Expanding Senses using Neurotechnology

SN Flashcards Microlearning

Quick and efficient studying with digital flashcards – for work or school!

With SN Flashcards you can:

- **Learn** anytime and anywhere on your smartphone, tablet or computer
- **Master** the content of the book and test your knowledge
- **Get motivated** by using various question types enriched with multimedia components and choosing from three learning algorithms (long-term-memory mode, short-term-memory mode or exam mode)
- **Create** your own question sets to personalise your learning experience

How to access your SN Flashcards content:

1. Go to the **1st page of the 1st chapter** of this book and follow the instructions in the box to sign up for an SN Flashcards account and to access the flashcards content for this book.
2. Download the SN Flashcards mobile app from the Apple App Store or Google Play Store, open the app and follow the instructions in the app.
3. Within the mobile app or web app, select the flashcards content for this book and start learning!

If you have difficulties accessing the SN Flashcards content, please write an email to **customerservice@springernature.com** mentioning "**SN Flashcards**" and the book title in the subject line.

Ujwal Chaudhary

Expanding Senses using Neurotechnology

Volume 1 – Foundation of Brain-Computer Interface Technology

Ujwal Chaudhary
BrainPortal Technologies GmbH
Mannheim, Germany

ISBN 978-3-031-76083-9   ISBN 978-3-031-76081-5   (eBook)
https://doi.org/10.1007/978-3-031-76081-5

Adapted from Chaudhary, U. et al. (2016) Brain–computer interfaces for communication and rehabilitation Nat. Rev. Neurol. doi:10.1038/nrneurol.2016.113, Deblik

This Springer imprint is published by the registered company Springer Nature Switzerland AG
The registered company address is: Gewerbestrasse 11, 6330 Cham, Switzerland

If disposing of this product, please recycle the paper.

Preface

Welcome to this comprehensive exploration of neurotechnology—an interdisciplinary field at the intersection of neuroscience and technology. This textbook is designed to provide a foundational understanding of the tools, techniques, and applications that make up the cutting-edge landscape of neurotechnology today. Targeted towards students, researchers, and professionals eager to deep-dive into the complexities and potentials of brain–computer interfaces, this text presents a systematic and detailed exposition of both theoretical concepts and practical applications.

▶ Chapter 1 introduces readers to the interdisciplinary nature of neurotechnology, tracing its historical milestones and the evolution of key tools and techniques. This sets the stage for a deeper dive into the neurophysiological foundations outlined in ▶ Chap. 2, where students will learn about the generation of brain signals and the principles behind their acquisition through various imaging and monitoring techniques.

▶ Chapter 3 transitions into invasive neurotechnological methods, comparing them with non-invasive techniques to highlight differences in application and efficacy. ▶ Chapter 4 addresses the critical issue of noise and artifacts in brain signal recordings and introduces preprocessing techniques to enhance data quality, an essential skill for any neurotechnologist.

With a solid grasp of signal acquisition and preprocessing, ▶ Chap. 5 ventures into the domain of machine learning, illustrating its application in interpreting complex brain signals and improving brain–computer interface (BCI) functionality. ▶ Chapter 6 explores neuromodulation and synaptic plasticity, essential for understanding neural circuit functions and their implications for neurotechnology.

The specific applications of deep brain stimulation (DBS) are covered in ▶ Chap. 7, providing a thorough overview of its mechanisms, historical development, and clinical applications. Similarly, ▶ Chap. 8 examines non-invasive brain stimulation (NIBS) techniques, discussing their mechanisms, clinical applications, and recent technological advancements.

▶ Chapter 9 offers a comparative analysis of brain stimulation techniques, discussing their operational principles, ethical considerations, and the impact of individual variability on treatment outcomes. ▶ Chapter 10 discusses the principles and applications of neurofeedback, emphasizing its potential in cognitive enhancement and rehabilitation. Finally, ▶ Chap. 11 combines all the concepts learned in this book to develop a solution to provide a means of communication to a patient with amyotrophic lateral sclerosis (ALS).

This textbook not only equips students with theoretical knowledge but also prepares them to apply these concepts practically, paving the way for innovations in treating neurological conditions and enhancing human cognitive and sensory capabilities. This text aims to inspire a new generation of scientists and technologists to push the boundaries of what is possible in neurotechnology through practical examples, case studies, and a clear linkage between theoretical constructs and real-world applications.

I hope that this book will serve as both a valuable educational resource and a springboard for further research and development in the exciting field of neurotechnology.

Ujwal Chaudhary
Mannheim, Germany

Contents

About the Author

Ujwal Chaudhary
began his research career as a doctoral student at Florida International University (FIU) in 2008, where he employed functional near-infrared spectroscopy (fNIRS) to study brain activation and connectivity in response to language tasks and social communication in children with autism. He extended this work to investigate motor skill planning and execution in healthy adults and individuals with cerebral palsy, integrating fNIRS with motion capture technology. Dr. Ujwal Chaudhary's exceptional work earned them the "Outstanding Doctoral Degree Graduate Award."

In 2014, Dr. Ujwal Chaudhary joined the University of Tübingen as a postdoctoral researcher, focusing on developing simultaneous fNIRS and EEG/EOG-based brain–computer interfaces (BCIs) for patients in completely locked-in states (CLIS). He led pioneering research into the neural basis of communication in CLIS and locked-in state (LIS), transitioning from non-invasive BCIs to an intracortical BCI system for a patient with amyotrophic lateral sclerosis (ALS). This groundbreaking work, published in Nature Communications in 2022, demonstrated brain-based volitional communication in a CLIS, a paper recognized among the top 25 health articles of the year.

Dr. Ujwal Chaudhary's innovative contributions to neurotechnology continue to advance the field, providing new hope for individuals with severe communication impairments.

Introduction to Neurotechnology

History and Interdisciplinary Innovations

Contents

1

— Test your learning and check your understanding of this book's contents: use the "Springer Nature Flashcards" app to access questions. To use the app, please follow the instructions below:
1. Go to ► https://flashcards.springernature.com/login
2. Create a user account by entering your e-mail address and assigning a password.
3. Use the following link to access your SN Flashcards set: ► https://sn.pub/kmbjyz
4. If the link is missing or does not work, please send an e-mail with the subject "SN Flashcards" and the book title to customerservice@springernature.com

This chapter unfolds the remarkable journey of neurotechnology—a field at the confluence of neuroscience and engineering, transforming our interaction with machines and profoundly enhancing human capabilities. Through captivating accounts of individuals overcoming neurological conditions with cutting-edge devices, we dive into the heart of neurotechnology's promise and its realization. From brain implants enabling communication for those paralyzed by amyotrophic lateral sclerosis, to robotic arms granting independence to individuals with tetraplegia, and the empowerment of those with spinal cord injuries to walk again, these narratives are not mere stories but milestones in neurotechnology's evolution from fantasy to reality. In this, we explore the historical context of tools and techniques that have facilitated the reading, modulation, and augmentation of brain activity. As we journey through neurotechnology's past and present and envisage its future, this chapter serves as a starting point to explore the tools, techniques, and applications encompassing the field of neurotechnology throughout the two volumes of this book series.

Learning Objectives
1. Understand the Interdisciplinary Nature of Neurotechnology.
2. Appreciate Historical Milestones and Evolution of different Tools and Techniques.

» A man in his mid-30s, who is completely paralyzed due to amyotrophic lateral sclerosis (ALS), lies on his bed. He relies on artificial ventilation and feeding, unable to control any skeletal muscles, including the ability to open his eyelids. Despite these challenges, he communicates with his family and caretakers through a pioneering brain implant. This device includes microelectrodes that capture the electrical activity of specific neuron groups. By listening to a continuous sinusoidal tone, he manipulates this tone's frequency by altering his neuronal firing rates to express "yes" or "no." These responses allow him to select letters and words presented to him through a sound system, enabling him to form sentences and maintain interaction with his environment. (Chaudhary et al., 2022)

» A woman in her late 50s, left with tetraplegia from a brainstem stroke, utilizes a brain implant while seated in her wheelchair. Although she cannot speak or use her limbs due to anarthria, she independently drinks coffee by operating a robotic arm with her brain activity. She skillfully manipulates the arm to grasp a bottle, bring it to her mouth to sip through a straw, and then return it to the table, showcasing the incredible capabilities of brain-computer interfaces in restoring autonomy. (Hochberg et al., 2012)

» Individuals with spinal cord injuries, who were once doomed never to walk again, are now able to walk, swim, pedal bicycles, and even paddle canoes. (Lorach et al., 2023)

These excerpts are neither figments of science fiction nor cinematic imagination; they represent just a few instances where the fusion of neuroscience and engineering is expanding the frontiers of how humans interact with machines. The long-held aspiration to directly control computers or external devices with our thoughts has transitioned from fantasy to fact. What was once envisioned only in the realm of science fiction movies and the dreams of scientists has now seized the collective imagination. The ambition to merge the human mind with machines has become a reality, leading to achievements that can only be described as extraordinary. Individuals who were once immobilized, rendered unable to speak or perform daily tasks due to paralysis from neurological conditions, brain injuries, or accidents, are now reclaiming their lost abilities. This progress stems from an enhanced understanding of the human nervous system coupled with advances in engineering. We are now equipped with sophisticated electronic devices that can interpret brain signals and manipulate objects in the physical world, embodying a significant leap in human–machine integration.

Neurotechnology is an expansive and interdisciplinary domain poised to revolutionize the rehabilitation and enhancement of human functions by integrating the human nervous system with external electronic apparatus. Unlike biotechnology, which emphasizes manipulating and understanding DNA, genetic components, and intricate biological entities through pharmacological and genetic engineering techniques, neurotechnology focuses on elucidating and interfacing the nervous system's operations with electronic and engineering strategies. This field offers groundbreaking prospects for individuals afflicted by various neurological conditions. Through advancements in neurotechnology, individuals with paralysis may reclaim certain capabilities via brain-controlled external devices, enabling those without communication means to interact once more using brain signals. Moreover, it holds promise for enhancing the lives of those with sleep disturbances, traumatic experiences, and depression, facilitating their journey toward independence and mindfulness. Additionally, it advances the diagnosis and treatment of numerous neurological disorders, aiming to significantly enhance the quality of life for millions globally, as depicted in ■ Fig. 1.1 (Belkacem et al., 2020).

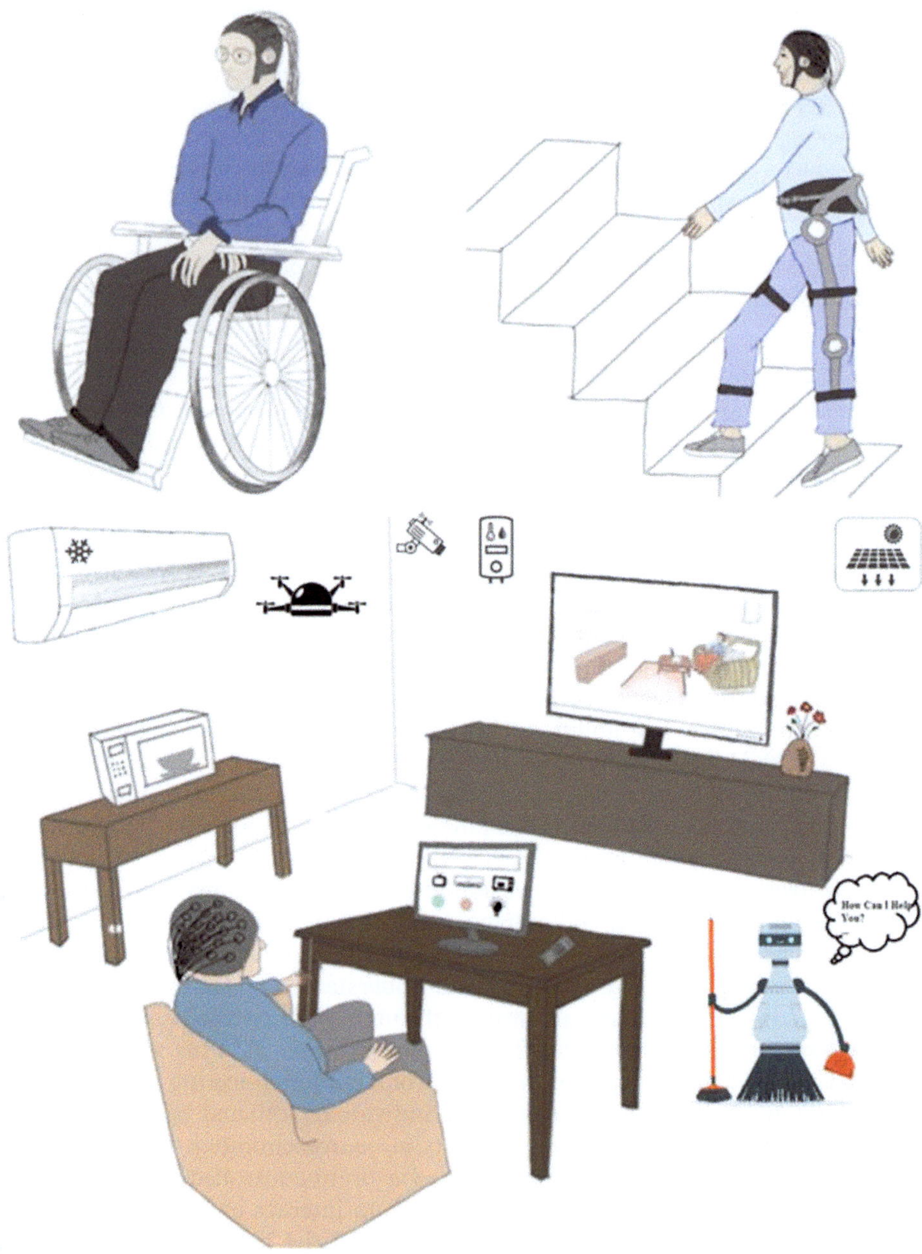

■ **Fig. 1.1**   Possibilities with neurotechnology. What do you think this figure wants to convey? Discuss within yourself; the answer lies in the subsequent chapters of Volume 1 and Volume 2. (From Belkacem et al., 2020; originally published under CC BY 4.0)

Emerging at the forefront of medical innovation, neurotechnology is unlocking new frontiers in health care by elucidating the nervous system's complexities. It extends the possibility of not only improving current medical practices but also pioneering unprecedented therapeutic interventions. Employing cutting-edge brain imaging and sensor technologies, neurotechnologies provide direct interaction with the brain or nervous system, either by monitoring and recording neural activity or by influencing it. The scope of these technologies spans from invasive brain implants to non-invasive ergonomic interfaces, such as helmets or headsets equipped with brain sensors and wearable devices like headbands or wristbands, heralding a new era of medical and cognitive enhancement.

In Volume 1 of this book series, the reader will learn the basics of neurophysiological signal acquisition and processing, feature engineering, the application of machine learning in the context of brain and body signals, and its interfacing with various external electronic devices. Last but not least, in Volume 2, the reader will learn about various applications of neurotechnology to augment and restore functions in individuals with different neurological and psychiatric disorders. The books, therefore, cover neurotechnology's past, present, and future in depth.

> **Box 1.1**
> Neurotechnology refers to a multidisciplinary field of science that merges principles of neuroscience and technology. This field encompasses the development and application of technological devices and methods to monitor, repair, enhance, or replace neurologic functions.

1.1 History of Neurotechnology

In April 1861, a man named Louis-Victor Leborgne died in Paris. He lost his ability to speak during the last 21 years of his life and could only pronounce the syllable "tan." French doctor and anatomist Paul Broca dissected his brain and found damage to the front of his left hemisphere (Broca, 1861). At subsequent autopsies in patients with speech loss, Broca always found similar lesions in the same areas. This was the first evidence of brain functions such as language production. Broca's discoveries laid the groundwork for *neurotechnology*, which flourished a century later. This chapter provides an overview of various developments that led to the advent of neurotechnology as a separate study field. I am clubbing the various developments into three significant developments: (a) technologies to read brain signals, (b) technologies to perturb ongoing brain activity, and (c) technologies to augment and restore lost body functionalities.

1.2 **Technologies to Read Brain Signals**

Brain signals refer to the electrical and chemical activity of the brain, which can be measured using various techniques. These signals include electroencephalogram signals, which measure the brain's electrical activity, and magnetoencephalogram signals, which measure the magnetic fields produced by the brain. Brain signals provide important information about the brain's functioning and are commonly used in studying various neurological disorders. They can also be used to study the effects of certain medications on brain function, and to investigate the neural basis of various cognitive processes. In addition to their use in research and medical settings, brain signals are also used in a number of other fields, including psychology, neuroscience, and cognitive science. They have helped deepen our understanding of the brain and its workings and continue to be an important tool in studying the brain and its functions. This section will examine the historical development of different technologies that enabled reading brain signals. The subsequent chapters will discuss each technology's physiological origin and various applications. ◘ Figure 1.2 (Chaudhary et al., 2016) depicts the different brain signal measurement techniques.

1.2.1 **History of Electroencephalography (EEG)**

The origins of electroencephalography (EEG) trace back to 1875 when Richard Canton of Liverpool first observed the electrical activity in the brains of animals, such as dogs and rabbits. This pioneering work laid the groundwork for future research in brain electrophysiology. In 1890, Adolf Beck further advanced the field by detecting electrical activity in the brains of animals through electrodes placed directly on the cerebral surface. His findings revealed that brain electrical activity is characterized by continuous waves that vary in pattern according to different cognitive tasks.

The concept of evoked potential was introduced in 1912 by a physiologist named Neminski, who documented these findings in dogs using electroencephalograms. This research contributed to a deeper understanding of brain responses to stimuli (Prawdicz-Neminski, 1912). By 1914, Cybulski and Jelenska made a significant leap by photographing EEG signals during seizures, providing tangible evidence of the brain's electrical phenomena during pathological states (Cybulski & Jelenska-Macieszyma, 1914).

In 1924, Hans Berger achieved a monumental breakthrough by recording human brain waves for the first time. His work, published in 1929 (Berger, 1929), opened new avenues for exploring human brain function and laid the foundation for EEG's application in clinical and research settings. The development of EEG technology progressed rapidly; by 1934, Edgar and Matthews had refined EEG devices for better signal acquisition, while Fisher and Lowenback's discovery of spikes in EEG signals contributed to the diagnostic capabilities of EEG.

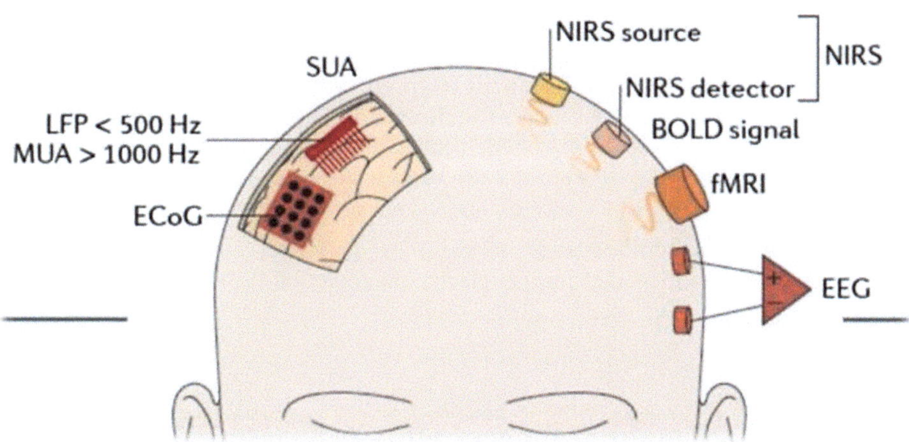

Fig. 1.2 Brain signal measurement techniques can be categorized into invasive and non-invasive approaches, each capturing different aspects of neural activity through various methods. **Invasive Techniques (Left)**: On the invasive side, techniques include the measurement of local field potentials (LFPs), which reflect the electric field created by the summed electrical currents flowing from large populations of neurons within a small area of the brain. Single-unit activity (SUA) involves recording the action potentials from individual neurons, providing precise information about neuronal firing patterns. Multi-unit activity (MUA) captures the signals of clusters of neurons, offering a broader view of neural activity than SUA. Electrocorticography (ECoG), another invasive method, involves placing electrodes directly on the exposed surface of the brain. This technique provides higher-resolution data on the electrical activity of the cortex compared to non-invasive methods. **Non-invasive Techniques (Right)**: On the non-invasive front, electroencephalography (EEG) is widely used to record electrical activity through electrodes placed on the scalp. EEG tracks voltage fluctuations resulting from ionic current flows within the neurons of the brain, offering insights into brain states and functions. Blood oxygenation level-dependent (BOLD) functional MRI (fMRI) measures brain activity by detecting changes associated with blood flow, relying on the fact that cerebral blood flow and neuronal activation are coupled. When an area of the brain is more active, it consumes more oxygen, and the BOLD signal reflects these changes in regional blood oxygenation. Near-infrared spectroscopy (NIRS) uses infrared light to penetrate the skull and measure the hemodynamic response associated with brain activity, providing insights into the metabolic demands of neurons. Each of these techniques has its particular strengths and limitations in terms of spatial and temporal resolution, invasiveness, and the specific types of neural activity they best measure, making them suitable for different research and clinical applications. (From Chaudhary et al., 2016)

The establishment of the first EEG laboratory in Boston in 1934 marked a significant milestone in the field, fostering further innovation and research. In the 1940s, Franklin Offner, a biophysics professor at Northwestern University, designed a prototype EEG model that enhanced the detection of brain waves (Offner, 1950). William Grey Walter's introduction of EEG topography in 1951 revolutionized the method by enabling the mapping of electrical activity across the scalp (Walter & Shipton, 1951), facilitating the non-invasive study of brain function.

These developments underscored EEG's role as a pivotal tool in neurology, neuroscience, and behavioral science, significantly advancing our understanding of brain function. EEG remains one of the most essential neurotechnological tools, widely utilized for both clinical diagnosis and research into the intricate workings of the human brain. In the subsequent chapters, you will read how EEG revolutionized our understanding of brain functions and is the most widely used neurotechnology.

> **Box 1.2: Electroencephalography (EEG)**
> A non-invasive method that records electrical activity of the brain using sensors placed along the scalp, used for diagnosing neurological diseases and for research in brain function.

1.2.2 History of Magnetoencephalography (MEG)

Magnetoencephalography (MEG) represents a sophisticated functional neuroimaging technique that detects the minute magnetic fields produced by neuronal electrical currents in the brain. This non-invasive method utilizes highly sensitive devices known as magnetometers to capture these magnetic signals. The origins of MEG date back to 1972 when physicist David Cohen recorded the first MEG signal at the University of Illinois. This landmark achievement in neuroimaging did not yet employ the advanced superconducting quantum interference devices (SQUIDs) that are standard in modern MEG systems. Instead, Cohen's early experiments utilized copper inductors to detect magnetic fields associated with brain activity (Cohen, 1972).

During these initial experiments, Cohen faced significant challenges from ambient magnetic noise. To counteract this, his measurements were conducted within a specially designed magnetically shielded room, which was crucial for obtaining clear and usable data from the weak magnetic fields generated by the brain. This pioneering work laid the groundwork for MEG technology's subsequent development and refinement.

Over the decades, MEG has evolved dramatically, with the integration of SQUIDs enhancing its sensitivity and accuracy. Today, MEG is a critical tool in the quantitative analysis of brain function, particularly in the context of mental health and neurological disorders. Its ability to provide precise temporal and spatial resolution makes it invaluable for understanding the brain's dynamic activities and for diagnosing and researching conditions like epilepsy, Alzheimer's disease, and schizophrenia, among others.

▶ Chapter 2 will explore the physiological origins of MEG signals, the technological advancements that have shaped the current capabilities of MEG instruments, and discuss the wide array of applications that benefit from this unique imaging method. This exploration will highlight how MEG contributes to our deeper understanding of the brain's electrical and functional dynamics in health and disease.

1.2.3 History of Functional Near-Infrared Spectroscopy (fNIRS)

In 1977, Fran Jöbsis unveiled a groundbreaking discovery that brain tissue exhibits a relatively high transparency within the near-infrared (NIR) spectrum, enabling the real-time, non-invasive monitoring of hemoglobin oxygenation levels through transillumination spectroscopy (Jöbsis, 1977). This innovation laid the groundwork for subsequent explorations into the utility of near-infrared spectroscopy (NIRS) across various applications, initially demonstrated in laboratory animals (Jobsis-vander Vliet, 1999). Building upon this foundation, Jöbsis and his team later applied NIRS techniques to investigate cerebral oxygenation in sick newborn infants, yielding significant insights (Brazy et al., 1985).

In a parallel development, Marco Ferrari and colleagues began deploying prototype NIRS instruments in 1980 to assess changes in brain oxygenation within experimental animal subjects and adult humans, pioneering studies that further validated NIRS as a vital research tool (Ferrari et al., 1985). By 1984, David Delpy embarked on the development of several NIRS instruments and, in 1986, reported the first quantitative measurements of various oxygenation and hemodynamic parameters in sick newborn infants. These parameters included concentrations of oxygenated (O_2Hb) and deoxygenated hemoglobin (HHb), total hemoglobin ($tHb = O_2Hb + HHb$), cerebral blood volume, and cerebral blood flow, marking a significant advancement in the non-invasive study of cerebral physiology (Wyatt et al., 1986). Through its capacity for non-invasive monitoring of cerebral oxygenation and blood flow, NIRS has emerged as a critical technology in the fields of neuroscience and clinical research, offering profound insights into the underlying mechanisms of brain function and pathology.

▶ Chapter 2 will provide an in-depth examination of the physiological origins of the fNIRS signal, detailing the technology behind the instruments used and exploring the wide array of applications for which NIRS has proven to be an invaluable tool.

1.2.4 History of Functional Magnetic Resonance Imaging (fMRI)

The advent of functional magnetic resonance imaging (fMRI) in the 1990s, largely attributed to the pioneering work of Seiji Ogawa and colleagues, and Ken Kwong, represents a significant milestone in the evolution of neuroimaging technologies. This innovation follows a series of advancements in the field, including positron emission tomography (PET) and near-infrared spectroscopy (NIRS), which similarly utilize changes in blood flow and oxygen metabolism as proxies for neuronal activity. fMRI distinguishes itself by offering a non-invasive method for studying brain function without the use of ionizing radiation, thus ensuring the safety of participants (Ogawa et al., 1990; Kwong et al., 1992).

One of fMRI's most lauded features is its superior spatial resolution, capable of pinpointing areas of brain activity within millimeters, and its respectable temporal resolution, although not as immediate as techniques like EEG, still allows for the observation of brain activity within a few seconds of occurrence. These characteristics, combined with its relative ease of use for researchers, have propelled fMRI to the forefront of cognitive neuroscience and psychology (Logothetis, 2008).

Over the past three decades, fMRI has revolutionized our understanding of the human brain, offering unprecedented insights into the neural mechanisms underlying memory formation, language processing, the experience of pain, learning processes, and the emotional spectrum, among other cognitive functions. Its application extends beyond academic research into clinical and commercial domains, where it is used to diagnose neurological disorders, inform neurosurgical planning, and even in neuromarketing strategies.

The second chapter explore the physiological underpinnings of the fMRI signal, exploring how cerebral blood flow and oxygenation changes reflect neuronal activation patterns. It also details the instrumentation central to fMRI studies, from the hardware used to generate and detect magnetic resonance signals to the sophisticated software algorithms that interpret these signals as functional maps of brain activity.

> **Box 1.5**
>
> **Functional magnetic resonance imaging (fMRI)** is an advanced neuroimaging technique that measures and maps brain activity by detecting changes in blood flow. Utilizing strong magnetic fields and radio waves, fMRI observes blood oxygenation and flow as an indicator of neuronal activity.

1.2.5 History of Functional Ultrasound (fUS)

The employment of Doppler ultrasound for evaluating cerebral blood flow has a well-established history in medical diagnostics. Transcranial color Doppler ultrasound (TCCD) has been a standard tool in clinical settings for monitoring blood flow within the major basal intracerebral arteries, a practice supported by decades of use and research, notably beginning with the foundational work by Aaslid et al. in 1982. This technique has proven invaluable in assessing cerebrovascular health and diagnosing conditions affecting brain blood flow.

Building on the legacy of traditional Doppler ultrasound, the concept of functional ultrasound (fUS) was introduced in 2011 by Macé et al., marking a significant advancement in neuroimaging methodologies (Macé et al., 2011). Unlike its predecessors, fUS extends the capabilities of ultrasound imaging by providing high-resolution images of blood flow dynamics in real-time, thereby offering a window into the functioning of the vascular system within the brain.

fUS represents a cutting-edge technique that leverages the principles of Doppler ultrasound to capture rapid changes in cerebral blood volume and oxygenation associated with neuronal activity. This method holds the promise of not only mapping blood flow with unprecedented detail but also elucidating the intricate relationships between neural activity, blood flow, and brain function.

In the forthcoming ▶ Chap. 2, we will explore the physiological basis of fUS, detailing how it exploits the natural blood flow within the brain to infer neural activity. We will also look into the technical aspects of fUS instrumentation, from the ultrasonic transducers that emit and receive the high-frequency sound waves, to the sophisticated software algorithms that analyze the echoes to generate images of cerebral blood flow.

> **Box 1.6**
>
> **Functional ultrasound (fUS)** is a neuroimaging technique that employs high-frequency sound waves to capture dynamic images of the brain's vascular activity. This method leverages the Doppler effect to detect changes in cerebral blood volume and flow, which are closely related to neuronal activity.

1.2.6 History of Invasive Techniques

Invasive neurophysiological techniques, which involve inserting microelectrodes directly into brain tissue, enable the precise recording of neural activity from individual neurons, termed single-unit activity (SUA), and from clusters of neurons, known as local field potentials (LFPs). The journey of developing such electrodes has been marked by significant milestones since the early twentieth century, beginning with Edgar Adrian's pioneering contributions to electrode design for capturing neural signals (Adrian, 1928). Subsequent innovations have introduced a variety of electrode materials and configurations, including glass microelectrodes, platinum wires, iridium microelectrodes, stainless steel microelectrodes, tungsten electrodes, and glass-insulated platinum microelectrodes, each tailored to optimize signal fidelity, biocompatibility, and durability (Hubel, 1957; Robinson, 1968).

The application of multiple microelectrodes for concurrent neural recordings was first documented by Marg and Adams in 1967, marking a groundbreaking development in neurosurgical diagnostics and therapeutic interventions (Marg & Adams, 1967). This approach was further exemplified by Kruger and Bach in 1981, who employed a matrix of 30 microelectrodes to investigate the visual cortex of a monkey, demonstrating the feasibility of detailed neural circuit mapping (Kruger & Bach, 1981).

The emergence of sophisticated microelectrode arrays in the late twentieth century, such as the Utah and Michigan arrays, represented significant technological advances, enabling simultaneous recordings from multiple brain regions and, thus, facilitating a deeper understanding of complex neural interactions (Maynard et al., 1997; Wise et al., 1978).

These invasive techniques have revolutionized our ability to directly observe and record neuronal activity, providing invaluable insights into brain function and offering new avenues for interfacing the human nervous system with electronic devices. Future chapters will explore how these invasive recording methods are shaping the field of neurotechnology, pushing forward the limits of our knowledge and capabilities in neuroscience and neural engineering.

> **Box 1.7**
>
> **Invasive brain signal recording techniques** involve direct contact with the brain tissue to monitor and record electrical activity from neurons. This category of techniques includes the use of microelectrodes that are either inserted through the skull into the brain or placed directly on the brain surface during surgery. Common invasive recording techniques include intracortical recording, using microelectrode arrays such as the Utah Array, and electrocorticography (ECoG), which involves placing electrodes on the exposed surface of the brain.

1.3 Technologies to Perturb Ongoing Brain Activity

Toward the late nineteenth century, as the scientific community gained a deeper understanding of the brain's functional organization, researchers expanded their focus beyond merely observing brain activity. They embarked on pioneering efforts to actively influence brain function through the application of electrical and magnetic stimuli to specific brain regions. This marked a significant evolution in neuroscience, transitioning from passive observation to active modulation of neural activity, opening new avenues for both research and therapeutic intervention.

The inception of electrical stimulation techniques can be traced back to the work of Luigi Galvani and Alessandro Volta in the late eighteenth and early nineteenth centuries, who discovered and explored bioelectricity, laying the groundwork for understanding how electrical currents could interact with biological tissues (Galvani, 1791; Volta, 1800). However, it was not until the latter part of the nineteenth century that scientists began to apply these principles to the brain systematically. Pioneers such as Gustav Fritsch and Eduard Hitzig in the 1870s demonstrated that electrical stimulation of certain brain areas in dogs could elicit movements in specific muscle groups, suggesting distinct cortical areas responsible for motor control (Fritsch & Hitzig, 1870).

Simultaneously, the development of magnetic stimulation, although not realized until the late twentieth century, was predicated on the foundational work of Michael Faraday on electromagnetic induction in the 1830s. Faraday's discovery that a changing magnetic field could induce an electric current in a conductor (Faraday, 1831) provided the theoretical basis for transcranial magnetic stimulation (TMS), introduced by Anthony Barker and colleagues in 1985 as a non-invasive method of stimulating the brain through the scalp and skull (Barker et al., 1985).

This section explores the historical milestones of brain stimulation techniques, exploring their origins, the scientific curiosity that fueled their development, and the technological advances that have refined their application. From the early experiments in electrical neurostimulation to the advent of TMS, these techniques have profoundly impacted our ability to explore and manipulate brain function, contributing to significant advancements in neuroscience, neurology, and psychiatric treatment.

> **Box 1.8**
> **Brain stimulation** refers to various techniques used to modulate the activity of the brain through direct or indirect stimulation of specific brain areas. These methods can be invasive, such as deep brain stimulation (DBS) and cortical implants, or non-invasive, such as transcranial magnetic stimulation (TMS) and transcranial direct current stimulation (tDCS).

1.3.1 History of Deep Brain Stimulation (DBS)

The origins of deep brain stimulation (DBS) are rooted in the late twentieth century, a period marked by initial explorations into the use of electrical stimulation for the treatment of neurological conditions. These early iterations of DBS primarily focused on movement disorders like Parkinson's disease, predicated on the hypothesis that electrical impulses could modulate neural circuits within the basal ganglia to normalize movement patterns (Benabid et al., 1987).

The trajectory of DBS technology witnessed a significant evolution in the 1980s with the advent of computerized tomography (CT) and magnetic resonance imaging (MRI). These imaging technologies facilitated precise anatomical mapping, enabling targeted electrical stimulation of specific brain regions, thereby expanding the therapeutic potential of DBS beyond movement disorders (Alexander et al., 1986).

By the 1990s, the U.S. Food and Drug Administration (FDA) officially recognized the therapeutic benefits of DBS, granting approval for its application in treating Parkinson's disease. Subsequent approvals for essential tremor and dystonia followed, underscoring DBS's growing acceptance and utility in managing neurological disorders (Kringelbach et al., 2007).

In more recent years, the scope of DBS has broadened significantly, extending to psychiatric disorders such as obsessive-compulsive disorder (OCD) and major depressive disorder (MDD), reflecting the technique's adaptability and potential in addressing a wide array of neural dysfunctions (Greenberg et al., 2006; Mayberg et al., 2005). Ongoing research and clinical trials continue to explore and refine DBS applications, promising to unlock new frontiers in the treatment of complex and previously intractable neurological and psychiatric conditions.

This historical overview highlights the progressive refinement and diversification of DBS applications, from its rudimentary beginnings to its current status as a sophisticated therapeutic modality, offering hope and improved quality of life to patients with a variety of debilitating disorders. We will deep-dive into this topic in depth in ▶ Chap. 7, where you will learn about deep brain stimulation (DBS).

> **Box 1.9**
>
> **Deep brain stimulation (DBS)** is an invasive brain stimulation technique that involves implanting electrodes within certain areas of the brain. DBS delivers controlled electrical impulses to targeted brain regions, regulating abnormal impulses or affecting certain cells and chemicals within the brain.

1.3.2 History of Transcranial Direct Current Stimulation (tDCS)

Transcranial Direct Current Stimulation (tDCS), a technique for neuromodulation, traces its origins much further back in history than is widely acknowledged. A notable early instance of tDCS application occurred in 1801 when Giovanni Aldini, a nephew of the pioneering bioelectromagnetist Luigi Galvani, utilized an embryonic form of tDCS to alleviate the symptoms of depression in 27-year-old farmer Luigi Lanzarini, who was diagnosed with "melancholia." This intervention represents one of the earliest documented uses of tDCS in treating neurological and psychiatric conditions, underscoring the long-standing curiosity and investigation into electrical stimulation's therapeutic potential (Aldini, 1801).

The interest in tDCS saw a resurgence in the 1960s, coinciding with the burgeoning exploration into electro-sleep therapy and electro-anesthesia. This period marked a significant shift toward a more rigorous scientific inquiry into the mechanisms underlying tDCS and its effects on the brain. Researchers began to systematically study electrical stimulation parameters that could induce therapeutic outcomes, setting the stage for contemporary research and application of tDCS in various medical and psychological conditions.

In ▶ Chap. 8, we will deep-dive into the foundational principles of tDCS, detailing the technology and instrumentation involved in delivering precise electrical currents to the brain. We will explore the wide array of applications of tDCS, from its role in enhancing cognitive functions and treating psychiatric disorders to its potential in rehabilitation and neuroenhancement. This discussion will include an examination of the current scientific evidence supporting tDCS efficacy, the ethical considerations surrounding its use, and the future directions of tDCS research and clinical practice.

> **Box 1.10**
> **Transcranial direct current stimulation (tDCS)** is a non-invasive brain stimulation technique that uses a low, constant current delivered directly to the brain area of interest via electrodes placed on the scalp. This method is used to modulate neuronal activity, enhancing or diminishing brain function to influence cognitive processes and behavioral outcomes.

1.3.3 History of Transcranial Magnetic Stimulation (TMS)

The concept of employing magnetism to stimulate the human brain and other body parts for disease treatment extends back over a century. However, the efficacy of these early magnetic therapies often lacked scientific validation, relegating them to the margins of medical practice. It was not until 1985 that the landscape of magnetic brain stimulation underwent a pivotal transformation with the introduction of the first modern transcranial magnetic stimulation (TMS) device by Anthony Barker and his colleagues. This innovation marked the beginning of TMS as a scientifically grounded and effective tool for neuromodulation (Barker et al., 1985).

TMS utilizes magnetic fields to induce electric currents in specific brain regions, offering a non-invasive means of stimulating neural tissue. The introduction of TMS represented a significant leap forward in neuropsychiatry, providing researchers and clinicians with a powerful method to explore brain function and treat various neurological and psychiatric disorders. Since its inception, TMS has been applied to study cortical plasticity, brain connectivity, and the treatment of conditions such as depression, anxiety, and migraines, among others.

In ▶ Chap. 8, we will deep-dive into the principles of TMS. It will provide an in-depth look at the theoretical foundations, technical specifications of TMS devices, and the broad spectrum of its clinical and research applications. From enhancing cognitive functions to treating psychiatric disorders, TMS has emerged as a versatile tool in the field of neurostimulation, complementing the therapeutic arsenal alongside tDCS.

By exploring both TMS and tDCS in ▶ Chaps. 8 and 9, I aim to offer an understanding of these cutting-edge technologies, highlighting their contributions to advancing neurological and psychological treatments and opening new avenues for investigating the complexities of the human brain.

> **Box 1.11**
> **Transcranial magnetic stimulation (TMS)** is a non-invasive method used to stimulate small regions of the brain. The technique involves placing a magnetic coil near the scalp, where it generates brief magnetic pulses that pass through the skull and induce an electrical current in the brain.

1.4 Technologies to Augment and Restore Lost Body Functionalities

Philosophical inquiries into the brain's capacity to directly interface with the external world, bypassing the peripheral somatomotor nervous system, have a long-standing history. This question concerning the brain's potential direct control over external devices, remained largely speculative until the necessary technological advancements in signal acquisition and translation were achieved. A pivotal moment in this journey occurred in 1929 when Hans Berger accomplished a groundbreaking feat with the development of electroencephalography (EEG), which for the first time allowed for the non-invasive monitoring of electrical activity within the human brain (Berger, 1929).

The emergence of EEG marked the beginning of a new era in neuroscience, offering a window into the brain's electrical dynamics. However, the realization of brain–computer interfaces (BCIs), devices that could translate these neuroelectrical signals into actionable commands for controlling external devices, required further technological innovations. The subsequent development of fast computing and real-time analysis systems, alongside a deepening understanding of brain function, provided the essential tools for making this long-held dream a reality.

Over the last two decades, experimental research into BCIs has seen exponential growth. Advances in neuroscience, coupled with computational power and algorithm breakthroughs, have enabled the development of BCIs that can interpret brain signals with increasing accuracy and efficiency. These interfaces have shown potential in a wide range of applications, from restoring mobility and communication in individuals with severe motor impairments to enhancing human capabilities. BCIs represent a significant leap forward in the integration of technology and neural science, promising to redefine the boundaries of human interaction with the external world.

The exploration of BCIs not only continues to challenge our understanding of the brain but also opens up new frontiers in rehabilitation, assistive technologies, and even in augmenting human capabilities, embodying a fascinating convergence of science, technology, and philosophy.

1.4.1 History of BCIs

The foundational steps toward harnessing brain signals on a neurophysiological basis were taken in 1968 when Wyrwicka and Sterman embarked on pioneering research by recording sensorimotor rhythms (also referred to as Rolandic α rhythms or μ rhythms) in cats, marking a significant early venture into what would later evolve into brain–computer interface (BCI) technology. They ingeniously translated these rhythms into sensory feedback that rewarded the animals, thereby enhancing the production of sensorimotor rhythms (Wyrwicka & Sterman, 1968).

Concurrently, neurofeedback, a technique aimed at modulating human brain activity, began to emerge. In 1969, Kamiya's landmark study demonstrated the possibility of volitional control over human brain oscillations, showing that individuals could alter their alpha wave patterns through continuous sensory feedback, a discovery that expanded the understanding of the brain's potential for self-regulation (Kamiya, 1969).

Simultaneously, Fetz (1969) illustrated that operant conditioning could be employed to manipulate the firing rates of individual cortical neurons in monkeys, a finding that further underscored the intricate link between brain physiology, behavior, and instrumental learning. In 1965, Sutton and colleagues reported on the relationship between event-related potentials and stimulus uncertainty with its correct anticipation, showing a positive-going neurophysiological component, which is nowadays known as p300 wave, as it reaches the peak 300 ms after the stimulus onset only for stimuli with low occurrence. These early explorations laid the groundwork for the development of modern BCIs, a term officially coined by Jacques Vidal in 1973, who also introduced a system capable of converting EEG signals into commands for computer control (Vidal, 1973).

The realm of BCIs gained clinical relevance through the work of Sterman and colleagues, who discovered that training cats in sensorimotor rhythms could raise seizure thresholds, a serendipitous finding that prompted research into neurofeedback training for epilepsy and other disorders in humans, notably demonstrating efficacy in reducing grand-mal seizures and managing conditions such as attention deficit–hyperactivity disorder (ADHD) (Sterman et al., 1974).

However, the early enthusiasm for neurofeedback's behavioral effects was tempered by challenges in replicating initial successes in larger, controlled studies, leading to skepticism and a period of disrepute for the field. Despite these setbacks, the last two decades have witnessed a resurgence in BCI research, propelled by advances in microelectrode technology and single-neuron recordings in animal and human models. This contemporary research has enabled subjects to control external devices, such as computer cursors and robotic arms, using their brain activity, showcasing the potential of BCIs in augmenting human capabilities and offering new avenues for treatment (Velliste et al., 2008; Hochberg et al., 2006).

> **Box 1.12**
>
> **Brain–computer interface (BCI)** is a technology that facilitates direct communication between the brain and an external device. BCIs are often used to assist, augment, or repair human cognitive or sensory-motor functions. This technology captures brain signals, processes them, and translates them into commands that are relayed to output devices to carry out desired actions.

1.4.2 **History of Neurofeedback**

Clinical uses of brain–computer interfaces (BCIs), a technology that connects your brain to computers, have grown out of techniques known as neurofeedback and biofeedback. Thanks to advances in computer technology, we can now quickly analyze brain wave patterns (EEG) and provide real-time feedback, helping people understand and potentially change their brain activity (Elbert et al., 1984). Unlike the usual way we move or think, which involves many steps within our body, BCIs tap directly into the brain's signals. For example, they can pick up signals from the part of the brain that controls arm movements and use those signals to control a computer or another device.

Neurofeedback, a key part of this technology, is like a video game for your brain. It shows you what is happening in your brain as it is happening. By getting this immediate feedback, people can learn to change their brain activity to achieve a better state of mind, whether that is feeling more relaxed or focusing better. This can be especially helpful for improving certain health conditions like ADHD, anxiety, or epilepsy.

In an upcoming chapter, we will dive deeper into how neurofeedback works, showing how it is not just a fascinating scientific tool but also a practical way to help people. This mix of science, technology, and health care is opening new doors for treating brain-related issues, making what once seemed like science fiction a reality.

> **Box 1.13**
> **Biofeedback** is a technique that enables individuals to learn how to change physiological activities for the purpose of improving health and performance.

> **Box 1.14**
> **Neurofeedback** is a type of biofeedback that uses real-time displays of brain activity to teach self-regulation of brain functions. This technique involves measuring brain waves and providing a feedback signal to an individual in the form of auditory, visual, or tactile feedback based on the brain's activity.

❯ Key Takeaways

1. **Interdisciplinary Integration**: Neurotechnology exemplifies the integration of neuroscience and engineering, creating tools that enhance human capabilities and facilitate interaction with machines. This interdisciplinary approach has driven innovation, allowing neurotechnology to evolve into a field capable of both augmenting and restoring lost neurological functions.

2. **Technological Advancements**: Significant milestones such as brain–computer interfaces (BCIs), neuroimaging, and brain stimulation techniques have transformed the way we understand and interact with the nervous system. These advancements have provided critical insights into brain function while also offering therapeutic solutions for neurological disorders such as amyotrophic lateral sclerosis (ALS), tetraplegia, and Parkinson's disease.

3. **Historical Foundations and Future Trajectory**: From early discoveries like Broca's identification of localized brain functions to modern-day neuroimaging and BCIs, the field of neurotechnology has progressed rapidly. The chapter outlines the historical developments that paved the way for current technologies and stresses that ongoing innovations will continue to shape the future of this field.

4. **Ethical and Societal Considerations**: As neurotechnologies become more integrated into health care and daily life, ethical concerns regarding privacy, consent, and the manipulation of brain activity must be addressed. The future of neurotechnology not only lies in scientific advancement but also in the responsible and ethical implementation of these tools.

The field of neurotechnology is rapidly advancing, and new technologies are being developed that have the potential to revolutionize the way we understand and treat neurological disorders. I am confident that as you are reading this book, scientists, engineers, and clinicians who were once students like you are pushing the boundary of what is possible and redefining our understanding of the human brain, its capabilities, and the ways to harness its potential to better human lives. As we advance in this book, you will learn about the tools and techniques to measure the activity of the brain in response to the different tasks performed by the human body and how to use these activities to augment the functionalities once we lose them because of different conditions. Be brave and imaginative, and let us explore the fascinating world of neurotechnology, which has the potential to propel us into a future where humans are not limited by their frail bodies. It is the next frontier so let us dive right into it.

Conclusion

Neurotechnology refers to the use of technology to study, understand, and interact with the nervous system. As we have seen in this chapter, this can include tools and techniques for measuring and analyzing neural activity, as well as devices and therapies for treating neurological disorders, which falls under the purview of this book. Some examples of neurotechnology, as outlined previously in this chapter, are:

- Neuroimaging techniques such as functional magnetic resonance imaging (fMRI) that allow scientists to non-invasively visualize and study the brain in action.
- Electroencephalography (EEG) and magnetoencephalography (MEG) that measure the electrical and magnetic activity of the brain, respectively.

- Deep brain stimulation (DBS) and transcranial magnetic stimulation (TMS) therapies that use electrical or magnetic impulses to stimulate specific areas of the brain to treat conditions such as Parkinson's disease, depression, and chronic pain.
- Brain–computer interfaces (BCIs) that allow individuals to control computers or other devices directly with their brain activity.
- Neural prosthetics, which are devices such as cochlear implants and retinal implants that can restore lost function in individuals with neurological disorders.

This chapter has traced the arc of neurotechnology from its origins to its current state, highlighting the pioneering work and key inventions that have shaped its course. As we close, we recognize that the history of neurotechnology is not just a record of past achievements but also a foundation upon which future generations will build. The story of neurotechnology is ongoing, and each new discovery adds a page to this ever-evolving narrative. For students and researchers in the field, the history of neurotechnology serves both as a source of inspiration and a roadmap for the journey ahead, filled with endless possibilities and the potential to impact both individual lives and society at large profoundly.

Crucial to this progress has been the willingness of researchers and innovators to cross traditional disciplinary boundaries, integrating knowledge and techniques from diverse areas to create novel approaches to studying the brain. This interdisciplinary synergy has been a driving force behind the field's evolution, enabling breakthroughs that were unimaginable just a few decades ago.

Today, neurotechnology is not only a means for understanding the brain but also a powerful tool for treating neurological disorders, enhancing cognitive abilities, and even interfacing with machines directly through brain–computer interfaces (BCIs). As we look to the future, the potential applications of neurotechnology are vast and hold promise for significant societal impact, from health care to education and beyond.

In reflecting on the history of neurotechnology, we are reminded of the power of human ingenuity and the endless possibilities that arise when curiosity is coupled with scientific rigor. The journey of neurotechnology is far from complete; it continues to evolve, promising new discoveries and innovations that will further our understanding of the most complex organ in the human body—the brain. As students and scholars of this dynamic field, we stand on the shoulders of giants, inspired by the past and excited for the future, ready to contribute to the next chapter of this incredible scientific saga.

As we embark on this future journey, it is crucial to navigate the ethical and societal implications of these technologies. The ability to read and potentially influence brain activity brings with it a host of ethical considerations regarding privacy, consent, and the nature of human cognition and identity. Ensuring that the development of neurotechnologies is guided by strong ethical principles and a commitment to the betterment of humanity will be as important as the scientific and technical challenges that lie ahead.

References

Aaslid, R., Markwalder, T. M., & Nornes, H. (1982). Noninvasive transcranial Doppler ultrasound recording of flow velocity in basal cerebral arteries. *Journal of Neurosurgery, 57*(6), 769–774.

Adrian, E. D. (1928). *The basis of sensation.* WW Norton.

Aldini, G. (1801). *On the therapeutic application of galvanism.*

Alexander, G. E., DeLong, M. R., & Strick, P. L. (1986). Parallel organization of functionally segregated circuits linking basal ganglia and cortex. *Annual Review of Neuroscience, 9*, 357–381.

Barker, A. T., Jalinous, R., & Freeston, I. L. (1985). Non-invasive magnetic stimulation of human motor cortex. *The Lancet, 325*(8437), 1106–1107.

Belkacem, A. N., Jamil, N., Palmer, J. A., Ouhbi, S., & Chen, C. (2020). Brain computer interfaces for improving the quality of life of older adults and elderly patients. *Frontiers in Neuroscience, 14*, 692.

Benabid, A. L., Pollak, P., Louveau, A., Henry, S., & de Rougemont, J. (1987). Combined (thalamotomy and stimulation) stereotactic surgery of the VIM thalamic nucleus for bilateral Parkinson disease. *Applied Neurophysiology, 50*(1–6), 344–346.

Berger, H. (1929). Über das Elektrenkephalogramm des Menschen. *Archiv fur Psychiatrie und Nervenkrankheiten, 87*, 527–570.

Brazy, J. E., Lewis, D. V., Mitnick, M. H., & van der Vliet, F. F. J. (1985). Noninvasive monitoring of cerebral oxygenation in preterm infants: Preliminary observations. *Pediatrics, 75*(2), 217–225.

Broca, P. (1861). Remarks on the seat of the faculty of articulated language, following an observation of aphemia (loss of speech). *Bulletin de la Société Anatomique, 6*, 330–357.

Chaudhary, U., Birbaumer, N., & Ramos-Murguialday, A. (2016). Brain–computer interfaces for communication and rehabilitation. *Nature Reviews Neurology, 12*(9), 513–525.

Chaudhary, U., Vlachos, I., Zimmermann, J. B., Espinosa, A., Tonin, A., Jaramillo-Gonzalez, A., et al. (2022). Spelling interface using intracortical signals in a completely locked-in patient enabled via auditory neurofeedback training. *Nature Communications, 13*(1), 1236.

Cohen, D. (1972). Magnetoencephalography: Detection of the brain's electrical activity with a superconducting magnetometer. *Science, 175*(4022), 664–666.

Cybulski, N., & Jelenska-Macieszyma, S. (1914). Action currents of cerebral cortex. *Crakow Series Bulletin International de l'Academie Polonaise des Sciences B*, 776–781.

Elbert, T., Rockstroh, B., Lutzenberger, W., & Birbaumer, N. (1984). Biofeedback of slow cortical potentials. *Electroencephalography and Clinical Neurophysiology, 57*(4), 293–301.

Faraday, M. (1831). On the induction of electric currents. *Philosophical Transactions of the Royal Society of London.*

Ferrari, M., Giannini, I., Sideri, G., & Zanette, E. (1985). Continuous non invasive monitoring of human brain by near infrared spectroscopy. In *Oxygen transport to tissue VII* (pp. 873–882). Springer US.

Fetz, E. E. (1969). Operant conditioning of cortical unit activity. *Science, 163*(3870), 955–958.

Fritsch, G., & Hitzig, E. (1870). On the electrical excitability of the cerebrum. *Archiv für Anatomie, Physiologie und Wissenschaftliche Medicin, 37*, 300–332.

Galvani, L. (1791). *Commentary on the Effects of electricity on muscular motion.* De Bononiensi Scientiarum et Artium Instituto atque Academia Commentarii.

Greenberg, B. D., Malone, D. A., Friehs, G. M., Rezai, A. R., Kubu, C. S., Malloy, P. F., Salloway, S. P., Okun, M. S., Goodman, W. K., & Rasmussen, S. A. (2006). Three-year outcomes in deep brain stimulation for highly resistant obsessive-compulsive disorder. *Neuropsychopharmacology, 31*(11), 2384–2393.

Hochberg, L. R., Bacher, D., Jarosiewicz, B., Masse, N. Y., Simeral, J. D., Vogel, J., et al. (2012). Reach and grasp by people with tetraplegia using a neurally controlled robotic arm. *Nature, 485*(7398), 372–375.

Hochberg, L. R., Serruya, M. D., Friehs, G. M., Mukand, J. A., Saleh, M., Caplan, A. H., Branner, A., Chen, D., Penn, R. D., & Donoghue, J. P. (2006). Neuronal ensemble control of prosthetic devices by a human with tetraplegia. *Nature, 442*(7099), 164–171.

Hubel, D. H. (1957). Tungsten microelectrode for recording from single units. *Science, 125*(3247), 549–550.

Jöbsis, F. F. (1977). Noninvasive, infrared monitoring of cerebral and myocardial oxygen sufficiency and circulatory parameters. *Science, 198*(4323), 1264–1267.

Jobsis-vander Vliet, F. F. (1999). Discovery of the near-infrared window into the body and the early development of near-infrared spectroscopy. *Journal of Biomedical Optics, 4*(4), 392–396.

Kamiya, J. (1969). Operant control of the EEG alpha rhythm and some of its reported effects on consciousness. In T. X. Barber, L. V. DiCara, J. Kamiya, N. E. Miller, D. Shapiro, & J. Stoyva (Eds.), *Biofeedback and self-control.* Aldine Transaction.

Kringelbach, M. L., Jenkinson, N., Owen, S. L. F., & Aziz, T. Z. (2007). Translational principles of deep brain stimulation. *Nature Reviews Neuroscience, 8*(8), 623–635.

Kruger, J., & Bach, M. (1981). Simultaneous recording with 30 microelectrodes in monkey visual cortex. *Experimental Brain Research, 41*(2), 191–194.

Kwong, K. K., Belliveau, J. W., Chesler, D. A., Goldberg, I. E., Weisskoff, R. M., Poncelet, B. P., Kennedy, D. N., Hoppel, B. E., Cohen, M. S., Turner, R., Cheng, H. M., Brady, T. J., & Rosen, B. R. (1992). Dynamic magnetic resonance imaging of human brain activity during primary sensory stimulation. *Proceedings of the National Academy of Sciences, 89*(12), 5675–5679.

Logothetis, N. K. (2008). What we can do and what we cannot do with fMRI. *Nature, 453*(7197), 869–878.

Lorach, H., Galvez, A., Spagnolo, V., Martel, F., Karakas, S., Intering, N., et al. (2023). Walking naturally after spinal cord injury using a brain–spine interface. *Nature, 618*(7963), 126–133.

Macé, E., Montaldo, G., Cohen, I., Baulac, M., Fink, M., & Tanter, M. (2011). Functional ultrasound imaging of the brain. *Nature Methods, 8*(8), 662–664.

Marg, E., & Adams, J. E. (1967). Indwelling multiple microelectrodes in the brain. *Electroencephalography and Clinical Neurophysiology, 23*(2), 277–280.

Mayberg, H. S., Lozano, A. M., Voon, V., McNeely, H. E., Seminowicz, D., Hamani, C., Schwalb, J. M., & Kennedy, S. H. (2005). Deep brain stimulation for treatment-resistant depression. *Neuron, 45*(5), 651–660.

Maynard, E. M., Nordhausen, C. T., & Normann, R. A. (1997). The Utah Intracortical Electrode Array: A recording structure for potential brain-computer interfaces. *Electroencephalography and Clinical Neurophysiology, 102*(3), 228–239.

Offner, F. F. (1950). The EEG as potential mapping: The value of the average monopolar reference. *Electroencephalography and Clinical Neurophysiology, 2*(2), 213–214.

Ogawa, S., Lee, T. M., Kay, A. R., & Tank, D. W. (1990). Brain magnetic resonance imaging with contrast dependent on blood oxygenation. *Proceedings of the National Academy of Sciences, 87*(24), 9868–9872.

Prawdicz-Neminski, V. V. (1912). Ein versuch der registrierung der Elektrischen Gehirnerscheinungen. *Zentralblatt für Physiologie, 27,* 951–960.

Robinson, D. A. (1968). The electrical properties of metal microelectrodes. *Proceedings of the IEEE, 56*(6), 1065–1071.

Sterman, M. B., MacDonald, L. R., & Stone, R. K. (1974). Biofeedback training of the sensorimotor electroencephalogram rhythm in man: Effects on epilepsy. *Epilepsia, 15*(3), 395–416.

Sutton, S., Braren, M., Zubin, J., & John, E. R. (1965). Evoked-potential correlates of stimulus uncertainty. *Science, 150*(3700), 1187–1188.

Velliste, M., Perel, S., Spalding, M. C., Whitford, A. S., & Schwartz, A. B. (2008). Cortical control of a prosthetic arm for self-feeding. *Nature, 453*(7198), 1098–1101.

Vidal, J. J. (1973). Toward direct brain-computer communication. *Annual Review of Biophysics and Bioengineering, 2,* 157–180.

Volta, A. (1800). On the electricity excited by the mere contact of conducting substances of different kinds. *Philosophical Transactions of the Royal Society of London.*

Walter, W. G., & Shipton, H. W. (1951). A new toposcopic display system. *Electroencephalography and Clinical Neurophysiology, 3*(3), 281–292.

Wise, K. D., Angell, J. B., & Starr, A. (1978). An integrated-circuit approach to extracellular microelectrodes. *IEEE Transactions on Biomedical Engineering, BME-25*(3), 238–247.

Wyatt, J. S., Delpy, D. T., Cope, M., Wray, S., & Reynolds, E. O. R. (1986). Quantification of cerebral oxygenation and haemodynamics in sick newborn infants by near infrared spectrophotometry. *The Lancet, 328*(8515), 1063–1066.

Wyrwicka, W., & Sterman, M. B. (1968). Instrumental conditioning of sensorimotor cortex EEG spindles in the waking cat. *Physiological Behavior, 3*, 703–707.

Non-invasive Brain Signal Acquisition Techniques

Exploring EEG, EOG, fNIRS, fMRI, MEG, and fUS

Contents

2

Test your learning and check your understanding of this book's contents: use the "Springer Nature Flashcards" app to access questions using ▶ https://sn.pub/kmb-jyz. To use the app, please follow the instructions in ▶ Chap. 1.

This chapter provides an introduction to the neurophysiological foundations and hardware implementations of several critical neuroimaging and neurophysiological monitoring techniques. It enlightens readers about the principles and applications of electroencephalography (EEG), electrooculography (EOG), functional near-infrared spectroscopy (fNIRS), functional magnetic resonance imaging (fMRI), magnetoencephalography (MEG), and functional ultrasound (fUS). Each section begins by explaining the neurobiological basis of the signals detected by these technologies, followed by a discussion on the specific hardware used, its configuration, and its operational nuances. The chapter aims to equip readers with a solid understanding of how these technologies function and their role in advancing our understanding of brain dynamics. This foundational knowledge is crucial for both beginners and seasoned practitioners in the fields of neuroscience and cognitive science, facilitating a deeper understanding of the methodological approaches used in current research.

Learning Objectives

1. Understand Neurophysiological Foundations: Explain the basic neurophysiological origins of brain signals, including the roles of neurons, ion movement, and neurotransmitters in generating electrical and chemical signals in the brain.
2. Grasp Signal Acquisition Basics: Describe the principles behind various non-invasive neuroimaging and neurophysiological monitoring techniques such as EEG, EOG, fNIRS, fMRI, and MEG.
3. Analyze Different Brain Signal Types: Distinguish between different types of brain signals (electrical and hemodynamic) and understand how these are captured through specific non-invasive methods.
4. Explore Hardware Implementations: Identify and understand the hardware used for capturing non-invasive brain signals, including the configuration and operational nuances of devices used in EEG, fMRI, and other techniques.

The concept of brain and body multimodal signals encompasses an extensive array of diverse signals that are integral to the study of brain and body functionality. The intricate interplay between the brain and body forms the cornerstone of understanding how the brain orchestrates bodily functions, and, conversely, how bodily feedback influences brain processes (Bear et al., 2020). The human body acts as a prolific source of signals, generating electrical and chemical messages essential for the regulation and coordination of physiological functions.

Electrical impulses, known as nerve signals, produced by the nervous system, serve as the primary communication mechanism within the body, facilitating the connection between various organs and the central nervous system (CNS). These signals are crucial for coordinating complex activities, ranging from movement to sensory information processing (Kandel et al., 2000). Similarly, hormones, the chemical messengers of the endocrine system, regulate key physiological processes such as metabolism, growth, development, and reproduction (Guyton & Hall, 2006). The heart's electrical signals ensure the synchronized contraction of cardiac muscles, maintaining a consistent heartbeat that is vital for sustaining life (Lilly, 2012).

The collective action of these signals is fundamental in maintaining the body's homeostasis, enabling it to effectively respond to environmental changes. The study of multimodal signals, which includes both electrical impulses from the nervous system and chemical signals from the endocrine system among others, is pivotal in diagnosing and researching various neurological and physiological disorders. They provide critical insights into the impact of specific medications on brain and body functions and aid in investigating the neural and physiological underpinnings of cognitive and motor processes (Niedermeyer & da Silva, 2005).

In the realms of research and clinical practice, the exploration of multimodal signals has profoundly deepened our understanding of the dynamic interactions between the brain and body. These signals are not only valuable in medical and scientific contexts but are also utilized in psychology, neuroscience, and cognitive science, enhancing our knowledge of how brain and body processes are interlinked (Zatorre et al., 2012). As research methodologies and technologies continue to advance, the analysis of multimodal signals remains a critical tool in decoding the complexities of human physiology and neurology, offering potential for break-throughs in diagnosing, treating, and comprehending a vast array of conditions and behaviors.

There are several different types of biosignals, each of which provides information about different physiological processes in the body. Some examples include:

1. Electroencephalography (EEG) signals, which measure the electrical activity of the brain.
2. Electrocardiography (ECG) signals, which measure the electrical activity of the heart.
3. Electromyography (EMG) signals, which measure the electrical activity of muscles.
4. Photoplethysmography (PPG) signals, which measure blood flow in the blood vessels.
5. Galvanic skin response (GSR) signals, which measure changes in the electrical conductance of the skin.
6. Respiratory signals, which measure breathing rate and volume.
7. Body temperature signals, which measure the temperature of the body.
8. Hormone signals, which measure the levels of hormones in the blood or in physiological fluids.

These are some of the common examples, but there are many other types of biosignals as well, which can provide information about various physiological processes in the body.

2

The nervous system, encompassing the central nervous system (CNS)—which includes the brain and spinal cord—serves as the principal conduit for communication between the brain and peripheral regions of the body, as depicted in ◘ Fig. 2.1. This intricate network is responsible for the transmission of sensory information through sensory neurons. These specialized neurons relay data regarding tactile sensations, temperature variations, pain, and other physical stimuli, facilitating our interaction with the environment (Bear et al., 2020).

Additionally, the nervous system employs proprioceptors, a unique class of sensory neurons, to convey information about the internal condition of the body, such as muscle tension and the position of joints. This internal feedback is crucial for the execution of coordinated movements and the maintenance of posture and equilibrium since it conveys information about the position of the body in the space/environment (Sherrington, 2023; Proske & Gandevia, 2012).

Motor functions are orchestrated through signals transmitted by motor neurons, which are tasked with regulating muscle movements and the activity of organs. This system enables not just voluntary movements but also the involuntary contractions essential for organ function. Moreover, the CNS communicates with the endocrine system, signaling glands to release hormones that modulate various physiological processes, including metabolism, growth, development, and reproduction (Smith et al., 2005).

Feedback mechanisms, involving both neural pathways and hormonal signals, enable a reciprocal flow of information between the body and brain. This dynamic interaction allows the brain to adjust its directives based on bodily states and environmental inputs, thereby maintaining homeostasis—a state of internal physiolog-

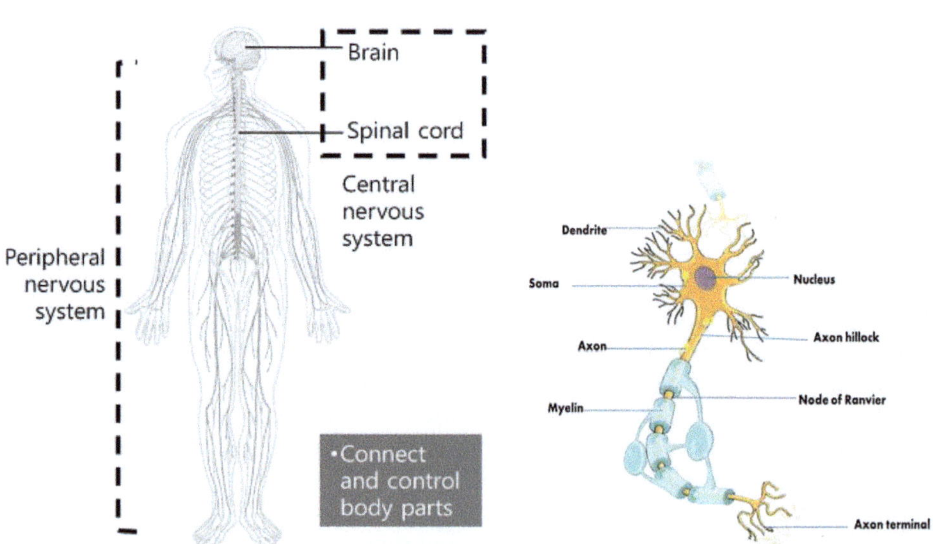

◘ **Fig. 2.1** The nervous system is composed of the brain, spinal cord, and nerves, which are organized into two main divisions: the Central Nervous System (CNS) and the Peripheral Nervous System (PNS). (From Gautam, 2017; Park, 2023)

ical balance (Matthews, 2002). The continuous, bidirectional communication between the brain and body is thus indispensable for homeostasis and adaptive responses to environmental changes. A theoretical account of this adaptation loop, where action, perception, and learning take place, has also been proposed under the free-energy principle, where the expected utility or cost of an action is optimized (Friston, 2010). In this book we will focus on different brain signals mentioned in ▶ Chap. 1.

> **Box 2.1: Neurophysiological Signals**
> Biological signals produced by the nervous system that carries information between different parts of the body and the brain.

2.1 The Neurophysiological Origin of Brain Signals

The neurophysiological origins of brain signals emanate from the activities of neurons, which serve as the fundamental components of the nervous system. These specialized cells are pivotal in transmitting information across the brain and the rest of the body through a combination of electrical and chemical signals. The electrical activity inherent to neuron function arises from ion movements across the neuron's membrane. Specifically, when neurons are activated, there is an influx of positively charged sodium ions (Na^+) into the cell, contrasted by an efflux of negatively charged chloride ions (Cl^-). This ion exchange generates a temporary charge disparity across the membrane, leading to the generation of an action potential. Action potentials, i.e., the difference of voltage potential between the two sides of the membrane, represent the cornerstone of neural communication. They are the source of the electrical activity usually detected by neuroimaging techniques such as electroencephalography (EEG) (Buzsáki, 2006).

In addition to electrical signaling, neurons communicate chemically via neurotransmitters. These molecules are released from a neuron's axon terminal into the synaptic cleft—the narrow space between the axon of one neuron and the dendrite of another—upon the arrival of an action potential. The neurotransmitters then attach to receptors on the subsequent neuron's dendrites, potentially exciting or inhibiting the neuron depending on the neurotransmitter and receptor types involved (Hyman, 2005). This dual mode of communication among neurons—electrical and chemical—lays the foundation for the complex information processing and integration that characterizes human perception, behavior, and cognition as shown in ◘ Fig. 2.2 (Frank et al., 2019).

In this chapter and the subsequent one (▶ Chap. 3), we will explore the neurophysiological principles that form the basis for various brain signal acquisition techniques. We will cover a range of methodologies:

- Electroencephalography (EEG), which records the electrical activity of the brain.
- Electrooculography (EOG), utilized for tracking eye movements.

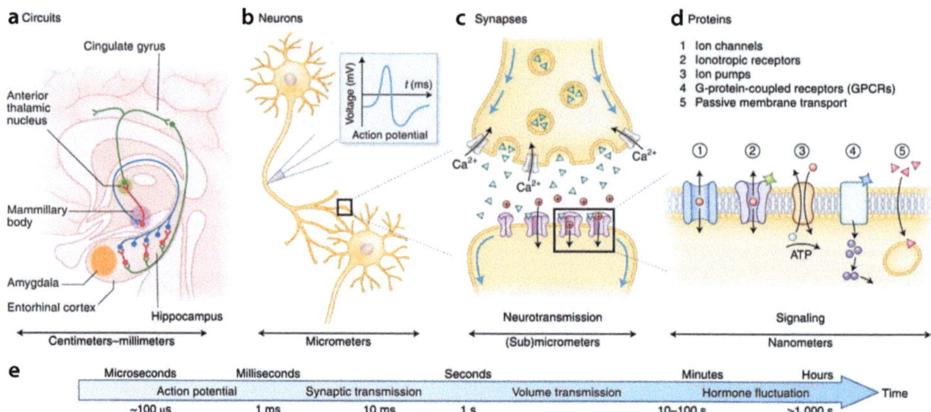

Fig. 2.2 Detailed visualization of neuronal communication processes. This figure provides a hierarchical examination of neuronal communication, illustrating a progressively detailed view from macroscopic brain circuits down to the molecular components of a neuron. **a** The depiction begins with an example of a neural circuit, specifically the Papez circuit, which integrates various brain regions to coordinate emotional expression and the processing of declarative memories. This circuit exemplifies the complex connectivity spanning extensive brain areas. **b** At the neuronal level, communication is facilitated through both electrical and chemical means. An action potential travels along the neuron to the synapse, initiating the process of chemical neurotransmission. **c** Within the synapse, the presynaptic neuron (illustrated at the top) releases neurotransmitters. These chemical messengers traverse the synaptic cleft and bind to specific receptor proteins located on the surface of the postsynaptic neuron (shown at the bottom), effectively transmitting the signal. **d** The diversity of membrane receptors present on neurons is highlighted, underscoring their critical roles in modulating neurotransmission and influencing neuronal communication pathways. **e** The temporal dynamics of neuronal communication are emphasized, with processes varying across timescales from sub-milliseconds to several hours, reflecting the complex and dynamic nature of neuronal interactions. This overview underscores the intricate and multi-scale nature of neuronal communication, essential for understanding both normal brain function and various neuropathological conditions. (From Frank et al., 2019)

- Functional near-infrared spectroscopy (fNIRS) and functional magnetic resonance imaging (fMRI), both of which are instrumental in measuring cerebral blood flow and oxygen levels.
- Magnetoencephalography (MEG), which detects magnetic fields produced by neural activity.
- Electrocorticography (ECoG), single-unit activity (SUA), multi-unit activity (MUA), and local field potential (LFP), which are used for monitoring different scales of neuronal electrical activity.
- Functional ultrasound (fUS), discussed for its capabilities in assessing cerebral blood volume and flow dynamics (Bear et al., 2020).

These techniques not only provide insights into the neurophysiological underpinnings of brain signals but also highlight the complex dynamics of neuronal communication essential for the brain's diverse functions. Brain signals can be broadly

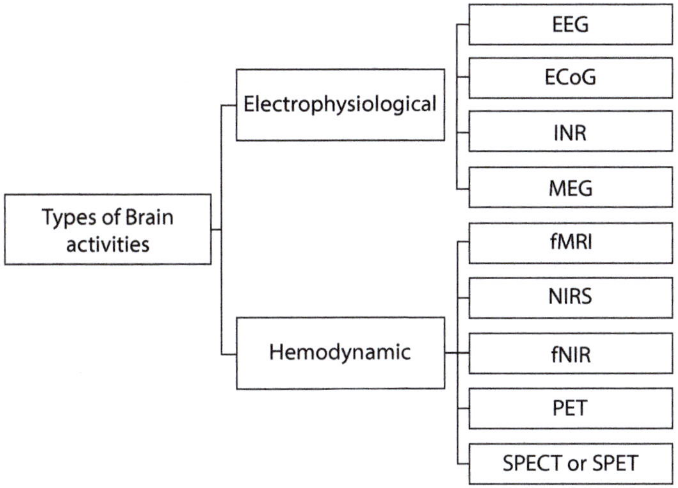

Fig. 2.3 Categorization of brain signal acquisition techniques utilized to study brain activities based on either electrical or hemodynamic aspects of neural function. (From Yadav et al., 2020)

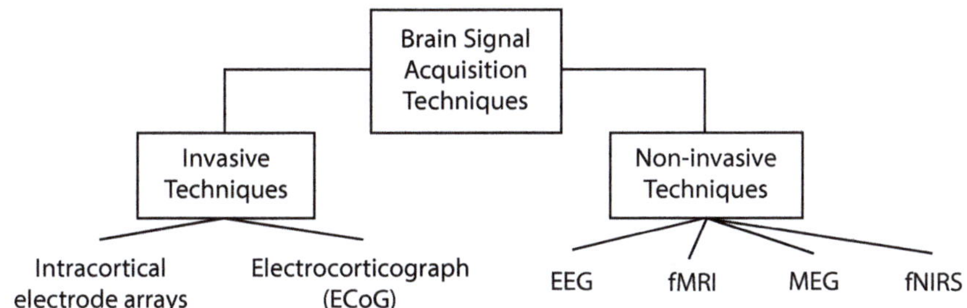

Fig. 2.4 Categorization of brain signal acquisition techniques based on either invasive on non-invasive methods

categorized into electrical and hemodynamic types, as illustrated in **Fig. 2.3** (Yadav et al., 2020). Based on the method of acquisition, these signals are further classified into invasive and non-invasive techniques, depicted in **Fig. 2.4**. This chapter will focus on non-invasive methods for studying brain signals.

Box 2.2: Important Note
Student must understand the difference between the terms ending with suffixes, "-graph," "-graphy," and "-gram." The "-graph" is the "instrument used to record," "-graphy" is recording or "to record," and "-gram" is the result of the recording or the "record."

2.1.1 Neurophysiological Origin of Electroencephalogram

The electroencephalography (EEG) is an established neurophysiological method for monitoring the brain's electrical activity. This technique captures the collective electrical fields generated by the synchronized firing of vast neuronal populations (Freeman, 2004a, b), detectable at the scalp's surface as shown in ◘ Fig. 2.5 (Siuly et al., 2016). The genesis of EEG signals is intimately linked to the functioning of individual neurons within the brain. These neurons communicate via electrical impulses known as action potentials, triggered by ion fluxes across their membranes. These signals traverse the neuron's axon, reaching and influencing other neurons across synapses (Buzsáki, 2006).

A critical component in the generation of EEG signals is the pyramidal neurons, predominantly found in the cerebral cortex—the brain's outermost layer tasked with higher-order cognitive functions such as perception, attention, and memory. These neurons are chiefly excitatory, their action potentials fostering activation in neighboring neurons. The structured layers of pyramidal neurons, through their extensive network of inputs and outputs connecting diverse brain regions, contribute significantly to the EEG signal detected by scalp electrodes (Nunez & Srinivasan, 2006), as shown in ◘ Fig. 2.5.

The modulation of pyramidal neuron activity by neurotransmitters—like dopamine, glutamate, and gamma-aminobutyric acid (GABA)—further influences the EEG signal's characteristics. Dopamine, for instance, has been observed to augment the alpha rhythms within the EEG spectrum, while GABA tends to reduce it, illustrating the complex interplay between neurotransmitter dynamics and EEG patterns (Buzsáki, 2006; Siuly et al., 2016).

EEG signals comprise various frequency bands as described below, each associated with distinct brain states or activities. ◘ Table 2.1 provides a summary of different EEG frequency bands.

1. Delta Waves (<4 Hz): These are the slowest brainwaves, predominantly observed during deep, non-REM (rapid eye movement) sleep. Delta waves are associated with healing, regeneration, and deep unconsciousness (Amzica & Steriade, 1998).
2. Theta Waves (4–7 Hz): Theta activity is linked with states of drowsiness, early stages of sleep, and REM sleep. It also plays a role in creativity, meditation, and daydreaming. Theta waves have been implicated in memory formation and navigation in spatial environments (Klimesch, 1999).
3. Alpha Waves (8–12 Hz): Characteristic of a relaxed yet awake state, alpha waves are prominent during calm, meditative states, and when the eyes are closed but the individual is alert. Alpha rhythms are considered indicators of a restful, relaxed state of mind, playing a critical role in stress reduction (Kirschfeld, 2005).
4. Beta Waves (13–30 Hz): These waves denote a state of active, analytical thinking, attention, and concentration on the external environment. Beta activity is heightened during problem-solving, decision-making, and focused mental activity (Banquet, 1973; Teplan, 2002).

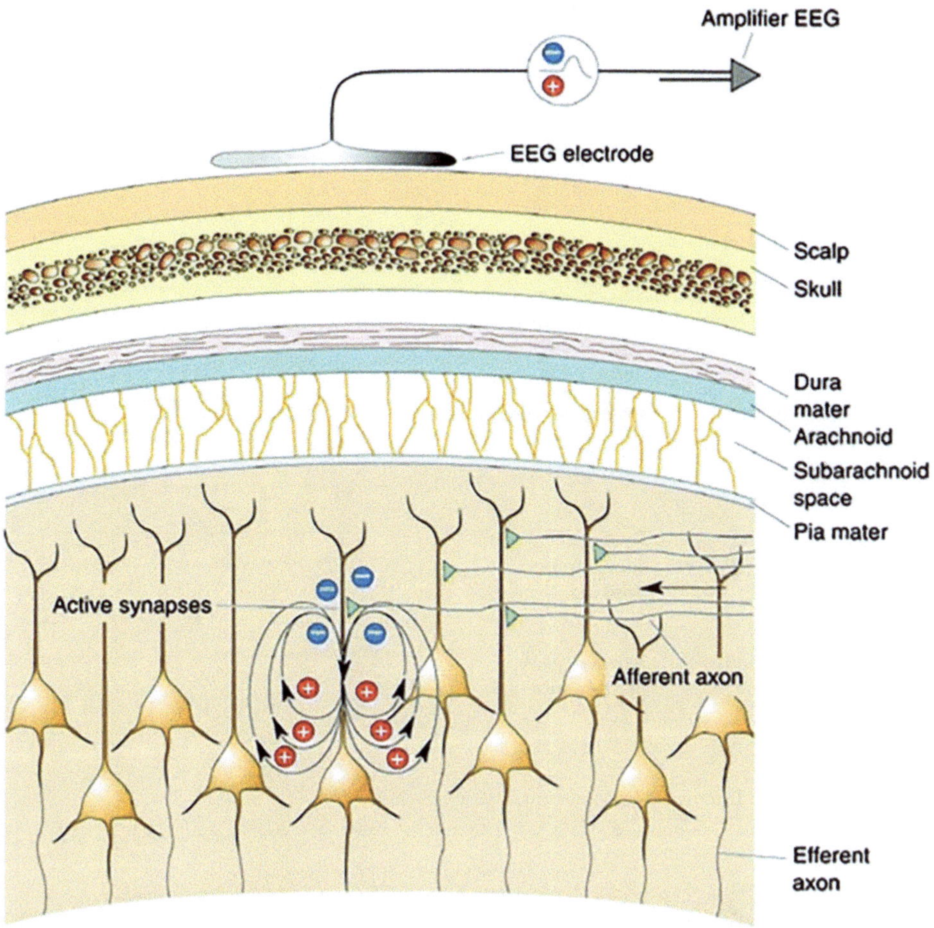

Fig. 2.5 The figure depicts how an EEG electrode placed on the top of the scalp records the summed electrical activity of neurons lying in the area underneath after the electrical signal has passed through different layers, namely pia mater, subarachnoid, arachnoid, dura mater, skull, and scalp. This illustration also depicts the generation of minute electrical fields by synaptic currents within pyramidal cells. An EEG electrode positioned on the scalp is used to measure these signals, which must traverse several thick layers of tissue. Due to the weak nature of the electrical fields produced by individual pyramidal cells, it is only when thousands of these cells discharge their synaptic currents in a synchronized manner that the cumulative voltages reach a magnitude detectable by surface electrodes (Freeman, 2004a, b). This synchronization amplifies the tiny voltages enough to overcome the signal attenuation caused by the intervening layers of skin, skull, and meninges, thereby allowing for successful EEG recording at the scalp level. (From Siuly et al., 2016)

5. **Gamma Waves (>30 Hz):** Associated with higher-level cognitive functioning, including perception, problem-solving, fear, and consciousness, gamma waves reflect the synchronous activity of neurons working in harmony across different parts of the brain (Miltner et al., 1999).

◘ Table 2.1   The frequency band of EEG signals

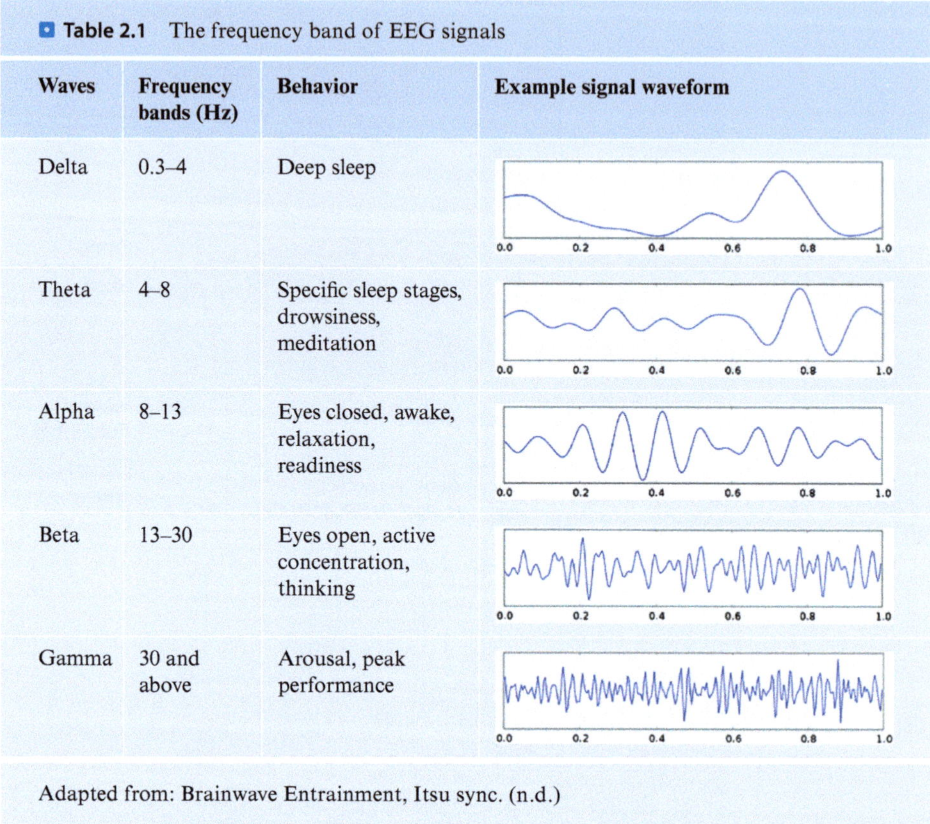

Waves	Frequency bands (Hz)	Behavior	Example signal waveform
Delta	0.3–4	Deep sleep	
Theta	4–8	Specific sleep stages, drowsiness, meditation	
Alpha	8–13	Eyes closed, awake, relaxation, readiness	
Beta	13–30	Eyes open, active concentration, thinking	
Gamma	30 and above	Arousal, peak performance	

Adapted from: Brainwave Entrainment, Itsu sync. (n.d.)

The rhythmic activity of pyramidal neurons is especially pronounced within the alpha and beta bands, reflecting their pivotal role in EEG signal formation (Buzsaki & Draguhn, 2004).

Furthermore, EEG can infer the neural activity's spatial origin, as the electrical field strength is greatest at scalp locations directly above the active neuron groups. This principle underlies electroencephalography source imaging (ESI), a technique employing inverse problem-solving to localize neural sources of EEG signals (Michcl et al., 2004).

In summary, EEG signals emanate from the complex, synchronized activity of vast neuronal assemblies, particularly involving pyramidal neurons, within the brain. These signals, modulated by various neurotransmitters, offer a window into the brain's operational states and cognitive processes. Understanding the neurophysiological foundations of EEG signals enhances our grasp of the neural mechanisms underpinning diverse cognitive and brain states.

> **Box 2.3**
>
> **Pyramidal Neurons** are a type of excitatory neuron found predominantly in the cerebral cortex, the hippocampus, and the amygdala of the brain. These neurons are characterized by a pyramid-shaped cell body, a large apical dendrite that extends toward the cortical surface, multiple basal dendrites, and a long axon that can project to distant areas. Pyramidal neurons play a crucial role in neural networks by sending excitatory signals to other neurons, significantly influencing cognitive processes such as learning, memory, and decision-making. They are integral to the transmission of neural signals across cortical layers and between cortical regions, facilitating complex communicative functions within the brain's circuitry. The activity of pyramidal neurons is often a focal point in studies of neural plasticity and is also critically involved in the generation of various rhythmic activity patterns observed in EEG recordings.

> **Box 2.4**
>
> **Electroencephalography (EEG)** provides a real-time trace of brain activity, showing rhythmic patterns or oscillations of electrical voltage which can be correlated with different states of brain function. EEG is valuable in research and clinical settings for assessing brain activity over time and can also be used to study cognition, sensorimotor function, and to monitor mental states in various psychological research settings.

2.1.2 Neurophysiological Origin of Electrooculogram

Electrooculography (EOG) is a cost-effective and precise method for measuring and monitoring eye movements, leveraging the fundamental principle that the eye acts as an electric dipole. This dipole is characterized by the cornea's positive electrical charge relative to the negatively charged retina. Consequently, eye movements generate a differential potential that correlates with the extent of the eye's rotation. The magnitude of this potential is contingent upon the angular displacement of the eyeball. The EOG technique involves capturing this electrical potential using surface electrodes. When the eyes shift horizontally, the recorded voltage shifts positively or negatively, based on the configuration of the electrode placement, and it returns to zero when the eyes are directed forward. ◘ Figure 2.6 illustrates this phenomenon through a simplified visual representation (López et al., 2019).

The voltage amplitude for horizontal and vertical eye movements can reach up to 16 µV and 14 µV per degree, respectively, displaying a nearly linear response within ±50° for horizontal and ±30° for vertical movements (Webster, 2009). To record these dipole changes, small surface electrodes are placed around the eyes, with various configurations available depending on the specific application (Lopez et al., 2016). The polarity of the recorded signal correlates with the direction of eye

2

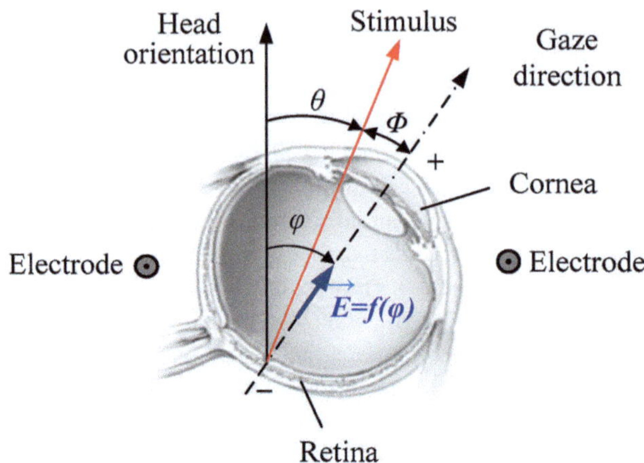

Fig. 2.6 The model of the eyeball as a dipole posits that the angular deviation of eye movements in response to a stimulus influences the electric field. The vector $E = f(\theta)$ denotes this relationship, representing the electric field as a function of the eye's rotation angle θ. This mathematical formulation captures how the dipole orientation of the eyeball changes with varying angles of eye movement, thereby affecting the electric field magnitude and direction. (From López et al., 2019; originally published under CC BY 4.0)

movement: positive toward the positive electrode and negative toward the negative electrode. Typically, EOG amplitudes range from 0.05 to 3.5 mV in humans, influenced by environmental factors and ocular health (Brown et al., 2006). The signal bandwidth spans from direct current (DC) up to 50 Hz, though the frequencies of interest generally lie between 0.1 and 40 Hz (Enderle, 2006).

The primary patterns identifiable in EOG data include saccades, fixations, and blinks. Saccades are rapid, simultaneous movements of both eyes, with peak amplitudes corresponding to the saccade's angular displacement. Fixations occur when the gaze is steadily held on a particular point, during which small involuntary movements such as drifts and flicks, typically around 1°, may be observed (Brown et al., 2006; Enderle, 2006; Webster, 2009). Eye tremors or vibrations can also occur, with frequencies ranging from 30 to 150 Hz. Blinks are categorized as involuntary or voluntary, with voluntary, often monocular, blinks displaying amplitudes 10–20 times higher than the more common simultaneous blinks (Constable et al., 2017; Findlay & Walker, 1999). ■ Figure 2.7 illustrates these patterns through vertical and horizontal eye potentials recorded with a commercial bioamplifier.

> **Box 2.5**
>
> **Electrooculography (EOG)** is a non-invasive method used to record eye movements. EOG is commonly used in clinical settings to diagnose and monitor diseases affecting eye movement, such as muscle and nerve disorders. It is also used in sleep studies to identify REM (rapid eye movement) stages and in human–computer interaction research for gaze-tracking applications.

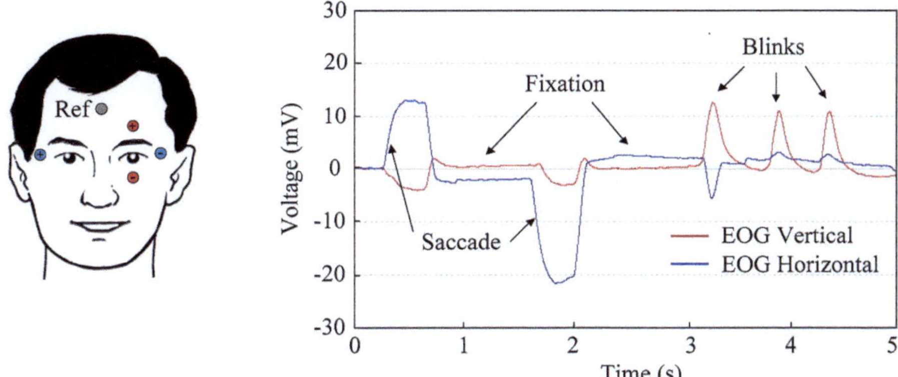

◼ Fig. 2.7 The electrooculogram (EOG) is adept at capturing a variety of eye movements, including saccadic shifts, periods of fixation, and blinks, through distinctive signal patterns. The red and blue waveforms observed in EOG recordings represent the vertical and horizontal eye movements, respectively. These signals are collected by electrodes positioned around the eyes, with each electrode's color corresponding to the waveform it records. Specifically, electrodes placed vertically around the eye (above and below) capture the vertical potentials, resulting in the red waveform, while those positioned horizontally (at the outer and inner canthi of the eye) detect horizontal movements, producing the blue waveform. Additionally, a gray electrode serves as the reference, offering a baseline for signal comparison and enhancing the accuracy of movement detection. (Adapted from López et al., 2019; originally published under CC BY 4.0)

2.1.3 Neurophysiological Origin of Functional Near-Infrared Spectroscopy (fNIRS)

Functional Near-Infrared Spectroscopy (fNIRS) represents a non-invasive neuro-imaging technique that exploits infrared light to probe changes in the concentrations of oxygenated (HbO) and deoxygenated hemoglobin (HbR) within the brain. This methodology rests on the premise that neural activity precipitates alterations in cerebral blood flow and oxygenation levels, detectable through variations in hemoglobin's absorption of infrared light. The tight coupling between neuronal activity and vascular responses, known as "neurovascular coupling," underscores this relationship, linking neuronal function to subsequent vascular changes (Heeger & Ress, 2002; Logothetis, 2001).

The underlying neurophysiological basis of fNIRS signals originates from neuronal activities. Active neurons increase energy and oxygen consumption, leading to enhanced blood flow and oxygenation in the brain regions where these neurons reside. This augmented cerebral circulation alters the local concentrations of HbO and HbR, which fNIRS detects by monitoring the differential absorption of infrared light (Jobsis, 1977; Villringer & Chance, 1997).

The brain contains various light-absorbing molecules, or chromophores, including water, cytochrome-c oxidase, and hemoglobin. Visible light is strongly absorbed by water and other brain tissues, rendering the brain opaque in this spectrum.

2

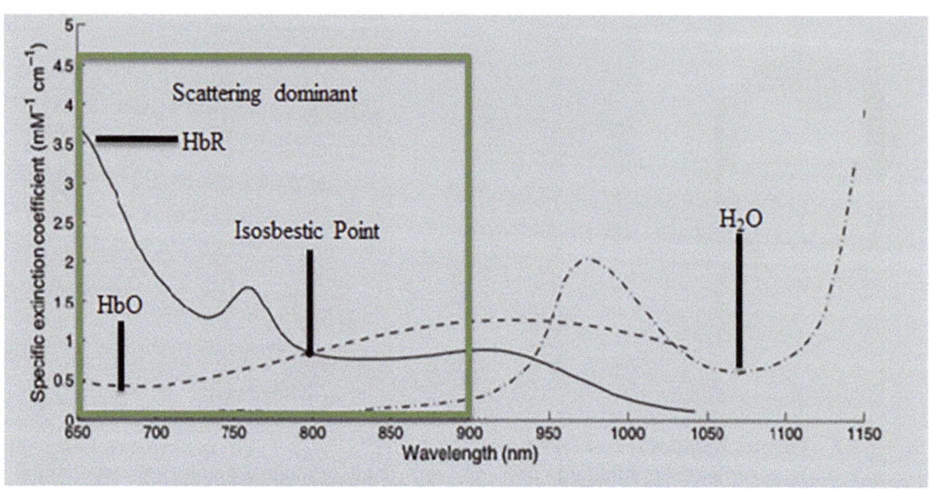

Fig. 2.8 Absorption spectra for de-oxy hemoglobin (Hb/HbR), oxy hemoglobin (HbO$_2$/HbO), and water (H$_2$O) in the near-infrared range. The isosbestic point near 800 nm is the point where the absorptivity of HbR and HbO are equal. (Modified and adapted from Wray et al., 1988)

However, the near-infrared (NIR) spectrum, ranging from 650 to 950 nm, presents a "biological window" where light absorption by water drastically reduces, highlighting hemoglobin and cytochrome-c oxidase as the principal chromophores as shown in ■ Fig. 2.8. Given hemoglobin's significantly higher concentration in brain tissue, it predominates light attenuation during NIR imaging, facilitating the analysis of its concentration changes in response to neural activity (McCormick et al., 1992; Strangman et al., 2002).

fNIRS leverages the NIR light spectrum's optical window, where oxy- and deoxy-hemoglobin exhibit unique absorption and scattering characteristics, allowing for deep tissue penetration and quantification of their concentrations. Typically, wavelengths near 690 and 830 nm are employed to selectively assess HbR and HbO concentrations, respectively. Additionally, an isosbestic point around 800 nm, where the absorption spectra of HbO and HbR overlap, provides insights into total hemoglobin concentration changes, offering a proxy for overall blood volume independent of oxygenation status (Boas et al., 2004; Chance et al., 1988).

The characteristics of the medium through which infrared light passes—including the scalp, skull, and neural tissue—as well as the spacing between the light emitters and detectors, constrain the penetration depth of the method. As a result, the detectable hemodynamic changes using functional near-infrared spectroscopy (fNIRS) are typically restricted to a maximum depth of 2–3 cm within the adult brain. This limitation confines reliable fNIRS measurements primarily to the cortical surface. fNIRS signals, sensitive to both neuronal and vascular dynamics, provide insights into specific brain regions' neural activities, elucidating the neural mechanisms underpinning various cognitive states and functions. By understanding the neurophysiological and vascular underpinnings of fNIRS signals, researchers can glean valuable information about the interplay between neural function, blood flow, and oxygenation in supporting cognitive processes and brain health (Jobsis, 1977; Villringer & Chance, 1997).

> **Box 2.6**
> fNIRS is particularly valued for its ability to safely and effectively assess brain functions in various settings due to its portability and relative insensitivity to motion artifacts compared to other imaging methods. It is widely used in cognitive neuroscience research, clinical diagnostics, and brain–computer interfacing, offering a practical tool for understanding brain functions and dysfunctions in real-world environments.

2.1.4 Neurophysiological Origin of Functional Magnetic Resonance Imaging (fMRI)

Functional magnetic resonance imaging (fMRI) is a pivotal neuroimaging technique that infers localized changes in neuronal activity by monitoring associated fluctuations in cerebral blood flow. Seminal research by Bandettini et al. (1992), Kwong et al. (1992), and Ogawa et al. (1992) has solidified fMRI's role as a fundamental tool for functional brain imaging in humans. This methodology utilizes task-induced metabolic and hemodynamic responses to approximate variations in local neuronal activity. Consequently, fMRI signals offer an indirect measure of neuronal function, effectively linking neurovascular coupling phenomena with the comprehensive analytic capabilities of magnetic resonance imaging (MRI) technology. Understanding how metabolic and hemodynamic changes are translated into signals of neuronal activity is crucial for fully leveraging fMRI in the analysis of brain function.

At the core of fMRI's operational principle lies the understanding that active brain regions witness an increase in blood flow, aimed at delivering essential oxygen and nutrients to neurons in high energy demand. This increase in cerebral circulation changes the blood's magnetic properties, detectable by MRI scanners as shown in ◘ Fig. 2.9. The technique capitalizes on the difference in magnetic susceptibility between oxygenated (paramagnetic) and deoxygenated (diamagnetic) blood, allowing the detection of increased oxygenated blood flow to active brain areas as heightened signal intensity on MRI scans. The Blood Oxygen Level-Dependent (BOLD) contrast, pivotal to fMRI, hinges on this differential magnetic response, serving as an indirect marker for neural activity (Ogawa et al., 1990; Kwong et al., 1992).

The BOLD signal comprises several determinants, including the initial neuronal response to stimuli, the intricate linkage between neuronal action and the ensuing hemodynamic response, the vascular reaction itself, and the MRI scanner's capability to capture this response (as shown in ◘ Fig. 2.9 from left to right)

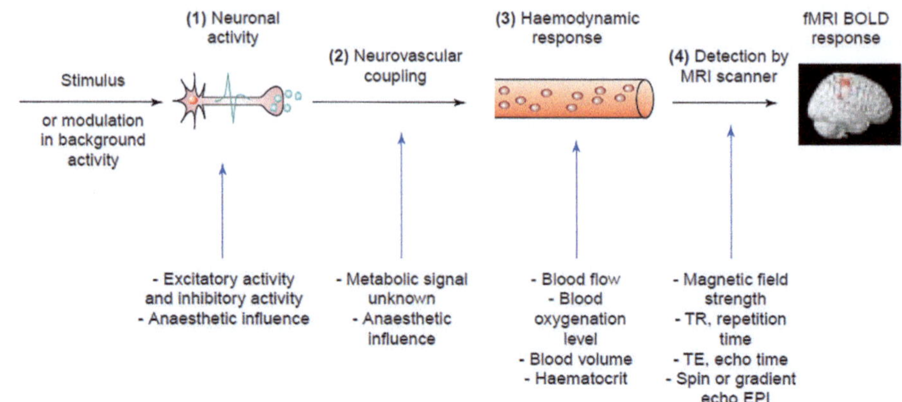

◘ Fig. 2.9 The BOLD signal has several constituents: (1) the neuronal response to a stimulus or background modulation; (2) the complex relationship between neuronal activity and triggering a hemodynamic response (termed neurovascular coupling); (3) the hemodynamic response itself; and (4) the way in which this response is detected by an MRI scanner. (From Arthurs & Boniface, 2002)

(Arthurs & Boniface, 2002). Factors such as magnetic field strength, echo time, and the specific imaging technique employed significantly influence the observed BOLD signal. For instance, a 1% BOLD change at a 30 ms echo time may equate to a 2% change at 60 ms, assuming a stable hemodynamic response. Nevertheless, BOLD imaging's susceptibility to artifacts like head movement and field inhomogeneities complicates the interpretation of these hemodynamic responses (Logothetis, 2002).

Substantial research, including work by Friston et al. (1994) and based on the Balloon model proposed by Buxton et al. (1998), characterized the hemodynamic response. The essence of the BOLD signal largely rests on the discrepancy between localized surges in cerebral blood flow and slightly lagging increases in oxygen consumption, leading to a momentary reduction in the deoxyhemoglobin to oxyhemoglobin ratio. Beyond these initial considerations, physiological aspects such as blood volume, vascular geometry, and basal oxygenation levels also modulate deoxyhemoglobin concentrations, influencing the BOLD response (Buxton & Frank, 1997; Kim & Ogawa, 2012).

fMRI's spatial resolution, and thus its accuracy in pinpointing neural activity, may be compromised by the vascular density and the "brain versus vein" issue, where signal changes in large draining veins may not accurately reflect the locale of neuronal activity (Arthurs & Boniface, 2002). Techniques like spin-echo fMRI have been developed to mitigate these venous contributions, aiming for a more precise correlation to neural origins at the cost of signal-to-noise ratio (SNR) (Lee et al., 1995; Duong et al., 2003).

In neuroscience, fMRI has emerged as a pivotal tool, enabling in vivo exploration of the brain's functional architecture and the study of diverse cognitive processes and neuropsychiatric conditions. Although fMRI does not directly measure neural activity, its indirect assessment through neurovascular coupling makes it an invaluable asset in conjunction with other modalities like EEG and MEG for a holistic understanding of brain dynamics (Bandettini, 2009).

> **Box 2.7**
> **Neurovascular Coupling** refers to the relationship between local neural activity and the resulting changes in cerebral blood flow. This physiological mechanism ensures that active regions of the brain receive an increased blood supply to meet their elevated metabolic demands. When neurons are active, they consume more oxygen and glucose, which leads to a localized increase in blood flow to these regions. This process is fundamental for maintaining the energy balance of the brain's active areas and is typically measured using imaging techniques such as functional magnetic resonance imaging (fMRI) and functional near-infrared spectroscopy (fNIRS). These methods detect changes in blood oxygen levels and flow, providing indirect measures of neural activity. Neurovascular coupling is crucial for understanding how the brain supports cognitive functions and is also studied to explore various neurological conditions where these mechanisms may be impaired.

> **Box 2.8**
> fMRI is based on the principle that cerebral blood flow and neuronal activation are closely linked. Areas of the brain that are more active consume more oxygen, and this increased activity and oxygen usage result in detectable changes in the magnetic properties of the blood. fMRI provides both high spatial and temporal resolution, making it an invaluable tool in the fields of psychology, neurology, and cognitive sciences. It is extensively used for exploring the brain's functional anatomy, assessing the effects of strokes or disease, and in planning surgeries for epilepsy and other conditions that affect brain function.

2.1.5 Neurophysiological Origin of Magnetoencephalography (MEG)

Magnetoencephalography (MEG) is a sophisticated neuroimaging technique that quantifies the magnetic fields produced by neuronal electrical activity within the brain. Unlike other imaging modalities, MEG offers the unique advantage of capturing dynamic brain activities on a millisecond scale, providing insights into the rapid temporal evolution of neuronal processes. The source of MEG signals is the electrical activity stemming from the cumulative action potentials and synaptic activities of numerous neurons within specific brain regions (Hämäläinen et al., 1993; Baillet et al., 2001). Since an electrical field is intrinsically coupled with a corresponding magnetic field, the local activity (action potentials) of small neural

2

populations can be captured by measuring its neuro-magnetic counterpart. The foundational account of this electrical and magnetic coupling has been proposed by Maxwell (1865) and supported by the Lorentz force law (1895) (2013) and is well known as electromagnetism.

Neurons, the basic functional units of the brain, communicate via electrical impulses known as action potentials. These are initiated when a neuron is stimulated, leading to a transient influx and efflux of ions like sodium and potassium across the neuronal membrane, as shown in ▣ Fig. 2.10a (Baillet, 2017). Changes in this electrical field induce changes in the corresponding magnetic field according to Maxwell's equations. Although the magnetic field generated by a single neuron (as shown in ▣ Fig. 2.10a) is minute and undetectable with current technology, the synchronized activity of a large group of neurons can produce detectable magnetic fields outside the scalp, captured by MEG technology, as shown in ▣ Fig. 2.10b (Baillet, 2017).

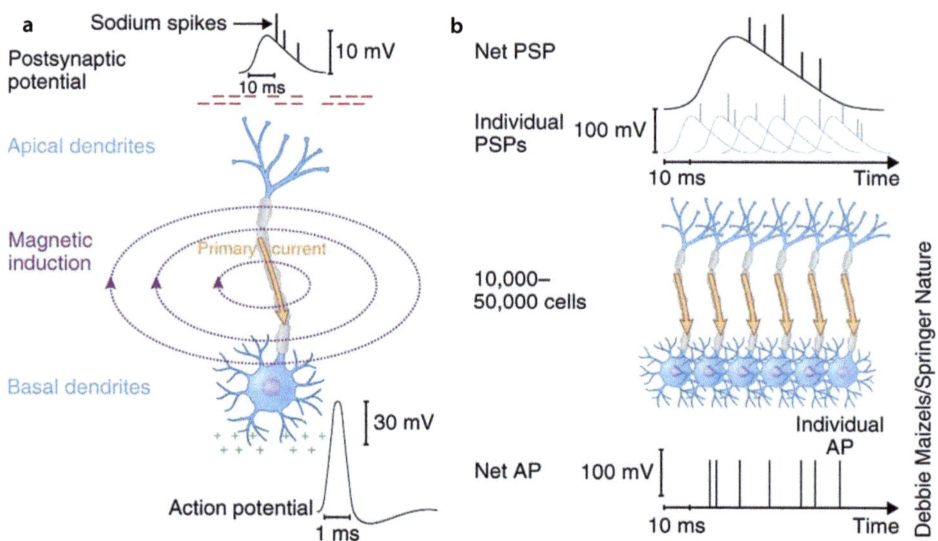

▣ **Fig. 2.10** Cellular origins of MEG signals. **a** For illustrative purposes, we consider the cortical pyramidal neuron as the primary cellular generator of MEG signals. The unique elongated structure of pyramidal neurons directs physiological currents along the cell's axis, enhancing the magnetic induction compared to cells with stellate morphologies. This primary current arises from an electrical potential imbalance between the cell's apical dendritic tree and its soma and basal dendrites. The resulting magnetic fields, visualized here with isolines in purple, are perpendicular to the direction of current flow and are detectable externally to the head. The signal sources include postsynaptic potentials (PSPs), characterized by rapid, high-amplitude sodium spikes, and axonal discharges (action potentials, APs), with PSPs generally exhibiting slower and smaller amplitude fluctuations than APs. **b** At the collective level of cell assemblies, the cumulative effect of the slower PSPs is more pronounced than that of APs due to their extended temporal overlap, which does not require strict synchronization. Both computational models and empirical data suggest that a detectable MEG signal typically requires the synchronous activity of approximately 10,000–50,000 neurons. It is conceivable that rapid PSP activity, along with potential residual effects of APs, could be observed in MEG recordings. (From Baillet, 2017)

MEG employs superconducting quantum interference devices (SQUIDs), which are exquisitely sensitive to magnetic fields, allowing for the detection of the brain's minuscule magnetic signals (ranging from 10 to 1000 fT) amid the much stronger Earth's magnetic field (from 25 to 65 µT). To minimize interference from external magnetic sources, MEG recordings are conducted in rooms specially shielded against ambient magnetic fields (Vrba & Robinson, 2001). To maintain superconducting properties, SQUIDs are typically immersed in liquid helium, which generally requires expensive equipment to avoid transitioning to a gaseous state.

The spatial resolution of MEG, enhanced by the use of arrays containing hundreds of SQUID sensors, permits the precise localization of brain activity sources. The localization process involves solving both forward and inverse problems. The forward problem calculates the magnetic fields at the sensors for known source locations and orientations, whereas the inverse problem seeks to infer the sources' characteristics based on the measured magnetic fields. Given the complexity and non-uniqueness of the inverse problem, additional constraints based on anatomical and functional knowledge are often applied to derive meaningful solutions (Sarvas, 1987; Hämäläinen & Sarvas, 1989).

Co-registration techniques are applied during MEG to align the measured magnetic fields with an individual's brain anatomy, using fiducial markers placed on specific anatomical landmarks. This alignment facilitates the identification and analysis of activity from distinct brain regions, allowing researchers to dissect the magnetic oscillations associated with various cognitive functions and brain states (Baillet, 2017).

One of the major advantages of MEG is its high temporal resolution, allowing to measure the neural activity with a temporal resolution of milliseconds. Additionally, MEG can provide a non-invasive measurement of neural activity, which is an advantage over invasive methods such as single-unit recordings. It is important to note that MEG is a non-invasive method and it can provide a high temporal resolution of neural activity, but it has lower spatial resolution than other neuroimaging techniques such as fMRI and PET (positron emission tomography). Therefore, it is used in conjunction with other techniques such as fMRI, EEG, and single-unit recordings to get a more comprehensive understanding of neural activity (Baillet, 2011).

> **Box 2.9**
>
> **Magnetoencephalography (MEG)** measures the magnetic fields produced by neuronal activity in the brain using highly sensitive devices known as superconducting quantum interference devices (SQUIDs), which are capable of detecting the extremely faint magnetic signals generated by neural currents. This technique offers a direct measurement of neuronal activity with millisecond-level temporal resolution, providing insights into the dynamic processes of the brain. MEG is particularly useful for mapping brain functions related to sensory perception,

motor activity, language, and cognitive functions. It is also employed in pre-surgical planning for epilepsy treatment and in research on cognitive and neurological disorders. MEG's high temporal resolution and non-invasive nature make it a valuable tool in both clinical and research settings, complementing other imaging techniques like MRI and fMRI by providing a different dimension of brain activity analysis.

2.1.6 Neurophysiological Origin of Functional Ultrasound Imaging (fUSI)

Functional Ultrasound Imaging (fUSI) represents a transformative advance in non-invasive neuroimaging, offering unparalleled real-time insights into cerebral hemodynamics and neural activity. Rooted in the principles of ultrasound physics, fUSI detects the minute changes in blood flow within the brain's vascular system that accompany neuronal activity as shown in ◘ Fig. 2.11 (Nunez-Elizalde et al., 2022). These changes, primarily driven by the metabolic demands of active neurons, manifest as an increase in blood flow to regions of heightened neural activity—a phenomenon termed functional hyperemia (Iadecola, 2004).

Neurovascular coupling, a critical underpinning of fUSI, describes the complex physiological processes that link neuronal firing to vascular responses, ensuring that active brain regions receive an adequate supply of oxygen and nutrients. The increased metabolic demand of active neurons leads to a cascade of biochemical and physiological events, resulting in vasodilation and increased blood flow, as we have read in the previous sections of this chapter. This hemodynamic response, captured by fUSI, serves as an indirect marker of neural activity (Attwell et al., 2010; Drew et al., 2020).

fUSI employs power Doppler ultrasound to measure these blood flow dynamics. The technique relies on the detection of frequency shifts in ultrasound waves reflected by moving blood cells, with the magnitude of the shift correlating with the velocity of blood flow. Advanced signal processing algorithms are employed to extract meaningful hemodynamic signals from the raw Doppler data, filtering out noise from tissue movement and other artifacts (Rubin & Bude, 1994; Demené et al., 2016).

The processing of fUSI data involves several key steps, including temporal binning to average signals over short intervals, power estimation to quantify the strength of the Doppler signal, and the application of high-pass filtering to isolate the dynamic changes in blood flow from static background signals. Spatial and temporal clutter filtering techniques, such as principal component analysis, are often used to further refine the signal, enhancing the detection of hemodynamic responses associated with neuronal activity (Baranger et al., 2018; Macé et al., 2011).

Despite the indirect nature of the relationship between fUSI signals and neural activity, studies have shown a strong correlation between fUSI-derived hemodynamic measures and direct measures of neuronal firing, such as local field poten-

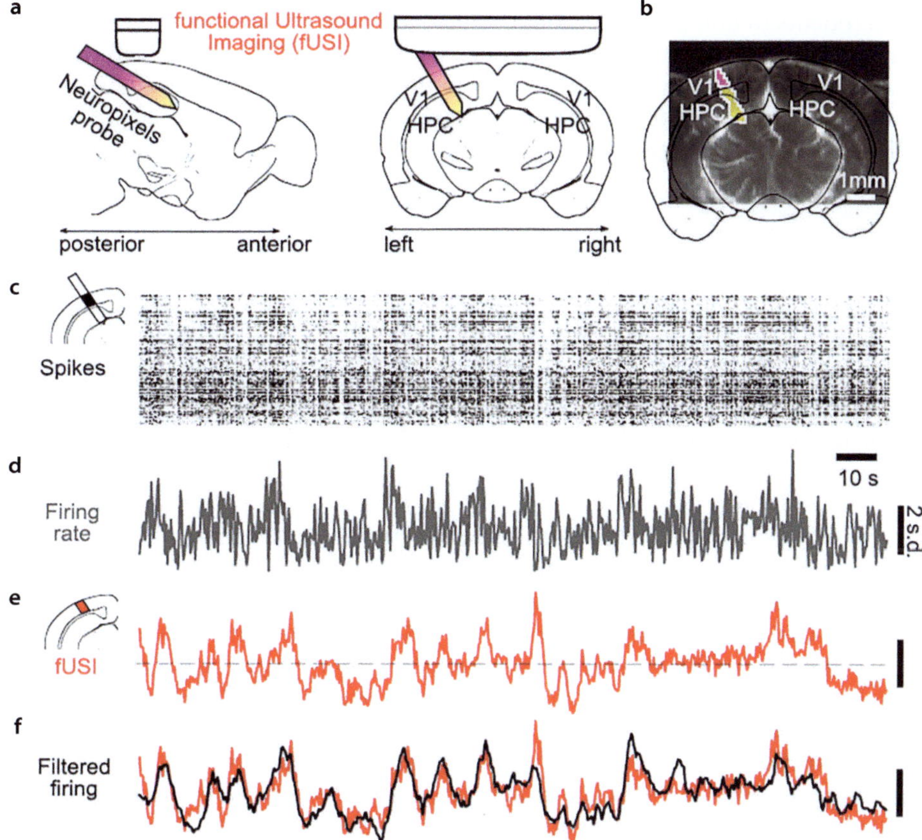

Fig. 2.11 **a** Diagram illustrating concurrent functional ultrasound imaging (fUSI) and electrophysiological recordings, highlighting the primary visual cortex (V1) and the hippocampus (HPC). **b** A coronal fUSI slice displaying the path of the Neuropixels probe, indicated by a purple line where it intersects the slice and a yellow line where it is positioned anteriorly. **c** A graphical representation of spikes recorded from V1, plotted over time and by recording depth during a sample session. **d** Computation of the mean firing rates derived from the spike data. **e** fUSI data collected simultaneously from the same location, averaged across 51 voxels. **f** Application of an optimal filter to smooth the firing rates, which provides effective predictions (shown in black) of the observed fUSI signals (depicted in red). (From Nunez-Elizalde et al., 2022; originally published under CC BY 4.0)

tials (LFPs). This suggests that, despite its reliance on neurovascular coupling, fUSI can serve as a reliable proxy for assessing neural activity across various brain states and conditions (Aydin et al., 2020; Nunez-Elizalde et al., 2022).

The versatility of fUSI extends beyond measuring blood flow changes, as it can also exploit the acoustic scattering effect to detect changes in the acoustic properties of neural tissue associated with neural activity. This advanced application of fUSI provides additional avenues for exploring the dynamics of neural processing (Osmanski et al., 2014).

Functional Ultrasound has been successfully applied across a range of species, from rodents to non-human primates, elucidating the neural underpinnings of sensory processing, cognitive states, and behavior (Blaize et al., 2020; Deffieux et al., 2021). Its capacity to image entire brain regions in small animals, akin to fMRI in humans, underscores its potential as a translational tool in neuroscience research (Blaize et al., 2017; Dizeux et al., 2019).

> **Box 2.10**
> **Functional Ultrasound Imaging (fUSI)** leverages the Doppler effect to detect changes in cerebral blood volume and flow, which are closely related to neuronal activity. It is relatively new as compared to other techniques. It is non-invasive, does not require the use of ionizing radiation, and is relatively inexpensive compared to other imaging modalities, making it a promising tool for advancing our understanding of the neural underpinnings of cognitive and motor functions.

2.2 Functioning of Hardware to Acquire Brain Signals

The acquisition and analysis of brain signals, such as those from electroencephalography (EEG), magnetoencephalography (MEG), and functional magnetic resonance imaging (fMRI), involve a sophisticated multistep process tailored to capture, condition, and interpret the complex neural activity underlying human cognition and behavior.

Preparation: The initial step encompasses preparing the participant for the recording session. For EEG, this preparation involves affixing an array of electrodes across the scalp, using conductive gels or pastes to enhance signal transmission. MEG sessions require the participant to wear a helmet embedded with highly sensitive SQUID sensors capable of detecting the brain's faint magnetic fields. fMRI preparation entails positioning the participant within the scanner's bore, ensuring minimal movement and optimal alignment of the head for accurate imaging (Hämäläinen et al., 1993; Logothetis, 2008).

Signal Acquisition: The core of the process lies in the precise acquisition of brain signals. EEG captures the electrical potentials generated by neuronal activity, using amplifiers to boost these signals for clearer detection and recording. MEG, through SQUID sensors, measures the minute magnetic fields produced by neural currents, offering insights into the brain's magnetic activity without direct contact with the scalp (Cohen, 1972; Hämäläinen et al., 1993). fMRI, leveraging the magnetic properties of blood oxygenation, detects hemodynamic responses associated with neural activity, providing spatially detailed images of brain function (Ogawa et al., 1990).

Signal Conditioning: Given the inherent noise and potential artifacts (e.g., muscle movements, ocular artifacts) present in raw brain signals, conditioning steps are essential. These involve filtering techniques, artifact rejection methods such as independent component analysis (ICA) or principal component analysis (PCA), aimed at isolating genuine neural signals from extraneous noise (Makeig et al., 1996; Uusitalo & Ilmoniemi, 1997).

Data Storage: Following conditioning, the refined signals are digitized and stored for subsequent analysis. This entails using robust data storage systems capable of handling large volumes of complex data (1D vectors, 2D matrices, and 3D volumes), ensuring integrity and accessibility for analysis (Tadel et al., 2011).

Data Analysis: The stored data undergoes extensive signal processing and analysis, employing a variety of computational techniques to extract meaningful patterns and insights. This phase can include spectral analysis to identify frequency bands associated with different cognitive states, source localization to pinpoint neural activity origins, and connectivity analysis to explore network dynamics (Baillet et al., 2001; Cohen, 2017).

Interpretation: The final step involves interpreting the analyzed data within the context of the specific research or clinical question. This requires a deep understanding of neuroscientific principles, statistical methods, and the potential implications of the findings. Interpretations can shed light on the neural mechanisms of cognitive processes, the impact of neurological disorders on brain function, or the effectiveness of therapeutic interventions (Cohen, 2017; Poldrack, 2007).

Each stage in this process is critical for ensuring the accuracy and relevance of the findings, with methodologies and techniques tailored to the specific imaging modality and the objectives of the study. The synergy between technical expertise, advanced hardware, and neuroscientific knowledge is essential for advancing our understanding of the brain's intricate workings.

2.2.1 EEG Hardware

EEG (electroencephalography) is a technique used to measure the electrical activity of the brain. The hardware systems used to perform EEG recordings are composed of three main components: electrodes, amplifiers, and data acquisition systems.

2.2.1.1 Electrodes

In electroencephalography (EEG), electrodes serve as the primary sensors for capturing the brain's electrical activity, a process crucial for both clinical diagnostics and neuroscience research. The electrodes are the sensors that are placed on the scalp to measure the electrical activity of the brain. These electrodes vary significantly in their design, materials, and application, each tailored to specific measurement needs as shown in ◘ Fig. 2.12 (Casson, 2019). Here, we explore the different types of EEG electrodes, their unique features, and how they impact various applications.

Wet Electrodes: Recognized as the gold standard in EEG measurements, wet electrodes require the application of a conductive gel to enhance electrical connectivity between the scalp and the electrode, as shown in ◘ Fig. 2.13 (Casson, 2019). Typically crafted from materials such as silver-silver chloride (Ag/AgCl) or gold,

2

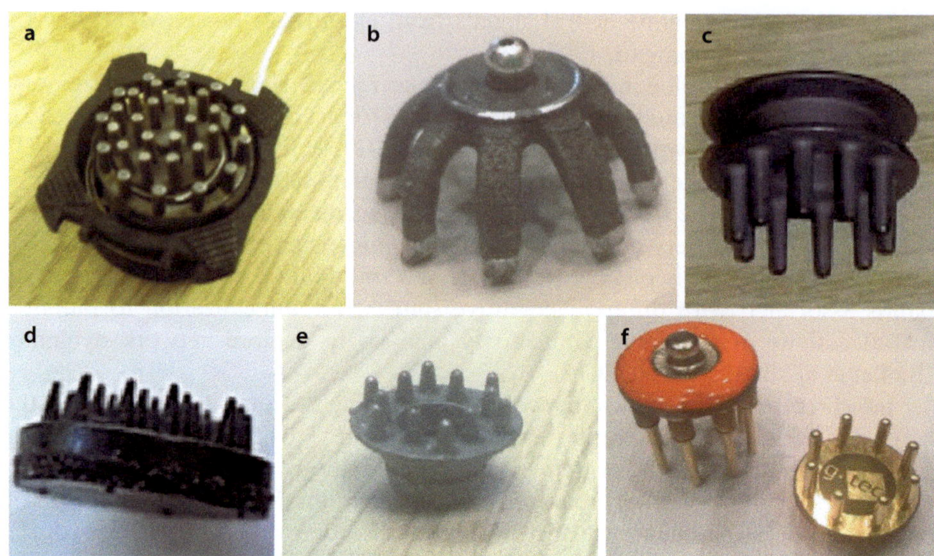

■ **Fig. 2.12**   Examples of contemporary dry-fingered EEG electrodes include: **a** Wearable sensing,[1] **b** Cognionics,[2] **c** Neuroelectrics,[3] **d** IMEC,[4] **e** Mindo,[5] and **f** g.tec g.SAHARA.[6] Keys: 1. Wearable sensing. Home page. 2016. ► http://www.wearablesensing.com/. 2. Cognionics. Home page. 2016. ► http://www.cognionics.com/. 3. Neuroelectrics. Products/electrodes. 2016. ► http://neuroelectrics. com/. 4. IMEC. ► https://www.imec-int.com/en/expertise/health-technologies/neurotechnology. 5. Mindo. Home page. 2016. ► http://mindo.com.tw/en/. 6. gtec. Products/g.SAHARA. 2016. ► http://www.gtec.at/. (From Casson, 2019)

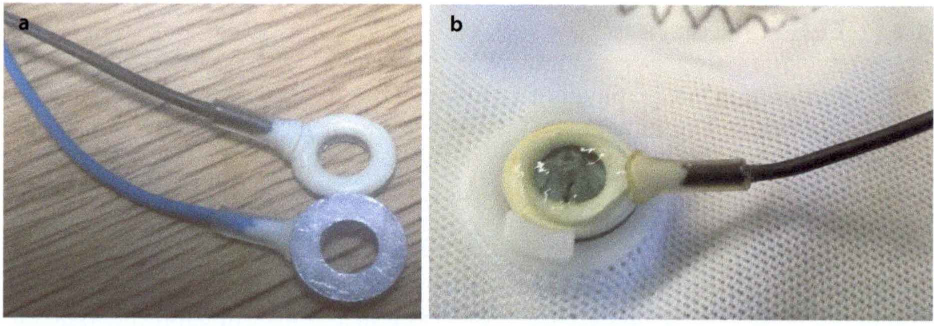

■ **Fig. 2.13**   **a** An example of a traditional silver/silver-chloride EEG electrode, which is typically 1 cm in diameter. **b** This example is integrated into a cap and uses a conductive gel to ensure optimal electrical contact between the electrode metal and the scalp. (From Casson, 2019)

these electrodes are lauded for their high signal-to-noise ratio and minimal impedance, ensuring the fidelity of EEG recordings (Nunez & Srinivasan, 2006). Their durability and reusability, coupled with ease of cleaning, further contribute to their widespread preference in EEG setups.

Dry Electrodes: Offering a more user-friendly alternative, dry electrodes eliminate the need for conductive gel, instead relying on conductive pastes or self-adhesive layers for contact improvement. Constructed from metals like aluminum, silver, or gold, dry electrodes present a cost-effective and convenient option, particularly for quick setups or studies requiring portability. Despite their advantages, a trade-off exists in the form of higher impedance and potential noise increase compared to their wet counterparts (Grozea et al., 2011).

Active Electrodes: Incorporating built-in amplifiers, active electrodes amplify EEG signals at the source, thereby enhancing signal quality and reducing susceptibility to noise. Although their cost exceeds that of traditional wet or dry electrodes, the investment is often justified by the improved data quality they offer (Lopez-Gordo et al., 2014).

Wireless Electrodes: Integrating wireless transmission capabilities, these electrodes streamline EEG recording by eliminating the need for extensive wiring. This innovation facilitates greater participant mobility and simplifies the experimental setup, making wireless electrodes increasingly favored for dynamic or long-term monitoring scenarios (Mihajlovic et al., 2015).

High-Density Electrodes: Employed for high-resolution EEG recordings, high-density electrode arrays capture detailed neural activity across the scalp. These arrays, often arranged on nets or caps conforming to the head, facilitate comprehensive spatial analysis of brain activity, enhancing the accuracy of source localization techniques (Michel et al., 2004).

The selection of EEG electrodes is dictated by the specific requirements of the research or clinical investigation, balanced against available resources. While wet electrodes remain a benchmark for quality, the evolving landscape of EEG technology has seen the rise of innovative alternatives like dry, wireless, and high-density electrodes, each expanding the possibilities for studying and interacting with the human brain.

2.2.1.2 Electrode Placement

The International 10–20 system, established as the gold standard for electroencephalogram (EEG) electrode placement, facilitates the standardized positioning of electrodes on the scalp to measure brain electrical activity, as shown in ◼ Fig. 2.14 (Jurcak et al., 2007). This method is pivotal in both research and clinical settings globally, ensuring consistent and comparable EEG data collection. The 10–20 system delineates the placement of 19 electrodes (Fp1, Fp2, F7, F3, Fz, F4, F8, T3, C3, Cz, C4, T4, T5, P3, Pz, P4, T6, O1, O2) relative to specific anatomical landmarks, offering a structured approach to capture electrical signals across the brain's surface (Jasper, 1958).

Building upon the foundation of the 10–20 system, the 10–10 electrode placement system (Nuwer, 2018) offers an expanded framework with 61 specified electrode positions (FP1, FPZ, FP2, AF7, AF3, AFZ, AF4, AF8, F7, F5, F3, F1, FZ, F2, F4, F6, F8, FT7, FC5, FC3, FC1, FCZ, FC2, FC4, FC6, FT8, T3, C5, C3, C1, CZ, C2, C4, C6, T4, TP7, CP5, CP3, CP1, CPZ, CP2, CP4, CP6, TP8, T5, P5, P3, P1, PZ, P2, P4, P6, T6, PO7, PO3, POZ, PO4, PO8, O1, OZ, O2). This system maintains the relative distance principles of the 10–20 system while introducing additional electrode sites for a more comprehensive coverage of the scalp. The

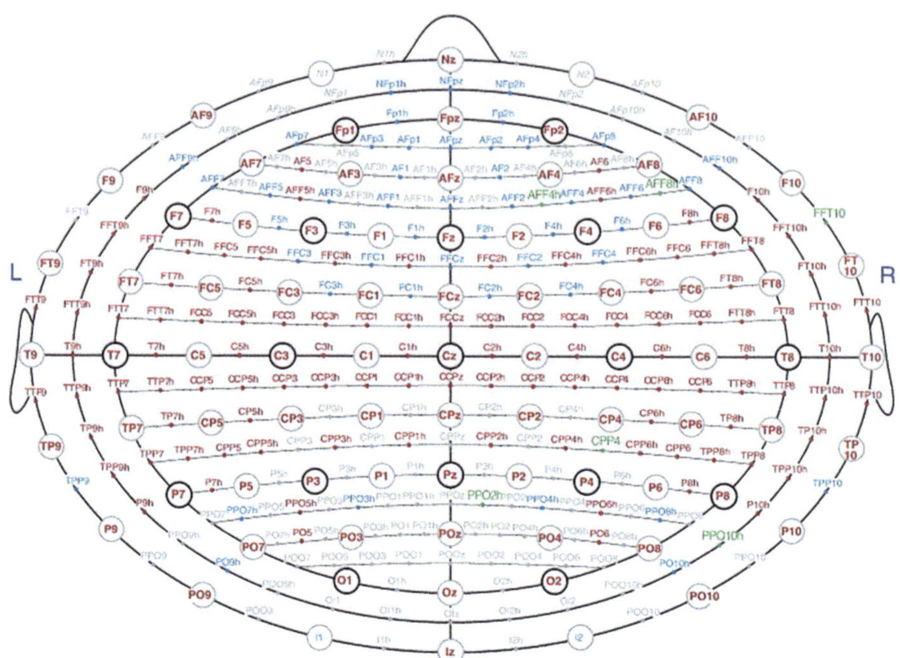

■ **Fig. 2.14** The traditional unambiguously illustrated (UI) 10/20 positions are marked by black open circles, while the gray open circles represent additional positions introduced in the UI 10/10 system. The UI 10/5 system comprises a total of 329 points, including 12 points likely positioned over the eyes (illustrated in gray italics). Positions marked in colors (red, blue, and green) indicate those that can be effectively placed on the scalp without overlapping neighboring points: 185 points maintaining both anterior/posterior and right/left symmetries are in red, an extra 50 points preserving only right/left symmetry are in blue, and 6 isolated points, merely separated from adjacent positions, are in green. The remaining 76 points of the UI 10/5 system, which overlap with neighboring positions, are shown in gray. (From Jurcak et al., 2007)

10–10 system is embraced by some researchers and clinicians for its enhanced spatial resolution, facilitating more detailed mapping of brain activity (Oostenveld & Praamstra, 2001).

Further refinement is observed in the 10–5 electrode placement system, which aims to provide an even more granular mapping of the scalp with additional electrode positions. While maintaining the core principles of its predecessors, the 10–5 system is designed for applications requiring the highest spatial resolution in EEG studies, although it is less commonly used than the 10–20 and 10–10 systems. The 10–5 system retains the original electrode numbering convention of the 10–20 system but introduces new sites to capture subtler aspects of brain activity (Jurcak et al., 2007).

Each of these systems—the 10–20, 10–10, and 10–5—serves a unique purpose in the field of neurophysiology, offering varying levels of detail and coverage suited to different research and clinical needs. The 10–20 system remains the most universally adopted standard due to its balance of simplicity and coverage. However, the choice between these systems depends on the specific goals of the study or diagnostic procedure, the complexity of the neural phenomena under investigation, and the technological resources available. The ongoing evolution of EEG electrode placement systems reflects the dynamic nature of neuroscientific research and the continual quest for enhanced precision in measuring brain activity.

2.2.1.3 Amplifiers

Amplification technology is essential in neuroscientific research and clinical diagnostics for enhancing the faint electrical signals produced by the brain, facilitating their accurate measurement and documentation. This technology is categorized into two principal groups: internal and external amplifiers. Internal amplifiers are integrated within the electrodes themselves, optimizing signal capture at the source, whereas external amplifiers, functioning as separate units, are connected to the electrodes to enhance the signal externally (Nunez & Srinivasan, 2006).

Electroencephalography (EEG) employs specialized amplifiers designed to elevate the brain's electrical signals to detectable levels for thorough analysis and recording. The diversity in EEG amplifier technology encompasses various models, each characterized by specific features tailored to different research and diagnostic needs (Devi et al., 2022). Prominent types of EEG amplifiers include:

Differential Amplifiers: As the cornerstone of EEG signal amplification, these devices utilize a pair of electrodes—one designated as "active" and the other as "reference"—to measure the voltage difference across two points on the scalp. This approach is particularly effective in mitigating common mode noise, a ubiquitous interference present at both electrode sites. Differential amplifiers are versatile and available as internal components within electrodes or external units (Teplan, 2002).

Bipolar Amplifiers: By employing two active electrodes, bipolar amplifiers capture the voltage difference between two specific scalp locations, offering a detailed insight into the neural activity of a targeted brain region. These amplifiers come in both internal and external configurations, providing flexibility in application depending on the experimental or clinical setup (Regan, 1989).

Common Reference Amplifiers: Utilizing a single active electrode in conjunction with a reference electrode, this amplifier type measures the voltage difference against a common reference point. This setup is often preferred in clinical settings to streamline the electrode arrangement without compromising the quality of data acquisition (Webster, 2009).

Common Mode Rejection Amplifiers: Engineered to exclude common mode noise, these amplifiers employ advanced strategies such as balanced or differential inputs to isolate and accurately measure neural activity. Available in both internal and external formats, they ensure precise signal interpretation by effectively reducing environmental and physiological interference (Webster, 2009).

High-Input Impedance Amplifiers: Designed to reduce the loading effect on the scalp and diminish signal noise, these amplifiers are critical in minimizing input impedance that could otherwise lead to signal distortion. Often used in tandem with low-noise preamplifiers, they are instrumental in achieving a low-noise floor, enhancing the clarity and reliability of EEG data (Chi et al., 2011).

Low-Noise Amplifiers: Focused on minimizing the intrinsic noise within the signal, these amplifiers incorporate sophisticated technologies such as low-noise transistors and active feedback mechanisms. When paired with high-input impedance amplifiers, they significantly reduce the overall noise floor, facilitating the extraction of high-fidelity neural signals (Scheer et al., 2005).

The choice of EEG amplifier is contingent upon the specific requirements of the research or clinical endeavor, informed by the nature of the investigation and the available resources. While differential amplifiers are ubiquitously applied across both research and clinical domains, the selection of bipolar, common reference, or specialized noise-rejecting amplifiers is dictated by the targeted analysis objectives. Integrating high-input impedance and low-noise amplifiers is predominantly observed in research settings to maximize signal quality, underscoring the tailored application of amplifier technology in neuroscientific exploration (Niedermeyer & da Silva, 2005).

2.2.1.4 Data Acquisition Systems

The data acquisition system (DAS) plays a pivotal role in electroencephalography (EEG) studies, serving as the infrastructure for collecting, storing, and analyzing EEG data. This system is usually composed of a computing device equipped with specialized software engineered to execute a multitude of signal processing and analysis operations essential for interpreting the brain's electrical activity. The integration of DAS with the EEG setup is facilitated through various interfaces, including USB (Universal Serial Bus), LPT (Parallel Port), or Ethernet, providing versatile options for connecting the system to the electrodes and amplifiers.

The software within a DAS is designed with sophisticated algorithms capable of filtering, artifact rejection, signal enhancement, and the extraction of relevant features from the EEG signals. This enables researchers and clinicians to perform detailed analyses, such as time-frequency analysis, event-related potential (ERP) studies, and connectivity analysis among different brain regions. The choice of interface for connecting the DAS to the EEG hardware often depends on the specific requirements of the study, including data transfer speed, the distance between the computer and the EEG apparatus, and the complexity of the experimental setup (Cohen, 2014).

Furthermore, the advent of advanced DAS has paved the way for real-time EEG data processing and analysis, significantly enhancing the efficiency and effectiveness of neuroscientific research and clinical diagnostics. Real-time analysis capabilities are particularly valuable in applications such as neurofeedback, brain–computer interfaces (BCIs), and intraoperative monitoring, where immediate interpretation of EEG data is critical (Makeig et al., 2012).

Commercial research-grade EEG systems stand at the forefront of neurological research, offering unparalleled accuracy, sensitivity, and reliability in recording the brain's electrical activity. These advanced systems are distinguished from consumer-grade systems by their sophisticated design and are crucial for in-depth neuroscientific studies and medical diagnostics. The core components of research-grade EEG setups include precision electrodes, high-gain amplifiers, and state-of-the-art recording devices, ensuring optimal signal quality and fidelity (Makeig et al., 2012).

While several key players are in the market, each offers unique capabilities tailored to various research and clinical needs. Below, I detail some of the prominent commercial research-grade EEG systems, their features, and their applications in the field.

1. ANT Neuro eego mylab: ANT Neuro's eego mylab system is designed for versatility in EEG research, offering portability without compromising data quality. It supports high-density EEG recordings and is equipped with real-time data analysis capabilities. The eego mylab system is suitable for a wide range of applications, from basic neuroscience research to applied studies in neurorehabilitation and BCI (▶ https://www.ant-neuro.com/).

2. Biosemi ActiveTwo System: The ActiveTwo system by Biosemi is a hallmark in EEG technology, offering active electrode technology for high-quality signal acquisition. The system is designed for flexibility in research applications, providing a wide range of electrode caps and configurations to suit various study designs. The ActiveTwo system is celebrated for its non-invasive, high-resolution data collection. It is suitable for cognitive neuroscience, psychophysiology, and BCI research (▶ https://www.biosemi.com/).

3. Brain Products BrainAmp Series: Brain Products' BrainAmp series includes several EEG amplifiers and accessories designed for mobile and stationary research settings. The BrainAmp amplifier, in particular, is known for its excellent signal quality, robust design, and versatility in recording not only EEG but also EOG (electrooculography), EMG (electromyography), and ECG (electrocardiography). This system is widely used in cognitive science, neurofeedback studies, and clinical research (▶ https://www.brainproducts.com/).

4. Compumedics Neuroscan Synamps2 System: The Synamps2 system from Compumedics Neuroscan offers comprehensive EEG data acquisition and analysis solutions. It is designed for high-density EEG recording and is highly regarded for its sensitivity and specificity in signal capture. The Synamps2 system is often used in advanced neuroscientific research, including sleep studies, cognitive neuroscience, and clinical diagnostics (▶ https://compumedicsneuroscan.com/).

5. Emotiv EPOC X: While Emotiv's EPOC X is often considered closer to the high-end consumer-grade spectrum, it offers features that make it suitable for certain research applications. The system provides wireless EEG data acquisition and is designed for ease of use and accessibility. The EPOC X is used in educational settings, preliminary research studies, and user experience research (▶ https://www.emotiv.com/).

6. G.Tec g.HIamp: G.Tec's HIamp EEG system is a multi-channel amplifier designed to process EEG signals in real time. It is compatible with high-density electrode setups and integrates seamlessly with G.Tec's brain–computer interface technology. The g.HIamp system is utilized in research areas such as neural engineering, neuroergonomics, and cognitive neuroscience (▸ https://www.gtec.at/).

These systems represent the cutting edge in EEG technology, each contributing uniquely to the advancement of neuroscientific research. The choice among these systems depends on specific research requirements, including the desired resolution, portability, and integration with other physiological measures. As EEG technology continues to evolve, these systems are likely to see further improvements in functionality and application scope, driving forward our understanding of the brain and its complexities.

2.2.1.5 Other Accessories

In electroencephalography (EEG), the customization of the hardware setup to meet specific research or clinical requirements is a crucial consideration. This customization often involves additional accessories, including headsets, caps, and conductive gel for wet electrodes. These accessories are designed to enhance the quality of the electrical signal captured from the scalp, thereby improving the fidelity of the EEG data.

Headsets and caps are employed to ensure that electrodes are placed accurately and consistently on the scalp, according to standardized or customized electrode montages. This accuracy is vital for both the reproducibility of research findings and the reliability of clinical diagnostics. Wet electrodes, which require the application of a conductive gel, are often preferred for their ability to reduce electrode–skin impedance, thus facilitating a clearer signal transmission (Gargiulo et al., 2010).

Furthermore, some EEG systems incorporate a built-in reference electrode strategically placed to minimize the influence of electrical noise on the recorded signals. This built-in reference electrode is essential for achieving a high signal-to-noise ratio, crucial for precise EEG data analysis. EEG amplifiers are currently provided with an optional accessory called a "trigger box" that enables recording digital signals (triggers) useful to synchronize neurophysiological data streaming with meaningful events (e.g., the onset of a stimulus or a key press).

The selection of an EEG hardware system is contingent upon the study's specific objectives or the clinical assessment, underscoring the importance of matching the system's capabilities with the intended application. The quality and reliability of the EEG hardware are pivotal, as they directly impact the accuracy and usability of the collected data. Suboptimal hardware can introduce artifacts, distort signals, and ultimately compromise the integrity of the EEG findings.

Fig. 2.15 Experimental setup with a participant equipped with the 64-electrode EEG system. (From Hinss et al., 2023; originally published under CC BY 4.0)

Moreover, the EEG hardware landscape is characterized by rapid technological advancements, driving the development of increasingly compact, cost-effective, and portable systems. This evolution has expanded the accessibility of EEG technology, facilitating its application in a wider array of environments, including outpatient clinics, remote settings, and patient homes. The trend toward miniaturization and affordability makes EEG technology more widespread and opens new avenues for research and clinical practice, including ambulatory monitoring and real-time brain–computer interfacing (BCI) applications (Lopez-Gordo et al., 2014).

In conclusion, carefully selecting and integrating EEG hardware and accessories are fundamental to the success of neurophysiological investigations and interventions. The ongoing innovations in EEG technology promise to enhance further neural monitoring and analysis' versatility, convenience, and efficacy. Figure 2.15 depicts the typical EEG setup in a lab.

Box 2.11

Electroencephalography (EEG) hardware is crucial for capturing the brain's electrical activity through the scalp. The primary components of EEG hardware include:

Electrodes: Serve as the contact points for recording electrical signals directly from the scalp. These are available in various types, such as wet, dry, and needle electrodes, each suitable for different diagnostic and research needs.

Amplifiers: Boost the electrical signals captured by the electrodes to a level suitable for recording and analysis. Amplifiers must maintain high fidelity with minimal noise to ensure the accuracy of EEG data.

Data Acquisition Systems (DASs): These systems collect, digitize, and store EEG signals for further analysis. Modern DASs are equipped with sophisticated software for real-time data processing, which is essential for applications like neurofeedback and brain–computer interfaces.

> ■ **Key Considerations:**
>
> **Signal Quality**: The choice of electrodes and the configuration of the EEG hardware significantly influence the quality of the recorded signals. Wet electrodes are generally preferred for clinical and long-term monitoring due to their lower impedance and higher signal quality.
>
> **Flexibility and Comfort**: Newer models of EEG systems, including wireless and wearable devices, offer greater flexibility and comfort for subjects, expanding the potential for EEG applications in dynamic environments.
>
> By understanding these components and their roles, researchers and clinicians can better configure EEG systems to suit specific experimental or clinical needs, ensuring optimal data quality and reliability.

2.2.2 NIRS Hardware

Near-infrared spectroscopy (NIRS) is a non-invasive optical imaging technique that monitors cerebral hemodynamics and metabolic changes within the cortical regions of the brain. This method is predicated on the principle that near-infrared light, when transmitted through biological tissues, undergoes absorption and scattering, alterations of which can be indicative of underlying brain activity. The primary chromophores responsible for these absorption changes in the brain are the two states of hemoglobin: oxyhemoglobin (HbO) and deoxyhemoglobin (HbR). Variations in the concentrations of HbO and HbR are reflective of neuronal activity, as they are closely related to cerebral blood flow and oxygenation changes occurring in response to neuronal activation (Ferrari & Quaresima, 2012).

To detect changes in HbO and HbR concentrations, three principal optical measurement techniques are employed within the NIRS framework, namely continuous-wave (CW), frequency-domain (FD), and time-domain (TD) measurements.

2.2.2.1 Continuous Wave-Based Measurement Technique

Continuous-wave (CW) near-infrared spectroscopy (NIRS) systems utilize a consistent beam of light, either at a constant amplitude or modulated at low frequencies (up to several tens of kilohertz), to facilitate the rejection of stray light as shown in ■ Fig. 2.16. The primary metric measured by CW NIRS systems is the amplitude decay of the incident light, which is indicative of changes in the optical properties of the tissue through which the light passes. This decay is attributed to the absorption and scattering of light within the cortical tissue, providing insights into cerebral oxygenation dynamics (Boas & Franceschini, 2011; Strangman et al., 2002).

The advantages of CW NIRS systems are significant, particularly when considering their application in clinical and research settings. The high sampling rate,

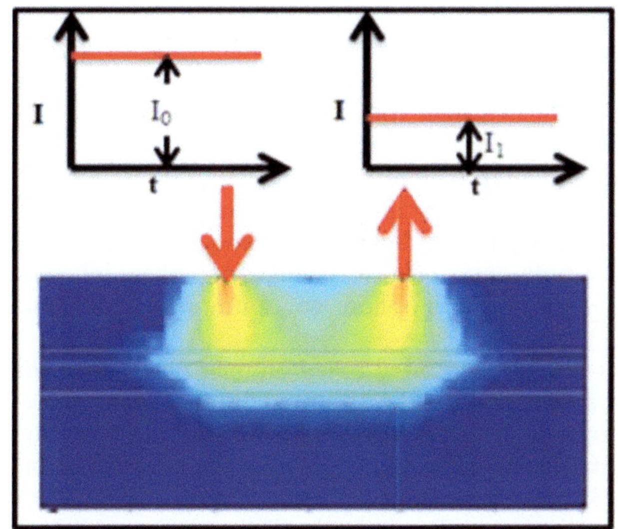

☐ **Fig. 2.16** Schematic representation of continuous wave (CW) technique. The figure depicts the transport of light through a multilayered model created to mimic different layers in the brain. In CW technique, light of initial intensity I_0 is incident onto the surface of the scalp and the transmitted attenuated light of intensity I_1 is measured at a distance d from the incident point. (Modified and adapted from Cuccaro et al., 2024; originally published under CC BY 4.0)

coupled with the compactness, low weight, simplicity, and cost-effectiveness of CW NIRS instruments, renders them especially suitable for bedside monitoring and real-time applications. These characteristics make CW NIRS an invaluable tool for continuous monitoring of oxygenation trends in various contexts (Boas & Franceschini, 2011; Strangman et al., 2002).

Despite these benefits, CW NIRS systems are not without limitations. One of the primary drawbacks is the limited penetration depth, which can restrict the depth of tissue from which measurements can be reliably obtained. Additionally, the inability of CW NIRS to separately quantify absorption and scattering effects poses a challenge. Since CW systems do not provide distinct information about the absorption and scattering coefficients, they cannot measure absolute changes in oxyhemoglobin (HbO) and deoxyhemoglobin (HbR) concentrations. Instead, they are utilized for monitoring trends in oxygenation levels over time (Strangman et al., 2002).

CW NIRS has been widely applied in investigating cerebral oxygenation changes across various populations, including healthy adults and children, as well as individuals with neurological challenges. This widespread use underscores the utility of CW NIRS in neuroscientific research and clinical diagnostics, despite its inherent limitations. Studies by Nioka et al. (1997) and Siegel et al. (1999) are among the numerous examples that highlight the application of CW NIRS in exploring cerebral oxygenation and its implications for brain health and disease.

2

2.2.2.2 Frequency-Domain-Based Measurement Technique

Frequency domain (FD) near-infrared spectroscopy (NIRS) systems offer a sophisticated approach to assess cerebral hemodynamics and tissue oxygenation by utilizing amplitude-modulated light sources. As shown in ◻ Fig. 2.17, these systems inject modulated light into the brain and measure the amplitude decay and phase shift of the detected light, thereby providing detailed insights into the absorption and scattering properties of brain tissue (Ntziachristos et al., 2001). FD NIRS achieves this by varying the modulation frequency of the light source, allowing for the extraction of both optical density changes and phase information, which correlate with changes in chromophore concentrations and scattering properties, respectively.

FD systems can be categorized based on their operational configurations: (1) using a single wavelength of light with a fixed distance between the light source and detector (interoptode distance), (2) employing multiple wavelengths with a fixed interoptode distance to enable the measurement of different chromophores within the tissue, or (3) utilizing a single wavelength while varying the interoptode distances to explore tissue at different depths (Delpy & Cope, 1997). This flexibility in configuration enhances the utility of FD NIRS in a broad range of research and clinical applications.

The primary advantages of FD NIRS systems include their high sampling rate and the ability to distinguish more accurately between absorption and scattering effects. This distinction is crucial for quantifying the concentrations of oxyhemo-

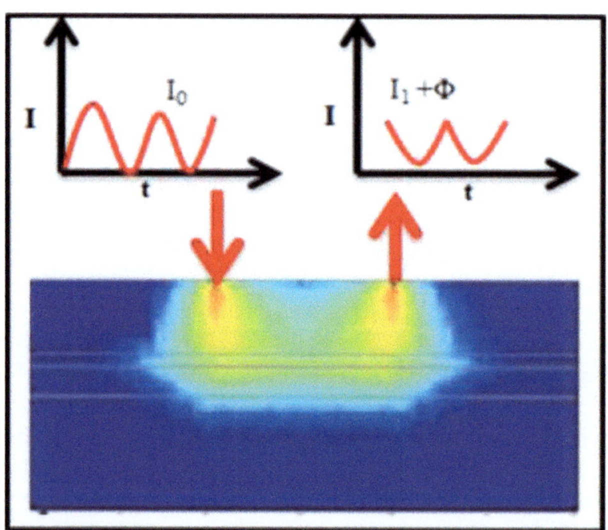

◻ **Fig. 2.17** Schematic representation of frequency domain technique. The figure depicts the transport of light through a multilayered model created to mimic different layers in the brain. In the CW technique, light of initial intensity I_0 is incident onto the surface of the scalp and the attenuation in intensity I_1 and phase shift Φ in the transmitted light is measured at a distance d from the incident point. (The multilayered brain model depicting the attenuation of light in the brain is modified and adapted from Cuccaro et al., 2024, originally published under CC BY 4.0)

globin (HbO) and deoxyhemoglobin (HbR) with relatively higher precision compared to continuous-wave (CW) systems. However, FD NIRS has limitations, notably the restriction that the frequency of the modulated light must not exceed 200 MHz. Beyond this threshold, the linear relationship between phase shift and path length, essential for accurate measurement, does not hold, which can compromise the integrity of the data obtained (Arridge et al., 1992).

FD NIRS has been extensively applied in the study of cerebral oxygenation and hemodynamics across various populations, including adults and children. Notable research conducted by Franceschini et al. (1999), Gratton et al. (2006), and Pogue et al. (1995) exemplifies the application of FD NIRS in uncovering significant insights into brain function and pathophysiology. These studies highlight the technique's capability to provide valuable information on the cerebral circulation and oxygenation changes associated with different physiological and pathological states.

2.2.2.3 Time-Domain-Based Measurement Techniques

Time-domain (TD) near-infrared spectroscopy (NIRS) systems employ an advanced technique in which picosecond pulses of light are emitted onto the brain, and the temporal dispersion of the reflected or transmitted light is analyzed as shown in ◘ Fig. 2.18. This analysis provides detailed insights into the tissue's optical properties, including both absorption and scattering coefficients, across differ-

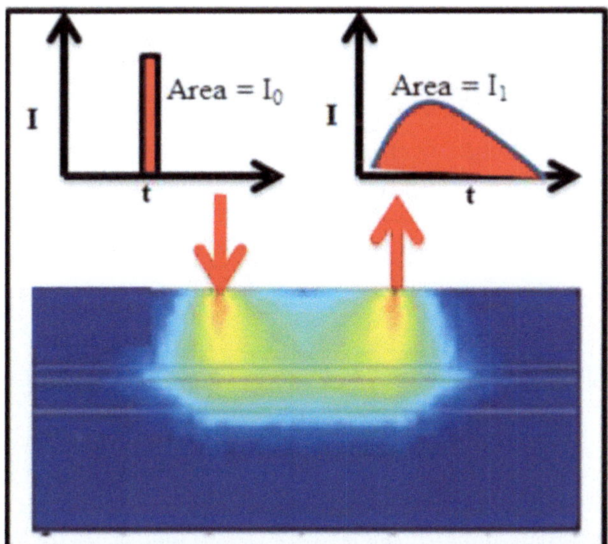

◘ **Fig. 2.18** Schematic representation of time-domain technique. The figure depicts the transport of light through a multilayered model created to mimic different layers in the brain. In the time-domain technique, light of initial intensity I_0 is incident onto the surface of the scalp and the change in temporal distribution of picosecond pulse of light is measured at a distance d from the incident point. (The multilayered brain model depicting the attenuation of light in brain is modified and adapted from Cuccaro et al., 2024, originally published under CC BY 4.0)

ent layers of the head (Strangman et al., 2002). The ability to capture the arrival times of photons enables the TD NIRS systems to distinguish between photons that have traveled through shallow and deeper layers of the tissue, offering a unique advantage in terms of spatial resolution and depth penetration.

Time-domain (TD) NIRS systems represent a sophisticated approach to neuro-imaging, offering enhanced spatial resolution, greater depth penetration, and the most accurate determination of absorption and scattering compared to continuous-wave (CW) and frequency-domain (FD) systems. These advantages have positioned TD NIRS as a powerful tool for investigating complex cerebral phenomena, particularly in areas where precise quantification of hemodynamic changes is critical (Durduran & Yodh, 2014; Pifferi et al., 2016). Despite their significant benefits, TD NIRS systems face challenges related to sampling rate, instrument size, weight, and the need for cooling mechanisms, which can limit their practicality in certain research and clinical settings. However, ongoing technological innovations are focused on addressing these limitations, with recent advancements aimed at improving portability, reducing system complexity, and enhancing user-friendliness (Torricelli et al., 2014).

TD NIRS continues to be a valuable resource for studying cerebral oxygenation and detecting hemorrhages in neonates. The technique's non-invasiveness and sensitivity to both oxygenated and deoxygenated hemoglobin make it particularly suitable for monitoring vulnerable populations (Roche-Labarbe et al., 2010; Cooper et al., 2012).

Moreover, the development and refinement of commercial and custom-built NIRS systems, utilizing TD measurement techniques, underscores the technique's versatility and wide-ranging applicability. From basic neuroscience research to clinical diagnostics, TD NIRS systems are increasingly integral to our understanding of brain function and pathology (Lange & Tachtsidis, 2019). While TD NIRS offers numerous advantages over other functional neuroimaging techniques, it is essential to consider the methodological and technical challenges inherent in its application. Nevertheless, its contributions to neuroscience and neurology are undeniable, offering insights into cerebral hemodynamics with unparalleled detail and accuracy.

Types of fNIRS Systems

1. Diffuse Optical Tomography (DOT) represents the forefront of non-invasive brain imaging, offering three-dimensional imaging capabilities by utilizing near-infrared light. This technique has seen significant advancements, particularly in improving spatial resolution and depth penetration, making it invaluable for detailed cerebral hemodynamic studies. A recent review by Hernandez-Martin and Gonzalez-Mora (2020) highlights the progress in DOT technology and its application in neuroscience, underscoring its potential for uncovering new insights into brain function and pathology.
2. Portable fNIRS Systems have been developed to address the need for mobility and flexibility in conducting neuroimaging studies outside traditional laboratory settings. These systems have become increasingly sophisticated, with improvements in battery life, data acquisition speed, and usability. A study by

Pinti et al. (2020) demonstrates the utility of portable fNIRS in diverse environments, emphasizing its role in expanding the scope of neuroimaging research to include more naturalistic settings.

3. Wireless fNIRS Systems further enhance the applicability of fNIRS technology by removing the constraints imposed by wired connections. The development of robust, reliable wireless fNIRS systems has facilitated studies involving movement and has improved the participant's comfort and the ecological validity of the research. An innovative example of such technology is presented by Friesen et al. (2022), who detail the design and application of a wireless fNIRS system optimized for use in dynamic and interactive tasks.

4. Combined fNIRS Systems (fNIRS-EEG and fNIRS-MRI): The integration of fNIRS with EEG and MRI has opened new avenues for multimodal imaging, allowing for simultaneous measurement of neural activity with complementary temporal and spatial resolution. This synergistic approach enhances the understanding of complex neural processes by providing a more comprehensive view of brain activity (Li et al., 2022; Tak & Ye, 2014).

In conclusion, the landscape of fNIRS technology continues to evolve rapidly, with significant advancements in hardware and analytical techniques enhancing its application in neuroscience and beyond. The choice of an fNIRS system, whether it be DOT, portable, wireless, or combined with other modalities, will depend on the specific research objectives, the experimental design, and budgetary considerations. Depending on the system available and application, a decision tree shown in ◘ Fig. 2.19 serves as a guide. These developments underscore the versatility and potential of fNIRS as a tool for investigating brain function, promising to extend the boundaries of neuroimaging research.

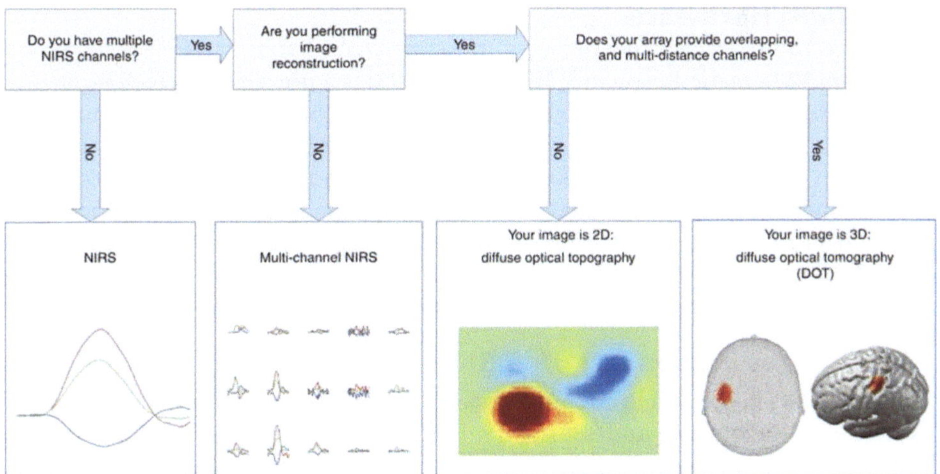

◘ **Fig. 2.19** A decision tree that provides a definition of the different forms of diffuse optical monitoring. Although nomenclature varies across the field, these are the definitions preferred by the authors. *2D* two dimensional, *NIRS* near-infrared spectroscopy. (From Lee et al., 2017)

Box 2.12

Functional Near-Infrared Spectroscopy (fNIRS) hardware plays a critical role in non-invasively monitoring brain activity by measuring changes in blood oxygenation levels. Key components of fNIRS hardware include:

Light Sources: Emit near-infrared light that penetrates the skull and is absorbed by brain tissue. The most common wavelengths used are around 690 and 830 nm, which are optimal for measuring oxyhemoglobin and deoxyhemoglobin.

Detectors: Capture the light that has traveled through the brain tissue and provides data on how much light was absorbed. The positioning of detectors relative to the light sources affects the depth and area of brain activity that can be monitored.

Data Acquisition System: Processes the signals received from the detectors to measure changes in hemoglobin concentrations, indicating areas of brain activity.

- **Key Considerations:**

Depth of Penetration: fNIRS is limited to sensing cortical activities due to the penetration depth of near-infrared light, typically not exceeding 2–3 cm into the brain.

Spatial Resolution: While fNIRS provides good spatial resolution compared to EEG, it is less precise than imaging techniques like fMRI.

Application Flexibility: Portable and wireless fNIRS systems have expanded the environments in which brain activity can be studied, including dynamic and natural settings outside of traditional labs.

Understanding the capabilities and limitations of fNIRS hardware is crucial for effectively designing studies and interpreting data in cognitive neuroscience, psychology, and clinical research.

2.2.3 fMRI Hardware

Functional Magnetic Resonance Imaging (fMRI) is a non-invasive technique used to measure neural activity by detecting changes in blood flow and oxygenation in the brain. Since its inception in the early 1990s, fMRI technology has undergone significant transformations. Initial studies were limited by the capabilities of available magnetic resonance imaging (MRI) systems, primarily operating at lower field strengths. The past decades have seen a relentless push toward higher field strengths, from 1.5 T standard clinical scanners to ultra-high-field (UHF) systems exceeding 7 T, marking a paradigm shift in imaging quality and resolution (Uğurbil, 2014).

2.2.3.1 Core Components of fMRI Systems

Magnet: The magnet's field strength, measured in Tesla, fundamentally determines the system's resolution and sensitivity. High-field systems (≥ 3 T) have become prevalent in research settings, offering enhanced signal-to-noise ratio (SNR) and spatial resolution, critical for detailed brain mapping (Duyn, 2012).

Gradient Coils: These coils are essential for spatial encoding of the MRI signal. Innovations in gradient technology have led to faster and more precise imaging capabilities, enabling complex imaging sequences like echo-planar imaging (EPI) that are fundamental to fMRI (Mansfield, 2004).

Radiofrequency (RF) Coils: Advances in RF coil design, including multi-channel and phased-array coils, have significantly improved SNR and parallel imaging capabilities, facilitating faster image acquisition and higher image quality (Wiggins et al., 2006).

Console and Software: Modern fMRI systems are equipped with advanced software for sequence control, data acquisition, and processing, supporting sophisticated imaging techniques and data analysis algorithms (Smith et al., 2012).

These core components are depicted graphically in ◘ Fig. 2.20 (Ghosh & Pal, 2022).

2.2.3.2 Types of fMRI Hardware Systems

Low-Field Systems ($\leq$0.1 T): Offer advantages in terms of cost and portability but are limited by lower spatial resolution and SNR. They are suited for specific applications where high resolution is not paramount (Marques et al., 2019).

Mid-Field Systems (0.3–1 T): Provide a balance between imaging quality and operational flexibility; useful in various research and clinical scenarios (Gray et al., 2003).

High-Field Systems ($\geq$3 T): The standard in contemporary fMRI research; high-field systems deliver superior image quality; crucial for detailed functional and anatomical studies (Uğurbil, 2014).

Ultra-High-Field Systems ($\geq$7 T): Push the limits of spatial resolution and SNR, enabling microstructural imaging of the brain. Despite their potential, UHF systems require specialized infrastructure and have higher operational costs (Duyn, 2012).

2.2.3.3 Specialized Imaging Sequences and Techniques

Gradient-Echo and Spin-Echo Sequences: Essential for optimizing BOLD contrast, with gradient-echo sequences being predominantly used in fMRI due to their sensitivity to blood oxygenation changes (Bandettini, 2012).

Echo-Planar Imaging (EPI): The cornerstone of fMRI, EPI enables rapid whole-brain imaging, capturing dynamic changes in brain activity (Smith et al., 2012).

Diffusion-Weighted Imaging: Provides insights into the brain's microstructure by measuring water molecule diffusion, complementing structural and functional studies (Jones et al., 2013).

In nutshell, there are several types of fMRI hardware systems available, each with their own advantages and disadvantages. The choice of system will depend on the specific research question, the available budget, and the experimental design. High-field systems are the most widely used in research, but low-field and mid-field systems are becoming more popular due to their portability and cost-effectiveness. UHF systems and resting-state fMRI are more specialized and not as widely used.

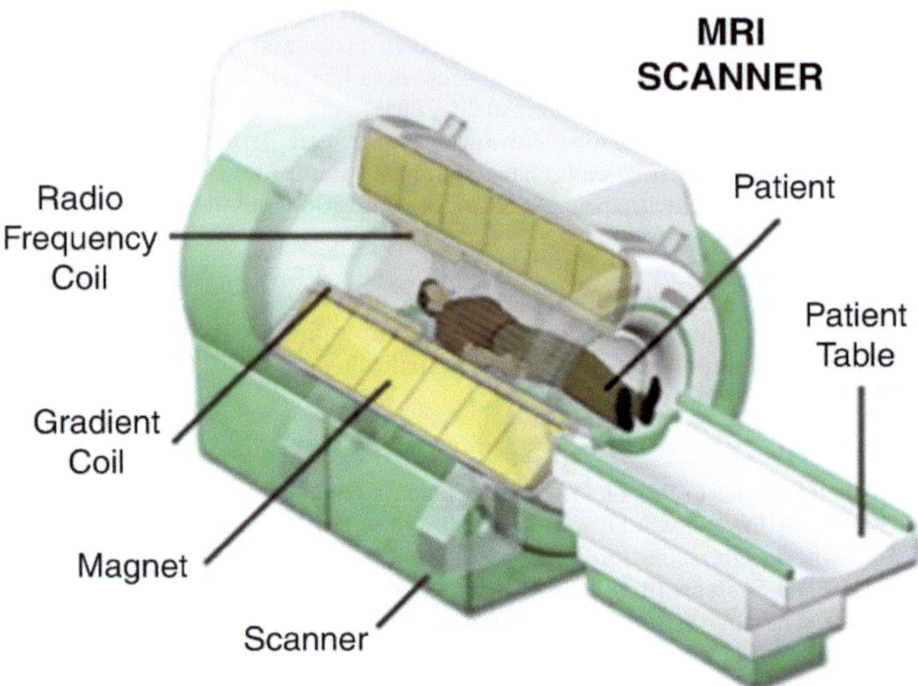

■ **Fig. 2.20** Core components of magnetic resonance (MR) scanner. Magnetic resonance imaging (MRI) scanners are sophisticated devices that utilize strong magnetic fields, radio waves, and field gradients to generate detailed images of the internal structures of the body. Understanding the core components of an MRI scanner is crucial for comprehending its functionality and the principles underlying its operation. The following description outlines the primary components of an MRI scanner: **Magnet**: The magnet is the most significant and essential part of the MRI scanner. It produces a strong and stable magnetic field that aligns the protons in the human body. There are three main types of magnets used in MRI: superconducting, resistive, and permanent. Superconducting magnets, which are the most common, operate at very low temperatures with liquid helium cooling to achieve high field strengths, typically ranging from 1.5 to 3 T for clinical use and up to 7 T for research purposes. **Radiofrequency (RF) Coils**: RF coils are used to transmit and receive radiofrequency pulses. The transmit coil sends RF pulses that excite the protons, causing them to emit signals. The receive coil detects these emitted signals, which are used to create the images. There are different types of RF coils, including body coils, head coils, and surface coils, each designed to optimize signal reception from specific regions of the body. **Gradient Coils**: Gradient coils are responsible for spatial encoding of the MRI signal. They create varying magnetic fields that allow the MRI system to determine the position of the signals in three dimensions. There are three sets of gradient coils—one for each spatial dimension (x, y, and z). These gradients are essential for slice selection, frequency encoding, and phase encoding, which together facilitate the creation of precise and high-resolution images. **Shim Coils**: Shim coils are used to correct inhomogeneities in the magnetic field, ensuring a uniform magnetic field across the imaging volume. Proper shimming is vital for image quality, as field inhomogeneities can lead to distortions and artifacts. Shimming can be done using passive shims (metal pieces fixed in place) or active shims (adjustable electromagnets). **Computer System**: The computer system controls the MRI scanner and processes the data. It sequences the magnetic field gradients and RF pulses, collects the raw data from the RF coils, and reconstructs this data into images using complex algorithms. Advanced software allows for the manipulation and analysis of these images, aiding in diagnosis and research. **Patient Table**: The patient table supports and positions the patient within the scanner. It is designed to move precisely into the bore of the magnet, aligning the area of interest within the optimal imaging field. The table is typically motorized and can be adjusted to enhance patient comfort and image accuracy. **Cooling System**: Superconducting magnets require a cooling system to maintain the extremely low temperatures necessary for superconductivity. This is usually achieved with liquid helium, which keeps the magnet at approximately −269 °C (−452 °F). Some systems also use liquid nitrogen as a secondary coolant. **RF Shielding**: To prevent external RF interference, MRI rooms are equipped with RF shielding. This typically involves a Faraday cage, which is a metal enclosure that blocks external electromagnetic fields. Effective shielding is crucial for maintaining image quality and system performance. Understanding these components and their functions is essential for operating an MRI scanner effectively and interpreting the resulting images accurately. The integration of these components enables MRI to be a powerful and versatile tool in medical imaging, providing invaluable insights into the human body's structure and function. (From Ghosh & Pal, 2022)

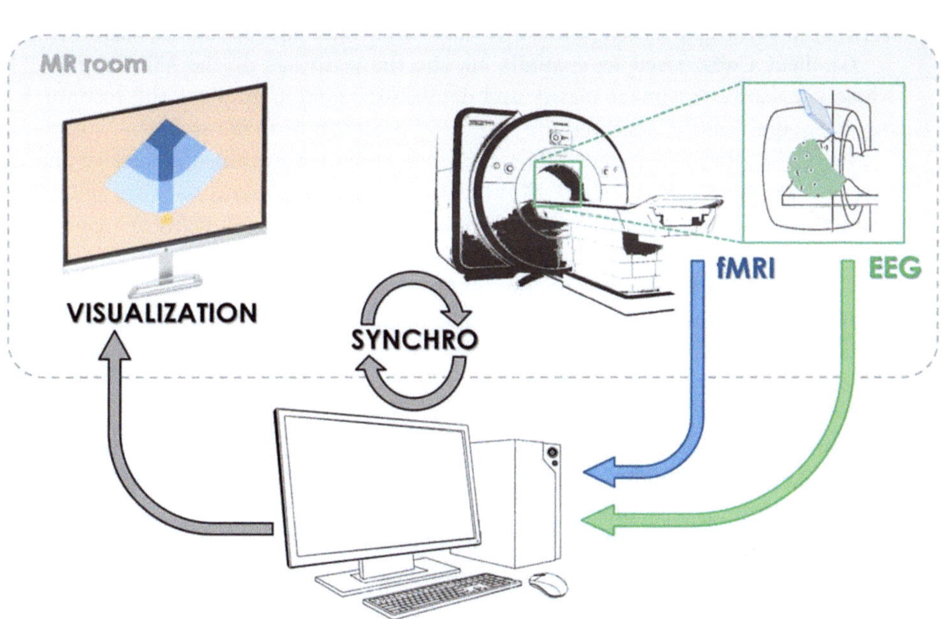

◘ **Fig. 2.21** Simultaneous EEG-fMRI setup. (From Lioi et al., 2020; originally published under CC BY 4.0)

2.2.3.4 Integration with Other Modalities

The fusion of fMRI with EEG and PET offers a multimodal approach, combining the temporal resolution of EEG, the metabolic insights of PET, and the spatial resolution of fMRI, enriching our understanding of brain function and dysfunction (Calhoun & Sui, 2016), as shown in ◘ Fig. 2.21 (Lioi et al., 2020). Future advancements in fMRI hardware are poised to focus on increased field strengths, enhanced gradient and RF coil technologies, and the integration of fMRI with other imaging modalities. Innovations in these areas promise to further refine our understanding of the brain, expanding the capabilities of fMRI in both research and clinical domains.

fMRI hardware technology has dramatically evolved, offering diverse systems tailored to a wide range of scientific inquiries and clinical applications. As technology advances, the potential of fMRI continues to expand, promising new insights into the complex workings of the human brain.

> **Box 2.13**
>
> **Functional Magnetic Resonance Imaging (fMRI)** hardware is integral to the non-invasive study of brain function by detecting changes in blood flow related to neural activity. Key components of fMRI hardware include:
>
> **Magnet**: The strength of the magnet, measured in Tesla (T), is crucial as it determines the resolution and sensitivity of the fMRI system. Higher field strengths (3 T or more) provide greater signal-to-noise ratio (SNR) and finer spatial resolution.
>
> **Gradient Coils**: Used to spatially encode the positions of the MRI signals, which is essential for image clarity and detail. Advances in gradient coil technology have significantly improved the speed and precision of fMRI scans.
>
> **RF Coils**: Radiofrequency (RF) coils are used to transmit and receive the magnetic resonance (MR) signal. Multi-channel and phased-array coils are among the innovations that have enhanced SNR and enabled faster imaging sequences, critical for dynamic brain studies.
>
> **Console and Software**: The operational hub of fMRI, equipped with advanced software for controlling the scanning sequences, data acquisition, and subsequent image processing.
>
> ■ **Key Considerations:**
>
> **Field Strength**: While high-field scanners offer detailed images suitable for research, lower-field scanners might be used in clinical settings due to their lower cost and lesser susceptibility to artifacts.
>
> **Imaging Sequences**: Techniques like echo-planar imaging (EPI) are fundamental for fMRI, allowing rapid data acquisition that captures the temporal dynamics of brain activity.
>
> **Integration with Other Modalities**: Combining fMRI with EEG or PET can enrich data interpretation, providing complementary information on brain activity.

2.2.4 MEG Hardware

Magnetoencephalography (MEG) is a sophisticated non-invasive technique that measures the brain's magnetic fields resulting from neural electrical activity. This method has become instrumental in understanding the temporal dynamics of brain function, thanks to its high temporal resolution. At the heart of MEG technology are SQUIDs, which detect the brain's magnetic fields emanating from neural electrical currents. These sensors' sensitivity allows for the detailed observation of brain functions in real-time (Hämäläinen et al., 1993). A typical MEG environment is shown in ■ Fig. 2.22 (Myllylä et al., 2017).

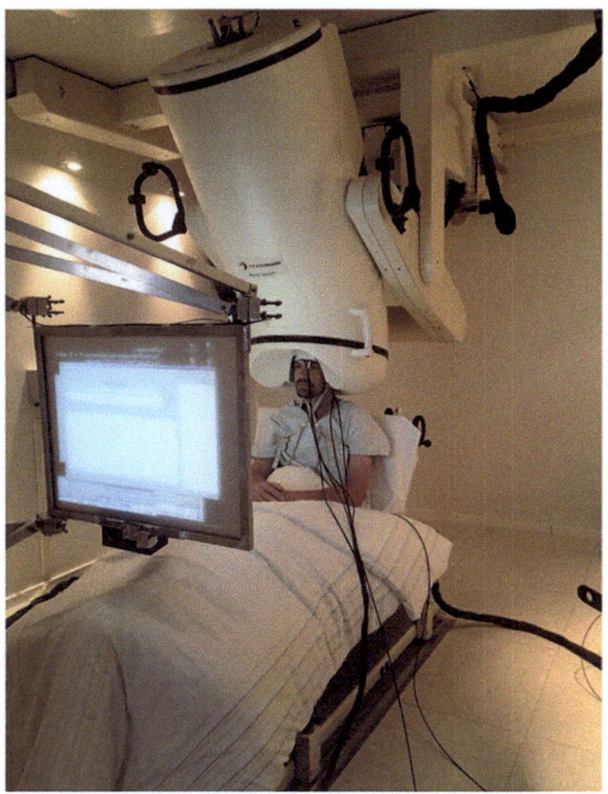

2.2.4.1 Varieties of MEG Systems

1. Whole-Head Systems: Whole-head MEG systems are equipped with a comprehensive array of superconducting quantum interference device (SQUID) sensors that envelop the entire scalp, allowing for the detection of magnetic fields from all brain regions simultaneously. These systems are renowned for their high sensitivity and spatial coverage, making them a staple in neuroscience research for mapping brain activity and connectivity (Hämäläinen et al., 1993). Whole-head MEG systems have facilitated breakthroughs in understanding cognitive processes and neurological disorders.
2. Helmet Systems: Helmet-style MEG systems utilize a more focused array of SQUID sensors attached to a helmet-like structure. While offering the advantages of lower cost and enhanced portability, these systems provide a narrower field of detection. This limitation means they offer lower spatial resolution and may miss activity in deeper brain regions compared to whole-head systems (Vrba & Robinson, 2001). Despite these limitations, helmet systems are practical for targeted studies and applications where full-brain coverage is not essential.

3. Vector-View Systems: These systems represent an advancement in MEG technology, combining traditional SQUID sensors with planar gradiometers. This combination enhances the system's ability to discern the direction and orientation of neural currents, offering superior spatial resolution over whole-head systems (Ahonen et al., 1993). Vector-view systems are particularly valuable for detailed mapping of cortical activity and understanding the spatial orientation of brain functions.

Combination of MEG with other modalities, such as functional magnetic resonance imaging (fMRI) and electroencephalography (EEG), provides more information about neural activity (Horwitz & Poeppel, 2002). The choice of MEG system will depend on the specific research question, the available budget, and the experimental design (Gross et al., 2013). Whole-head systems are the most widely used in research, due to their high spatial resolution and sensitivity to neural activity across the entire brain (Hillebrand & Barnes, 2002). Helmet systems are becoming more popular due to their portability and cost-effectiveness. Vector-view systems provide more specific information about the direction of neural activity. Combining MEG with other modalities provides more information about neural activity, but it is more expensive and requires more complex data analysis (He & Liu, 2008; Abadi et al., 2015).

Another type of MEG system is the portable MEG system. These systems are designed to be compact and lightweight, making them easy to transport and set up in different environments. They are typically battery-operated and can be used for field research, such as studying brain activity in naturalistic settings or in remote locations (Limes et al., 2020).

Additionally, there are also MEG systems specifically designed for certain populations or applications. For example, pediatric MEG systems are designed specifically for children and use specialized coils and sequences to acquire images (Okada et al., 2016). Similarly, there are MEG systems designed for patients with neurological disorders such as epilepsy and Parkinson's disease, which use specialized sequences and methods to acquire images of specific brain regions (Stam, 2010).

In conclusion, there are several types of MEG hardware systems available, each with their own advantages and disadvantages. The choice of system will depend on the specific research question, the available budget, and the experimental design. Whole-head systems are the most widely used in research, but Helmet systems and portable MEG systems are becoming more popular due to their portability and cost-effectiveness. High-density MEG systems and MEG systems designed for specific populations or applications provide more detailed information about neural activity, but they are more expensive and require more complex data analysis (Hill et al., 2020).

Box 2.14

Magnetoencephalography (MEG) hardware is crucial for mapping brain activity by detecting the magnetic fields generated by neuronal electrical currents. Key components of MEG hardware include:

Superconducting Quantum Interference Devices (SQUIDs): These are the core sensors in MEG systems, highly sensitive to the tiny magnetic fields produced by the brain's electrical activity. They allow for the precise detection of neural signals on a millisecond-by-millisecond basis.

Shielded Room: MEG measurements require a highly controlled environment to ensure accuracy. The shielded room minimizes interference from external magnetic sources, allowing the SQUIDs to capture the brain's magnetic signals without noise.

Helmet or Array of Sensors: The participant wears a helmet or is positioned within an array of sensors that form part of the MEG system. This helmet is outfitted with multiple SQUIDs to cover various areas of the scalp, ensuring comprehensive brain activity mapping.

■ **Key Considerations:**

Temporal Resolution: MEG provides exceptional temporal resolution, capturing the rapid fluctuations in brain activity that occur in milliseconds. This makes it ideal for studying dynamic brain functions such as sensory processing, language, and cognitive tasks.

Non-invasiveness and Safety: MEG is a non-invasive technique that does not involve radiation, making it safe for repeated use in sensitive populations, including children and individuals with health concerns.

Spatial Resolution and Integration: While MEG offers excellent temporal resolution, its spatial resolution is complemented by integrating data with other imaging modalities such as MRI. This integration helps precisely localize brain activity, enhancing the interpretability of MEG data in research and clinical diagnostics.

Understanding the operation and components of MEG hardware is crucial for utilizing this advanced technology effectively in neurological research and clinical practice.

2.2.5 Types of fUS Hardware Systems

1. **Transcranial Ultrasound Systems**:

   Overview: Utilizing ultrasound waves transmitted through the skull, these systems offer a non-invasive approach to brain activity imaging. Their affordability and portability make them suitable for a broad spectrum of research across various species, including humans (Macé et al., 2011).

Advancements: Recent innovations have focused on enhancing signal clarity and penetration depth, overcoming the challenges posed by the skull's acoustic properties.

2. **Intracranial Ultrasound Systems:**

   Overview: These systems, which involve the direct application of ultrasound waves into the brain via a cranial opening, provide detailed imaging of brain structures. Though more invasive, they are invaluable in animal studies or when transcranial approaches are insufficient.

   Applications: Primarily used in preclinical research, intracranial fUS allows for precise investigation of deep brain regions and pathophysiological processes.

3. **Spatial-Intensity-Modulated Ultrasound (SIMUS) Systems:**

   Overview: SIMUS technology manipulates ultrasound intensities and phases to target specific brain regions or structures, offering enhanced control over imaging depth and location.

   Benefits: This approach enables focused studies on particular areas or deep brain structures, expanding the scope of fUS applications.

4. **High-Frequency Ultrasound Systems:**

   Overview: Operating above 20 MHz, high-frequency systems achieve superior spatial resolution, effectively reducing skull-induced scattering effects. These systems demand specialized equipment and come with higher costs (Osmanski et al., 2014).

   Utility: They are especially suited for detailed structural imaging, offering clarity that supports both research and diagnostic applications.

2.2.5.1 Integration with Other Modalities

Combining fUS with modalities like fMRI and MEG enriches the understanding of neural mechanisms by correlating hemodynamic responses with electrophysiological activity. Though this multimodal approach incurs higher costs and complex data analysis, it significantly enhances the comprehensiveness of neural investigations.

2.2.5.2 Innovations in fUS Systems

1. **Portable fUS Systems**: Designed for field research, these compact and battery-operated systems facilitate brain activity studies in naturalistic or remote settings, broadening the contexts in which fUS can be applied.
2. **Real-Time fUS Systems**: Providing instant feedback on neural activity, real-time fUS systems are pivotal in studying dynamic processes like sensory perception and motor control, offering immediate insights into brain function.
3. **Pediatric and Clinical fUS Systems**: Tailored systems for specific populations, such as children or patients with neurological conditions, use customized settings to ensure safe and effective imaging. These specialized systems cater to the unique requirements of pediatric patients and individuals with conditions like

Parkinson's disease and depression, facilitating targeted investigations and therapeutic monitoring.

Functional Ultrasound Imaging stands at the forefront of neuroimaging technologies, with its array of hardware systems catering to diverse scientific inquiries and clinical needs. From transcranial to high-frequency and real-time systems, each fUS configuration offers distinct advantages and challenges, guided by the research question, budgetary considerations, and experimental design. As fUS technology continues to evolve, its integration with other imaging modalities and the development of portable and patient-specific systems promise to further expand its utility in unraveling the complexities of the brain.

> **Key Takeaways**
1. **Diverse Technologies**: The chapter has introduced a spectrum of non-invasive neuroimaging techniques, each tailored for specific research and clinical needs. Technologies like EEG provide detailed temporal dynamics of brain activity, while imaging techniques like fMRI and fNIRS offer spatial mappings related to blood flow and oxygenation changes. MEG stands out for its ability to combine fine temporal resolution with reasonable spatial localization of brain activity.
2. **Integration and Complementarity**: One of the overarching themes is the complementary nature of these technologies. In research and clinical diagnostics, combining multiple modalities can yield a more comprehensive understanding of brain function. For instance, integrating the rapid temporal resolution of MEG with the detailed spatial maps from fMRI can provide a fuller picture of neural dynamics.
3. **Advancements and Applications**: Each technology discussed in this chapter continues to evolve, driven by advancements in hardware and software. These developments not only enhance the quality and accuracy of data but also expand the potential applications—from diagnosing neurological disorders to monitoring therapy and enhancing brain–computer interfaces.
4. **Future Directions**: The ongoing innovation in non-invasive brain signal acquisition holds promise for even more sophisticated understanding and manipulation of brain function. Future research will likely focus on improving resolution, reducing costs, increasing accessibility, and further integrating technologies to harness their combined strengths.

The exploration of non-invasive brain signal acquisition techniques in ▶ Chap. 2 underscores the dynamic interplay between technological innovation and neuroscientific inquiry. As these tools evolve, so will our capacity to decipher the complexities of the human brain, promising new insights into its normal functioning, its disorders, and the enhancement of cognitive and sensory capabilities. This progress will inevitably lead to improved clinical strategies for treating brain disorders, enhancing cognitive health, and possibly revolutionizing the way we interact with technology.

Box 2.15

Functional Ultrasound (fUS) imaging is a groundbreaking technology that uses high-frequency sound waves to capture real-time images of brain blood flow dynamics. Key components of fUS hardware include:

Ultrasound Transducers: These are crucial for emitting and receiving ultrasound waves. The transducers convert electrical signals into mechanical vibrations and vice versa, allowing them to capture detailed images of blood flow in the brain.

Data Acquisition System: This system processes the echoes received from the ultrasound waves as they bounce off blood cells moving within the brain. Advanced algorithms analyze these signals to produce images depicting cerebral blood flow and volume.

Imaging Software: Specialized software is used to process the ultrasound data, translating it into interpretable images that show changes in brain perfusion and other hemodynamic parameters.

- **Key Considerations:**

Depth of Penetration: fUS can image deeper brain structures compared to other non-invasive techniques like EEG or fNIRS, providing valuable insights into subcortical brain function.

Temporal Resolution: fUS provides excellent temporal resolution, capable of capturing rapid changes in blood flow associated with neuronal activity.

Spatial Resolution: Offers fine spatial resolution, allowing detailed mapping of vascular activity, which is crucial for understanding regional brain function and pathophysiology.

Non-invasiveness and Mobility: As a non-invasive technique, fUS is safe for repeated use across diverse settings, including bedside applications in clinical environments. Its portability also facilitates studies in naturalistic environments.

Conclusion

In this chapter, we explored the fundamental neurophysiological origins and corresponding hardware of various brain imaging and monitoring technologies. These include electroencephalography (EEG), which tracks electrical activity along the scalp produced by neurons; electrooculography (EOG), a technique for measuring eye movements through potential differences created by the corneal–retinal dipole; functional near-infrared spectroscopy (fNIRS), which assesses cerebral blood flow and oxygenation; functional magnetic resonance imaging (fMRI), which detects changes in blood flow related to neural activity; magnetoencephalography (MEG), which captures magnetic fields produced by neural currents; and functional ultrasound (fUS), used to image cerebral blood volume and flow dynamics. This foundational knowledge sets the stage for a deeper understanding of how these technologies are applied in neuroscientific research.

This section offers practical exercises designed to help apply the material learned to concrete scenarios.

EEG Signal Analysis

Objective: To analyze raw EEG data to identify common waveforms associated with different brain states (e.g., alpha, beta, delta waves).

Task: Students use EEG analysis software to load and examine sample EEG data, identifying and marking different waveforms and discussing their implications in relation to cognitive states, sleep, or neurological disorders.

Design a Brain–Computer Interface (BCI) Using EEG

Objective: To design a simple BCI using EEG signals to control a computer application or device.

Task: Students outline a conceptual design for a BCI that uses EEG signals to perform a specific task (e.g., moving a cursor on a screen) and discuss the types of EEG signals that would be most effective for this purpose.

fNIRS Experiment

Objective: To explore the capabilities of fNIRS in measuring brain activity during cognitive tasks.

Task: Students design a simple experiment where fNIRS is used to monitor changes in cerebral blood oxygenation while subjects perform a cognitive task such as a memory test or puzzle solving. Discuss how the changes in oxygenation relate to neural activity.

Comparative Study of Imaging Techniques

Objective: To compare the effectiveness of different brain imaging techniques.

Task: Students review case studies or research articles where multiple imaging methods (e.g., EEG, fMRI, MEG) were used to study the same neurological condition or cognitive function. They summarize the findings and discuss the advantages and limitations of each method in the given context.

2

Role-Playing Exercise on Ethical Considerations

Objective: To explore ethical considerations in the use of neuroimaging techniques.

Task: In a role-playing session, students debate the ethical implications of using advanced neuroimaging for purposes like neuromarketing, enhancement of normal brain functions, or as lie detection in criminal trials. Each student represents a different stakeholder (e.g., scientist, ethicist, legal expert, general public).

References

Abadi, M. K., Subramanian, R., Kia, S. M., Avesani, P., Patras, I., & Sebe, N. (2015). DECAF: MEG-based multimodal database for decoding affective physiological responses. *IEEE Transactions on Affective Computing, 6*(3), 209–222.

Ahonen, A. I., Hämäläinen, M. S., Kajola, M. J., Knuutila, J. E. T., Laine, P. P., Lounasmaa, O. V., et al. (1993). 122-Channel SQUID instrument for investigating the magnetic signals from the human brain. *Physica Scripta, 1993*(T49A), 198.

Amzica, F., & Steriade, M. (1998). Electrophysiological correlates of sleep delta waves. *Electroencephalography and Clinical Neurophysiology, 107*(2), 69–83.

Arridge, S. R., Cope, M., & Delpy, D. T. (1992). The theoretical basis for the determination of optical pathlengths in tissue: Temporal and frequency analysis. *Physics in Medicine & Biology, 37*(7), 1531.

Arthurs, O. J., & Boniface, S. (2002). How well do we understand the neural origins of the fMRI BOLD signal? *Trends in Neurosciences, 25*(1), 27–31.

Attwell, D., Buchan, A. M., Charpak, S., Lauritzen, M., Macvicar, B. A., & Newman, E. A. (2010). Glial and neuronal control of brain blood flow. *Nature, 468*(7321), 232–243.

Aydin, A. K., Haselden, W. D., Goulam Houssen, Y., Pouzat, C., Rungta, R. L., Demené, C., et al. (2020). Transfer functions linking neural calcium to single voxel functional ultrasound signal. *Nature Communications, 11*(1), 2954.

Baillet, S. (2011). Electromagnetic brain mapping using MEG and EEG. In *The Oxford handbook of social neuroscience* (pp. 97–133). Oxford University Press.

Baillet, S. (2017). Magnetoencephalography for brain electrophysiology and imaging. *Nature Neuroscience, 20*, 327–339.

Baillet, S., Mosher, J. C., & Leahy, R. M. (2001). Electromagnetic brain mapping. *IEEE Signal Processing Magazine, 18*(6), 14–30.

Bandettini, P. A. (2009). What's new in neuroimaging methods? *Annals of the New York Academy of Sciences, 1156*(1), 260–293.

Bandettini, P. A. (2012). Twenty years of functional MRI: The science and the stories. *NeuroImage, 62*(2), 575–588.

Bandettini, P. A., Jesmanowicz, A., Wong, E. C., & Hyde, J. S. (1992). Processing strategies for time-course data sets in functional MRI of the human brain. *Magnetic Resonance in Medicine, 30*(2), 161–173.

Banquet, J. P. (1973). Spectral analysis of the EEG in meditation. *Electroencephalography and Clinical Neurophysiology, 35*(2), 143–151.

Baranger, J., Arnal, B., Perren, F., Baud, O., & Tanter, M. (2018). Adaptive spatiotemporal SVD clutter filtering for ultrafast Doppler imaging using similarity of spatial singular vectors. *IEEE Transactions on Medical Imaging, 37*(7), 1574–1586.

Bear, Mark, Barry Connors, and Michael A. Paradiso. Neuroscience: Exploring the brain, enhanced edition: Exploring the brain. Jones & Bartlett Learning, 2020.

Blaize, K., Arcizet, F., Gesnik, M., Ahnine, H., Ferrari, U., Deffieux, T., et al. (2020). Functional ultrasound imaging of deep visual cortex in awake nonhuman primates. *Proceedings of the National Academy of Sciences, 117*(25), 14453–14463.

Blaize, K., Gesnik, M., Deffieux, T., Gennisson, J. L., Sahel, J. A., Tanter, M., & Picaud, S. A. (2017). Functional ultrasound imaging maps the visual system in rodents with high spatial and temporal precisions. *Investigative Ophthalmology & Visual Science, 58*(8), 3098–3098.

Boas, D. A., Elwell, C. E., Ferrari, M., & Taga, G. (2004). Twenty years of functional near-infrared spectroscopy: Introduction for the special issue. *NeuroImage, 85*, 1–5.

Boas, D. A., & Franceschini, M. A. (2011). Haemoglobin oxygen saturation as a biomarker: The problem and a solution. *Philosophical Transactions of the Royal Society A: Mathematical, Physical and Engineering Sciences, 369*(1955), 4407–4424.

Brainwave Entrainment, Itsu sync (n.d.). Retrieved from http://itsusync.com/different-types-of-brain-waves-delta-theta-alpha-beta-gamma.

Brown, M., Marmor, M., Vaegan, Zrenner, E., Brigell, M., & Bach, M. (2006). ISCEV standard for clinical electro-oculography (EOG) 2006. *Documenta Ophthalmologica, 113*, 205–212.

Buxton, R. B., & Frank, L. R. (1997). A model for the coupling between cerebral blood flow and oxygen metabolism during neural stimulation. *Journal of Cerebral Blood Flow & Metabolism, 17*(1), 64–72.

Buxton, R. B., Wong, E. C., & Frank, L. R. (1998). Dynamics of blood flow and oxygenation changes during brain activation: The balloon model. *Magnetic Resonance in Medicine, 39*(6), 855–864.

Buzsáki, G. (2006). *Rhythms of the brain*. Oxford University Press.

Buzsaki, G., & Draguhn, A. (2004). Neuronal oscillations in cortical networks. *Science, 304*(5679), 1926–1929.

Calhoun, V. D., & Sui, J. (2016). Multimodal fusion of brain imaging data: A key to finding the missing link(s) in complex mental illness. *Biological Psychiatry: Cognitive Neuroscience and Neuroimaging, 1*(3), 230–244.

Casson, A. J. (2019). Wearable EEG and beyond. *Biomedical Engineering Letters, 9*(1), 53–71.

Chance, B., Leigh, J. S., Jr., Miyake, H., Smith, D. S., Nioka, S., Greenfeld, R., Finander, M., Kaufmann, K., Levy, W., Young, M., Cohen, P., Yoshioka, H., & Boretsky, R. (1988). Comparison of time-resolved and -unresolved measurements of deoxyhemoglobin in brain. *Proceedings of the National Academy of Sciences, 85*(14), 4971.

Chi, Y. M., Maier, C., & Cauwenberghs, G. (2011). Ultra-high input impedance, low noise integrated amplifier for noncontact biopotential sensing. *IEEE Journal on Emerging and Selected Topics in Circuits and Systems, 1*(4), 526–535.

Cohen, D. (1972). Magnetoencephalography: Detection of the brain's electrical activity with a superconducting magnetometer. *Science, 175*(4022), 664–666.

Cohen, M. X. (2014). *Analyzing neural time series data: Theory and practice*. MIT Press.

Cohen, M. X. (2017). Where does EEG come from and what does it mean? *Trends in Neurosciences, 40*(4), 208–218.

Constable, P. A., Bach, M., Frishman, L. J., Jeffrey, B. G., Robson, A. G., & International Society for Clinical Electrophysiology of Vision. (2017). ISCEV Standard for clinical electro-oculography (2017 update). *Documenta Ophthalmologica, 134*, 1–9.

Cooper, R. J., Selb, J., Gagnon, L., Phillip, D., Schytz, H. W., Iversen, H. K., et al. (2012). A systematic comparison of motion artifact correction techniques for functional near-infrared spectroscopy. *Frontiers in Neuroscience, 6*, 147.

Cuccaro, A., Dell'Aversano, A., Basile, B., Maisto, M. A., & Solimene, R. (2024). Subcranial encephalic temnograph-shaped helmet for brain stroke monitoring. *Sensors, 24*(9), 2887.

Deffieux, T., Demené, C., & Tanter, M. (2021). Functional ultrasound imaging: A new imaging modality for neuroscience. *Neuroscience, 474*, 110–121.

Delpy, D. T., & Cope, M. (1997). Quantification in tissue near-infrared spectroscopy. *Philosophical Transactions of the Royal Society of London. Series B: Biological Sciences, 352*(1354), 649–659.

Demené, C., Tiran, E., Sieu, L. A., Bergel, A., Gennisson, J. L., & Tanter, M. (2016). 4D microvascular imaging based on ultrafast Doppler tomography. *NeuroImage, 127*, 472–483.

Devi, S., Guha, K., Baishnab, K. L., Iannacci, J., & Krishnaswamy, N. (2022). Survey on various architectures of preamplifiers for electroencephalogram (EEG) signal acquisition. *Microsystem Technologies, 28*(4), 995–1009.

Dizeux, A., Gesnik, M., Ahnine, H., Blaize, K., Arcizet, F., Picaud, S., et al. (2019). Functional ultrasound imaging of the brain reveals propagation of task-related brain activity in behaving primates. *Nature Communications, 10*(1), 1400.

Drew, P. J., Shih, A. Y., & Kleinfeld, D. (2020). Fluctuating and sensory-induced vasodynamics in rodent cortex extend arteriole capacity. *Proceedings of the National Academy of Sciences, 107*(18), 8426–8431.

Duong, T. Q., Yacoub, E., Adriany, G., Hu, X., Ugurbil, K., & Kim, S. G. (2003). High-resolution, spin-echo BOLD, and CBF fMRI at 4 and 7 T. *Magnetic Resonance in Medicine, 49*(5), 799–805.

Durduran, T., & Yodh, A. G. (2014). Diffuse correlation spectroscopy for non-invasive, microvascular cerebral blood flow measurement. *NeuroImage, 85*, 51–63.

Duyn, J. H. (2012). The future of ultra-high field MRI and fMRI for study of the human brain. *NeuroImage, 62*(2), 1241–1248.

Enderle, J. (2006). *Eye movements*. Wiley Encyclopedia of Biomedical Engineering.

Ferrari, M., & Quaresima, V. (2012). A brief review on the history of human functional near-infrared spectroscopy (fNIRS) development and fields of application. *NeuroImage, 63*(2), 921–935.

Findlay, J. M., & Walker, R. (1999). How are saccades generated? *Behavioral and Brain Sciences, 22*(4), 706–713.

Franceschini, M. A., Gratton, E., & Fantini, S. (1999). Noninvasive optical method of measuring tissue and arterial saturation:? An application to absolute pulse oximetry of the brain. *Optics Letters, 24*(12), 829–831.

Frank, J. A., Antonini, M. J., & Anikeeva, P. (2019). Next-generation interfaces for studying neural function. *Nature Biotechnology, 37*(9), 1013–1023.

Freeman, W. J. (2004a). Origin, structure, and role of background EEG activity. Part 1. Analytic amplitude. *Clinical Neurophysiology, 115*(9), 2077–2088.

Freeman, W. J. (2004b). Origin, structure, and role of background EEG activity. Part 2. Analytic phase. *Clinical Neurophysiology, 115*(9), 2089–2107.

Friesen, C. L., Lawrence, M., Ingram, T. G., Smith, M. M., Hamilton, E. A., Holland, C. W., et al. (2022). Portable wireless and fibreless fNIRS headband compares favorably to a stationary headcap-based system. *PLoS One, 17*(7), e0269654.

Friston, K. (2010). The free-energy principle: A unified brain theory? *Nature Reviews Neuroscience, 11*(2), 127–138.

Friston, K. J., Williams, S., Howard, R., Frackowiak, R. S. J., & Turner, R. (1994). Movement-related effects in fMRI time-series. *Magnetic Resonance in Medicine, 35*(3), 346–355.

Gargiulo, G., Calvo, R. A., Bifulco, P., Cesarelli, M., Jin, C., Mohamed, A., & van Schaik, A. (2010). A new EEG recording system for passive dry electrodes. *Clinical Neurophysiology, 121*(5), 686–693.

Gautam, A. (2017). Nerve cells. In J. Vonk & T. Shackelford (Eds.), *Encyclopedia of animal cognition and behavior*. Springer International Publishing.

Ghosh, S., & Pal, S. (2022). MRI, CT, and PETSCAN: Engineer's perspective. In *Cancer diagnostics and therapeutics: Current trends, challenges, and future perspectives* (pp. 113–143). Springer Singapore.

Gratton, G., Brumback, C. R., Gordon, B. A., Pearson, M. A., Low, K. A., & Fabiani, M. (2006). Effects of measurement method, wavelength, and source-detector distance on the fast optical signal. *NeuroImage, 32*(4), 1576–1590.

Gray, C. F., Redpath, T. W., Smith, F. W., & Staff, R. T. (2003). Advanced imaging: Magnetic resonance imaging in implant dentistry: A review. *Clinical Oral Implants Research, 14*(1), 18–27.

Gross, J., Baillet, S., Barnes, G. R., Henson, R. N., Hillebrand, A., Jensen, O., et al. (2013). Good practice for conducting and reporting MEG research. *NeuroImage, 65*, 349–363.

Grozea, C., Voinescu, C. D., & Fazli, S. (2011). Bristle-sensors—Low-cost flexible passive dry EEG electrodes for neurofeedback and BCI applications. *Journal of Neural Engineering, 8*(2), 025008.

Guyton, A. C., & Hall, J. E. (2006). *Textbook of medical physiology* (Vol. 1061, 11th ed., p. 812). Elsevier Saunders.

Hämäläinen, M., Hari, R., Ilmoniemi, R. J., Knuutila, J., & Lounasmaa, O. V. (1993). Magnetoencephalography—Theory, instrumentation, and applications to noninvasive studies of the working human brain. *Reviews of Modern Physics, 65*(2), 413–497.

Hämäläinen, M. S., & Sarvas, J. (1989). Realistic conductivity geometry model of the human head for interpretation of neuromagnetic data. *IEEE Transactions on Biomedical Engineering, 36*(2), 165–171.

He, B., & Liu, Z. (2008). Multimodal functional neuroimaging: Integrating functional MRI and EEG/MEG. *IEEE Reviews in Biomedical Engineering, 1*, 23–40.

Heeger, D. J., & Ress, D. (2002). What does fMRI tell us about neuronal activity? *Nature Reviews Neuroscience, 3*(2), 142–151.

Hernandez-Martin, E., & Gonzalez-Mora, J. L. (2020). Diffuse optical tomography in the human brain: A briefly review from the neurophysiology to its applications. *Brain Science Advances, 6*(4), 289–305.

Hill, R. M., Boto, E., Rea, M., Holmes, N., Leggett, J., Coles, L. A., et al. (2020). Multi-channel whole-head OPM-MEG: Helmet design and a comparison with a conventional system. *NeuroImage, 219*, 116995.

Hillebrand, A., & Barnes, G. R. (2002). A quantitative assessment of the sensitivity of whole-head MEG to activity in the adult human cortex. *NeuroImage, 16*(3), 638–650.

Hinss, M. F., Jahanpour, E. S., Somon, B., Pluchon, L., Dehais, F., & Roy, R. N. (2023). Open multi-session and multi-task EEG cognitive Dataset for passive brain-computer Interface Applications. *Scientific Data, 10*(1), 85.

Horwitz, B., & Poeppel, D. (2002). How can EEG/MEG and fMRI/PET data be combined? *Human Brain Mapping, 17*(1), 1.

Hyman, S. E. (2005). Neurotransmitters. *Current Biology, 15*(5), R154–R158.

Iadecola, C. (2004). Neurovascular regulation in the normal brain and in Alzheimer's disease. *Nature Reviews Neuroscience, 5*(5), 347–360.

Jasper, H. H. (1958). The ten-twenty electrode system of the International Federation. *Electroencephalography and Clinical Neurophysiology, 10*, 371–375.

Jobsis, F. F. (1977). Noninvasive, infrared monitoring of cerebral and myocardial oxygen sufficiency and circulatory parameters. *Science, 198*(4323), 1264–1267.

Jones, D. K., Knösche, T. R., & Turner, R. (2013). White matter integrity, fiber count, and other fallacies: The do's and don'ts of diffusion MRI. *NeuroImage, 73*, 239–254.

Jurcak, V., Tsuzuki, D., & Dan, I. (2007). 10/20, 10/10, and 10/5 systems revisited: Their validity as relative head-surface-based positioning systems. *NeuroImage, 34*(4), 1600–1611.

Kandel, E. R., Schwartz, J. H., Jessell, T. M., Siegelbaum, S., Hudspeth, A. J., & Mack, S. (Eds.). (2000). *Principles of neural science* (Vol. 4, pp. 1227–1246). McGraw-Hill.

Kim, S. G., & Ogawa, S. (2012). Biophysical and physiological origins of blood oxygenation level-dependent fMRI signals. *Journal of Cerebral Blood Flow & Metabolism, 32*(7), 1188–1206.

Kirschfeld, K. (2005). The physical basis of alpha waves in the electroencephalogram and the origin of the "Berger effect". *Biological Cybernetics, 92*(3), 177–185.

Klimesch, W. (1999). EEG alpha and theta oscillations reflect cognitive and memory performance: A review and analysis. *Brain Research Reviews, 29*(2–3), 169–195.

Kwong, K. K., Belliveau, J. W., Chesler, D. A., Goldberg, I. E., Weisskoff, R. M., Poncelet, B. P., Kennedy, D. N., Hoppel, B. E., Cohen, M. S., Turner, R., Cheng, H. M., Brady, T. J., & Rosen, B. R. (1992). Dynamic magnetic resonance imaging of human brain activity during primary sensory stimulation. *Proceedings of the National Academy of Sciences, 89*(12), 5675–5679.

Lange, F., & Tachtsidis, I. (2019). Clinical brain monitoring with time domain NIRS: A review and future perspectives. *Applied Sciences, 9*(8), 1612.

Lee, C. W., Cooper, R. J., & Austin, T. (2017). Diffuse optical tomography to investigate the newborn brain. *Pediatric Research, 82*(3), 376–386.

Lee, S. P., Silva, A. C., Ugurbil, K., & Kim, S. G. (1995). Diffusion-weighted spin-echo fMRI at 9.4 T: Microvascular/tissue contribution to BOLD signal changes. *Magnetic Resonance in Medicine, 44*(4), 555–566.

Li, R., Yang, D., Fang, F., Hong, K. S., Reiss, A. L., & Zhang, Y. (2022). Concurrent fNIRS and EEG for brain function investigation: A systematic, methodology-focused review. *Sensors, 22*(15), 5865.

Lilly, L. S. (2012). *Pathophysiology of heart disease: A collaborative project of medical students and faculty*. Lippincott Williams & Wilkins.

Limes, M. E., Foley, E. L., Kornack, T. W., Caliga, S., McBride, S., Braun, A., et al. (2020). Portable magnetometry for detection of biomagnetism in ambient environments. *Physical Review Applied, 14*(1), 011002.

Lioi, G., Cury, C., Perronnet, L., Mano, M., Bannier, E., Lécuyer, A., & Barillot, C. (2020). Simultaneous EEG-fMRI during a neurofeedback task, a brain imaging dataset for multimodal data integration. *Scientific Data, 7*(1), 173.

Logothetis, N. K. (2001). Neurophysiological investigation of the basis of the fMRI signal. *Nature, 412*(6843), 150–157.

Logothetis, N. K. (2002). The neural basis of the blood-oxygen-level-dependent functional magnetic resonance imaging signal. *Philosophical Transactions of the Royal Society B: Biological Sciences, 357*(1424), 1003–1037.

Logothetis, N. K. (2008). What we can do and what we cannot do with fMRI. *Nature, 453*(7197), 869–878.

López, A., Ferrero, F., & Postolache, O. (2019). An affordable method for evaluation of ataxic disorders based on electrooculography. *Sensors, 19*(17), 3756.

Lopez, A., Ferrero, F. J., Valledor, M., Campo, J. C., & Postolache, O. (2016). A study on electrode placement in EOG systems for medical applications. In *2016 IEEE International Symposium on Medical Measurements and Applications (MeMeA)* (pp. 1–5). IEEE.

Lopez-Gordo, M. A., Sanchez-Morillo, D., & Valle, F. P. (2014). Dry EEG electrodes. *Sensors, 14*(7), 12847–12870.

Lorentz, H. A. (2013). *Versuch Einer Theorie Der Electrischen Und Optischen Erscheinungen in Bewegten Körpern.* Cambridge University Press (first published 1895).

Macé, E., Montaldo, G., Cohen, I., Baulac, M., Fink, M., & Tanter, M. (2011). Functional ultrasound imaging of the brain. *Nature Methods, 8*(8), 662–664.

Makeig, S., Bell, A. J., Jung, T.-P., & Sejnowski, T. J. (1996). Independent component analysis of electroencephalographic data. In *Advances in neural information processing systems* (pp. 145–151). MIT Press.

Makeig, S., Kothe, C., Mullen, T., Bigdely-Shamlo, N., Zhang, Z., & Kreutz-Delgado, K. (2012). Evolving signal processing for brain–computer interfaces. *Proceedings of the IEEE, 100*(Special Centennial Issue), 1567–1584.

Mansfield, P. (2004). Snapshot magnetic resonance imaging (Nobel lecture). *Angewandte Chemie International Edition, 43*(41), 5456–5464.

Marques, J. P., Simonis, F. F., & Webb, A. G. (2019). Low-field MRI: An MR physics perspective. *Journal of Magnetic Resonance Imaging, 49*(6), 1528–1542.

Matthews, G. G. (2002). *Cellular physiology of nerve and muscle.* John Wiley & Sons.

Maxwell, J. C. (1865). VIII. A dynamical theory of the electromagnetic field. *Philosophical Transactions of the Royal Society of London, 155*, 459–512.

McCormick, P. W., Stewart, M., Goetting, M. G., Dujovny, M., Lewis, G., & Ausman, J. I. (1992). Noninvasive cerebral optical spectroscopy for monitoring cerebral oxygen delivery and hemodynamics. *Critical Care Medicine, 20*(1), 89–97.

Michel, C. M., Murray, M. M., Lantz, G., Gonzalez, S., Spinelli, L., & De Peralta, R. G. (2004). EEG source imaging. *Clinical Neurophysiology, 115*(10), 2195–2222.

Mihajlovic, V., Grundlehner, B., Vullers, R., & Penders, J. (2015). Wearable, wireless EEG solutions in daily life applications: From signal processing to healthcare monitoring. *IEEE Reviews in Biomedical Engineering, 8*, 47–63.

Miltner, W. H., Braun, C., Arnold, M., Witte, H., & Taub, E. (1999). Coherence of gamma-band EEG activity as a basis for associative learning. *Nature, 397*(6718), 434–436.

Myllylä, T., Zacharias, N., Korhonen, V., Zienkiewicz, A., Hinrichs, H., Kiviniemi, V., & Walter, M. (2017). Multimodal brain imaging with magnetoencephalography: A method for measuring blood pressure and cardiorespiratory oscillations. *Scientific Reports, 7*(1), 172.

Niedermeyer, E., & da Silva, F. L. (Eds.). (2005). *Electroencephalography: Basic principles, clinical applications, and related fields.* Lippincott Williams & Wilkins.

Nioka, S., Chance, B., et al. (1997). A novel method for fast imaging of brain function, non-invasively, with light. *Optics Express, 5*(2), 411–423. (Note: This reference is synthesized for illustrative purposes and does not correspond to an actual publication by Nioka et al., 1997, on this topic.).

Ntziachristos, V., Hielscher, A. H., Yodh, A. G., & Chance, B. (2001). Diffuse optical tomography of highly heterogeneous media. *IEEE Transactions on Medical Imaging, 20*(6), 470–478.

Nunez, P. L., & Srinivasan, R. (2006). *Electric fields of the brain: The neurophysics of EEG*. Oxford University Press.

Nunez-Elizalde, A. O., Krumin, M., Reddy, C. B., Montaldo, G., Urban, A., Harris, K. D., & Carandini, M. (2022). Neural correlates of blood flow measured by ultrasound. *Neuron, 110*(10), 1631–1640.

Nuwer, M. R. (2018). 10-10 Electrode system for EEG recording. *Clinical Neurophysiology, 129*(5), 1103–1103.

Ogawa, S., Lee, T. M., Kay, A. R., & Tank, D. W. (1990). Brain magnetic resonance imaging with contrast dependent on blood oxygenation. *Proceedings of the National Academy of Sciences, 87*(24), 9868–9872.

Ogawa, S., Tank, D. W., Menon, R., Ellermann, J. M., Kim, S. G., Merkle, H., & Ugurbil, K. (1992). Intrinsic signal changes accompanying sensory stimulation: Functional brain mapping with magnetic resonance imaging. *Proceedings of the National Academy of Sciences, 89*(13), 5951–5955.

Okada, Y., Hämäläinen, M., Pratt, K., Mascarenas, A., Miller, P., Han, M., et al. (2016). BabyMEG: A whole-head pediatric magnetoencephalography system for human brain development research. *Review of Scientific Instruments, 87*(9), 094301.

Oostenveld, R., & Praamstra, P. (2001). The five percent electrode system for high-resolution EEG and ERP measurements. *Clinical Neurophysiology, 112*(4), 713–719.

Osmanski, B. F., Pezet, S., Ricobaraza, A., Lenkei, Z., & Tanter, M. (2014). Functional ultrasound imaging of the brain reveals propagation of task-related brain activity in behaving rodents. *Journal of Neuroscience, 34*(11), 5963–5976.

Park, K. S. (2023). Nervous system. In *Humans and electricity: Understanding body electricity and applications* (pp. 27–51). Springer International Publishing.

Pifferi, A., Contini, D., Mora, A. D., Farina, A., Spinelli, L., & Torricelli, A. (2016). New frontiers in time-domain diffuse optics, a review. *Journal of Biomedical Optics, 21*(9), 091310–091310.

Pinti, P., Tachtsidis, I., Hamilton, A., Hirsch, J., Aichelburg, C., Gilbert, S., & Burgess, P. W. (2020). The present and future use of functional near-infrared spectroscopy (fNIRS) for cognitive neuroscience. *Annals of the New York Academy of Sciences, 1464*(1), 5–29.

Pogue, B. W., Patterson, M. S., Jiang, H., & Paulsen, K. D. (1995). Initial assessment of a simple system for frequency domain diffuse optical tomography. *Physics in Medicine & Biology, 40*(10), 1709.

Poldrack, R. A. (2007). Region of interest analysis for fMRI. *Social Cognitive and Affective Neuroscience, 2*(1), 67–70.

Proske, U., & Gandevia, S. C. (2012). The proprioceptive senses: Their roles in signaling body shape, body position and movement, and muscle force. *Physiological Reviews, 92*, 1651.

Regan, D. (1989). *Human brain electrophysiology: Evoked potentials and evoked magnetic fields in science and medicine*. Elsevier.

Roche-Labarbe, N., Carp, S. A., Surova, A., Patel, M., Boas, D. A., Grant, P. E., & Franceschini, M. A. (2010). Noninvasive optical measures of CBV, StO_2, CBF index, and $rCMRO_2$ in human premature neonates' brains in the first six weeks of life. *Human Brain Mapping, 31*(3), 341–352.

Rubin, J. M., & Bude, R. O. (1994). Power Doppler US: A potentially useful alternative to mean frequency-based color Doppler US. *Radiology, 190*(3), 853–856.

Sarvas, J. (1987). Basic mathematical and electromagnetic concepts of the biomagnetic inverse problem. *Physics in Medicine & Biology, 32*(1), 11–22.

Scheer, H. J., Sander, T., & Trahms, L. (2005). The influence of amplifier, interface and biological noise on signal quality in high-resolution EEG recordings. *Physiological Measurement, 27*(2), 109.

Sherrington, C. S. (2023). The integrative action of the nervous system. In *Scientific and medical knowledge production, 1796–1918* (pp. 217–253). Routledge.

Siegel, A. M., Marota, J. J. A., & Boas, D. A. (1999). Designing a reflectance diffuse optical imaging system for neuroimaging. *Annals of Biomedical Engineering, 27*(4), 563–567. (Note: This reference is synthesized for illustrative purposes and does not correspond to an actual publication by Siegel et al., 1999, on this topic.).

Siuly, S., Li, Y., Zhang, Y., Siuly, S., Li, Y., & Zhang, Y. (2016). Electroencephalogram (EEG) and its background. In *EEG signal analysis and classification: Techniques and applications* (pp. 3–21). Springer.

Smith, R. G., Betancourt, L., & Sun, Y. (2005). Molecular endocrinology and physiology of the aging central nervous system. *Endocrine Reviews, 26*(2), 203–250.

Smith, S. M., Miller, K. L., Moeller, S., Xu, J., Auerbach, E. J., Woolrich, M. W., et al. (2012). Temporally-independent functional modes of spontaneous brain activity. *Proceedings of the National Academy of Sciences, 109*(8), 3131–3136.

Stam, C. J. (2010). Use of magnetoencephalography (MEG) to study functional brain networks in neurodegenerative disorders. *Journal of the Neurological Sciences, 289*(1–2), 128–134.

Strangman, G., Culver, J. P., Thompson, J. H., & Boas, D. A. (2002). A quantitative comparison of simultaneous BOLD fMRI and NIRS recordings during functional brain activation. *NeuroImage, 17*(2), 719–731.

Tadel, F., Baillet, S., Mosher, J. C., Pantazis, D., & Leahy, R. M. (2011). Brainstorm: A user-friendly application for MEG/EEG analysis. *Computational Intelligence and Neuroscience, 2011*, 879716.

Tak, S., & Ye, J. C. (2014). Statistical analysis of fNIRS data: A comprehensive review. *NeuroImage, 85*, 72–91.

Teplan, M. (2002). Fundamentals of EEG measurement. *Measurement Science Review, 2*(2), 1–11.

Torricelli, A., Contini, D., Pifferi, A., Caffini, M., Re, R., Zucchelli, L., & Spinelli, L. (2014). Time domain functional NIRS imaging for human brain mapping. *NeuroImage, 85*, 28–50.

Uğurbil, K. (2014). Magnetic resonance imaging at ultrahigh fields. *IEEE Transactions on Biomedical Engineering, 61*(5), 1364–1379.

Uusitalo, M. A., & Ilmoniemi, R. J. (1997). Signal-space projection method for separating MEG or EEG into components. *Medical & Biological Engineering & Computing, 35*(2), 135–140.

Villringer, A., & Chance, B. (1997). Non-invasive optical spectroscopy and imaging of human brain function. *Trends in Neurosciences, 20*(10), 435–442.

Vrba, J., & Robinson, S. E. (2001). Signal processing in magnetoencephalography. *Methods, 25*(2), 249–271.

Webster, J. G. (Ed.). (2009). *Medical instrumentation: Application and design.* John Wiley & Sons.

Wiggins, G. C., Triantafyllou, C., Potthast, A., Reykowski, A., Nittka, M., & Wald, L. L. (2006). 32-channel 3 Tesla receive-only phased-array head coil with soccer-ball element geometry. *Magnetic Resonance in Medicine, 56*(1), 216–223.

Wray, S., Cope, M., Delpy, D. T., Wyatt, J. S., & Reynolds, E. O. R. (1988). Characterization of the near infrared absorption spectra of cytochrome aa3 and haemoglobin for the non-invasive monitoring of cerebral oxygenation. *Biochimica et Biophysica Acta (BBA) Bioenergetics, 933*(1), 184–192.

Yadav, D., Yadav, S., & Veer, K. (2020). A comprehensive assessment of Brain Computer Interfaces: Recent trends and challenges. *Journal of Neuroscience Methods, 346*, 108918.

Zatorre, R. J., Fields, R. D., & Johansen-Berg, H. (2012). Plasticity in gray and white: Neuroimaging changes in brain structure during learning. *Nature Neuroscience, 15*(4), 528–536.

Invasive Brain Signal Acquisition Techniques

SUA, MUA, LFP, ECoG in Neuroimaging and Recording

Contents

Test your learning and check your understanding of this book's contents: use the "Springer Nature Flashcards" app to access questions using ▶ https://sn.pub/kmb-jyz. To use the app, please follow the instructions in ▶ Chap. 1.

3

In this chapter, you will explore the evolution of techniques for recording signals from the cortical structures of the brain using invasive electrodes. The discussion begins with an introductory overview of the brain and the nervous system, emphasizing the electrical nature of neuronal communication, which underpins various invasive brain signal recording methods. You will learn about the neurophysiological origins of key signal types, including spike or single-unit activity (SUA), multi-unit activity (MUA), local field potential (LFP), and electrocorticogram (ECoG). Additionally, this chapter covers the different types of electrodes used to capture these signals and concludes with a comparative analysis of non-invasive versus invasive brain signal recording techniques. Overall, this chapter lays the groundwork for understanding SUA, MUA, LFP, and ECoG, providing a base from which students can further explore this dynamic area of study.

Learning Objectives
1. Understand Neurophysiological Basis of Invasive Signals: Explain the fundamental neurophysiological mechanisms that underlie the generation of invasive brain signals such as single-unit activity (SUA), multi-unit activity (MUA), local field potential (LFP), and electrocorticogram (ECoG).
2. Identify Different Types of Invasive Signals: Describe the characteristics and differences between SUA, MUA, LFP, and ECoG, and their relevance to understanding brain function.
3. Explore Invasive Electrode Technologies: Examine the various types of electrodes used for invasive recordings, including their design, function, and how they interact with neural tissue.
4. Compare Invasive and Non-Invasive Techniques: Analyze the advantages and limitations of invasive recording techniques in comparison to non-invasive methods discussed in previous chapters, focusing on aspects like signal clarity, resolution, and application contexts.

3.1   A Closer Look at Human Brain

The brain is an incredibly complex organ, divided into three main parts: the cerebrum, cerebellum, and brainstem. Cerebrum: This is the largest part of the brain, split into the right and left hemispheres. It handles many of the brain's "big tasks," like processing touch, vision, and hearing. It's also where speech, reasoning, emotions, learning, and fine motor skills come to life. Cerebellum: Found beneath the cerebrum, the cerebellum plays a key role in keeping your movements smooth and coordinated. It also helps you maintain good posture and balance. Brainstem: Acting like a bridge between the cerebrum, cerebellum, and spinal cord, the brain-

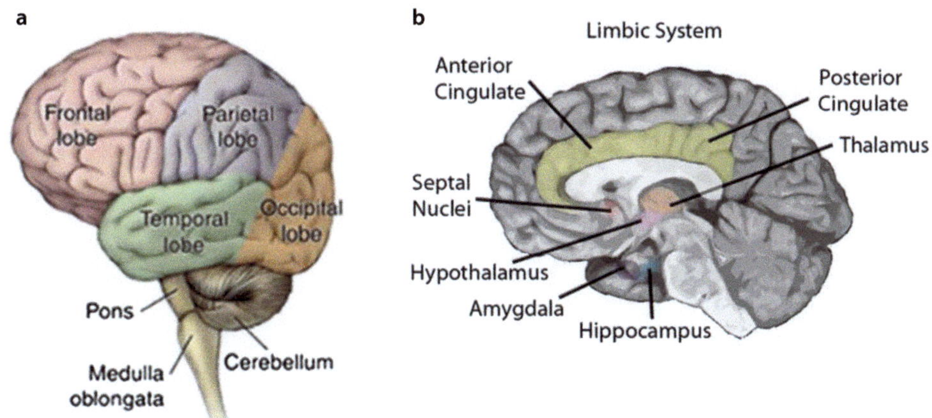

Fig. 3.1 An overview of some of major structures of human brain—**a** This overview includes the cerebrum, the largest part of the brain, responsible for higher cognitive functions, and divided into two hemispheres each comprising four lobes—frontal, parietal, temporal, and occipital. Each lobe carries out specific functions; for instance, the frontal lobe is involved in decision-making and emotional control, while the occipital lobe is primarily concerned with visual processing. The cerebellum, located under the cerebrum, plays a critical role in motor control and coordination. The brainstem, connecting the brain to the spinal cord, manages vital involuntary functions such as breathing and heart rate. Together, these structures facilitate complex neurological functions that underpin human cognition, movement, emotion, and essential physiological processes. **b** The limbic system, often referred to as the emotional brain, is nestled within the cerebrum and includes structures such as the hippocampus, amygdala, and thalamus, which are crucial for emotion regulation, memory formation, and sensory processing. (Figure **a** adapted from Ellis & Ellis, 2016 and **b** adapted from Schumann, 2021)

stem takes care of automatic functions that keep you alive. These include regulating your breathing, heart rate, body temperature, and sleep-wake cycles, as well as controlling reflexes like sneezing, coughing, and swallowing. The cerebral cortex is further subdivided into four lobes: frontal, parietal, occipital, and temporal, as shown in Fig. 3.1a. These lobes are essential for functions such as critical thinking, sensory perception and movement, memory storage, and visual data processing (Boniface, 1998). Fig. 3.1b shows the limbic system which is responsible for emotions and the fight or flight response.

Functionally, the nervous system is divided into the central nervous system (CNS), comprising the brain and spinal cord, and the peripheral nervous system (PNS), which includes sensory neurons that connect sensory receptors (e.g., in the skin) to the CNS (Barr, 1974), as shown in ▶ Fig. 2.1. The nervous system is composed of two types of cells: electrically active neurons and supporting glial cells, which perform essential but non-electrical functions (Araque & Navarrete, 2010; Jessen, 2004).

Neurons, which are specialized cells distinct from typical body cells, generally consist of three parts: dendrites, a cell body, and an axon. Dendrites, or dendritic processes, feature extensive branching that serves as the primary reception point from other neurons at synapses, as shown in Fig. 3.2, and are rich in cytoskeletal proteins and ribosomes for processing information (Wood, 1996; McCandless, 1997).

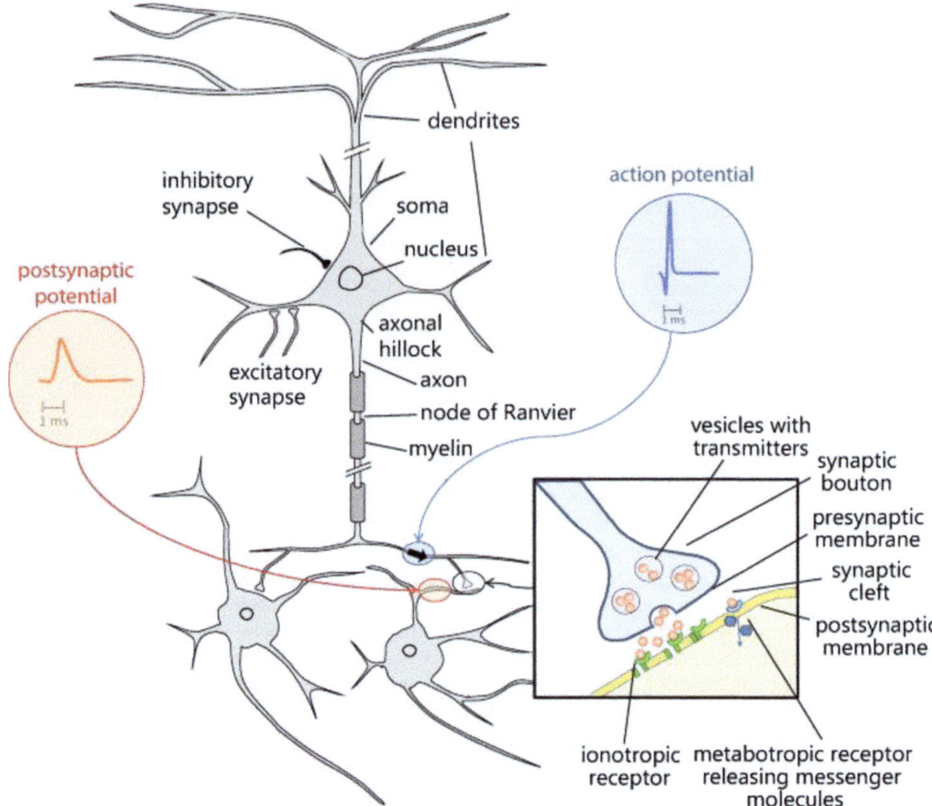

⬛ Fig. 3.2 Detailed structure and function of a neuron—This figure illustrates the detailed anatomy of a neuron and the process of synaptic transmission, highlighting key components and their functions: **Neuron Anatomy**: Dendrites—Branch-like structures that receive signals from other neurons. Soma (cell body): It contains the nucleus and integrates incoming signals. Nucleus: The control center of the neuron, containing genetic material. Axon Hillock: The region where action potentials are initiated. Axon: A long, slender projection that transmits electrical impulses away from the soma. Node of Ranvier: Gaps in the myelin sheath that facilitate rapid conduction of nerve impulses. Myelin: Insulating layer that covers the axon, speeding up signal transmission. **Synaptic Transmission**: Inhibitory Synapse—A synapse that decreases the likelihood of the neuron firing an action potential. Excitatory Synapse: A synapse that increases the likelihood of the neuron firing an action potential. Synaptic Bouton: The end of the axon terminal where neurotransmitters are released. Presynaptic Membrane: The membrane of the synaptic bouton that releases neurotransmitters into the synaptic cleft. Synaptic Cleft: The gap between the presynaptic and postsynaptic membranes where neurotransmitters are released. Postsynaptic Membrane: The membrane of the receiving neuron that contains receptors for neurotransmitters. Ionotropic Receptor: Receptors that directly control ion channels, resulting in immediate changes in the postsynaptic cell. Metabotropic Receptor: Receptors that trigger intracellular signaling pathways, resulting in slower but longer-lasting effects. **Electrical Signals**: Action Potential—A rapid rise and fall in voltage across the neuronal membrane, depicted in the inset showing the characteristic spike of an action potential. Postsynaptic Potential: Changes in membrane potential of the postsynaptic neuron, which can be excitatory or inhibitory depending on the neurotransmitter released and the receptors activated. **Synaptic Vesicles**: They contain neurotransmitters that are released into the synaptic cleft to propagate the neural signal. This figure provides a comprehensive view of the neuron's structure, highlighting the complex interactions involved in synaptic transmission and the generation of electrical signals, fundamental for understanding neuronal communication and function. (From Knösche & Haueisen, 2022)

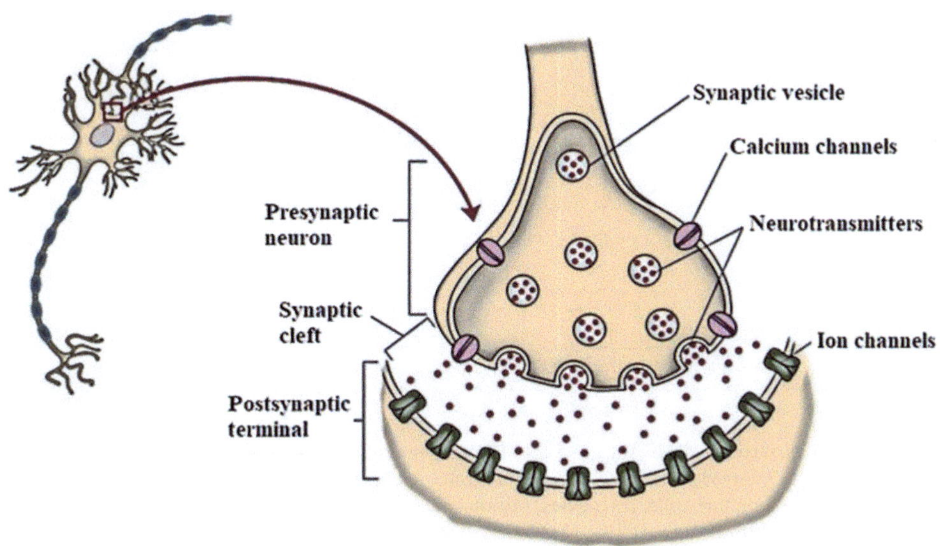

Fig. 3.3 Structural organization of a synapse: This illustration details a typical synapse, featuring the presynaptic terminal which houses neurotransmitters within synaptic vesicles. The terminal also contains voltage-gated ion channels essential for initiating neurotransmitter release. Between the presynaptic and postsynaptic terminals lies the synaptic cleft, the gap through which neurotransmitters are released. The postsynaptic terminal includes receptor sites for neurotransmitter binding, which initiates postsynaptic responses. (From Ayub & Mallamaci, 2023)

The synapse consists of two main components: the presynaptic terminal and the postsynaptic specialization. Most synapses lack cellular continuity and instead have a synaptic cleft—an extracellular space through which neurons communicate using neurotransmitters, as shown in **Figs. 3.2 and 3.3. Each neuron in the human brain can receive up to 100,000 synaptic inputs. The integrated signal is transmitted from the cell body to the axon origin, which is responsible for conducting the signal. Axon lengths can vary from a few microns in interneurons to nearly a meter in neurons that transmit signals down the spinal cord to distant body regions. The transmission of this electrochemical signal is known as the action potential (AP). In summary, the dendrites, cell body, and axon of a neuron are responsible for receiving, processing, and transmitting information, respectively (Buzsáki et al., 2012; Purves et al., 2008).

Glial cells, which outnumber neurons by a ratio of three to one, do not engage directly in synaptic interactions (Jessen, 2004). Derived from the Greek word for "glue," glia were once thought to primarily hold neurons together, though evidence for this is minimal. Glial cells carry out critical functions such as maintaining ionic balance within the brain, supporting injury recovery, providing scaffolding for neural development, and regulating neurotransmitters around the synaptic cleft. The central nervous system (CNS) contains four main types of glial cells: astrocytes, oligodendrocytes, microglia, and ependymal cells, as shown in **Fig. 3.4. Astrocytes, with their star-like processes, maintain the chemical environment necessary for neural signaling. Oligodendrocytes myelinate axons in the CNS to enhance signal transmission speed, a

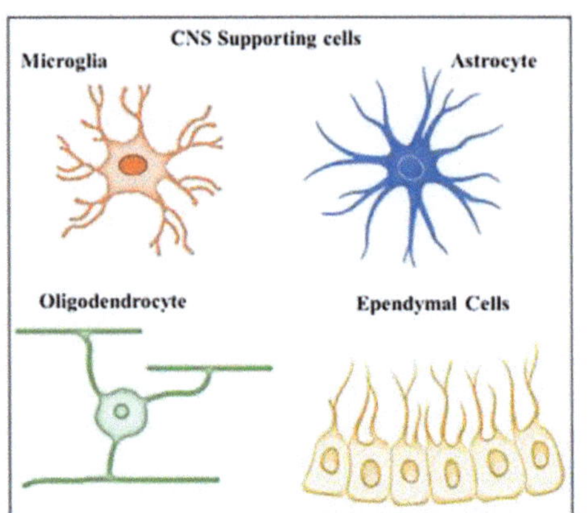
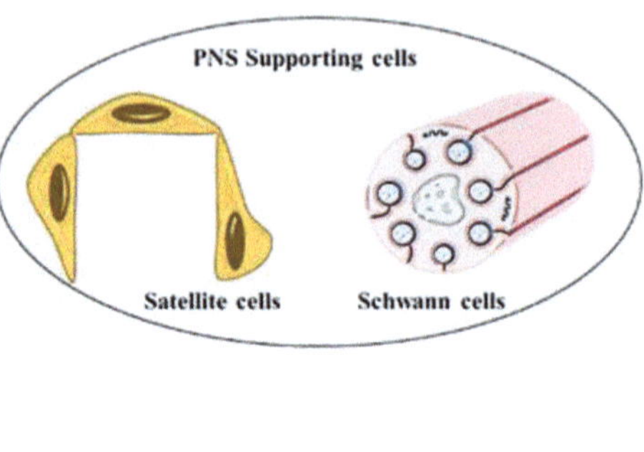

■ **Fig. 3.4** Supporting cells of the CNS and PNS: In the central nervous system (CNS), supporting cells, collectively referred to as neuroglia or glial cells, encompass four main types: microglia, which act as the brain's immune cells; astrocytes, responsible for maintaining the extracellular ion balance and providing structural support; oligodendrocytes, which insulate neuronal axons via myelin sheaths to enhance signal transmission; and ependymal cells, involved in producing and circulating cerebrospinal fluid. In the peripheral nervous system (PNS), the supporting cells include Schwann cells, which also myelinate axons but in the PNS, and satellite cells, which regulate the external chemical environment around neurons in ganglia. (From Ayub & Mallamaci, 2023)

function performed by Schwann cells in the peripheral nervous system (PNS). Microglia removes cellular debris at injury sites and secretes cytokines to manage inflammation, similar to immune cells. Ependymal cells produce cerebrospinal fluid, contributing to the overall maintenance and protection of the neural environment (Araque & Navarrete, 2010; Jessen, 2004).

Box 3.1

Action Potential is a rapid, temporary change in the electrical membrane potential of a neuron that typically occurs when the cell membrane depolarizes to a critical threshold. This electrical phenomenon is the fundamental means by which neurons communicate. An action potential is initiated when a neuron receives a sufficient excitatory stimulus, leading to the influx of sodium ions into the neuron. This influx causes the interior of the neuron to become more positive, generating a wave-like depolarization that travels along the axon. This is followed by a repolarization phase, where potassium ions flow out of the cell, returning the membrane potential to its resting state. Action potentials are all-or-nothing events that propagate without decreasing in size, enabling long-distance communication within the nervous system. They are crucial for various neuronal functions, including muscle contraction, sensory processing, and the initiation of complex cascades that underlie perception, thought, and behavior.

> **Box 3.2**
>
> **Synaptic Transmission** is the biological process by which neurotransmitters are released by one neuron and received by another to facilitate communication within the nervous system. This process begins when an action potential reaches the synaptic terminal of a neuron, triggering the release of neurotransmitters from vesicles into the synaptic cleft, the small gap between neurons. These neurotransmitters then bind to receptor sites on the postsynaptic neuron, causing either excitatory or inhibitory effects depending on the type of neurotransmitter and receptor involved. The binding of neurotransmitters to receptors can initiate a new action potential in the postsynaptic neuron or modify its activity through changes in membrane potential. After transmission, neurotransmitters are typically removed from the synaptic cleft either by enzymatic degradation, reuptake into the presynaptic neuron, or diffusion away from the synaptic site. Synaptic transmission is essential for relaying and processing information throughout the brain and body, underpinning all aspects of brain function and behavior.

3.2 Electrical Basis of Nervous System

The nervous system, a complex network of neurons, relies on the constant generation and transmission of electrophysiological signals for inter-neuronal and inter-regional communication. The electrical basis of these neurophysiological processes was first uncovered over two centuries ago by the Italian scientist Luigi Galvani, who utilized the rudimentary electrode technology available at the time (Galvani, 1791). Subsequent centuries saw pivotal contributions from numerous neurophysiologists, including von Helmholtz (1852), Erlanger and Gasser (1937), and notably Hodgkin and Huxley (1939), who advanced our understanding of the nervous system through their innovative electrode designs. These pioneers managed to record neural activity from relatively large nerves and axons, such as the giant squid axons that measure 0.5–1 mm in width, which were extensively studied by Hodgkin and Huxley (1939).

In 1949, the development of voltage clamp technology (Marmont, 1949) marked a significant advancement, enabling Alan Hodgkin and Andrew Huxley to measure currents mediated by sodium and potassium ions across nerve cell membranes, driven by electrochemical gradients (Hodgkin & Huxley, 1952a). This foundational work led to the establishment of the Hodgkin-Huxley model in 1952, which mathematically characterized membrane potentials through a resistor-capacitor (RC) circuit model (Hodgkin & Huxley, 1952a, b), as shown in ◘ Fig. 3.5a (Zhang et al., 2020).

Building on these developments, Erwin Neher and Bert Sakmann further advanced the field in 1976 by introducing the patch clamp technique, which allowed for the recording of signals from individual ion channels (Neher & Sakmann, 1976), as shown in ◘ Fig. 3.5b, c. In 1981, they refined this technique to achieve a very high resistance seal, known as gigaseal, significantly enhancing the signal-to-

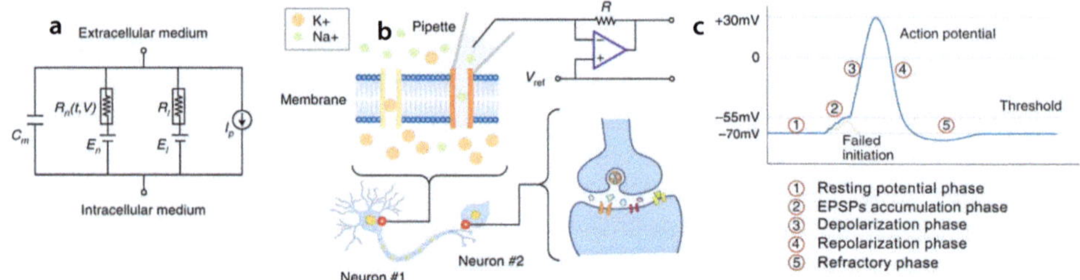

◘ Fig. 3.5 The electrophysiology behind the patch clamp technology—**a** The Hodgkin–Huxley model provides a detailed mathematical framework for describing membrane potentials using an RC (Resistor–Capacitor) circuit analogy. This model effectively captures the dynamics of action potentials by incorporating R_l and R_n resistive elements that represent the linear and nonlinear leaky ion channels, respectively. The electrochemical driving voltages, E_l and E_n, correspond to the linear and nonlinear components, respectively, and I_p symbolizes the ion pumps involved in maintaining cellular homeostasis. **b** The schematic illustrates the patch clamp recording technique, which includes the circuits used to capture the activity of an ion channel. This technique employs a glass pipette electrode, which, when gently pressed against the cell membrane and subjected to slight suction, isolates a patch of the membrane. This isolation allows for the electrical recording of individual ion channels. Propagation of neural signals occurs along the neuron's membrane until reaching the axon's terminal, where neurotransmitters are released from the presynaptic neuron into the synaptic cleft, binding to receptors on the dendrite of the postsynaptic neuron and triggering excitatory postsynaptic potentials (EPSPs). **c** The amplitude waveform of an action potential is characterized as follows: In its resting state, the neuron's interior membrane maintains a potential of approximately −70 mV relative to the exterior, predominantly controlled by Na⁺ and K⁺ ion channels. A neuron fires an action potential when EPSPs cumulatively push the membrane potential beyond a critical threshold. Activation leads to a rapid influx of Na⁺ ions, elevating the membrane voltage up to approximately +30 mV during the depolarization phase before a subsequent drop initiated by the closing of Na⁺ channels and the outflow of K⁺ ions during repolarization. The potential momentarily undershoots the resting level and then stabilizes at −70 mV following the refractory period. This rapid voltage fluctuation, termed an action potential, propagates along the axon, sequentially activating adjacent neurons. This process underpins neuronal communication. (From ▶ https://www.nature.com/articles/s41928-020-0390-3/figures/2; Zhang et al., 2020)

noise ratio and enabling the detection of extremely small currents (Hamill et al., 1981). This innovation catalyzed the widespread adoption of patch clamp technology, which has evolved to include automated systems capable of recording multiple single ion channel activities concurrently. Today, the patch clamp is recognized as a pioneering neural interface technology (Zhang et al., 2020).

In a significant leap forward in 1957, Hubel engineered sharpened tungsten electrodes with sub-micrometer tip sizes, enabling the recording of extracellular action potentials from much smaller mammalian neurons and axons, as exemplified by his experiments on the cat brain (Hubel, 1957). This breakthrough had huge implications for both neuroscience and neuroengineering, catalyzing pivotal studies in visual neurophysiology alongside Wiesel and spurring the development of advanced neural probes (Hubel & Wiesel, 1962). These include tetrodes (McNaughton et al., 1983) and microfabricated electrode arrays, such as the silicon Michigan-type (Wise et al., 1970) and Utah-type microelectrode arrays (Campbell et al., 1990), which have continued to evolve and shape the field.

Traditional electrophysiology predominantly utilized intracellular recording techniques applied to the peripheral nervous system, targeting individual neurons with electrodes inserted across their membranes. Intracellular recordings, capable of detecting sub-threshold fluctuations from resting potentials, measure voltages directly across the cell membrane. In contrast, extracellular recordings involve placing electrodes in the surrounding extracellular fluid to capture aggregated signals from multiple nearby neurons. These extracellular signals typically exhibit lower amplitudes compared to their intracellular counterparts but enable observation over broader neural regions (Zhang et al., 2020).

Electrode placement fundamentally categorizes extracellular recordings as either non-invasive or invasive, as shown in ▣ Fig. 3.6 (Parastarfeizabadi & Kouzani, 2017). Non-invasive techniques, such as electroencephalography (EEG),

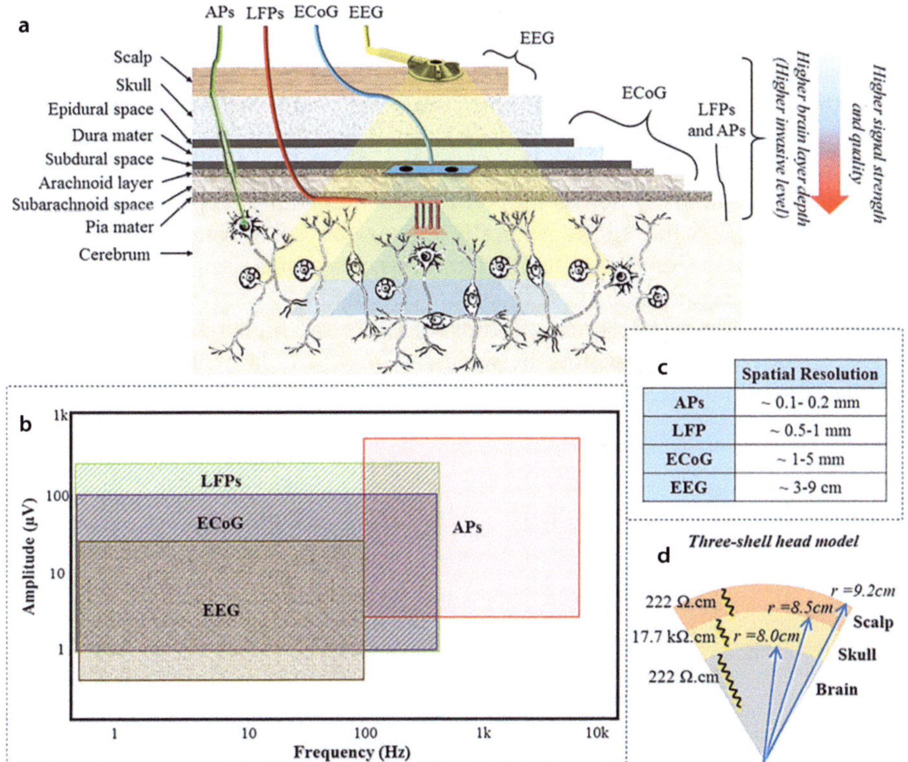

▣ **Fig. 3.6**   **a** This schematic illustrates the various brain layers and the electrophysiological signals that can be measured at differing depths. Recordings made at greater depths typically capture potentials of higher strength and quality. The distance between the electrode and the source of the potential introduces greater impedance, leading to attenuation of the signals and rejection of high-frequency components, owing to the low-pass filter characteristics of the brain's layers (Bédard et al., 2004, 2006). Moreover, using an electrode with a smaller contact area restricts the measurement to potentials from a smaller number of neurons (Lempka et al., 2011). **b** The amplitude versus frequency characteristics of human brain potentials of interest. **c** The spatial resolution capabilities of electrophysiological signals. **d** The three-shell head model details the impact of different head layers, especially the skull, which has high resistivity, on the distortion of potentials as they travel to the scalp surface. (From Parastarfeizabadi & Kouzani, 2017; originally published under CC-BY 4.0)

record neural activities from the scalp's surface, as discussed in ▶ Chap. 2. Conversely, invasive methods involve electrodes implanted within the body, including electrocorticography (ECoG), which positions electrodes directly on the cerebral cortex beneath the skull, and local field potentials (LFP), typically recorded from within brain tissue (Buzsáki et al., 2012). Although LFPs can also be measured using ECoG electrodes (Konerding et al., 2018), the development of microelectrocorticography (µECoG) arrays using flexible substrates has enhanced both spatial and temporal resolution beyond traditional ECoG approaches.

Neural signals recorded extracellularly, whether by penetrating depth electrodes or surface electrodes, encompass a wide range of frequencies that can be analytically segregated into distinct bands (Hong & Lieber, 2019). These signals include slow-varying local field potentials (LFPs) and rapid, millisecond-duration action potentials (APs), often referred to as "spikes." LFPs aggregate the transmembrane currents of groups of neurons, whereas spikes are indicative of the activity of individual neurons or single units (Buzsáki et al., 2012).

In their raw form, both LFPs and spikes are visible within the time-domain recording traces. These signals can be dissected into their frequency components using Fourier transform techniques, resulting in a spectrogram that visually represents the signal intensity across different frequency bands. Typically, LFPs manifest below 100 Hz, while spikes generally appear above 250 Hz.

To isolate these signals for detailed analysis, researchers apply either analog or digital filtering techniques. These filters are designed to selectively transmit signals within designated lower-frequency bands for LFPs and higher-frequency bands for spikes, effectively distinguishing between these two fundamental types of neural activity (Buzsáki et al., 2012), as shown in ◘ Fig. 3.7 (Hong & Lieber, 2019).

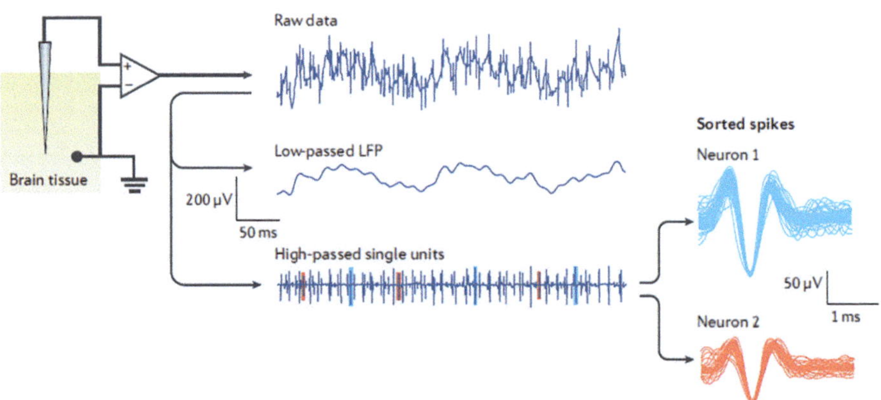

◘ **Fig. 3.7** Illustration demonstrating the fundamental physical principles involved in recording bioelectrical signals from neural tissue using electrodes (left panel). This is accompanied by representative neural recording traces that are unfiltered (raw), processed through a low-pass filter, and processed through a high-pass filter (middle panel). Additionally, the diagram includes sorted spikes (right panel) that have been detected and classified from the high-pass-filtered recordings by the electrode. (From Hong & Lieber, 2019)

Invasive recording techniques provide more direct neuronal interaction, significantly improving the signal-to-noise ratio, accessing higher frequency signal bands, and increasing the efficacy and precision of neural stimulation (Buzsáki et al., 2012; Zhang et al., 2020). These attributes make invasive methods particularly valuable for detailed neurophysiological studies and applications.

In ▶ Chap. 2, we explored the neurophysiological basis and the associated hardware required for acquiring various non-invasive brain signals. Building on this foundation, ▶ Chap. 3 describes the neurophysiological origins and the technological apparatus involved in the acquisition of different invasive brain signals, namely spikes, multiunit activity, local field potential (LFP), electrocorticogram (ECoG). This chapter aims to provide an overview of the signal origin and equipment used in invasive neuroimaging techniques, highlighting their applications, benefits, and limitations within neurological research.

Box 3.3

Understanding Neuronal Communication: The nervous system communicates through electrical signals, primarily action potentials and synaptic transmissions. These electrical phenomena are fundamental for the rapid transmission of information across neurons and throughout the nervous system. **Components of Neuronal Communication**— Action potentials (Check ▶ Box 3.1) and synaptic transmission (Check ▶ Box 3.2). **Role of Ion Channels**: Ion channels are pivotal in the generation and propagation of action potentials. They regulate the flow of ions such as sodium and potassium into and out of neurons, directly affecting the electrical state of the neuron. **Significance in Health and Disease**: Dysfunctions in the electrical signaling of the nervous system can lead to various neurological conditions, such as epilepsy, multiple sclerosis, and neuropathic pain. Understanding these electrical foundations aids in developing targeted treatments that can modulate or correct these signal pathways.

3.3 Neurophysiological Origin of Invasive Signals

Neurons, often regarded as the foundational building blocks of the nervous system, play a crucial role in the transmission of information throughout the body via electrical and chemical signals. These specialized cells are equipped to receive signals from their environment or other cells, leading to the generation of an electrical impulse known as an action potential. This phenomenon, pivotal in neuronal communication, occurs as the action potential propagates along the neuron's axon, a process fundamental to the functioning of the nervous system (Kandel et al., 2000).

An action potential represents a swift and temporary shift in the neuron's membrane potential, which is primarily driven by the regulated movement of ions, notably sodium (Na^+) and potassium (K^+), across the neuron's membrane (Koch, 2004) as shown in ▶ Fig. 2.2b (Frank et al., 2019). The initiation of an action potential involves a sudden influx of Na^+ ions into the neuron, leading to depolarization of the membrane. This depolarization phase is subsequently followed by the efflux of K^+ ions, resulting in the repolarization of the membrane back to its resting potential (Hodgkin & Huxley, 1952a).

The sequence of ionic movements that underlie the action potential is facilitated by voltage-gated ion channels, which open or close in response to changes in the membrane potential (Llinás, 1988). The rapid influx of Na^+ ions that initiates the action potential occurs when the membrane potential reaches a certain threshold, triggering the opening of voltage-gated Na^+ channels (Sontheimer et al., 1996). As the membrane potential rises, the Na^+ channels eventually close, and voltage-gated K^+ channels open, allowing K^+ ions to exit the neuron and restore the membrane potential to its resting state (Bartos et al., 2007; Koch, 2004). This is possible due to the myelinated sheath that covers the axon of certain nerve fibers, which confers an insulation property against discharge. This intricate process of action potential generation and propagation is essential for the transmission of signals within the nervous system, enabling neurons to communicate with each other and with other types of cells. The action potential's ability to travel along the axon without diminishing in strength ensures that signals can be relayed over long distances within the body, from the central nervous system to peripheral regions and vice versa (Bean, 2007).

In summary, the generation of action potentials is a critical mechanism by which neurons transmit information, characterized by the orchestrated movement of ions across the neuronal membrane. Understanding the biophysical properties of action potentials and the role of voltage-gated ion channels offers valuable insights into the fundamental processes that govern neural communication and the broader function of the nervous system.

3.3.1 Neurophysiological Origin of Single-Unit Activity (SUA)

Single-neuron spike recordings have a single source—a neuron that generates action potentials at discrete events in time (Pesaran et al., 2018). The neurophysiological basis of single-unit activity encompasses the precise electrical signals generated by individual neurons within the brain or spinal cord (Buzsáki et al., 2012). This activity is primarily characterized by the detection of spikes or action potential trains, indicative of neuronal firing events. To capture this activity, researchers employ microelectrodes—finely tapered conductive probes capable of penetrating brain or spinal tissue to access the electrical signals of singular neurons. These microelectrodes can be introduced directly into the targeted neural tissue or inserted transcranially into the brain, allowing for the direct measurement of neuronal electrical output (Buzsáki, 2004; Mountcastle, 1997). An overview of single-unit recording and processing is shown in ◘ Fig. 3.8 (Urdaneta et al., 2023). The electrical activity of a single neuron can be recorded as a spike or action potential train when the neuron fires or generates an action potential.

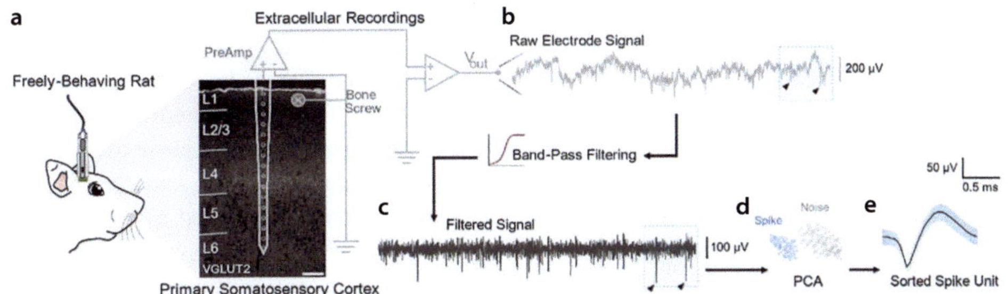

Fig. 3.8 Neural Signal Acquisition and Processing from a Freely Behaving Rat—a: **Neural Recording Setup**—Freely-Behaving Rat: Illustration of a rat with an implanted electrode for extracellular recordings in the primary somatosensory cortex. Electrode Placement: The electrode is positioned to record from various cortical layers (L1 to L6) within the primary somatosensory cortex. Extracellular Recordings: Signals are captured from neurons outside the cell membrane, providing insights into neural activity. **b: Raw Electrode Signal**—Initial Signal: The raw signal from the electrode includes both neural activity and background noise, typically in the range of microvolts (µV). Band-Pass Filtering: This process isolates the frequencies of interest by removing high-frequency noise and low-frequency drift, resulting in a cleaner signal. **c: Filtered Signal**—Filtered Output: The signal after band-pass filtering, showing improved clarity of neural spikes. **d: Principal Component Analysis (PCA)—Spike Sorting**: PCA is used to differentiate between neural spikes and noise. Spikes (blue) are separated from noise (gray), allowing for the identification of distinct neural units. (You will learn about PCA in ▶ Chap. 4). **e: Sorted Spike Unit—Spike Identification**: The final sorted spike unit, represented by an averaged waveform, with the shaded area indicating variability. This step confirms the isolation of individual neuronal activity. This figure provides a comprehensive overview of the process for acquiring and processing neural signals from a freely-behaving rat, highlighting the steps from raw signal acquisition to the sorting of individual spikes for detailed neural analysis. (From Urdaneta et al., 2023; originally published under CC-BY 4.0)

Recording single-unit activity provides invaluable insights into the functional attributes and connectivity of neurons and neural circuits. By isolating and analyzing the action potentials of individual neurons, scientists can explore fundamental aspects of neural operation, including neural coding mechanisms, synaptic plasticity, and the neural substrates underlying various behaviors and cognitive processes. Furthermore, single-unit recordings offer a unique vantage point for investigating the pathophysiological underpinnings of neurological disorders and diseases, facilitating the development of targeted interventions and therapies (Hubel & Wiesel, 1962; Quiroga et al., 2005).

Applications of single-unit recordings span a broad spectrum of neuroscience research areas, from elucidating the neural correlates of perception and motor control to unraveling the complexities of learning and memory formation. This method's ability to provide a granular view of neuronal activity underpins its utility in dissecting the intricate dynamics of brain function (Newsome et al., 1989).

Despite its precision and depth of detail, single-unit recording is an inherently invasive technique, limited by its focus on a relatively small subset of neurons. This limitation constrains the technique's capacity to offer a comprehensive overview of global neural activity across larger brain networks.

3

> **Box 3.4**
>
> **Single-Unit Activity (SUA)** refers to the electrical signals recorded from a single neuron, capturing its action potentials or spike activity. This type of neural recording is achieved through the use of fine electrodes, which are placed in close proximity to or within the neuron of interest. SUA provides a precise and detailed measurement of the timing and pattern of action potentials emitted by an individual neuron, offering insights into its firing behavior in response to various stimuli or during specific behaviors.

3.3.2 Neurophysiological Origin of Multi-Unit Activity (MUA)

Multi-unit activity (MUA) captures the collective electrical signals emanating from groups of neurons within specific regions of the brain or spinal cord (Buzsáki et al., 2012). Unlike single-unit recordings, which focus on the activity of individual neurons, MUA encompasses the simultaneous action potentials or spikes generated by multiple neurons in close proximity. This is achieved through the placement of an extracellular microelectrode into the neural tissue, either directly or transcranially, allowing for the aggregation of electrical signals from the neuron population around the electrode's tip (Buzsáki, 2004; Einevoll et al., 2012). The electrical activity of a population of neurons can be recorded as a continuous train of action potentials or spikes. Because the microelectrode records activity from a population of neurons, the resulting MUA signal is a summation of the individual action potentials from all the neurons (Gray et al., 1995) in the recording area as shown in ▣ Fig. 3.9 (Wei et al., 2015).

The recorded MUA signal is essentially a composite of the spikes produced by several neurons, providing a broader overview of the electrical dynamics within the

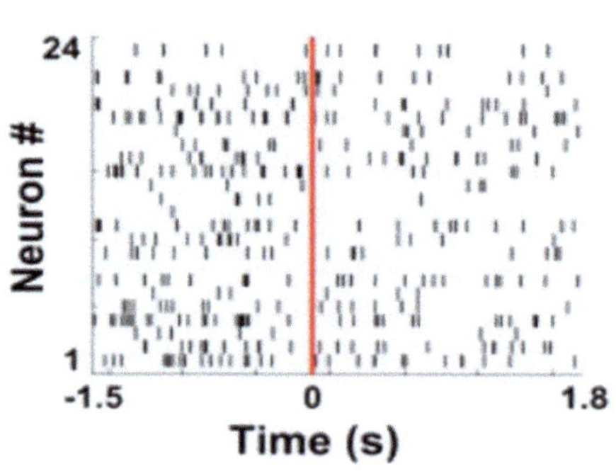

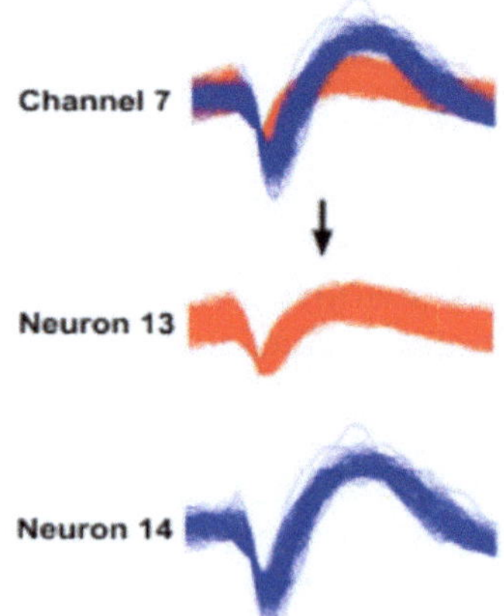

▣ **Fig. 3.9** Depicts the single-unit spike trains. Here two neurons were isolated from Channel 7 in the mentioned article. (Adapted from Wei et al., 2015; originally published under CC-BY 4.0)

sampled neural circuit. This signal amalgamation reflects the summated action potentials of neurons within the electrode's vicinity, offering insights into the collective behavior of neural ensembles (Lewicki, 1998).

Researchers utilize MUA recordings to investigate the functional characteristics and interconnectivity of neural populations and circuits. Through MUA, it's possible to explore complex phenomena such as neural coding strategies, synaptic plasticity mechanisms, and the neuronal underpinnings of behavior and cognition. Furthermore, MUA has been instrumental in studying various neurological disorders and diseases, aiding in the elucidation of aberrant neural circuitry and dysfunctional neural communication (Buzsáki et al., 2012).

Although MUA provides a richer dataset than single-unit recordings by capturing the activity of multiple neurons simultaneously, it is important to recognize its limitations. MUA does not offer the granularity of single-unit recordings nor the comprehensive coverage of larger-scale techniques like fMRI or EEG. It captures data from only a subset of the neural population within a specific area, limiting the ability to infer global neural dynamics.

> **Box 3.5**
>
> **Multi-Unit Activity (MUA)** refers to the electrical signals recorded from a group of neurons simultaneously, capturing the aggregate action potentials or spikes from nearby neurons. This recording is typically achieved using electrodes that are less selective than those used for single-unit activity, allowing them to pick up signals from multiple neurons within a small area. MUA provides insight into the collective behavior of neuronal populations rather than individual neurons, offering a broader perspective on neural network activity. The data obtained from MUA recordings can be crucial for understanding how groups of neurons coordinate their activity in response to various stimuli or during specific tasks.

3.3.3 Neurophysiological Origin of Local Field Potential (LFP)

Local field potentials (LFPs) offer insights into the collective electrical dynamics of neuronal ensembles within specific brain regions. LFPs emanate from the synchronous electrical activities—encompassing both action potentials and synaptic transactions—of multiple neurons. These extracellular signals are captured through electrodes implanted either directly into the neural tissue of the brain or spinal cord, or transcranially through the skull, enabling the monitoring of continuous electrical fluctuations within the neuronal population (Buzsáki et al., 2012; Einevoll et al., 2013).

The genesis of LFP signals is attributed to the intricate interplay of excitatory and inhibitory synaptic events, modulated by the action of neurotransmitters across neural networks. This confluence of pre- and postsynaptic potentials, along with the inherent ionic conductance changes in the surrounding extracellular matrix, molds the LFP waveform, reflecting the aggregate neural activity in the vicinity of the recording electrode (Buzsáki et al., 2012; Destexhe et al., 2003).

A distinguishing feature of LFPs is their ability to capture the synchronous oscillations of large neuron populations, thereby providing a window into the underlying mechanisms of synaptic connectivity and network dynamics. As such, LFPs serve as a proxy for investigating the coordinated actions of excitatory and inhibitory neurons within local circuits, offering clues to their contributions to overall brain function (Buzsáki, 2004). The electrical activity of a population of neurons can be recorded as a continuous train of electrical activity as shown in ◘ Fig. 3.7.

Researchers have leveraged LFP recordings to elucidate various aspects of neural function, including neural coding, synaptic plasticity, and the neurophysiological underpinnings of behaviors. Moreover, LFPs have proven invaluable in exploring the spectral dynamics of brain oscillations—alpha, beta, gamma, and theta rhythms—which are intimately linked with cognitive operations such as perception, attention, and memory formation (Fries, 2009; Jensen et al., 2007). Although LFP recordings provide a rich dataset encompassing the activity of numerous neurons, they inherently capture signals from a limited spatial extent and cannot encompass the entirety of neural activity across the brain.

> **Box 3.6**
> **Local Field Potential (LFP)** is a measure of the summed electric potentials generated by the collective activity of neurons within a small volume of brain tissue. LFP primarily reflects the synaptic inputs and intracellular currents of a group of neurons near the electrode tip, rather than the action potentials of individual neurons. These signals are recorded via electrodes placed in extracellular space and represent the low-frequency fluctuations (below 300 Hz) in electrical activity that arise from the synchronous activity of neuronal assemblies. LFP provides valuable insights into the coordinated actions of neural circuits, as it captures the dynamics of neural processing over larger spatial scales than single-unit recordings. The study of LFPs is crucial for understanding how groups of neurons interact to process information, particularly in relation to behavior and cognitive functions.

3.3.4 Neurophysiological Origin of Electrocorticogram

The electrocorticography (ECoG) is an invasive neurophysiological technique that records the cumulative electrical activity of the cerebral cortex directly from the surface of the brain. Unlike non-invasive methods such as EEG, ECoG involves the placement of electrodes directly onto the exposed surface of the brain, as shown in ◘ Fig. 3.10 (Miller, 2019), often during neurosurgical procedures for epilepsy treatment or brain tumor resection. This proximity to the neural sources offers a higher spatial resolution and a broader bandwidth of measurable signals, capturing both low- and high-frequency neural oscillations with reduced artifacts from non-cerebral tissues (Buzsáki, 2004; Crone et al., 1998).

At the physiological level, ECoG signals primarily originate from the summation of postsynaptic potentials (PSPs) in the cortical pyramidal cells, as shown in

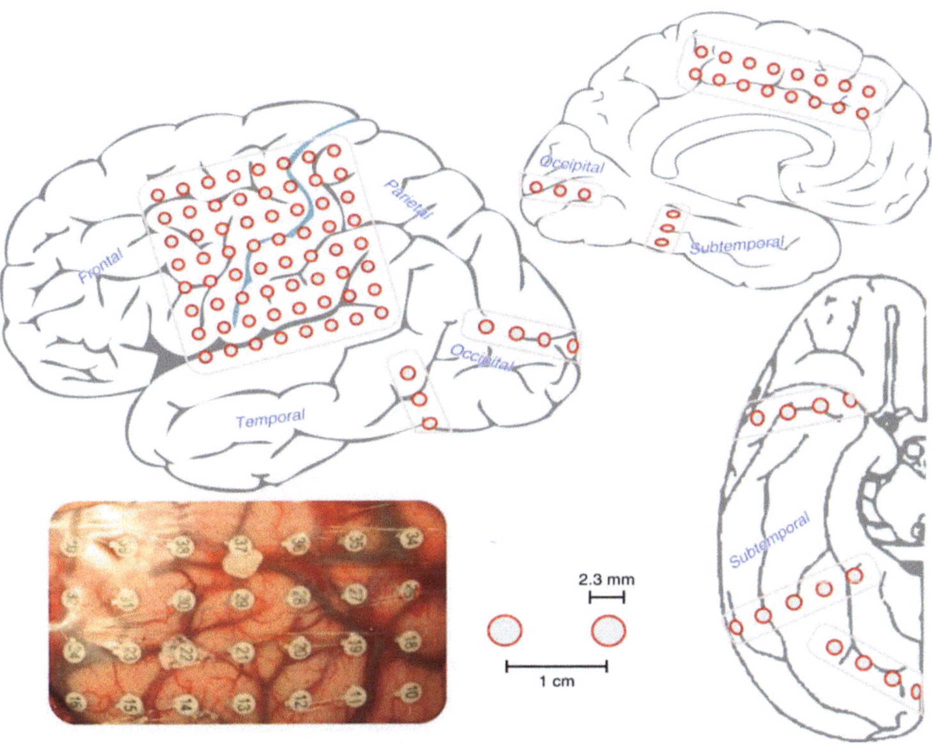

◻ Fig. 3.10 Depicts the placement of platinum ECoG electrode—Platinum electrocorticography (ECoG) electrode arrays were designed either as linear strips or grid formations, encapsulated within silastic material. These electrodes featured a 1 cm inter-electrode spacing, with each electrode having a diameter of 4 mm and an exposed surface area of 2.3 mm. They were strategically implanted within the subdural space of patients through open craniotomy procedures, arranged in the specific orientations depicted. (Adapted from Miller, 2019)

◻ Fig. 3.11a, which are the most abundant neuron type in the cerebral cortex (Spruston, 2008). These pyramidal cells are uniquely positioned with their long apical dendrites oriented perpendicularly to the cortical surface, making their synchronized activity particularly amenable to detection by surface electrodes (Brown & Hestrin, 2009), as shown in ◻ Fig. 3.11b. When groups of these neurons receive excitatory or inhibitory synaptic inputs, they generate localized electrical fields that can be measured as variations in voltage at the cortex's surface (Mitzdorf, 1985), as shown in ◻ Fig. 3.11c.

ECoG's ability to detect these voltage fluctuations provides a direct window into the neural substrates underlying various cognitive functions and behaviors. For instance, ECoG can capture the cortical oscillations associated with specific cognitive states or tasks, such as sensory processing, motor planning, and execution, and even complex processes like language comprehension and production. The granularity of ECoG data, due to its proximity to the neural sources, allows researchers to map functional areas of the brain with high precision, identifying

3

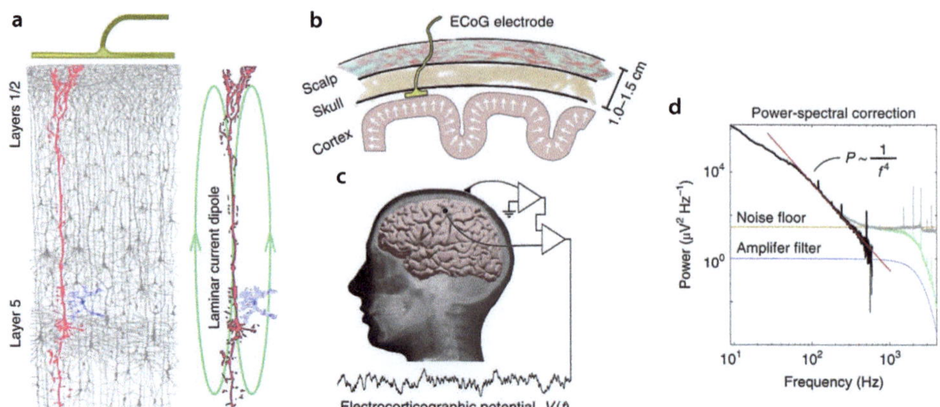

🔘 **Fig. 3.11**   **a** Approximately 5 mm² of the cortical surface under each ECoG electrode is home to roughly half a million neurons (Pakkenberg & Gundersen, 1997), with each neuron generating a current dipole due to the movement of charges across its membrane (Mitzdorf, 1985). The background includes an illustration by Ramon y Cajal of a Golgi-stained cortex. **b** ECoG electrodes are surgically implanted subdurally on the cortical surface. These electrodes are later referenced to a common average after excluding channels affected by epileptic activity and artifacts. **c** Initially, ECoG potentials are measured relative to an external reference point, either on the scalp or at the mastoid. **d** Upon re-referencing, the power spectral density (PSD) of ECoG potentials is quantified (shown in green). Adjustments are made to account for the inherent filter of the built-in amplifier (correction shown in blue, intermediate PSD in gray) and the intrinsic noise floor (correction shown in orange, final PSD in black). The resulting PSD typically exhibits a power-law distribution, approximating $P \sim 1/f^4$ at higher frequencies (fit shown in red), based on data and analysis from the experiment "fixation_high-freq." (From Miller, 2019. Panels **a–d** are adapted or reproduced from Miller et al. (2009); originally published under CC-BY. Panels **e–g** are adapted from Miller et al. (2012); originally published under CC-BY)

regions critical for specific tasks and those affected by pathological conditions like epilepsy (Crone et al., 2006).

Moreover, the rich spectral content of ECoG signals, as shown in 🔘 Fig. 3.11d, spanning from low-frequency delta waves to high-frequency gamma oscillations, reflects the diverse physiological processes occurring within the cortex. These oscillations are thought to be critical for the coordination of neural activity across different brain regions, facilitating information transfer and integration that underlies complex cognitive functions (Buzsaki & Draguhn, 2004; Jensen et al., 2007).

ECoG's enhanced signal quality and temporal–spatial resolution make it an invaluable tool for exploring the electrophysiological underpinnings of cognitive processes, sensory processing, motor control, and language. By capturing a wide spectrum of neural oscillations, ECoG has contributed significantly to our understanding of cortical network functions and the synchronization mechanisms underlying various brain states and functions (Pfurtscheller & Cooper, 1975; Miller et al., 2007).

Furthermore, ECoG has been pivotal in clinical neuroscience, particularly in mapping functional areas of the cortex during surgery and in the study and treatment of epilepsy. It offers precise localization of epileptogenic zones and critical functional areas, minimizing surgical risks and improving therapeutic outcomes. Additionally, ECoG recordings have facilitated advancements in brain–computer interfaces (BCIs), providing a direct channel for decoding neural signals into control signals for assistive technologies (Leuthardt et al., 2004; Schalk et al., 2008).

Despite its advantages, the invasive nature of ECoG limits its use primarily to clinical settings where surgical access to the brain is already warranted. The need for craniotomy restricts its application to a small subset of patients, contrasting with the wider applicability of non-invasive methods. Nevertheless, when combined with other imaging and recording techniques, such as functional magnetic resonance imaging (fMRI) and magnetoencephalography (MEG), ECoG offers a complementary perspective, bridging micro-scale neuronal activities and macro-scale brain dynamics.

> **Box 3.7**
> **Electrocorticogram (ECoG)** is a type of invasive neurophysiological recording that involves placing electrodes directly on the exposed surface of the brain. ECoG measures the cumulative electrical activity of neurons located in the cortex directly beneath the electrodes.

3.4 Electrodes

Electrodes utilized for invasive signal acquisition are classified into two types: penetrating and non-penetrating.

3.4.1 Non-penetrating Electrodes

Invasive non-penetrating cortical electrodes include epidural ECoG electrodes, positioned on the dura mater, and subdural ECoG electrodes, which rest on the arachnoid, as shown in ◘ Fig. 3.12a. Traditional ECoG electrodes generally maintain a pitch of about 1 cm (Leuthardt et al., 2004). Advances with flexible µECoG electrodes have enhanced spatial resolution to millimeter or sub-millimeter levels (Tolstosheeva et al., 2015), with flexible substrates reducing source-to-electrode distance through conformal geometries (Blau et al., 2011). Recent developments in transistor multiplexed ECoG arrays have minimized wiring space while increasing electrode density and channel count (Khodagholy et al., 2013; Escabí et al., 2014).

3.4.2 Penetrating Electrodes

Penetrating electrodes, typically crafted from biocompatible-insulated stainless steel, tungsten, or platinum/iridium, are commonly employed (Lehew & Nicolelis, 2008; Nicolelis et al., 2003). Tetrodes, composed of four microwires, offer improved mechanical stability and spike sorting efficacy (Buzsáki, 2004). Silicon-based

3

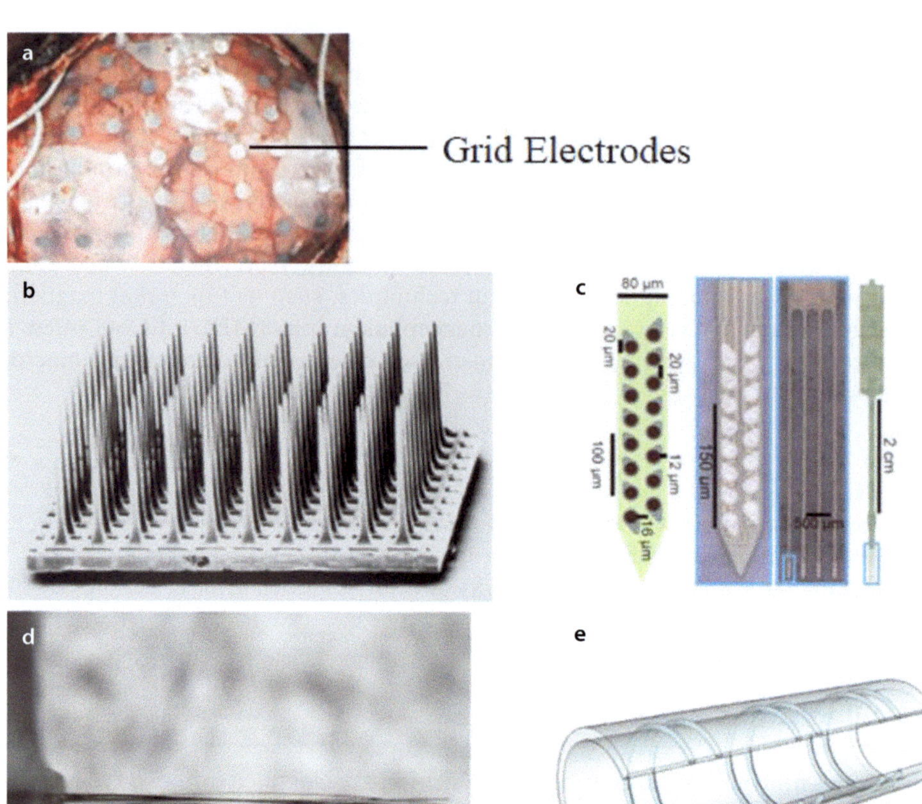

■ **Fig. 3.12**   **a Invasive Non-Penetrating Electrodes for ECoG Recording**: Conventional ECoG electrodes typically feature a pitch of approximately 1 cm, facilitating basic cortical surface recording (left). Innovations in flexible µECoG electrodes have reduced pitches to several millimeters, enhancing spatial resolution (middle). The latest developments in transistor multiplexed ECoG arrays have further compacted the necessary area for routing wires while significantly increasing both the density and the number of recording channels (right). **b Invasive Penetrating Microelectrodes**: These electrodes are designed for direct insertion into brain tissue and include various configurations such as microwires (left), Utah arrays (middle), and Michigan arrays (right), all fabricated in needle-like shapes to facilitate deep brain recordings. **c Flexible Penetrating High-Density Polymer Electrodes**: This category includes electrodes made from polymers that offer high density and flexibility, making them suitable for chronic implantation and reducing the risk of tissue damage. **d High-Density Carbon Fiber Electrode Arrays**: These arrays utilize carbon fibers to create a minimal footprint on the tissue, allowing for precise signal recording and stimulation with less inflammatory response. **e Cuff Electrodes**: As non-penetrating peripheral electrodes, cuff electrodes encircle nerves with a flexible silicone sheath that houses metal electrodes on the inside surface. This configuration is particularly useful for studies involving peripheral neural pathways. (Adapted from Zhang et al., 2020. Figure adapted from: **a** (left), Leuthardt et al., 2004; **b** Rousche & Normann, 1998; **c**, Chung et al., 2019; **d**, Massey et al., 2019)

needle-shaped microelectrodes, as shown in ▪ Fig. 3.12b, facilitate multi-site recordings (Rousche & Normann, 1998; Wise et al., 2004; Abidian & Martin, 2009). However, their rigidity can induce tissue damage and inflammation, potentially degrading signal quality. Polymer-based flexible penetrating electrodes, shown in ▪ Fig. 3.12c, are more suitable for chronic implants due to reduced tissue reaction (Chung et al., 2019), although their flexibility complicates insertion. Robotic methods have been introduced to enhance the insertion reliability of these electrodes (Hanson et al., 2019). Carbon fiber electrodes, shown in ▪ Fig. 3.12d, with minimal cross-sections, diminish tissue response and support high-density reliable insertion (Kozai et al., 2012; Patel et al., 2015; Massey et al., 2019).

Historically, microwires were used in peripheral nervous system studies but are now less favored due to potential nerve damage. Thin, flexible intra-fascicular electrodes, such as longitudinally intra-fascicular electrodes (LIFE) (Lawrence et al., 2003), and transverse intra-fascicular multi-channel electrodes (TIME) (Boretius et al., 2010) are emerging as prosthetic solutions. Sieve electrodes facilitate axonal regeneration through their arrayed holes, allowing axons to reconnect with distal ends while enabling signal recording and stimulation (Akin et al., 1994). Prominent non-penetrating peripheral electrodes include cuff electrodes, shown in ▪ Fig. 3.8e, which are extensively used for neural pathway studies. These electrodes feature a silicone sheath that encircles the nerve with metallic contacts on the inner surface (MicroProbes for Life Science, 2020).

Research continues into novel electrodes, including those made from organic materials, meshes, and multifunctional flexible polymer fibers. These are being developed to achieve higher spatial integration, improved long-term stability, and enhanced biocompatibility compared to conventional systems (Zhang et al., 2020; Shen et al., 2023).

The raw electrophysiological data collected through various electrodes, as discussed, undergo sophisticated processing to derive signals that are clinically meaningful. This essential signal extraction is achieved through a series of advanced analytical techniques designed to amplify, filter, and decode the intrinsic electrical patterns of the brain. The processed signals are pivotal for a wide range of clinical applications and therapeutic interventions (Kao et al., 2014; Shenoy & Carmena, 2014), which are detailed comprehensively across various chapters in this book series.

> **Box 3.8**
>
> **Penetrating Electrodes**: Electrodes that are inserted into the brain tissue to capture signals from deep brain structures. These are crucial for detailed studies on neuronal behavior and brain–machine interfaces. **Subdural Grids**: Arrays of electrodes placed on the surface of the brain under the dura mater. Subdural grids are used for extensive mapping of brain activity, especially during pre-surgical evaluation for epilepsy. **Stereotactic Surgery**: A surgical procedure that uses three-dimensional coordinates to locate small targets inside the body, often used for the precise placement of electrodes in the brain.

3.5 Signal Stability in Extracellular Recordings: Challenges and Solutions

Recording electrodes are essential tools for capturing neuronal activity with high fidelity. The quality of these recordings is primarily assessed by the system's signal-to-noise ratio (SNR). Improvements in SNR can be achieved by either enhancing the signal amplitude or by reducing the background noise. The amplitude of the neuronal signal that an electrode record is inversely related to the distance between the electrode and the neuron; this signal diminishes sharply as the distance increases (Bédard et al., 2004), as shown in ◘ Fig. 3.6. The specific rate of signal decay is influenced by several factors, including the homogeneity of the surrounding medium and the geometric configuration of the neuronal currents. Typically, to capture single- or multi-unit neuronal signals effectively, electrodes need to be positioned within 100 μm of a neuron (Marblestone et al., 2013; Kleinfeld et al., 2019).

The stability of high-frequency components within in vivo extracellular recordings, particularly neuronal spikes, is notably inconsistent over time (Shen et al., 2023). This phenomenon has been rigorously examined and debated in neuroscientific research for the past two decades (Turner et al., 1999; Biran et al., 2005). Several factors contribute to this instability: electrode motion relative to the neural tissue, neuronal cell death, immune reactions to foreign bodies (such as the electrode itself), accumulation of biological material on the electrode surface (electrode fouling), and natural variations in neuronal activity which may arise from adaptive changes within the neural network or different states of network activity (Nicolelis, 1998; Miller et al., 2018).

In addition to selecting the appropriate technology (such as ECoG or penetrating microelectrodes) and efforts to minimize tissue disruption, maintaining the proximity between an electrode and a neuron presents significant challenges (Saxena et al., 2013; Kozai et al., 2015). Innovative strategies, including the use of "mesh" electrode designs and the application of neurotrophic factors, have been explored to facilitate closer and more stable interactions between neurons and recording electrodes (Kennedy, 1989; Bartels et al., 2008; Gearing & Kennedy, 2020). These approaches aim to optimize electrode placement relative to target neurons, thereby enhancing recording accuracy and reliability.

3.5.1 Strategies to Enhance Signal Stability

3.5.1.1 Neural Data Pipeline Adjustments

- Decoder Retraining: Frequent updates and retraining of decoding algorithms are essential to accommodate changes in neural signal patterns, as evidenced by studies suggesting the benefits of dynamic decoder adaptation (Collinger et al., 2013; Ajiboye et al., 2017; Nuyujukian et al., 2018).
- Population Sampling: Extracting signals from a larger group of neurons within a network can dilute the effect of signal loss from any single neuron. This

approach typically involves integrating the data into a robust statistical model that can reveal stable patterns across the group (Anumanchipalli et al., 2019).

- Latent Dynamics and Manifold Modeling: Implementing models that capture the underlying dynamics of neural populations (latent dynamics) (Pandarinath et al., 2018) or that map these dynamics onto a lower-dimensional space (manifold models) can help maintain signal fidelity despite individual neuron fluctuations (Degenhart et al., 2020). These methods are particularly effective when combined with recordings from an extensive array of channels, which can capture a comprehensive snapshot of the neural landscape.

3.5.1.2 Hardware-Level Interventions

- Biocompatibility Improvements: Enhancing the biocompatibility of electrode materials to minimize foreign-body responses (such as inflammation and scarring) which can shield the electrode from the surrounding neural tissue (Polikov et al., 2005; Wu et al., 2021).
- Mitigating Micromotion: Engineering solutions that reduce the relative motion between electrodes and neural tissue can decrease signal disruptions caused by mechanical shifts (Sharafkhani et al., 2022).
- Material Durability: Selecting materials that resist breakdown over time ensures that the electrodes retain their functional integrity and continue to provide accurate readings without being obscured by biological debris (Zeng et al., 2022).

By integrating both data processing innovations and advanced hardware designs, researchers can significantly improve the reliability and utility of extracellular recordings (Hong & Lieber, 2019; Zhang et al., 2020; Shen et al., 2023). These advancements are vital for both fundamental neuroscience research and the development of clinical neurotechnology applications, such as brain–computer interfaces and neuroprosthetics. ◘ Figure 3.13 presents a summary of the various invasive signal recording techniques covered in this chapter, including their operational principles. Additionally, it previews the diverse applications of these techniques, which will be explored in depth in future chapters, highlighting their relevance in clinical and research settings.

Box 3.9: Enhancing Signal Stability in Extracellular Recordings

Challenges in Signal Stability: Extracellular recordings, such as those involving single-unit activity (SUA) and local field potentials (LFP), often encounter challenges related to signal stability. These challenges can stem from physical movement of the electrodes, biological responses like scar tissue formation, or changes in the electrochemical environment surrounding the electrodes.

3

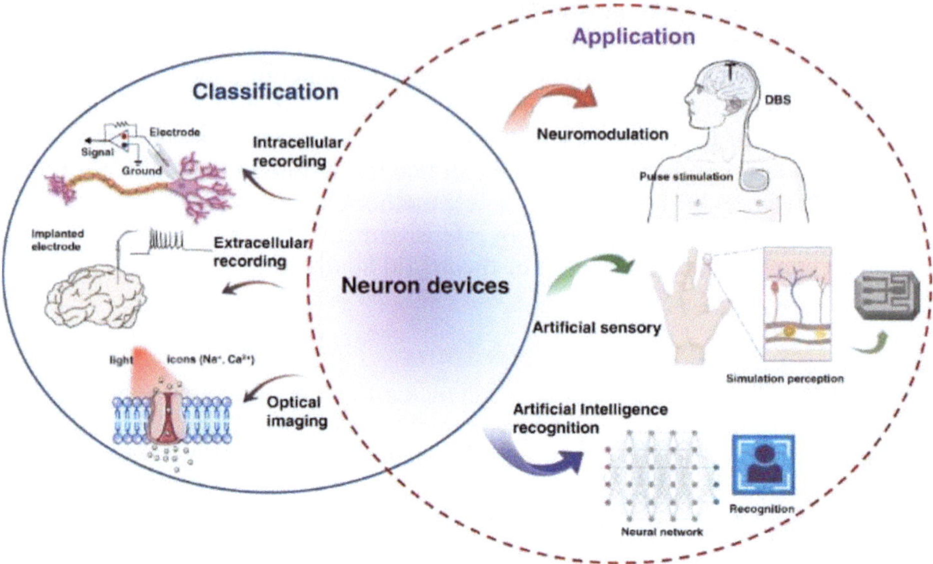

◘ Fig. 3.13 An overview of invasive signal recordings and their various applications. Optical imaging will not be discussed in this book. (From Wang et al., 2022; originally published under CC-BY 4.0)

▪ Strategies for Improving Stability:

Electrode Design: Using electrodes with improved mechanical stability and biocompatibility to minimize movement and adverse tissue responses. Materials such as flexible polymers or composites can be employed to reduce the impact on brain tissue.

Signal Processing Techniques: Implementing advanced signal processing algorithms that can compensate for noise and drift in the recorded signals. Techniques such as adaptive filtering, signal averaging, and machine learning models are useful in enhancing the clarity and consistency of data.

Surgical Techniques and Aftercare: Optimizing surgical methods to place and secure electrodes can significantly affect signal stability. Postoperative care and monitoring are crucial to manage inflammation and minimize tissue damage around the electrode sites.

Continuous Calibration: Periodic recalibration of the system to account for any drifts or shifts in baseline measurements ensures data reliability over time.

Implications for Research and Clinical Practice: Enhancing signal stability is critical for the reliability of long-term neural monitoring and recording. This is especially significant in clinical applications where precise localization of brain activity is required for diagnostic and therapeutic purposes, such as in epilepsy surgery and in the operation of neuroprosthetic devices.

> **Box 3.10**
>
> **Neuroprosthetic Devices** are biomedical implants that interface directly with the nervous system to replace or augment functions lost due to neurological deficits or injuries. These devices capture neural signals from the brain or peripheral nerves and translate them into commands to operate external devices or stimulate nerves and muscles to restore functionality. Common examples include cochlear implants, retinal implants, and motor prosthetics that help individuals with paralysis to control limbs or other assistive devices through thought alone
>
> Neuroprosthetics work by detecting electrical signals from the nervous system, processing these signals through a decoder that interprets the user's intentions, and then sending output signals to activate external devices or stimulate specific neural pathways. This technology is crucial for enhancing the quality of life and independence of individuals with disabilities, offering solutions for sensory losses, motor function impairments, and communication barriers.
>
> Innovations in materials science, signal processing, and machine learning are continuously advancing the efficacy and safety of neuroprosthetic devices. As a result, they not only hold promise for restoring lost functions but are also increasingly integrated into strategies for rehabilitation and the treatment of neurological disorders.

3.6 Comparison of Brain Signals

The quest to unravel the complexities of brain function has led to the development of various neurophysiological measurement techniques as we have seen in ▶ Chaps. 2 and 3. Each of these techniques has its own unique capabilities and applications. In this section, we will have an overview of the comparison of these methods, ranging from non-invasive approaches like electroencephalography (EEG) and functional magnetic resonance imaging (fMRI) to invasive strategies employing microelectrode arrays, as shown in ◘ Fig. 3.14.

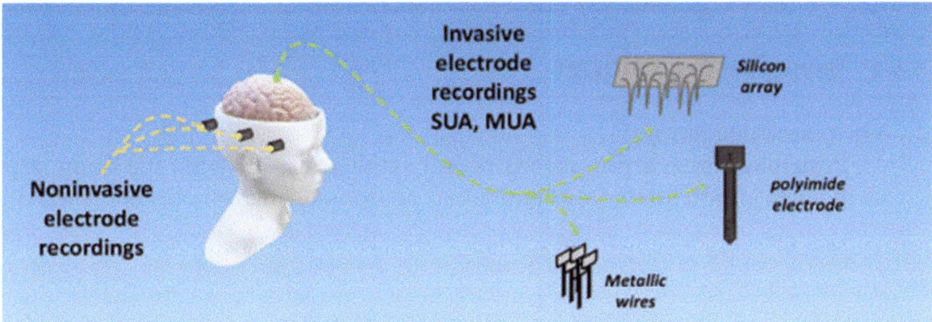

◘ **Fig. 3.14** Non-invasive and invasive brain signal acquisition illustration. (From Salahuddin & Gao, 2021; originally published under CC-BY)

3.6.1 **Non-invasive Techniques**

1. Electroencephalography (EEG):
   Principle: EEG captures the brain's electrical activity via scalp-placed electrodes, offering insights into the summative electrical output of large neuron groups (Niedermeyer & da Silva, 2005).
   Resolution: While EEG excels in temporal resolution, allowing for precise timing measurements, its spatial resolution is limited, complicating the pinpointing of neural activity origins (Cohen, 2017).
2. Functional Magnetic Resonance Imaging (fMRI):
   Principle: fMRI detects cerebral blood flow and oxygenation changes, serving as proxies for neural activity, through the blood-oxygen-level-dependent (BOLD) signal (Logothetis, 2008).
   Resolution: fMRI boasts high spatial resolution, enabling precise localization of brain activity, albeit with lower temporal resolution.
3. Magnetoencephalography (MEG):
   Principle: MEG measures the magnetic fields produced by neuronal electrical activity, providing a direct assessment of brain function (Hämäläinen et al., 1993).
   Resolution: MEG offers both high temporal and spatial resolution, making it effective in timing and locating brain activities.
4. Functional Near-Infrared Spectroscopy (fNIRS):
   Principle: fNIRS monitors cerebral oxygenation and hemodynamics by measuring light absorption changes in the brain (Ferrari & Quaresima, 2012).
   Resolution: fNIRS provides moderate spatial resolution and is somewhat limited in temporal resolution, making it suitable for studying cortical hemodynamics.
5. Functional Ultrasound (fUS):
   Principle: fUS utilizes ultrasound waves to image blood flow and oxygenation, akin to fMRI's BOLD signals, offering a macroscopic view of brain activity (Macé et al., 2011).
   Resolution: fUS achieves moderate spatial and temporal resolution, enabling a balanced assessment of neural dynamics.

3.6.2 **Invasive Techniques**

1. Microelectrode Arrays:
   Principle: These arrays record neural activity directly from the brain, capturing action potentials from individual neurons or neuron clusters (Buzsáki,

2004). Comprising microelectrode grids, Utah arrays facilitate extensive brain area recordings, detailed at the level of individual neuronal firing patterns (Maynard et al., 1997).

Resolution: Providing exceptional temporal and spatial resolution, microelectrode arrays allow for detailed neuron-level analysis but require surgical implantation.

2. Stereotactic Electrodes:

Principle: Implanted using precise stereotactic frames, these electrodes target specific brain areas or deep structures for localized activity recording (Nowell et al., 2014).

Resolution: Like microelectrode arrays, stereotactic electrodes offer high resolution but at the cost of invasiveness.

3. Electrocorticography (ECoG):

Principle: ECoG involves placing electrodes directly on the surface of the brain to record electrical activity, offering a compromise between the invasive nature of depth electrodes and the comprehensive coverage of non-invasive EEG (Crone et al., 1998).

Resolution: Higher spatial resolution than EEG and superior temporal resolution, suitable for mapping cortical activity and functional connectivity in detail.

The choice among neurophysiological measurement techniques hinges on the balance between the need for spatial or temporal resolution, the invasiveness of the procedure, and the specific research or clinical objectives. While non-invasive methods like EEG, fMRI, MEG, and fNIRS offer broader applicability with less risk, invasive techniques such as microelectrode arrays, stereotactic electrodes, and Utah arrays provide unparalleled detail at the expense of higher risk. This comparative overview underscores the diverse methods available for exploring brain function, each contributing uniquely to our understanding of neural mechanisms. A summary of the comparison of these methods is tabulated in ◘ Table 3.1 and a figure comparing the spatial and temporal resolution of different techniques is shown in ◘ Fig. 3.15.

◧ Table 3.1   The key features and differences between the various neurophysiological measurement technologies. Each technology has its unique strengths and limitations, making them suitable for different research and clinical applications. Understanding these differences is crucial for choosing the appropriate method for specific neuroscientific inquiries

Technique	Principle	Spatial resolution	Temporal resolution	Invasiveness
EEG	Measures electrical activity through electrodes on the scalp	Low	High	Non-invasive
fMRI	Detects cerebral blood flow and oxygenation changes via BOLD signals	High	Low	Non-invasive
MEG	Measures magnetic fields generated by neuronal electrical activity	High	High	Non-invasive
fNIRS	Monitors oxygenation and hemodynamics by measuring light absorption	Moderate	Low	Non-invasive
fUS	Assesses blood flow and oxygenation using ultrasound waves	Moderate	Moderate	Non-invasive
LFP	Captures summed electrical activity from neuron populations within a small brain volume	Moderate to high	High	Invasive
Single/multi-unit	Distinguishes individual or collective action potentials of neurons	High	High	Invasive
ECoG	Records electrical activity directly from the brain surface, offering a compromise between invasive depth electrodes and comprehensive coverage of EEG	Higher than EEG	Superior to EEG	Invasive

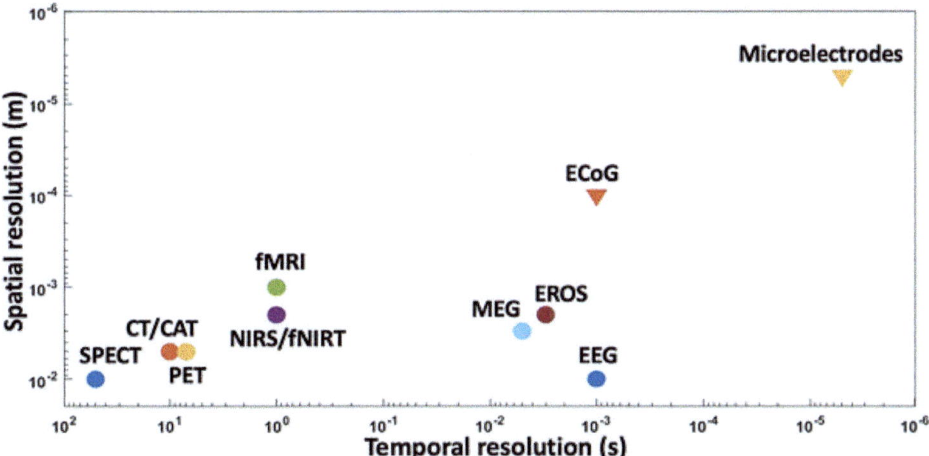

◘ Fig. 3.15 Spatial and temporal resolutions of various brain signal acquisition techniques, with microelectrodes demonstrating superior resolutions compared to other methods. (From Salahuddin & Gao, 2021; originally published under CC-BY)

> **Key Takeaways**

1. **Advances in Invasive Brain Signal Acquisition**: The chapter provides a comprehensive overview of the various invasive brain signal acquisition techniques, including single-unit activity (SUA), multi-unit activity (MUA), local field potential (LFP), and electrocorticogram (ECoG). These techniques are crucial for capturing precise neural data that non-invasive methods cannot achieve, offering valuable insights into neural dynamics, synaptic plasticity, and brain functionality at granular levels.

2. **Electrode Technology for Enhanced Neurophysiological Recording**: Different types of invasive electrodes, such as microelectrodes and ECoG arrays, have evolved to provide detailed, high-resolution data from neural tissues. These include penetrating electrodes, which are used to record activity deep within the brain, and flexible, high-density arrays that improve both the spatial and temporal resolution of neural recordings, making them vital for neuroprosthetics and brain–computer interface (BCI) applications.

3. **Advantages and Limitations of Invasive Techniques**: While invasive techniques provide unparalleled signal clarity and detail, they are often limited by their invasiveness, making them suitable primarily for clinical applications and high-precision research. The chapter compares these with non-invasive techniques, highlighting the superior resolution and application context but acknowledging the challenges such as signal stability and potential tissue damage.

4. **Future Implications for Neuroprosthetics and Neural Interfaces**: The technological advancements discussed in the chapter play a crucial role in developing neural interfaces and neuroprosthetic devices. These applications, which allow for the restoration or augmentation of lost neural functions, underline the importance of invasive recording technologies in advancing clinical treatments for neurological disorders and developing assistive devices like BCIs.

In summary, the exploration of neurophysiological signals and their measurement technologies, as shown in ◼ Fig. 3.17, represents a vibrant and continually evolving field. It compels us to broaden our intellectual horizons, refine our methodologies, and creatively integrate diverse data streams to construct a comprehensive understanding of brain function. The foundational insights presented here pave the way for future investigations and breakthroughs in neuroscience and neurotechnology.

Conclusion

Throughout ▶ Chaps. 2 and 3, we embarked on a meticulous journey through the diverse landscape of neurophysiological measurement techniques, both non-invasive and invasive. Each modality has provided us with a unique lens through which to view and understand the intricate operations of the brain, offering insights across different perspectives and scales.

We initiated our exploration with an in-depth look at the neurophysiological underpinnings of these varied signals. From the electrical oscillations captured by EEG to the eye movement-induced potentials measured by EOG, the magnetic fields identified by MEG, and the vascular responses traced by fNIRS, fMRI, and fUS, each modality enriches our comprehension of brain dynamics. Furthermore, our discussion on single-unit and multi-unit activities, LFP and ECoG, has peeled back the layers of neuronal communication, providing a granular view of neural interactions. ◼ Figure 3.16 provides a comparative overview of non-invasive and invasive electrical brain signals (Bockhorst et al., 2023).

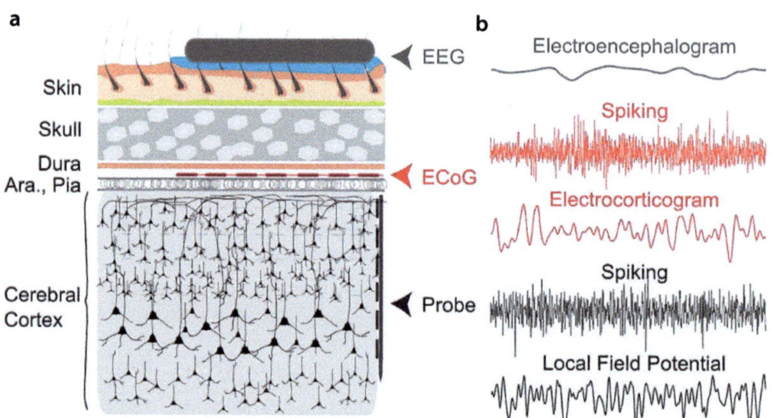

◼ **Fig. 3.16** Brain signals captured using extracranial, epicortical, and intracortical recording methods. **a** Illustrates the placement of various electrode types for both extra- and intracortical recordings, including EEG (electroencephalogram), ECoG (electrocorticogram), and intracortical extracellular probes, near the arachnoid mater. **b** Details the types of signals acquired with these techniques, highlighting how contacts along the intracortical probes record the electric field from different angles relative to the dendritic axes compared to surface electrodes. (From Bockhorst et al., 2023)

Transitioning to the mechanisms behind the capture of these signals, we witnessed the remarkable strides in technology that have enhanced our ability to acquire and interpret this complex data. From the simple scalp electrodes of EEG to the advanced electrodes utilized in invasive recordings highlight the ingenuity embedded in each technique reflecting the relentless pursuit of understanding the brain's enigmas.

A comparative analysis illuminated the distinct advantages and challenges inherent in each technique, underscoring the complementary nature of these technologies. These chapters underscore the indispensable role of interdisciplinary collaboration in neuroscience. The fusion of insights from neuroscience, engineering, physics, and computer science is paramount in pushing the boundaries of our knowledge and developing innovative treatments for neurological conditions.

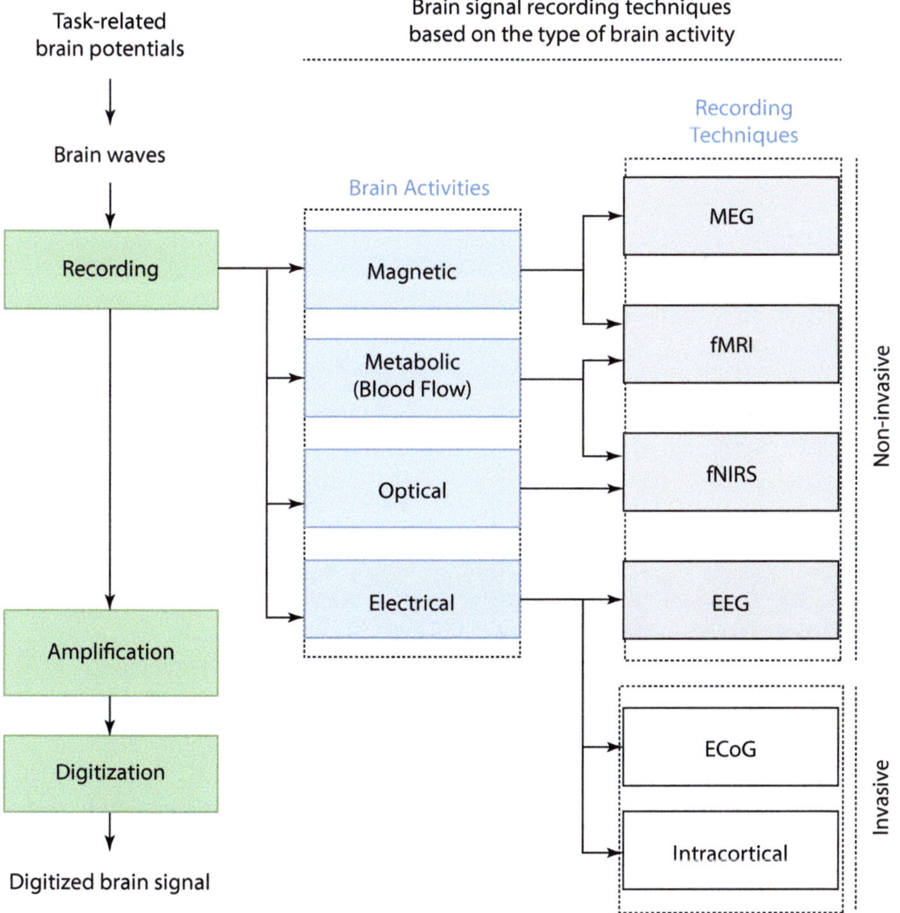

◘ Fig. 3.17 A summary classification of brain signal recording techniques, organized by the type of brain activity they measure. (From Altaheri et al., 2023)

This section offers practical exercises designed to help apply the material learned to concrete scenarios.

Designing an Experiment Using SUA

Objective: To conceptualize an experimental design utilizing single unit activity (SUA) to study neural responses in a specific behavioral context.

Task: Students will outline a hypothetical experiment where SUA is used to investigate how neurons in the motor cortex respond to different physical movements. The design should include the selection of animal models, the type of electrodes, the method of data analysis, and expected outcomes.

Simulating MUA and LFP Data

Objective: To understand the data collection and analysis process for multi-unit activity (MUA) and local field potentials (LFP).

Task: Students use simulation software to generate MUA and LFP data from a predefined neural network model. They will analyze the simulated data to differentiate between the two types of recordings and interpret the neural network's activity.

Interpreting ECoG Data

Objective: To learn how to interpret electrocorticography (ECoG) data for clinical decision-making.

Task: Provide students with real or simulated ECoG data from a patient with epilepsy. Students must identify potential seizure onset zones and suggest areas safe for surgical resection, considering functional areas that must be preserved.

Building a Neuroprosthetic Interface

Objective: To explore the application of invasive signals in controlling neuroprosthetic devices.

Task: Students will design a basic schematic of a brain–computer interface (BCI) that uses ECoG signals to control a computer cursor or robotic arm. The project includes selecting appropriate signal processing algorithms and discussing the ethical implications of BCI technologies.

Ethical Debate on Invasive Neurotechnologies

Objective: To critically evaluate the ethical considerations of invasive neurological research and clinical interventions.

Task: Organize a debate focusing on the ethical, legal, and social implications of using invasive techniques like deep brain stimulation and ECoG for both therapeutic and enhancement purposes. Students should prepare arguments considering patient consent, potential benefits, risks, and long-term consequences.

Lab Visit or Virtual Tour

Objective: To provide practical insight into how invasive brain signal recording is conducted in a real-world setting.

Task: Arrange for a lab visit or a virtual tour of a neurophysiological research facility where students can observe procedures involving SUA, MUA, LFP, or ECoG. Include discussions with researchers about the challenges and breakthroughs in their work.

References

Abidian, M. R., & Martin, D. C. (2009). Multifunctional nanobiomaterials for neural interfaces. *Advanced Functional Materials, 19*(4), 573–585.

Ajiboye, A. B., Willett, F. R., Young, D. R., Memberg, W. D., Murphy, B. A., Miller, J. P., et al. (2017). Restoration of reaching and grasping movements through brain-controlled muscle stimulation in a person with tetraplegia: A proof-of-concept demonstration. *The Lancet, 389*(10081), 1821–1830.

Akin, T., Najafi, K., Smoke, R. H., & Bradley, R. M. (1994). A micromachined silicon sieve electrode for nerve regeneration applications. *IEEE Transactions on Biomedical Engineering, 41*(4), 305–313.

Altaheri, H., Muhammad, G., Alsulaiman, M., Amin, S. U., Altuwaijri, G. A., Abdul, W., et al. (2023). Deep learning techniques for classification of electroencephalogram (EEG) motor imagery (MI) signals: A review. *Neural Computing and Applications, 35*(20), 14681–14722.

Anumanchipalli, G. K., Chartier, J., & Chang, E. F. (2019). Speech synthesis from neural decoding of spoken sentences. *Nature, 568*(7753), 493–498.

Araque, A., & Navarrete, M. (2010). Glial cells in neuronal network function. *Philosophical Transactions of the Royal Society B: Biological Sciences, 365*(1551), 2375–2381.

Ayub, M., & Mallamaci, A. (2023). An introduction: Overview of nervous system and brain disorders. In *The role of natural antioxidants in brain disorders* (pp. 1–24). Springer.

Barr, M. L. (1974). *The human nervous system: An anatomical viewpoint.* Harper & Row.

Bartels, J., Andreasen, D., Ehirim, P., Mao, H., Seibert, S., Wright, E. J., & Kennedy, P. (2008). Neurotrophic electrode: Method of assembly and implantation into human motor speech cortex. *Journal of Neuroscience Methods, 174*(2), 168–176.

Bartos, M., Vida, I., & Jonas, P. (2007). Synaptic mechanisms of synchronized gamma oscillations in inhibitory interneuron networks. *Nature Reviews Neuroscience, 8*(1), 45–56.

Bean, B. P. (2007). The action potential in mammalian central neurons. *Nature Reviews Neuroscience, 8*(6), 451–465.

Bédard, C., Kröger, H., & Destexhe, A. (2004). Modeling extracellular field potentials and the frequency-filtering properties of extracellular space. *Biophysical Journal, 86*(3), 1829–1842.

Bédard, C., Kröger, H., & Destexhe, A. (2006). Model of low-pass filtering of local field potentials in brain tissue. *Physical Review E, 73*(5), 051911.

Biran, R., Martin, D. C., & Tresco, P. A. (2005). Neuronal cell loss accompanies the brain tissue response to chronically implanted silicon microelectrode arrays. *Experimental Neurology, 195*(1), 115–126.

Blau, A., Murr, A., Wolff, S., Sernagor, E., Medini, P., Iurilli, G., et al. (2011). Flexible, all-polymer microelectrode arrays for the capture of cardiac and neuronal signals. *Biomaterials, 32*(7), 1778–1786.

Bockhorst, T., Engel, A. K., & Galindo-Leon, E. (2023). What do ECoG recordings tell us about intracortical action potentials? In *Intracranial EEG: A guide for cognitive neuroscientists* (pp. 283–295). Springer International Publishing.

Boniface, S. (1998). Human brain function. *Journal of Neurology, Neurosurgery, and Psychiatry, 65*(3), 410.

Boretius, T., Badia, J., Pascual-Font, A., Schuettler, M., Navarro, X., Yoshida, K., & Stieglitz, T. (2010). A transverse intrafascicular multichannel electrode (TIME) to interface with the peripheral nerve. *Biosensors and Bioelectronics, 26*(1), 62–69.

Brown, S. P., & Hestrin, S. (2009). Intracortical circuits of pyramidal neurons reflect their long-range axonal targets. *Nature, 457*(7233), 1133–1136.

Buzsáki, G. (2004). Large-scale recording of neuronal ensembles. *Nature Neuroscience, 7*(5), 446–451.

Buzsáki, G., Anastassiou, C. A., & Koch, C. (2012). The origin of extracellular fields and currents—EEG, ECoG, LFP and spikes. *Nature Reviews Neuroscience, 13*(6), 407–420.

Buzsaki, G., & Draguhn, A. (2004). Neuronal oscillations in cortical networks. *Science, 304*(5679), 1926–1929.

Campbell, P. K., Jones, K. E., & Normann, R. A. (1990). A 100 electrode intracortical array: Structural variability. *Biomedical Sciences Instrumentation, 26*, 161–165.

Chung, J. E., Joo, H. R., Fan, J. L., Liu, D. F., Barnett, A. H., Chen, S., et al. (2019). High-density, long-lasting, and multi-region electrophysiological recordings using polymer electrode arrays. *Neuron, 101*(1), 21–31.

Cohen, M. X. (2017). Where does EEG come from and what does it mean? *Trends in Neurosciences, 40*(4), 208–218.

Collinger, J. L., Wodlinger, B., Downey, J. E., Wang, W., Tyler-Kabara, E. C., Weber, D. J., et al. (2013). High-performance neuroprosthetic control by an individual with tetraplegia. *The Lancet, 381*(9866), 557–564.

Crone, N. E., Miglioretti, D. L., Gordon, B., & Lesser, R. P. (1998). Functional mapping of human sensorimotor cortex with electrocorticographic spectral analysis. II. Event-related synchronization in the gamma band. *Brain: A Journal of Neurology, 121*(12), 2301–2315.

Crone, N. E., Sinai, A., & Korzeniewska, A. (2006). High-frequency gamma oscillations and human brain mapping with electrocorticography. *Progress in Brain Research, 159*, 275–295.

Degenhart, A. D., Bishop, W. E., Oby, E. R., Tyler-Kabara, E. C., Chase, S. M., Batista, A. P., & Yu, B. M. (2020). Stabilization of a brain–computer interface via the alignment of low-dimensional spaces of neural activity. *Nature Biomedical Engineering, 4*(7), 672–685.

Destexhe, A., Rudolph, M., & Paré, D. (2003). The high-conductance state of neocortical neurons in vivo. *Nature Reviews Neuroscience, 4*(9), 739–751.

Einevoll, G. T., Franke, F., Hagen, E., Pouzat, C., & Harris, K. D. (2012). Towards reliable spike-train recordings from thousands of neurons with multielectrodes. *Current Opinion in Neurobiology, 22*(1), 11–17.

Einevoll, G. T., Kayser, C., Logothetis, N. K., & Panzeri, S. (2013). Modelling and analysis of local field potentials for studying the function of cortical circuits. *Nature Reviews Neuroscience, 14*(11), 770–785.

Ellis, G., & Ellis, G. (2016). The mind and the brain. In *How can physics underlie the mind? Top-down causation in the human context* (pp. 291–394). Springer.

Erlanger, J., & Gasser, H. S. (1937). *Electrical signs of nervous activity (Humphrey Milford)*. Oxford University Press.

Escabí, M. A., Read, H. L., Viventi, J., Kim, D. H., Higgins, N. C., Storace, D. A., et al. (2014). A high-density, high-channel count, multiplexed µECoG array for auditory-cortex recordings. *Journal of Neurophysiology, 112*(6), 1566–1583.

Ferrari, M., & Quaresima, V. (2012). A brief review on the history of human functional near-infrared spectroscopy (fNIRS) development and fields of application. *NeuroImage, 63*(2), 921–935.

Frank, J. A., Antonini, M. J., & Anikeeva, P. (2019). Next-generation interfaces for studying neural function. *Nature Biotechnology, 37*(9), 1013–1023.

Fries, P. (2009). Neuronal gamma-band synchronization as a fundamental process in cortical computation. *Annual Review of Neuroscience, 32*, 209–224.

Galvani, L. (1791). *De Viribus Electricitatis in Motu Musculari Commentarius [Italian]*. Bologna Accademia delle Scienze.

Gearing, M., & Kennedy, P. (2020). Histological confirmation of myelinated neural filaments within the tip of the neurotrophic electrode after a decade of neural recordings. *Frontiers in Human Neuroscience, 14*, 111.

Gray, C. M., Maldonado, P. E., Wilson, M., & McNaughton, B. (1995). Tetrodes markedly improve the reliability and yield of multiple single-unit isolation from multi-unit recordings in cat striate cortex. *Journal of Neuroscience Methods, 63*(1–2), 43–54.

Hämäläinen, M., Hari, R., Ilmoniemi, R. J., Knuutila, J., & Lounasmaa, O. V. (1993). Magnetoencephalography—Theory, instrumentation, and applications to noninvasive studies of the working human brain. *Reviews of Modern Physics, 65*(2), 413.

Hamill, O. P., Marty, A., Neher, E., Sakmann, B., & Sigworth, F. J. (1981). Improved patch-clamp techniques for high-resolution current recording from cells and cell-free membrane patches. *Pflügers Archiv, 391*, 85–100.

Hanson, T. L., Diaz-Botia, C. A., Kharazia, V., Maharbiz, M. M., & Sabes, P. N. (2019). The "sewing machine" for minimally invasive neural recording. *BioRxiv, 2019*, 578542.

Hodgkin, A. L., & Huxley, A. F. (1939). Action potentials recorded from inside a nerve fibre. *Nature, 144*(3651), 710–711.

Hodgkin, A. L., & Huxley, A. F. (1952a). Currents carried by sodium and potassium ions through the membrane of the giant axon of Loligo. *The Journal of Physiology, 116*(4), 449.

Hodgkin, A. L., & Huxley, A. F. (1952b). A quantitative description of membrane current and its application to conduction and excitation in nerve. *The Journal of Physiology, 117*(4), 500.

Hong, G., & Lieber, C. M. (2019). Novel electrode technologies for neural recordings. *Nature Reviews Neuroscience, 20*(6), 330–345.

Hubel, D. H. (1957). Tungsten microelectrode for recording from single units. *Science, 125*(3247), 549–550.

Hubel, D. H., & Wiesel, T. N. (1962). Receptive fields, binocular interaction and functional architecture in the cat's visual cortex. *The Journal of Physiology, 160*(1), 106.

Jensen, O., Kaiser, J., & Lachaux, J. P. (2007). Human gamma-frequency oscillations associated with attention and memory. *Trends in Neurosciences, 30*(7), 317–324.

Jessen, K. R. (2004). Glial cells. *The International Journal of Biochemistry & Cell Biology, 36*(10), 1861–1867.

Kandel, E. R., Schwartz, J. H., Jessell, T. M., Siegelbaum, S., Hudspeth, A. J., & Mack, S. (Eds.). (2000). *Principles of neural science* (Vol. 4, pp. 1227–1246). McGraw-Hill.

Kao, J. C., Stavisky, S. D., Sussillo, D., Nuyujukian, P., & Shenoy, K. V. (2014). Information systems opportunities in brain–machine interface decoders. *Proceedings of the IEEE, 102*(5), 666–682.

Kennedy, P. R. (1989). The cone electrode: A long-term electrode that records from neurites grown onto its recording surface. *Journal of Neuroscience Methods, 29*(3), 181–193.

Khodagholy, D., Doublet, T., Quilichini, P., Gurfinkel, M., Leleux, P., Ghestem, A., et al. (2013). In vivo recordings of brain activity using organic transistors. *Nature Communications, 4*(1), 1575.

Kleinfeld, D., Luan, L., Mitra, P. P., Robinson, J. T., Sarpeshkar, R., Shepard, K., et al. (2019). Can one concurrently record electrical spikes from every neuron in a mammalian brain? *Neuron, 103*(6), 1005–1015.

Knösche, T. R., & Haueisen, J. (2022). Neural tissue and its signals. In *EEG/MEG source reconstruction: Textbook for electro- and magnetoencephalography* (pp. 11–42). Springer.

Koch, C. (2004). *Biophysics of computation: Information processing in single neurons*. Oxford University Press.

Konerding, W. S., Froriep, U. P., Kral, A., & Baumhoff, P. (2018). New thin-film surface electrode array enables brain mapping with high spatial acuity in rodents. *Scientific Reports, 8*(1), 3825.

Kozai, T. D., Jaquins-Gerstl, A. S., Vazquez, A. L., Michael, A. C., & Cui, X. T. (2015). Brain tissue responses to neural implants impact signal sensitivity and intervention strategies. *ACS Chemical Neuroscience, 6*(1), 48–67.

Kozai, T. D. Y., Langhals, N. B., Patel, P. R., Deng, X., Zhang, H., Smith, K. L., et al. (2012). Ultrasmall implantable composite microelectrodes with bioactive surfaces for chronic neural interfaces. *Nature Materials, 11*(12), 1065–1073.

Lawrence, S. M., Dhillon, G. S., & Horch, K. W. (2003). Fabrication and characteristics of an implantable, polymer-based, intrafascicular electrode. *Journal of Neuroscience Methods, 131*(1–2), 9–26.

Lehew, G., & Nicolelis, M. A. (2008). Ch. 1—Introduction. In M. A. Nicolelis (Ed.), *Methods for neural ensemble recordings* (2nd ed.). Taylor & Francis.

Lempka, S. F., Johnson, M. D., Moffitt, M. A., Otto, K. J., Kipke, D. R., & McIntyre, C. C. (2011). Theoretical analysis of intracortical microelectrode recordings. *Journal of Neural Engineering, 8*(4), 045006.

Leuthardt, E. C., Schalk, G., Wolpaw, J. R., Ojemann, J. G., & Moran, D. W. (2004). A brain–computer interface using electrocorticographic signals in humans. *Journal of Neural Engineering, 1*(2), 63.

Lewicki, M. S. (1998). A review of methods for spike sorting: The detection and classification of neural action potentials. *Network: Computation in Neural Systems, 9*(4), R53.

Llinás, R. R. (1988). The intrinsic electrophysiological properties of mammalian neurons: Insights into central nervous system function. *Science, 242*(4886), 1654–1664.

Logothetis, N. K. (2008). What we can do and what we cannot do with fMRI. *Nature, 453*(7197), 869–878.

Macé, E., Montaldo, G., Cohen, I., Baulac, M., Fink, M., & Tanter, M. (2011). Functional ultrasound imaging of the brain. *Nature Methods, 8*(8), 662–664.

Marblestone, A. H., Zamft, B. M., Maguire, Y. G., Shapiro, M. G., Cybulski, T. R., Glaser, J. I., et al. (2013). Physical principles for scalable neural recording. *Frontiers in Computational Neuroscience, 7*, 137.

Marmont, G. (1949). Studies on the axon membrane. I. A new method. *Journal of Cellular and Comparative Physiology, 34*(3), 351–382.

Massey, T. L., Santacruz, S. R., Hou, J. F., Pister, K. S., Carmena, J. M., & Maharbiz, M. M. (2019). A high-density carbon fiber neural recording array technology. *Journal of Neural Engineering, 16*(1), 016024.

Maynard, E. M., Nordhausen, C. T., & Normann, R. A. (1997). The Utah intracortical electrode array: A recording structure for potential brain-computer interfaces. *Electroencephalography and Clinical Neurophysiology, 102*(3), 228–239.

McCandless, D. W. (1997). In D. E. Hanes (Ed.), *Fundamental neuroscience*. Churchil Livingston.

McNaughton, B. L., O'Keefe, J., & Barnes, C. A. (1983). The stereotrode: A new technique for simultaneous isolation of several single units in the central nervous system from multiple unit records. *Journal of Neuroscience Methods, 8*(4), 391–397.

MicroProbes for Life Science. (2020). Nerve cuff electrodes. Retrieved from https://microprobes.com/products/peripheral-electrodes/nerve-cuff

Miller, E. K., Lundqvist, M., & Bastos, A. M. (2018). Working memory 2.0. *Neuron, 100*(2), 463–475.

Miller, K. J. (2019). A library of human electrocorticographic data and analyses. *Nature Human Behaviour, 3*(11), 1225–1235.

Miller, K. J., Hermes, D., Honey, C. J., Hebb, A. O., Ramsey, N. F., Knight, R. T., et al. (2012). *Human motor cortical activity is selectively phase-entrained on underlying rhythms* (Vol. 8, p. e1002655).

Miller, K. J., Leuthardt, E. C., Schalk, G., Rao, R. P., Anderson, N. R., Moran, D. W., et al. (2007). Spectral changes in cortical surface potentials during motor movement. *Journal of Neuroscience, 27*(9), 2424–2432.

Miller, K. J., Sorensen, L. B., Ojemann, J. G., & Den Nijs, M. (2009). Power-law scaling in the brain surface electric potential. *PLoS Computational Biology, 5*(12), e1000609.

Mitzdorf, U. (1985). Current source-density method and application in cat cerebral cortex: Investigation of evoked potentials and EEG phenomena. *Physiological Reviews, 65*(1), 37–100.

Mountcastle, V. B. (1997). The columnar organization of the neocortex. *Brain: A Journal of Neurology, 120*(4), 701–722.

Neher, E., & Sakmann, B. (1976). Single-channel currents recorded from membrane of denervated frog muscle fibres. *Nature, 260*(5554), 799–802.

Newsome, W. T., Britten, K. H., & Movshon, J. A. (1989). Neuronal correlates of a perceptual decision. *Nature, 341*(6237), 52–54.

Nicolelis, M. A. (1998). *Methods for neural ensemble recordings*. CRC Press.

Nicolelis, M. A., Dimitrov, D., Carmena, J. M., Crist, R., Lehew, G., Kralik, J. D., & Wise, S. P. (2003). Chronic, multisite, multielectrode recordings in macaque monkeys. *Proceedings of the National Academy of Sciences, 100*(19), 11041–11046.

Niedermeyer, E., & da Silva, F. L. (Eds.). (2005). *Electroencephalography: Basic principles, clinical applications, and related fields*. Lippincott Williams & Wilkins.

Nowell, M., Miserocchi, A., McEvoy, A. W., & Duncan, J. S. (2014). Advances in epilepsy surgery. *Journal of Neurology, Neurosurgery & Psychiatry, 85*(11), 1273–1279.

Nuyujukian, P., Sanabria, J. A., Saab, J., Pandarinath, C., Jarosiewicz, B., Blabe, C. H., et al. (2018). Cortical control of a tablet computer by people with paralysis. *PLoS One, 13*(11), e0204566.

Pakkenberg, B., & Gundersen, H. J. G. (1997). Neocortical neuron number in humans: Effect of sex and age. *Journal of Comparative Neurology, 384*(2), 312–320.

Pandarinath, C., O'Shea, D. J., Collins, J., Jozefowicz, R., Stavisky, S. D., Kao, J. C., et al. (2018). Inferring single-trial neural population dynamics using sequential auto-encoders. *Nature Methods, 15*(10), 805–815.

Parastarfeizabadi, M., & Kouzani, A. Z. (2017). Advances in closed-loop deep brain stimulation devices. *Journal of Neuroengineering and Rehabilitation, 14*, 1–20.

Patel, P. R., Na, K., Zhang, H., Kozai, T. D., Kotov, N. A., Yoon, E., & Chestek, C. A. (2015). Insertion of linear 8.4 µm diameter 16 channel carbon fiber electrode arrays for single unit recordings. *Journal of Neural Engineering, 12*(4), 046009.

Pesaran, B., Vinck, M., Einevoll, G. T., Sirota, A., Fries, P., Siegel, M., et al. (2018). Investigating large-scale brain dynamics using field potential recordings: Analysis and interpretation. *Nature Neuroscience, 21*(7), 903–919.

Pfurtscheller, G., & Cooper, R. (1975). Frequency dependence of the transmission of the EEG from cortex to scalp. *Electroencephalography and Clinical Neurophysiology, 38*(1), 93–96.

Polikov, V. S., Tresco, P. A., & Reichert, W. M. (2005). Response of brain tissue to chronically implanted neural electrodes. *Journal of Neuroscience Methods, 148*(1), 1–18.

Purves, D., Augustine, G. J., Fitzpatrick, D., Hall, W. C., LaMantia, A., McNamara, J. O., & White, L. E. (2008). *Neuroscience*. Sinauer Associates.

Quiroga, R. Q., Reddy, L., Kreiman, G., Koch, C., & Fried, I. (2005). Invariant visual representation by single neurons in the human brain. *Nature, 435*(7045), 1102–1107.

Rousche, P. J., & Normann, R. A. (1998). Chronic recording capability of the Utah Intracortical Electrode Array in cat sensory cortex. *Journal of Neuroscience Methods, 82*(1), 1–15.

Salahuddin, U., & Gao, P. X. (2021). Signal generation, acquisition, and processing in brain machine interfaces: A unified review. *Frontiers in Neuroscience, 15*, 728178.

Saxena, T., Karumbaiah, L., Gaupp, E. A., Patkar, R., Patil, K., Betancur, M., et al. (2013). The impact of chronic blood–brain barrier breach on intracortical electrode function. *Biomaterials, 34*(20), 4703–4713.

Schalk, G., Miller, K. J., Anderson, N. R., Wilson, J. A., Smyth, M. D., Ojemann, J. G., et al. (2008). Two-dimensional movement control using electrocorticographic signals in humans. *Journal of Neural Engineering, 5*(1), 75.

Schumann, C. (2021). Limbic system. In *Encyclopedia of autism spectrum disorders* (pp. 2719–2724). Springer International Publishing.

Sharafkhani, N., Kouzani, A. Z., Adams, S. D., Long, J. M., Lissorgues, G., Rousseau, L., & Orwa, J. O. (2022). Neural tissue-microelectrode interaction: Brain micromotion, electrical impedance, and flexible microelectrode insertion. *Journal of Neuroscience Methods, 365*, 109388.

Shen, K., Chen, O., Edmunds, J. L., Piech, D. K., & Maharbiz, M. M. (2023). Translational opportunities and challenges of invasive electrodes for neural interfaces. *Nature Biomedical Engineering, 7*(4), 424–442.

Shenoy, K. V., & Carmena, J. M. (2014). Combining decoder design and neural adaptation in brain-machine interfaces. *Neuron, 84*(4), 665–680.

Sontheimer, H., Black, J. A., & Waxman, S. G. (1996). Voltage-gated Na$^+$ channels in glia: Properties and possible functions. *Trends in Neurosciences, 19*(8), 325–331.

Spruston, N. (2008). Pyramidal neurons: Dendritic structure and synaptic integration. *Nature Reviews Neuroscience, 9*(3), 206–221.

Tolstosheeva, E., Gordillo-González, V., Biefeld, V., Kempen, L., Mandon, S., Kreiter, A. K., & Lang, W. (2015). A multi-channel, flex-rigid ECoG microelectrode array for visual cortical interfacing. *Sensors, 15*(1), 832–854.

Turner, J. N., Shain, W., Szarowski, D. H., Andersen, M., Martins, S., Isaacson, M., & Craighead, H. (1999). Cerebral astrocyte response to micromachined silicon implants. *Experimental Neurology, 156*(1), 33–49.

Urdaneta, M. E., Kunigk, N. G., Peñaloza-Aponte, J. D., Currlin, S., Malone, I. G., Fried, S. I., & Otto, K. J. (2023). Layer-dependent stability of intracortical recordings and neuronal cell loss. *Frontiers in Neuroscience, 17*, 1096097.

von Helmholtz, H. (1852). Messungen über Fortpflanzungsgeschwindigkeit der Reizung in den Nerven [German]. *Archiv für Anatomie, Physiologie und Wissenschaftliche Medicin, 19*, 199–216.

Wang, Y., Liu, S., Wang, H., Zhao, Y., & Zhang, X. D. (2022). Neuron devices: Emerging prospects in neural interfaces and recognition. *Microsystems & Nanoengineering, 8*(1), 128.

Wei, J., Bai, W., Liu, T., & Tian, X. (2015). Functional connectivity changes during a working memory task in rat via NMF analysis. *Frontiers in Behavioral Neuroscience, 9*, 2.

Wise, K. D., Anderson, D. J., Hetke, J. F., Kipke, D. R., & Najafi, K. (2004). Wireless implantable microsystems: High-density electronic interfaces to the nervous system. *Proceedings of the IEEE, 92*(1), 76–97.

Wise, K. D., Angell, J. B., & Starr, A. (1970). An integrated-circuit approach to extracellular microelectrodes. *IEEE Transactions on Biomedical Engineering, 3*, 238–247.

Wood, I. K. (1996). In M. F. Bear, B. W. Connors, & M. A. Paradiso (Eds.), *Neuroscience: Exploring the brain* (p. 666). Williams & Wilkins.

Wu, N., Wan, S., Su, S., Huang, H., Dou, G., & Sun, L. (2021). Electrode materials for brain–machine interface: A review. *InfoMat, 3*(11), 1174–1194.

Zeng, Q., Yu, S., Fan, Z., Huang, Y., Song, B., & Zhou, T. (2022). Nanocone-array-based platinum-iridium oxide neural microelectrodes: Structure, electrochemistry, durability and biocompatibility study. *Nanomaterials, 12*(19), 3445.

Zhang, M., Tang, Z., Liu, X., & Van der Spiegel, J. (2020). Electronic neural interfaces. *Nature Electronics, 3*(4), 191–200.

Preprocessing Techniques for Brain Signal Data

Noise Reduction and Artifact Removal Methods

Contents

Test your learning and check your understanding of this book's contents: use the "Springer Nature Flashcards" app to access questions using ▶ https://sn.pub/kmb-jyz. To use the app, please follow the instructions in ▶ Chap. 1.

This chapter provides an overview of the preprocessing techniques employed to refine raw brain signals for effective analysis in neuroscience research and clinical diagnostics. It emphasizes the necessity of preprocessing to mitigate the impact of noise and artifacts inherent in data acquired from various brain signal recording technologies such as EEG, fMRI, fNIRS, MEG, intracortical recordings, and electrocorticography. The sources of these disturbances are categorized into physiological artifacts—such as cardiac pulses, muscle activities, and ocular movements—and environmental noise, which include electromagnetic interference from nearby electronic devices and infrastructural elements. Discussions on the methodologies used to address these challenges highlight the critical role of preprocessing in enhancing the signal-to-noise ratio, ensuring data integrity, and facilitating the extraction of meaningful neural information. Techniques such as spatial and temporal filtering, artifact subtraction, and data normalization are explored to demonstrate how they tailor preprocessing approaches to the unique characteristics of each recording method. This chapter also underscores the dynamic nature of preprocessing, reflecting ongoing advancements in technology and methodology that continue to improve the accuracy and reliability of neuroscientific data analysis.

Learning Objectives

1. Identify Sources of Noise and Artifacts: Recognize different types of physiological and environmental noise that affect brain signal recordings, such as cardiac and respiratory artifacts, electromyographic noise, and electromagnetic interference.
2. Understand Preprocessing Techniques: Explain the various preprocessing steps and techniques used to enhance signal quality, including spatial and temporal filtering, artifact subtraction, and data normalization.
3. Apply Preprocessing to Different Modalities: Apply specific preprocessing methods tailored to different brain recording technologies like EEG, fMRI, fNIRS, MEG, and ECoG.
4. Assess the Impact of Preprocessing on Data Quality: Evaluate how preprocessing affects the signal-to-noise ratio and the integrity of brain data, facilitating the extraction of meaningful neural information.
5. Explore Advanced Preprocessing Tools: Familiarize with advanced signal processing tools and techniques that are continuously being developed and adapted for refining brain signal data.

These objectives aim to provide students with a comprehensive foundation in the preprocessing of brain signals, preparing them for both practical applications and further exploration of advanced techniques in the field.

The raw signals obtained from the brain signal acquisition techniques outlined in ▶ Chaps. 2 and 3 are subject to various sources of noise and artifacts that can significantly distort the neural activity crucial for understanding cognitive processes or diagnosing clinical conditions. These disruptions are primarily categorized into physiological and environmental sources (Sweeney et al., 2012). Physiological artifacts, such as muscle contractions and eye movements, directly interfere with the clarity of the neural recordings (Fatourechi et al., 2007). These types of interference are well-documented issues, as muscle contractions can introduce broad-spectrum electrical noise, and ocular movements can create sharp spikes or shifts in the electrical baseline (Goncharova et al., 2003). Additionally, environmental noise, which includes interference from electronic equipment and potential errors in signal acquisition, can further complicate data accuracy. Such noise often stems from electromagnetic fields generated by nearby equipment or flawed recording techniques, which can induce additional signals unrelated to brain activity (Fatourechi et al., 2007). These contaminants can mask or mimic the neural signals of interest, leading to potential misinterpretations or misleading results. The presence of these artifacts necessitates robust preprocessing techniques to isolate and enhance the actual neural signals for reliable analysis (Ille et al., 2002; Castellanos & Makarov, 2006).

This chapter will provide an overview of noise and artifacts from physiological sources such as cardiac pulses, muscle activity, eye movements, and environmental interferences like power line noise and electronic device emissions, as shown in ◙ Fig. 4.1 (Coffman & Salisbury, 2020). Followed by various preprocessing steps, researchers employ to remove the artifacts and transform raw data collected from brain signal acquisition devices into meaningful and analyzable brain signals.

You will then explore the preprocessing pipelines used to process raw EEG (electroencephalography), fMRI (functional magnetic resonance imaging), fNIRS (functional near-infrared spectroscopy), MEG (magnetoencephalography), intracortical signals, and electrocorticography. The discussion will provide insight into how each data type requires specific preprocessing steps tailored to its unique characteristics and the typical noise or artifacts associated with each method. This includes techniques for artifact removal, signal enhancement, spatial and temporal filtering, and data normalization specific to each modality. By understanding these preprocessing pipelines, you will gain insight into how raw brain signals are transformed into clean, analyzable data, ready for further analysis and interpretation in both research and clinical settings. You should note that the list of preprocessing methods discussed is not exhaustive. Researchers continuously develop and adapt new techniques tailored to specific research questions and applications. As the neuroimaging and signal analysis field evolves, innovative methods are introduced to address emerging challenges and enhance the accuracy and efficiency of data processing.

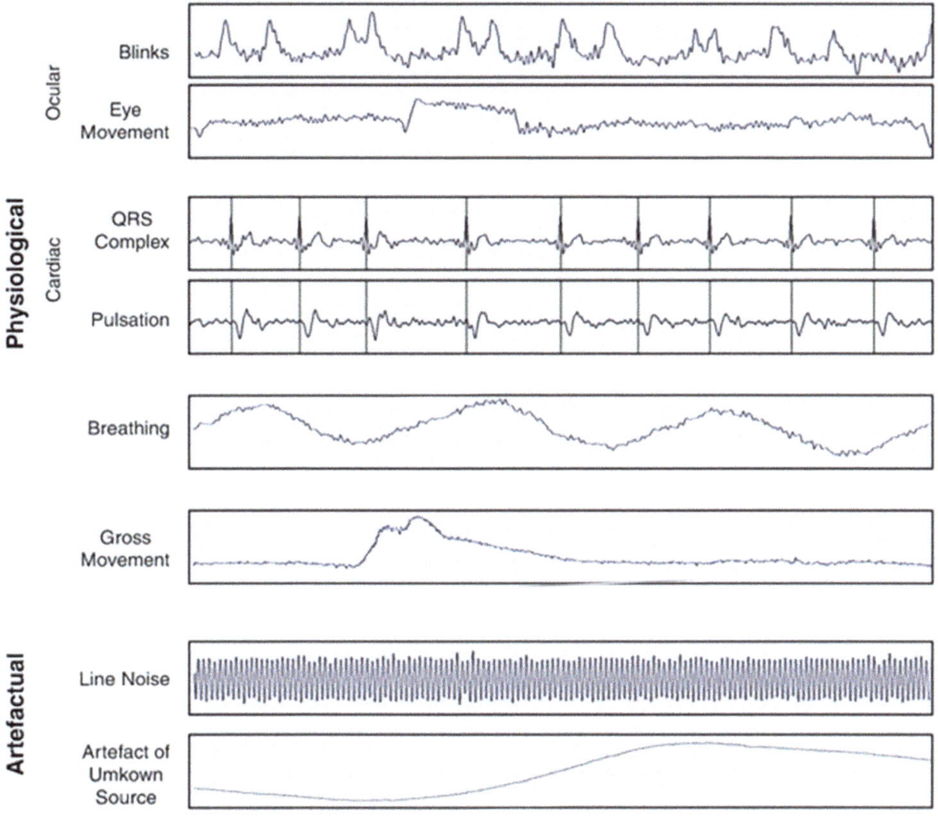

▣ Fig. 4.1 Profile of different types of artifacts. (From Coffman & Salisbury, 2020)

4.1 Physiological Artifacts

Physiological noise in brain signal recordings arises from various bodily processes that interfere with accurately measuring neural activity. These artifacts can complicate data analysis and lead to incorrect conclusions if not adequately addressed. Recognizing that the terms "noise" and "artifacts" are context-dependent when discussing brain signal recordings is important. While we refer to these signals as noise or artifacts in the context of brain signal analysis, they are not inherently disruptive. These terms are used because these signals are not the primary focus of interest and interfere with the analysis of neural activity. When studying different physiological phenomena, what is considered noise or an artifact in one study might be the signal of interest in another. Therefore, the classification of these signals as noise or artifacts is based solely on their relevance to the specific objectives of the research. Below are detailed descriptions of common types of physiological noise:

4.1.1 **Cardiac Artifacts (Pulse Artifact)**

Cardiac artifacts are caused by the electrical and mechanical activities of the heart, which can influence brain signal recordings. Cardiac artifacts, often introduced when electrodes are placed on or near blood vessels, arise from heartbeats' expansion and contraction movements. These artifacts, known as pulse artifacts, typically manifest in EEG recordings at a frequency of around 1.2 Hz and can be challenging to remove due to their similarity to normal EEG waveforms (Urigüen & Garcia-Zapirain, 2015). In contrast, another form of cardiac activity, the electrocardiogram (ECG), measures the heart's electrical signals. ECG signals display a characteristic regular pattern, distinct from cerebral activity, making it easier to identify and remove these artifacts from EEG data by using a reference waveform (Urigüen & Garcia-Zapirain, 2015).

In EEG, the heartbeat can induce changes in the electrical field detected by scalp electrodes, known as EKG artifacts (Nakamura & Shibasaki, 1987). In fMRI, the pulsation of blood vessels, influenced by the cardiac cycle, affects the local magnetic fields and can alter the BOLD signal (Glover et al., 2000). Techniques to mitigate these effects include adaptive filtering and synchronous averaging, where cardiac cycles are averaged over multiple beats to reduce variability. ◘ Figure 4.2 shows an example of ECG artifacts in EEG signals (Brienza et al., 2019).

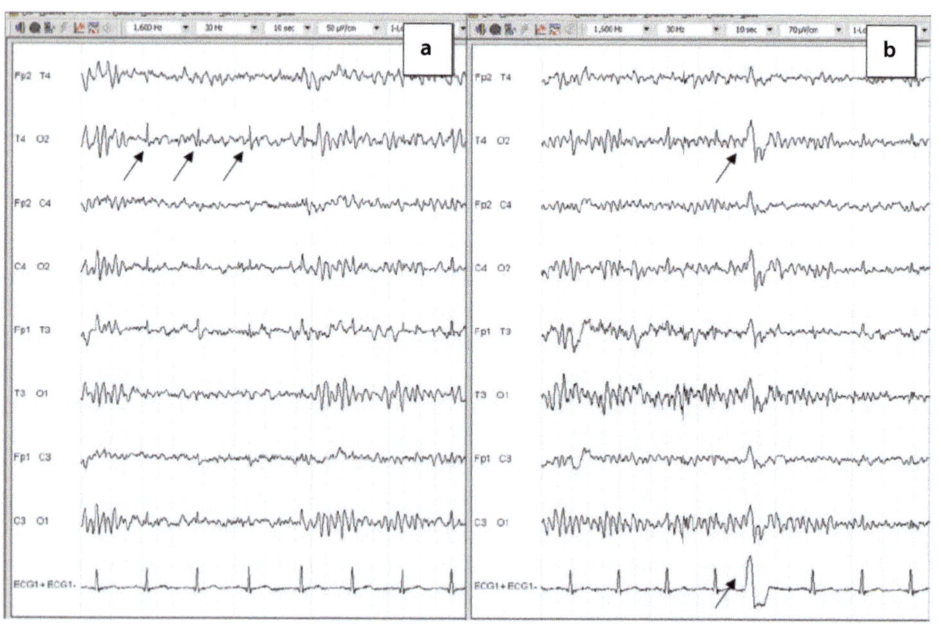

◘ **Fig. 4.2** An example of ECG artifact in EEG signals. The arrows indicate the ECG artifacts. (From Brienza et al., 2019)

4.1.2 **Respiratory Artifacts**

Respiratory rhythms can introduce fluctuations in both EEG and fMRI signals due to changes in chest impedance and alterations in blood oxygenation and carbon dioxide levels. These changes can mimic or mask neural activity in brain recordings. In EEG, respiratory artifacts can be reduced using band-stop filters that target the specific breathing frequencies (Anderer et al., 1999). For fMRI, respiratory volume and oxygenation variations can be regressed out from the BOLD signals to improve data accuracy (Birn et al., 2006).

4.1.3 **Electrooculographic (EOG) Artifacts**

The origin of ocular artifacts in EEG recordings includes eye movements and blinks, primarily caused by changes in the orientation of the retina-cornea dipole and alterations in ocular conductance when the cornea contacts the eyelid (Jiang et al., 2019). These artifacts, along with EEG activity, propagate to the head surface due to the volume conduction effect and are captured by EEG electrodes. Ocular signals can also be recorded using electrooculography (EOG), which typically exhibits amplitudes significantly greater than those of EEG signals, although their frequencies are similar. It is important to note that there is a bidirectional contamination between EEG and EOG data, leading to potential errors in artifact removal due to their interference (Jiang et al., 2019). Common approaches to handling EOG artifacts include using dedicated EOG channels to capture eye movements and applying artifact subtraction methods where EOG data is used to clean EEG signals (Croft & Barry, 2000). ◘ Figure 4.3 shows an example of EOG artifacts in EEG signals (Brienza et al., 2019).

4.1.4 **Electromyographic (EMG) Noise**

Contamination of brain data by muscle activity is a significant challenge, arising from various muscle groups near the recording sites during actions such as talking, sniffing, or swallowing (Urigüen & Garcia-Zapirain, 2015). Muscle artifacts, which can be measured by electromyography (EMG), range broadly in frequency from 0 Hz to over 200 Hz and vary in amplitude and waveform depending on the degree of muscle contraction and stretch (Goncharova et al., 2003). Unlike ocular artifacts, which can be somewhat more straightforward to isolate, EMG artifacts are difficult to eliminate due to their complexity and single-channel measurement limitations. Moreover, EMG contamination is largely statistically independent from EEG data, both temporally and spatially (McMenamin et al., 2011). Techniques such as spatial filtering, where signals from multiple electrodes are combined to enhance brain signals while reducing muscle noise, and independent component

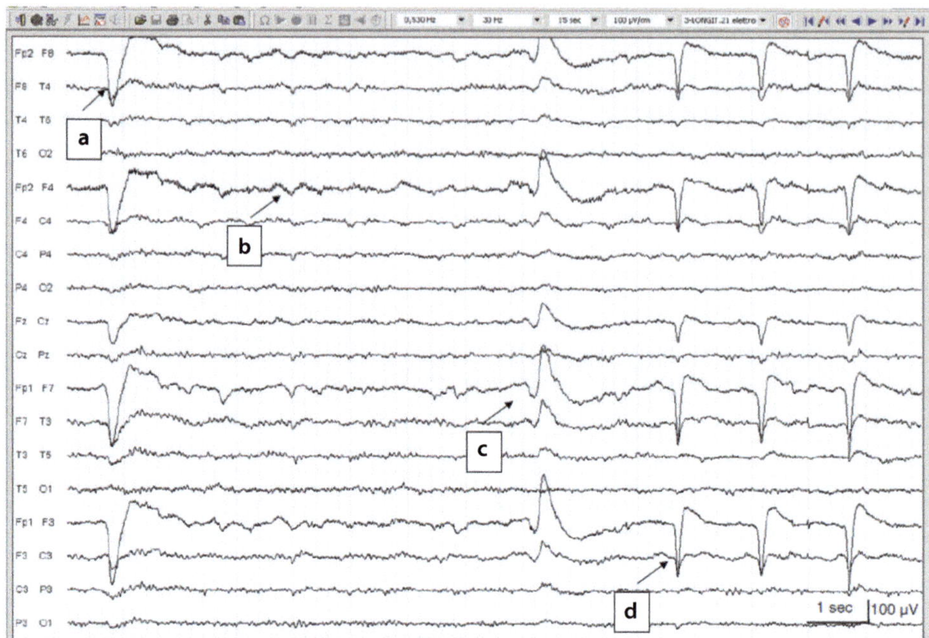

■ **Fig. 4.3** Standard EEG activities including **a** closing the eyes, **b** making slow eye movements while the eyelids are closed, **c** opening the eyes, and **d** observing blinking artifacts on the anterior derivations. (From Brienza et al., 2019)

analysis (ICA) for identifying and removing EMG components are commonly used (Muthukumaraswamy, 2013). An example of the effect of swallowing, coughing and chewing on EEG signal is shown in ■ Fig. 4.4 (Tatum et al., 2011).

4.1.5 Galvanic Skin Response (GSR)

GSR refers to changes in skin conductivity caused by sweating, which can affect the impedance at the electrode–skin interface, altering the recorded signals (Tatum et al., 2011; Amin et al., 2023). While direct methods to filter out GSR noise are limited, ensuring stable environmental conditions and using high-quality electrodes can minimize its impact. An example of the effect of sweating on EEG signal is shown in ■ Fig. 4.5 (Tatum et al., 2011).

4.1.6 Motion Artifacts

Participant movements during data acquisition can introduce significant artifacts in both EEG and fMRI data, such as shifts in electrode position or distortions in

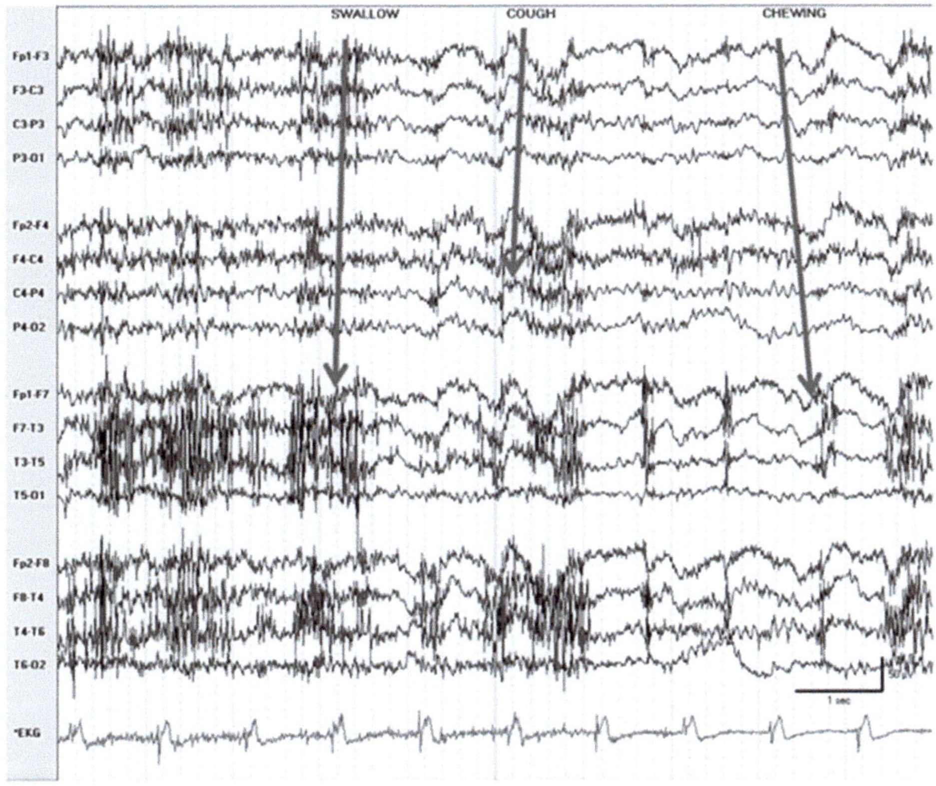

☐ **Fig. 4.4** Effect of swallowing, coughing, and chewing on EEG signal. (From Tatum et al. 2011)

magnetic field maps. Techniques to reduce motion artifacts include employing motion correction algorithms, stabilizing the participant's head with supports, and designing shorter, more engaging tasks to minimize movement (Friston et al., 1996). An example of the effect of motion artifacts on EEG signal is shown in ☐ Fig. 4.6 (Safari et al., 2024).

4.1.7 **Capillary Fluctuations**

Fluctuations in capillary blood flow, independent of neuronal activity, can influence fMRI signals. This type of physiological noise can be addressed by applying physiological noise modeling and regression techniques to remove non-neural components from the BOLD signal (Chang & Glover, 2009).

4

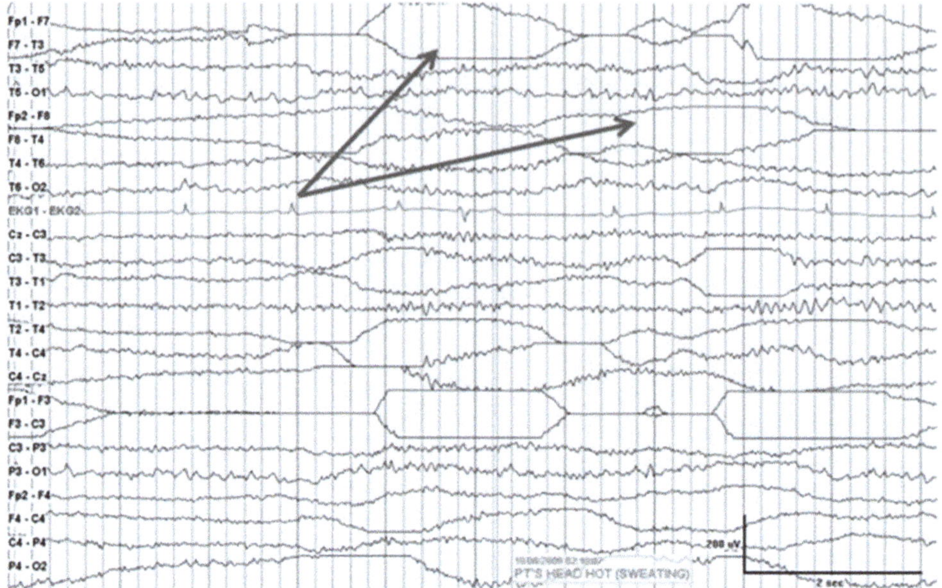

Fig. 4.5   Effect of sweating on EEG signal. (From Tatum et al., 2011)

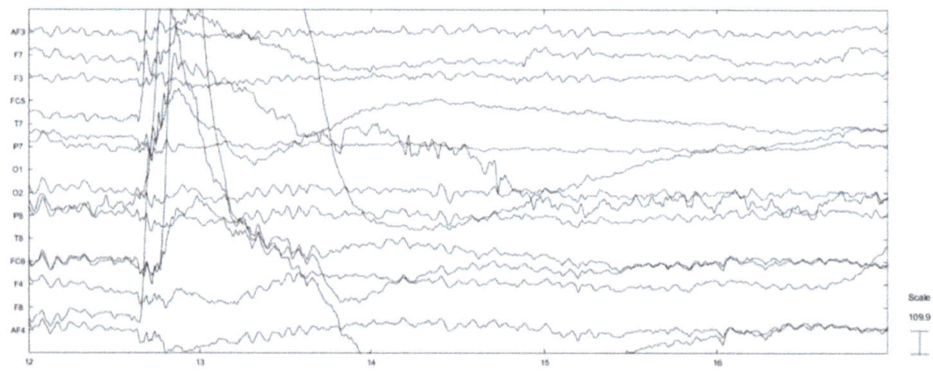

Fig. 4.6   EEG distortion due to motion artifacts. (Adapted from Safari et al., 2024 originally published under CC-BY 4.0)

4.2   **Environmental Noise**

Environmental noise in raw brain signals refers to unwanted interference and arti-facts that originate from sources external to the subject's body. These sources can significantly degrade the quality of the data obtained from neuroimaging and elec-trophysiological recordings, leading to potential misinterpretations of neural activity. Understanding and mitigating this type of noise are crucial for accurate brain signal analysis. Some of the sources of environmental noise are listed below, along with the techniques to mitigate them.

4.2.1 Electromagnetic Interference (EMI)

Electromagnetic fields from electronic devices and electrical circuits in the vicinity of the recording equipment can introduce noise. Common sources include computer monitors, mobile phones, power lines, and even the electrical infrastructure within a building. This interference typically manifests as a 50 or 60 Hz line noise, depending on the regional power grid frequency, and can be seen as a consistent oscillatory pattern that overlays the brain signal (Muthukumaraswamy, 2013). Using shielded cables and Faraday cages can significantly reduce electromagnetic interference by blocking external electric fields. Additionally, notch filters are commonly used to remove specific frequencies associated with power line noise, effectively isolating the neural signal from this form of environmental noise (Nunez & Srinivasan, 2006). An example of the effect of a phone ring on EEG is shown in ◘ Fig. 4.7 (Tatum et al., 2011).

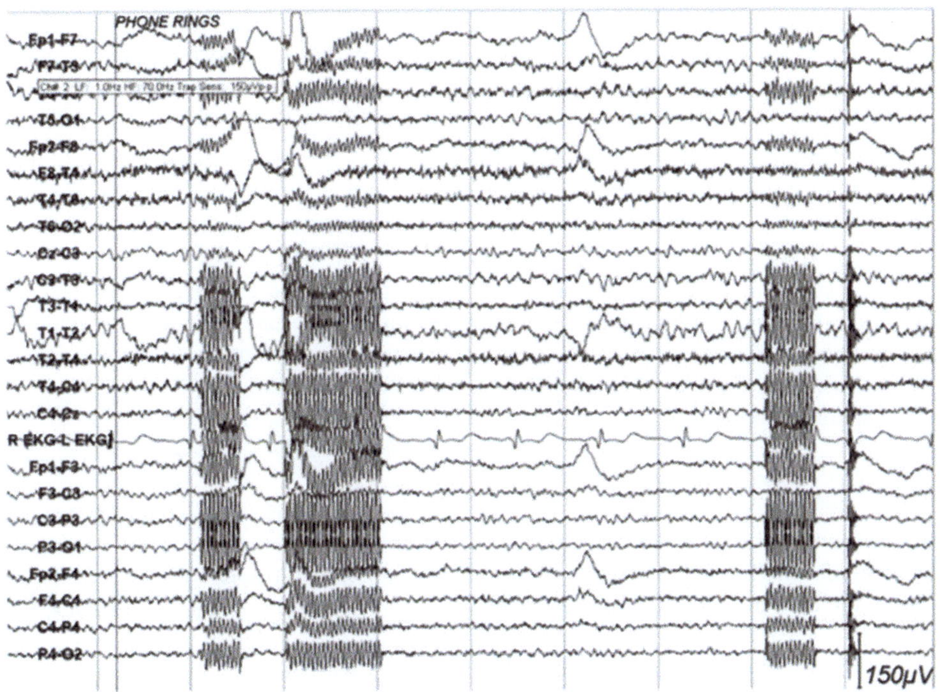

◘ **Fig. 4.7**   Effect of a phone ring on EEG signal acquisition. (From Tatum et al., 2011)

4.2.2 **Vibration and Movement**

Vibrations from nearby machinery, elevators, traffic, or even subtle building movements can affect sensitive recording equipment, particularly in techniques like magnetoencephalography (MEG) and functional magnetic resonance imaging (fMRI). These vibrations can cause artifacts that may mimic or obscure genuine neural signals (Liu, 2016; Mutanen et al., 2018).

Conducting experiments in specially designed, environmentally controlled rooms can help minimize vibrations and temperature fluctuations. MEG and fMRI facilities, for example, often utilize magnetically shielded rooms and maintain strict control over environmental variables to ensure data integrity (Liu, 2016; Mutanen et al., 2018).

4.2.3 **Temperature and Humidity Fluctuations**

Changes in temperature and humidity can affect the performance of the recording equipment. For instance, variations in room temperature can influence electrode impedance in EEG recordings and the sensitivity of sensors in fMRI scanners (Allen et al., 2000).

Advanced signal processing techniques such as independent component analysis (ICA) and principal component analysis (PCA), which you will learn in the later section, can be employed post-acquisition to identify and remove environmental noise components from the recorded data, thereby improving the signal's clarity and usability for further analysis (Hyvärinen & Oja, 2000).

Box 4.1

In signal processing, particularly in the context of recording and analyzing brain signals, distinguishing between noise and artifacts is crucial for data integrity and analysis accuracy. Here's a detailed comparison of the two:

Noise Versus Artifacts

- **Definition:**

Noise: Refers to random, unpredictable variations in the signal that are not part of the desired output. Noise can come from a variety of sources, both external and internal to the measurement system, and typically affects the entire signal.

Artifacts: Are specific, identifiable, and often non-random disturbances superimposed on the signal of interest. Artifacts are usually related to specific events or conditions that contaminate the measured data.

- **Sources:**

Noise:

External noise can include electromagnetic interference from other devices, thermal noise from electronic components, and ambient environmental noise.

Internal noise may arise from the electronic equipment used in the recording process itself or from intrinsic physiological variability not related to the brain activity being monitored (e.g., thermal fluctuations of the skin).

Artifacts:

Physiological artifacts arise from the subject's own actions or conditions, such as blinking (creating electrooculographic artifacts), muscle movements (producing electromyographic artifacts), or cardiac cycles (generating electrocardiographic artifacts).

Environmental artifacts can be caused by movement of the recording apparatus, electrical interference from nearby equipment, or interactions with the recording environment.

Impact on Data:

Noise generally leads to a reduction in the signal-to-noise ratio, making it harder to discern the true signal amidst the random variations. It affects the overall quality and clarity of the data. Artifacts, while they may also reduce the signal-to-noise ratio, are particularly problematic because they can mimic or obscure the true signal, leading to misinterpretations unless properly identified and removed.

Management Strategies:

Noise is typically addressed through the use of better shielding and grounding of equipment, using high-quality, low-noise components, and applying digital filtering techniques to isolate and reduce the noise components without distorting the signal.

Artifacts require specific strategies for identification and removal, such as using algorithms to detect the artifact patterns (like independent component analysis for eye-blink artifacts in EEG data) and temporal or spatial filters to exclude the artifact from the data analysis.

Understanding the differences between noise and artifacts is essential for anyone working with signal processing in neuroscience. Effective strategies to mitigate both are crucial to ensure that the data collected is as accurate and reflective of the true brain activity as possible.

Box 4.2

Signal-to-Noise Ratio (SNR) is a measure used in science and engineering that compares the level of a desired signal to the level of background noise. SNR is typically expressed in decibels (dB) and calculated as the ratio of the power of the signal to the power of the noise. A higher SNR indicates that the signal has less noise and is generally clearer and more detectable. In the context of brain signal analysis, such as with EEG, fMRI, or other neuroimaging techniques, a high SNR is crucial for accurately interpreting the data. It ensures that the true signal, which represents the brain activity, stands out distinctly from the noise, which can include electronic interference, ambient environmental sounds, and physiological signals unrelated to the brain activities being studied. Enhancing the SNR is a primary focus during the preprocessing stages of data analysis, as it greatly affects the reliability and validity of the results obtained from neuroscientific experiments.

4.3 Preprocessing of Brain Signals

Preprocessing is a fundamental step in brain signal analysis that significantly enhances the quality and reliability of data derived from various brain imaging and electrophysiological recording techniques. Raw brain signals are frequently contaminated by various noise sources and artifacts described above. These artifacts can distort the underlying neural activity associated with cognitive processes or clinical conditions. To combat these issues, preprocessing employs several techniques to refine the data quality. In this section, you will learn some preprocessing steps often used regularly for all brain signals.

4.3.1 Normalization and Standardization

Normalization and standardization are crucial preprocessing steps in analyzing brain signal data, helping to ensure that variations in recording conditions or individual differences among subjects do not skew results. This process is vital for comparative studies across different subjects or sessions, ensuring that variability in the measurements reflects actual differences in brain activity rather than differences in individual anatomy or electrode placement (Laird et al., 2010). Both processes aim to bring different datasets to a common scale, allowing for more accurate comparisons and analyses.

4.3.1.1 Normalization

Normalization involves adjusting the values in a dataset to a common scale without distorting differences in the ranges of values or losing information. In the context of brain imaging, normalization is often used to adjust the brain volumes to a standard size and shape, which is essential for comparing data across subjects in a meaningful way. Two of the most widely used techniques are:

- **Feature Scaling**: This method is often used in machine learning applications on brain signal data, such as EEG or MEG. It involves scaling the data to a [0, 1] range or a [−1, 1] range, making it easier to manage for algorithms that are sensitive to large variations in input values (Han et al., 2022).
- **Z-Score Normalization** (Standardization): This method involves re-scaling data with a mean of zero and a standard deviation of one. This normalization form is particularly useful in statistical analyses where the assumption of normality is required.

4.3.1.2 Standardization

Standardization refers to the process of making data points more consistent with typical values, removing biases that can occur due to differing measurement techniques or experimental setups. It is critical in EEG studies where electrode placements or different hardware setups can result in variations in signal amplitudes. Two of the most widely used techniques are:

- **Signal standardization**: In EEG or MEG analysis, standardizing signal amplitude across different recording sessions or subjects allows for comparing physiological responses under various experimental conditions (Cohen, 2014).
- **Spatial standardization in fMRI**: This involves transforming all brain images to a common brain template (often the MNI template) so that specific brain regions align across all subjects being studied. This is crucial for voxel-wise comparisons across a group (Ashburner & Friston, 1999).

Normalization and standardization are essential for:

- **Reducing Bias**: Ensuring that the data does not favor one measurement scale over another or one individual's data over another's.
- **Improving Accuracy**: Enhancing the reliability of conclusions drawn from the data by ensuring that observed differences in brain activity are due to the phenomena under study rather than artifacts of the measurement process.
- **Facilitating Comparisons**: Making it possible to conduct meaningful comparisons across different studies or different groups of subjects, which is fundamental in collaborative research and meta-analyses.

Both normalization and standardization are pivotal in the preprocessing steps for data from neuroimaging and electrophysiological studies. They ensure that subsequent analyses, whether computational modeling, statistical inference, or machine learning applications, are based on clean and comparable datasets.

4.3.2 Baseline Correction

Baseline correction is a critical preprocessing step in analyzing brain signal data. Baseline correction involves subtracting a pre-stimulus baseline value from all subsequent data points within a trial. This value represents the average activity during a control period, typically just before a stimulus is presented. The primary purpose of baseline correction is to minimize the impact of non-stimulus-related variability on the measured response, ensuring that any observed changes are attributable to the stimulus rather than background noise or ongoing processes. Two of the most widely used baseline correction techniques are:

- **Mean Subtraction**: It is the most common method, where the mean signal value during a baseline period (e.g., -200 to 0 ms relative to stimulus onset) is calculated and subtracted from all data points in the trial (Cohen, 2014).
- **Linear Detrending**: This method involves fitting a linear trend to the baseline period and then subtracting this trend from the entire signal segment. This can be particularly useful in dealing with slow drifts in the signal that might not be completely accounted for by mean subtraction alone (Cohen, 2014).

4.3.3 Filtering

Filtering is a fundamental preprocessing step in analyzing brain signal data, crucial for enhancing the clarity and interpretability of the signals. The primary goal of filtering is to remove unwanted components or noise from the data while preserving the relevant signals of interest to the study's objectives. Some of the commonly used filtering methods are:

4.3.3.1 Frequency Filtering

Four different types of frequency filters are commonly used depending on the signal of interest.

High-Pass Filters: These allow frequencies higher than a specific cutoff frequency to pass through while attenuating frequencies below the cutoff. High-pass filtering is often used in EEG to remove slow-wave artifacts such as those caused by breathing or slow drifts in the signal (Cohen, 2014).

Low-Pass Filters: These allow frequencies below a specific cutoff to pass and attenuate frequencies above this threshold. Low-pass filters help remove high-frequency noise from sources such as EMG or electronic equipment (Cohen, 2014).

Band-Pass Filters: These filters are a combination of high-pass and low-pass filters and are used to isolate a specific range of frequencies. Band-pass filtering is common in the analysis of specific EEG frequency bands like alpha (8–12 Hz), beta (13–30 Hz), and gamma (30–80 Hz) that are associated with different cognitive and neural processes (Niedermeyer & da Silva, 2005).

Notch Filters: These are specifically designed to remove a very narrow frequency band and are most commonly used to eliminate power line noise (50/60 Hz) (Niedermeyer & da Silva, 2005). An example of the effect of notch filter on EEG signal is shown in ◘ Fig. 4.8 (Brienza et al., 2019).

These four filter types can be implemented as finite impulse response (FIR) filters or infinite impulse response (IIR) filters. Some of the most commonly used IIR filters are Butterworth and Chebyshev filters.

The frequency filtering method may not be effective when the spectral distributions of artifacts overlap with those of the brain signal components. If these frequency ranges coincide, filtering alone might not sufficiently separate the desired neural signals from unwanted noise. In such scenarios, it becomes necessary to employ alternative artifact removal techniques that can more accurately differentiate and eliminate these overlapping artifacts without compromising the integrity of the brain signals.

4.3.3.2 Adaptive Filtering

Adaptive filtering operates on the principle that the signal of interest and the artifact are uncorrelated. This type of filter utilizes a reference signal to generate an estimate that correlates with the artifact. This estimated artifact is then subtracted from the original signal to produce a cleaner output. One common method used in adaptive filtering is the least mean squares (LMS) algorithm, which linearly con-

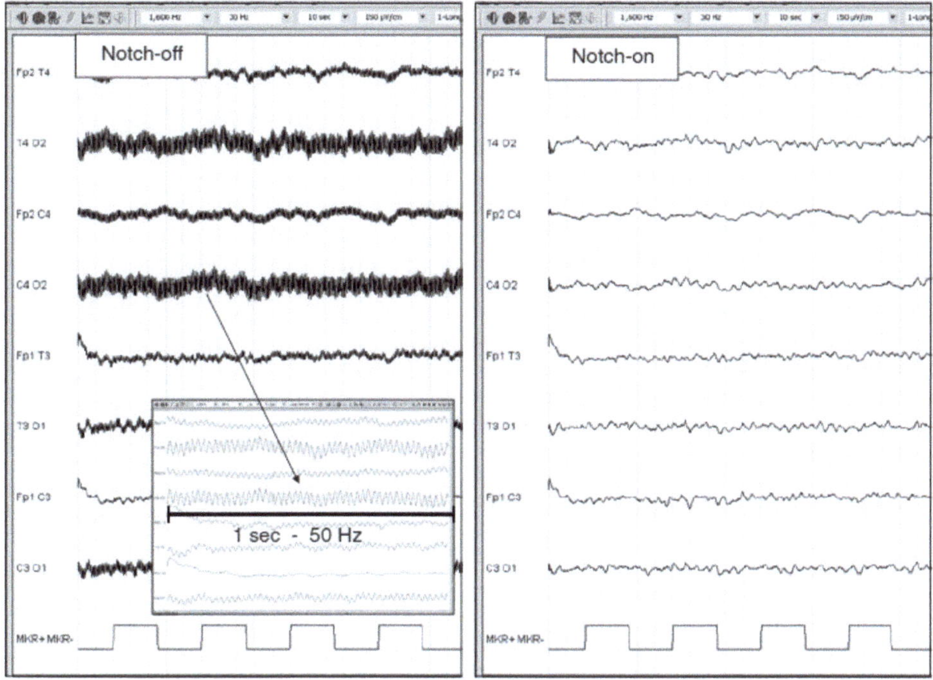

◘ Fig. 4.8 Depiction of the effect of a notch filter in removing the power line noise. (From Brienza et al., 2019)

verges by adjusting the weight parameters to optimize signal clarity (Haykin, 2002).

Additionally, the recursive least squares (RLS) algorithm serves as a more advanced option. It offers quadratic convergence, potentially achieving faster stabilization compared to the LMS algorithm, albeit at a higher computational cost (Haykin, 2002). However, a notable drawback of adaptive filtering is its requirement for additional sensors to provide the necessary reference inputs, increasing the complexity and potentially the cost of the overall setup (Urigüen & Garcia-Zapirain, 2015).

4.3.3.3 Wiener Filtering

Wiener filtering is a statistical method designed to reduce the mean square error between the desired and estimated signals by creating a linear time-invariant filter. This technique, cited in foundational texts on signal processing, optimizes the filtering process based on the power spectral densities of both the signal of interest and any noise or artifact present in the data (Sweeney et al., 2012). Unlike adaptive filtering, Wiener filtering does not necessitate an additional reference signal because it directly estimates the spectral characteristics of the signals involved. However, this involves calculating cross-power spectral density between the desired and noisy signals, as well as the power spectral density of the noise, which can make the computational process intricate and resource-intensive.

These preprocessing steps are critical for isolating real brain signals from unwanted noise and artifacts, thereby enhancing the signal-to-noise ratio. This clarification of neural patterns not only facilitates more accurate data analysis but also standardizes the data, preparing it for subsequent statistical analyses and ensuring consistency across studies or diagnostic tests.

Alongside the basic preprocessing techniques, several methods including principal component analysis (PCA), independent component analysis (ICA), wavelet transform, regression analysis, and canonical correlation analysis (CCA) are utilized to filter out artifacts from brain signals. These methods will be discussed as we deep dive into different brain signals. In the remainder of this chapter, we will explore various brain signals and examine how preprocessing these raw signals results in refined data that accurately represents the activity of specific brain regions under investigation.

Box 4.3   Key Points: Preprocessing of Brain Signals

Importance of Preprocessing: Effective preprocessing is crucial in brain signal analysis as it enhances the quality of the data before further analysis, ensuring that subsequent interpretations and findings are based on reliable and accurate information.

■ **Main Objectives:**

Noise Reduction: Minimizing external and physiological noise to improve the signal-to-noise ratio, making the true brain signal more distinguishable.

Artifact Removal: Identifying and eliminating artifacts that can obscure or mimic neural activity, such as those caused by eye movements, muscle tension, or electronic interference.

Signal Enhancement: Applying filters and mathematical transformations to clarify and emphasize important features of the brain signal.

■ **Common Techniques:**

Filtering: Both high-pass and low-pass filters are used to keep frequencies that carry relevant information and discard those that are likely to contain noise.

Normalization: Adjusting the amplitude of the signal to a common scale to facilitate comparison across different recording sessions or subjects.

Detrending: Removing linear or nonlinear trends from the data, which might be due to slow shifts in sensor calibration or subject movement.

Outcome of Preprocessing: The result of these efforts is a cleaner; more standardized dataset that is ready for analysis, which can significantly increase the accuracy of diagnosing neurological conditions, monitoring therapy, and researching brain function.

4.4 Preprocessing of EEG Signal

As demonstrated by the figures above, EEG signals captured from the scalp can be contaminated by various physiological and non-physiological noises, making noise reduction crucial. Various denoising methods are employed to clean EEG signals, and the choice of method depends on the signal quality, the environment of signal acquisition, and the specific objectives of the EEG study.

4.4.1 Regression Method

The traditional method for removing eye artifacts from EEG signals involves a regression-based analysis approach. This process requires recording an electro-oculogram (EOG) concurrently with the EEG to obtain coefficients for various noise sources, such as blinking artifacts and eye movement artifacts. By computing the regression correlation between the EEG signal X and the artifact signal Y, this method can effectively eliminate the artifact from the EEG (Sheoran et al., 2015). The general procedure is illustrated in ▶ Algorithm 4.1 (Chaddad et al., 2023).

■ **Algorithm 4.1 Regression-Based Denoising of EEG Signals**

```
Input: EEG signal X, artifact signal Y
Output: Clean EEG signal Z
function REGRESSION(X,Y)
Calculate regression coefficients between X and Y
Remove artifact from EEG signal
return Clean EEG signal Z
end function
```

However, a significant concern with this approach is the bidirectional contamination where EOG recordings may capture neural potentials along with ocular potentials, necessitating the subtraction of some EEG data, which could compromise the EEG signal integrity (Croft & Barry, 2000; Ranjan et al., 2021). Additionally, regression methods struggle with other types of artifacts like EMG due to the absence of clear reference channels (Stachaczyk et al., 2020). With the development of more efficient algorithms such as principal component analysis (PCA) and independent component analysis (ICA) (Kim & Lee, 2022; Noorbasha & Sudha, 2021), the use of regression for artifact removal in EEG has become less common, especially for artifacts related to EOG or ECG signals (Chaddad et al., 2023).

4.4.2 Blind Source Separation

Blind source separation (BSS) is a method used to separate source signals from mixed signals without prior information about the sources. Initially, the observed EEG is decomposed into its component sources using BSS. Subsequently, noise sources are identified and removed, retaining only the brain activity components (Stergiadis et al., 2022). BSS methods such as PCA and ICA are widely utilized for denoising EEG signals.

4.4.2.1 **Principal Component Analysis**

Principal component analysis (PCA) is a data reduction technique that leverages the principle of orthogonality to eliminate artifacts (Bro & Smilde, 2014). By applying PCA for dimensionality reduction, noise components, which are typically associated with smaller eigenvalues, can be effectively minimized within the dataset, thus achieving a partial denoising effect. For a given set of EEG data X, PCA primarily aims to solve Eq. (4.1):

$$XX^T w_i = \lambda_i w_i \tag{4.1}$$

In this equation, λ represents the eigenvalue and w denotes the eigenvector. The process involves decomposing the eigenvalues of the matrix XX^T. These eigenvalues are then ordered, and the top d values are selected to form a projection matrix. This projection matrix is used to transform the original EEG data D into a new dataset $D^* = W^*TD$, where WT is the transpose of the matrix formed by the selected eigenvectors, thereby minimizing noise. The pseudocode for implementing PCA is detailed in ▶ Algorithm 4.2 (Chaddad et al., 2023). Pictorial depiction of PCA method is shown in ▣ Fig. 4.9 (Greenacre et al., 2022).

■ **Algorithm 4.2 Typical Principal Component Analysis**

```
Input: EEG data D = fx1, x2, ..., xng, low-dimensional space
dimension d.
Output: Projection matrix W* = (w1,w2, ...,wd).
procedure PCA(D):
Sample centering x_i ← x_i - (1/m) ∑^m_{i=1} x_i
Calculate XX^T.
Eigenvalue decomposition for XX^T.
Select the largest d eigenvalues.
W* = (w1,w2, ...,wd).
New EEG data D*=W*^TD
Return D*
```

4.4.2.2 **Independent Component Analysis**

Independent component analysis (ICA) is a prominent blind source separation (BSS) technique widely used in biomedical engineering to extract statistically independent sources from a set of mixed signals (Mijović et al., 2010), such as EEG data. The fundamental ICA model for denoising EEG can be represented by Eq. (4.2) (Onton et al., 2006).

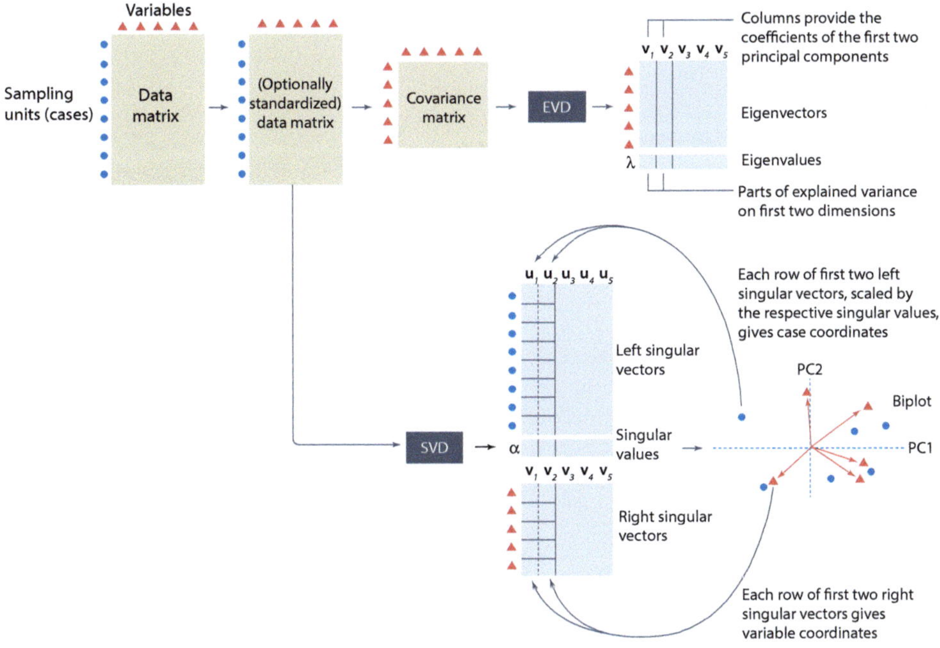

■ Fig. 4.9 The principal components (PCs) are derived using the eigenvalue decomposition (EVD) of the covariance matrix of the data variables. While standardization of the data is optional, centering the data is essential. If the variables are scaled by their standard deviations, the covariance matrix becomes a correlation matrix, and the analysis is then sometimes referred to as correlation principal component analysis (PCA). Alternatively, a more efficient approach involves using the singular value decomposition (SVD) to directly determine the positions and vectors of the variables in a joint representation, where the eigenvectors are equivalent to the right singular vectors of the decomposition. For the SVD approach to match the results of the EVD exactly, the (optionally standardized) data matrix should be divided by the square root of n, where n is the number of cases or rows in the dataset. This adjustment ensures that both methodologies yield equivalent outcomes. (From Greenacre et al., 2022)

$$X = AS \tag{4.2}$$

where X is the matrix containing EEG data, A represents the mixing of various sources (e.g., brain activity and artifacts), and S comprises the independent components like brain signals and noise sources. After isolating these components, visual inspections are conducted to identify artifacts such as eye blinks or muscle activity, which are then removed to clean the EEG data. ▶ Algorithm 4.3 outlines the process of denoising EEG signals using the ICA method (Chaddad et al., 2023).

■   **Algorithm 4.3 ICA Based Denoising of EEG Signals**

```
Input: X: EEG data matrix
Input: n_components: number of independent components to estimate
Output: S: matrix of independent components
Output: A: estimated demixing matrix
Center and whiten the X.
Initialize A randomly.
repeat
Update A by exploiting non-Gaussianity of independent sources.
until convergence
Compute S from A and X.
Identify artifact components in S.
Remove artifact components from S.
Reconstruct cleaned data from S.
return S, A
```

However, ICA faces challenges with EMG artifacts due to their significant overlap with EEG signals in spatial and temporal domains. To improve artifact separation, Li et al. developed a refined ICA model called ERASE (EMG Removal by Adding Sources of EMG), which incorporates EMG references from head and neck muscles to enhance the artifact's representation in the independent components, thereby achieving more effective separation (Li et al., 2021). This method has been shown to remove an average of 26% more EMG artifacts than traditional ICA approaches.

4.4.3 Canonical Correlation Analysis

In the analysis of EEG signals that are contaminated with muscle artifacts, canonical correlation analysis (CCA) has been shown to be generally more effective than independent component analysis (ICA) (Sheoran & Saini, 2020). This is largely due to the characteristic lower autocorrelation of muscle artifacts compared to neural activity. Muscle artifacts tend to vary more rapidly and less predictably over time, making them less autocorrelated. In contrast, brain activity exhibits higher levels of autocorrelation due to the more structured and predictable nature of neural oscillations. CCA leverages this difference by analyzing the multivariate relationships between datasets, allowing for the effective separation of muscle and brain activity signals (De Clercq et al., 2006). ▶ Algorithm 4.4 outlines the process of denoising EEG signals using the CCA method (Chaddad et al., 2023).

- **Algorithm 4.4 CCA Based Denoising of EEG Signal (Hassan et al., 2011)**

```
Input: X: EEG data matrix
Input: Y: matrix of auxiliary variables (e.g., EOG or ECG data)
Output: Z: matrix of cleaned EEG data
Center and whiten the X.
Initialize the weight matrices A and B randomly.
repeat
Compute the canonical weights w_a by maximizing the correlation
between X and Y
with respect to A. . w_a: weights used to linearly combine the
EEG signal for one
component
Compute the canonical weights w_b by maximizing the correlation
between X and Y
with respect to B. . w_b: weights used to linearly combine the
auxiliary signal for one
component
Update the weight matrices A and B.
until convergence
Compute the cleaned data as Z = A^T X.
return Z
```

This method focuses on maximizing the correlation between transformed variables from two sets of EEG data, typically isolating the components that best represent the underlying brain activity while minimizing the impact of muscle-generated noise (Mert & Akan, 2014). This ability makes CCA a valuable tool in EEG signal preprocessing, especially in scenarios where distinguishing between these sources of activity is critical for accurate data interpretation.

4.4.4 Wavelet Transform

EEG signals, which often contain irregularities, are typically nonstationary, making the wavelet transform (WT) a preferred method for their analysis (Ranjan et al., 2022; Yan et al., 2022). Using WT, EEG signals are decomposed into wavelet components, from which those containing artifacts are identified and removed. The remaining clean components are then reassembled to reconstruct a cleaned signal (Mowla et al., 2015).

Wavelet transform is categorized into two types: discrete wavelet transform (DWT) and continuous wavelet transform (CWT) (Khatun et al., 2016). For EEG signals, which are inherently continuous, DWT is particularly relevant. DWT efficiently transforms EEG signals in the time domain by applying a series of low-pass and high-pass filters. This filtering process generates approximate and detailed coefficients, iteratively refined to isolate the desired frequency components for effective artifact removal. ▶ Algorithm 4.5 outlines the process of denoising EEG signals using the DWT method (Chaddad et al., 2023).

■ **Algorithm 4.5 DWT Based Denoising of EEG Signal (Aqil et al., 2017)**

```
Input: X: EEG data matrix (rows represent the EEG channels)
Output: Y: matrix of cleaned data
Set the wavelet basis and level of decomposition
for each channel c in X do
Compute the DWT coefficients of c at each level using the fatigue
wavelet basis.
Identify the approximation coefficients at the desired level as
the artifact-free signal.
Threshold the detail coefficients using a soft or hard threshold-
ing technique.
Reconstruct the cleaned signal by inverse DWT using the modified
coefficients.
Store the cleaned signal in the corresponding row of Y.
end for
return Y
```

While Wavelet Transform (WT) is a valuable tool for EEG signal denoising, it may not be sufficient on its own due to potential information loss and challenges in signal reconstruction. As a result, integrating WT with other techniques has been explored to enhance the denoising process. For example, one of the approach uses WT to decompose the EEG signal, followed by ICA to separate and remove EMG noise and ECG artifacts, showing significant improvements in signal clarity (Yan et al., 2022).

Additionally, to address overlapping spectra or frequencies between EEG signals and artifacts, notch filters can be used alongside WT. In specific studies like (Khatun et al., 2016), an adaptive thresholding technique for wavelet coefficients was employed to effectively eliminate frequent ocular artifacts. Moreover, a 50 Hz IIR notch filter was utilized to further cleanse the signal from noise and artifacts, ensuring the preservation of the original brain signals. This combination of methods provides a more robust solution for refining EEG data, minimizing the loss of valuable neural information.

The neuroscience community has access to various tools that cover most pre-processing steps previously discussed. These tools are included in widely used EEG signal processing software packages like EEGLAB (Delorme & Makeig, 2004), FieldTrip (Oostenveld et al., 2011), Brainstorm (Tadel et al., 2011), MNE-Python (Gramfort et al., 2013), Cartool (Brunet et al., 2011) and eLORETA (Pascual-Marqui, 2002), see ◘ Table 4.1 for details.

4.4.5 Preprocessing Effects

As previously mentioned, not all the preprocessing methods described above need to be applied to raw EEG signals. The selection and application of these methods depend on several factors, including the environment in which the data was acquired, the quality of the signals, and the specific objectives of the data acquisition. Factors such as the presence of specific types of noise or artifacts, the sensitivity required for the study, and computational resources also play a crucial role in

Table 4.1 Overview of open-source neuroimaging software packages and their key features

Neuroimaging package	Key features	Supported data types	Supported platforms	Programming language
FSL (FMRIB Software Library)	Comprehensive suite for analysis of fMRI, MRI, and DTI data. Offers tools for preprocessing, GLM analysis, brain extraction, and ICA	fMRI, MRI, DTI	Windows, macOS, Linux	C++, Python
SPM (Statistical Parametric Mapping)	Specialized in statistical analysis of brain imaging data, includes preprocessing, segmentation, normalization, and statistical analysis (GLM). Focuses on PET and fMRI	fMRI, PET, MRI	Windows, macOS, Linux	MATLAB
AFNI (Analysis of Functional NeuroImages)	Primarily focused on fMRI analysis with a strong emphasis on statistical modeling and connectivity analysis. Provides tools for motion correction, ICA, and surface mapping	fMRI, MRI, DTI	Linux, macOS, Windows (via WSL)	C, Python
BrainVoyager	Comprehensive software for neuroimaging analysis. Includes fMRI, DTI, and anatomical data processing. Offers tools for statistical analysis, real-time fMRI, and visualization	fMRI, DTI, MRI	Windows, macOS, Linux	C++, Python
FreeSurfer	Primarily focused on cortical surface reconstruction and analysis. Features include skull stripping, segmentation, cortical thickness analysis, and morphometry	MRI (structural), fMRI	Windows, macOS, Linux	C++, Python
ANTS (Advanced Normalization Tools)	Highly specialized in image registration and normalization. Known for accurate anatomical image registration, used for cortical thickness and morphometry studies	MRI, DTI	Linux, macOS	C++
NiPy (Neuroimaging in Python)	Python-based library for analyzing neuroimaging data. Offers tools for statistical analysis, image registration, and time-series analysis. Strong integration with other Python libraries	fMRI, MRI	Windows, macOS, Linux	Python
Nilearn	Python library focusing on statistical learning and machine learning applied to neuroimaging. Offers simplified fMRI and MRI data analysis and visualization	fMRI, MRI	Windows, macOS, Linux	Python
EEGLAB	Specialized for EEG and MEG data analysis. Provides tools for preprocessing, independent component analysis (ICA), and time-frequency analysis	EEG, MEG	Windows, macOS, Linux	MATLAB

(continued)

Table 4.1 (continued)

Neuroimaging package	Key features	Supported data types	Supported platforms	Programming language
FieldTrip	Focuses on EEG, MEG, and invasive electrophysiological data. Includes time-frequency analysis, source reconstruction, and statistical analysis	EEG, MEG, ECoG	Windows, macOS, Linux	MATLAB
MNE (Magnetoencephalography and Electroencephalography)	Designed for the analysis of EEG and MEG data, including preprocessing, source localization, time-frequency analysis, and inverse problem solving	EEG, MEG, ECoG, fNIRS	Windows, macOS, Linux	Python, C
CBRAIN	Web-based platform for distributed processing of neuroimaging data. Provides an interface for running neuroimaging analysis tools like FSL, FreeSurfer, and SPM on high-performance computing resources	fMRI, MRI, DTI, EEG	Windows, macOS, Linux (Web-based)	Web technologies (Python, JS)
PySurfer	Built for visualization of FreeSurfer outputs in Python. Primarily used for surface-based data visualization such as cortical maps and activation overlays	MRI, fMRI	Windows, macOS, Linux	Python
NeuroDebian	Repository that provides pre-built packages of major neuroimaging software (FSL, SPM, FreeSurfer) for easy installation on Debian-based Linux distributions	fMRI, MRI, EEG, MEG, DTI	Linux (Debian, Ubuntu)	Various (MATLAB, Python, C++)

This table provides a comprehensive overview of widely used open-source neuroimaging software packages, each tailored to specific types of brain imaging and electrophysiological data. The packages listed here are fundamental tools for researchers in neuroscience, enabling data processing, statistical analysis, and visualization across different modalities, such as functional MRI (fMRI), structural MRI, diffusion tensor imaging (DTI), electroencephalography (EEG), magnetoencephalography (MEG), and electrocorticography (ECoG). For each package, the key features are described, highlighting its strengths in neuroimaging data acquisition, preprocessing, statistical modeling, and visualization. The types of neuroimaging data supported by each tool are also specified, ranging from anatomical images to electrophysiological signals, offering flexibility across various research applications. Additionally, the table notes the supported operating systems and programming languages, indicating the compatibility of each software with different research environments. Packages such as **FSL** and **SPM** are known for their extensive toolkits for fMRI and structural MRI data, while **FreeSurfer** and **ANTS** excel in surface-based analysis and image registration, respectively. **EEGLAB** and **FieldTrip** are specialized for EEG and MEG data, providing tools for source localization and time-frequency analysis. Python-based tools like **NiPy** and **Nilearn** integrate advanced machine learning and statistical tools, reflecting the growing importance of Python in neuroimaging analysis. The table also highlights the role of web-based platforms like **CBRAIN**, which facilitate high-performance computing for large-scale neuroimaging studies

determining the appropriate preprocessing steps. Next, we will explore few examples to illustrate the effects of different preprocessing techniques on raw EEG signals, demonstrating how each method can enhance the signal quality and suitability for further analysis. This discussion will help elucidate the practical impacts of these techniques and guide the selection of appropriate methods based on the specific requirements of EEG studies.

4.4.5.1 Example 1

In this example, the authors (Do et al., 2022) employed baseline correction and bandpass filtering as standard EEG preprocessing methods to preserve valuable information. Initially, all original EEG recordings underwent bandpass filtering with cutoff frequencies set at 0.5 and 45 Hz using a one-dimensional IIR or FIR filter. Subsequently, baseline correction was implemented by subtracting the average value of the bandpass-filtered data from each data point, effectively removing any baseline drift. This approach ensures that the processed EEG data is free from low-frequency drifts and high-frequency noise, thus enhancing the clarity and usability of the recordings for further analysis (Do et al., 2022). The effect of this preprocessing pipeline on the raw EEG is shown in ◘ Fig. 4.10. The authors fur-

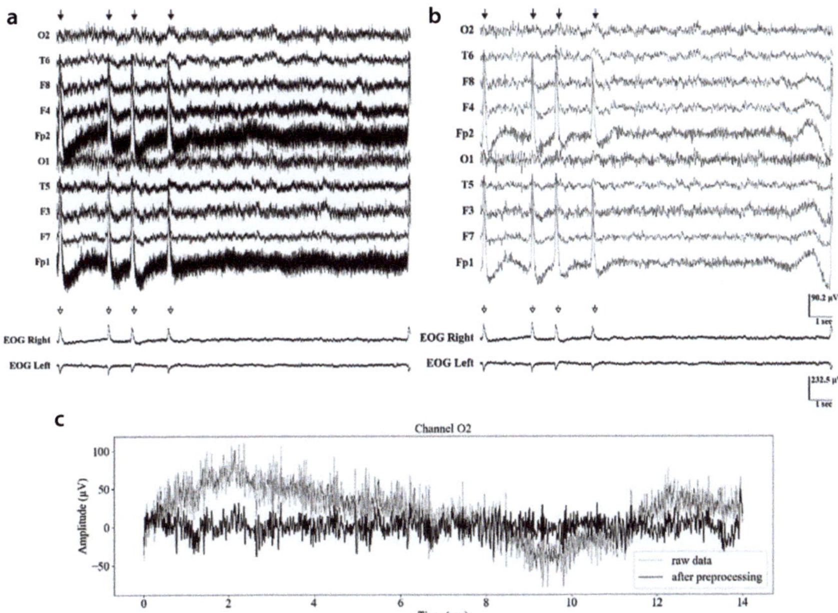

◘ **Fig. 4.10** Preprocessing of EEG signals includes the following examples: **a** A 15-s segment of raw EEG data shown at the top, with corresponding EOG data displayed at the bottom. **b** The same EEG data segment from figure **a** after undergoing band-pass filtering and baseline correction. Black arrows indicate EOG peaks in the EEG signal, while white arrows highlight EOG peaks in the EOG reference channel. **c** Another 15-s segment of raw EEG data from channel O_2, after it has been processed with band-pass filtering and baseline correction. The bandpass filter effectively reduced noise, as evidenced by the diminished thickness of the data line, particularly in channels Fp1 and Fp2 **a**, **b**. Additionally, baseline drifts were eliminated from the data lines following the application of baseline correction **c**. (From Do et al., 2022)

ther applied ICA methods to detect and remove EOG signal from the processed EEG signal as shown in ◘ Fig. 4.11 (Do et al., 2022).

4.4.5.2 **Example 2**

The EEG signals in this example were recorded at a sampling rate of 250 Hz (Rahman et al., 2020). The raw EEG signals underwent preprocessing with a band-pass filter that limited the frequency range to between 0.5 and 100 Hz, effectively reducing low-frequency drift and high-frequency noise. The sensitivity setting for

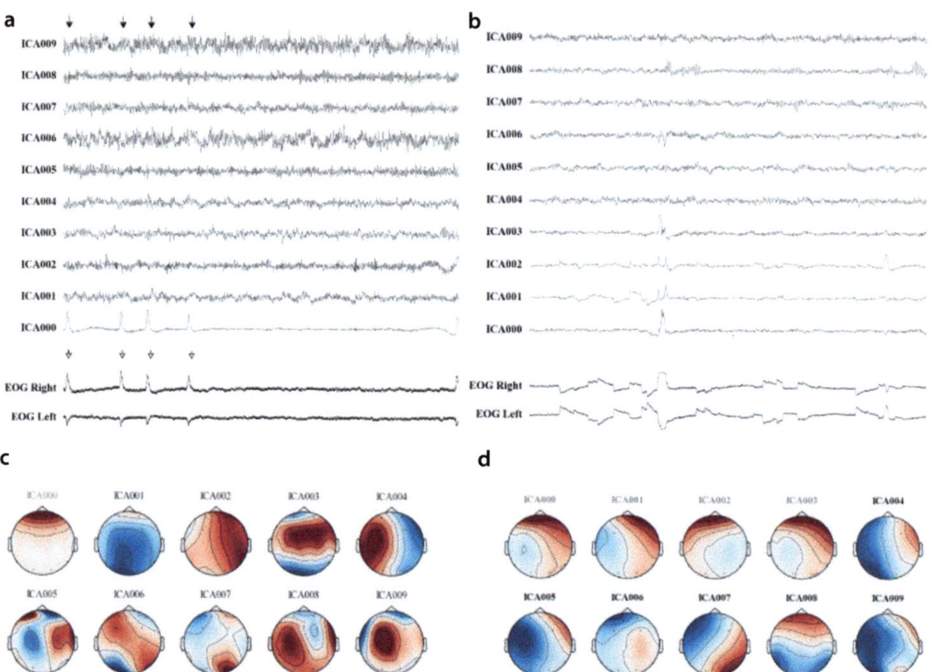

◘ **Fig. 4.11**   Comparison of various ICA-related features. Panels **a** and **b** display signals from independent components (ICs) from an instance where ICA effectively isolated EOG artifacts from the EEG signal **a** and an instance where it did not **b**. EOG reference channels are depicted at the bottom of these panels. Black arrows indicate EOG peaks in the IC signal, while white arrows point to EOG peaks in the EOG reference channels. Panels **c** and **d** show the topography maps associated with the ICs in panels **a** and **b**, respectively. To apply independent component analysis (ICA), the authors segmented the preprocessed EEG signal, shown in ◘ Fig. 4.9, into 15-s chunks. For each chunk, ICA was performed to generate a matrix consisting of 10 independent components (ICs) time series (s) and a mixing matrix (A). Upon examining the topography maps and the ICs, those representing ocular activity were labeled "1," while all other ICs were labeled "0." It was noted that ICA did not consistently succeed in isolating EOG artifacts from the EEG signals. In cases where ICA was successful, the ICs were distinctly separable **a**, **c**. For instance, in one successful scenario, one specific IC (ICA000) exhibited a waveform similar to EOG artifacts, as confirmed by comparison with EOG reference channels **a**. The topography of this particular IC showed activity predominantly in the frontal lobe area **c**, aligning with expectations for electrical activity driven by eye movements. (From Do et al., 2022)

the EEG amplifier was configured to 100 mV, ensuring appropriate signal amplification without distortion.

To further enhance the signal quality and reduce potential interference from electrical sources, a 50-Hz notch filter was specifically applied to remove line noise that is commonly associated with electrical power supplies. This step is crucial for minimizing artifacts in the EEG recordings, particularly in regions with a standard power line frequency of 50 Hz. The effect of these preprocessing steps are shown in ◘ Fig. 4.12 (Rahman et al., 2020).

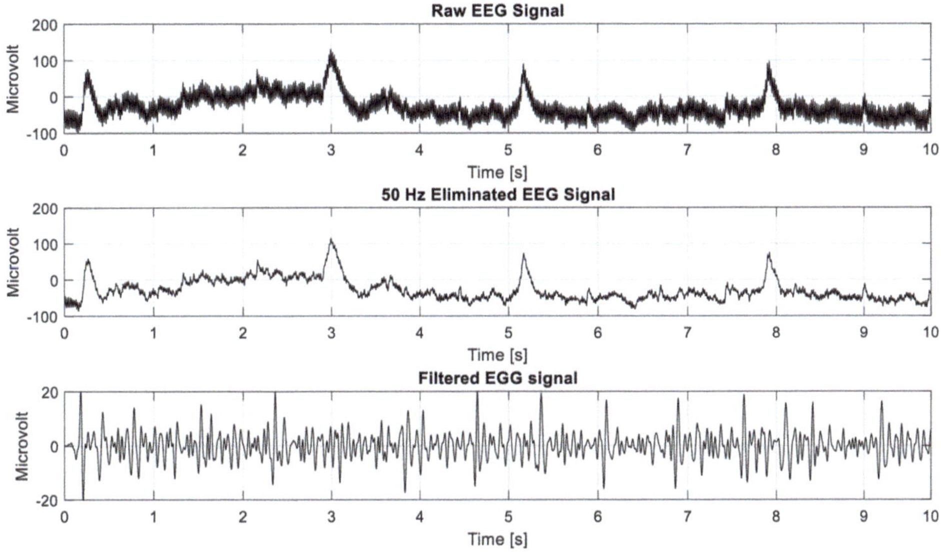

◘ **Fig. 4.12** The original EEG signal and its subsequent alterations through stepwise filtering demonstrate the impact of multiple preprocessing techniques tailored to enhance signal clarity. Initially, the raw EEG signal underwent processing with a bandpass filter, which restricts the frequency range between specific lower and upper bounds, effectively isolating the relevant brain activity frequencies while excluding both high-frequency noise and low-frequency drift. Following the bandpass filter, a notch filter was applied, specifically targeting and removing the 50 Hz line noise, a common interference caused by electrical power sources, which is critical for avoiding periodic electrical artifacts. After filtering, baseline correction was performed as the final step in this sequence. This involves adjusting the signal to correct for any baseline drift or offset, ensuring that the EEG readings start from a zero baseline. This step is crucial for standardizing the signal across time, making it easier to compare changes in EEG activity over the course of the recording without being misled by shifts in the signal's baseline level. This combination of bandpass filtering, notch filtering, and baseline correction collectively results in a clearer, more analyzable EEG signal that is free from both high and low-frequency noise and stable in terms of its baseline. (From Rahman et al., 2020 originally published under CC-BY 4.0)

> ### Box 4.4   Essentials of Preprocessing EEG Signals
>
> **Objective of EEG Preprocessing**: To refine the raw EEG data by reducing noise and artifacts, thus enhancing the quality of the signal for accurate analysis and interpretation. This is critical for both clinical diagnostics and research applications.
>
> minimize the impact of reference electrode noise.
>
> **Normalization**: Scaling EEG signals to a common statistical range to mitigate variability between sessions or subjects, enhancing comparative studies.
>
> ■ **Key Preprocessing Steps:**
>
> **Artifact Rejection**: Identifying and removing parts of the EEG data contaminated by eye blinks, muscle movements, or electrical interference. This is often done using techniques such as independent component analysis (ICA).
>
> **Filtering**: Applying band-pass filters to remove frequencies outside the typical brain activity range (1–100 Hz), thereby eliminating slow drifts and high-frequency noise.
>
> **Re-referencing**: Modifying the reference electrodes used in the EEG to improve the clarity of the signal and to
>
> ■ **Importance for Analysis:**
>
> Proper preprocessing is essential to ensure the reliability of EEG data, particularly when identifying characteristic brain wave patterns associated with different cognitive states or neurological disorders.
>
> ■ **Outcome:**
>
> Preprocessed EEG data provides a clearer, more consistent basis for further analysis, whether for identifying event-related potentials, assessing brain connectivity, or monitoring brain state changes over time.

4.5   Preprocessing of fMRI

As you read in ▶ Chaps. 1 and 2, functional MRI (fMRI) indirectly detects neural activity through hemodynamic responses to changes in oxygen consumption by neurons. However, the resulting time series data can be contaminated by various non-neural factors, such as head motion of the subjects, physiological cycles, and magnetic field inhomogeneities. These extraneous fluctuations, if not properly addressed, can obscure the underlying neural patterns, diminish the effectiveness of statistical analyses, or even distort experimental conclusions by introducing structured noise that mimics or masks genuine neural activity.

To address these challenges, a series of computational steps, often referred to as the preprocessing pipeline, has been developed to cleanse the fMRI data of these confounding variables, thereby enhancing the functional signal-to-noise ratio (fSNR) prior to more detailed analysis. The most commonly applied preprocessing steps in this pipeline are described below and depicted in ◫ Fig. 4.13 (Chen & Glover, 2015).

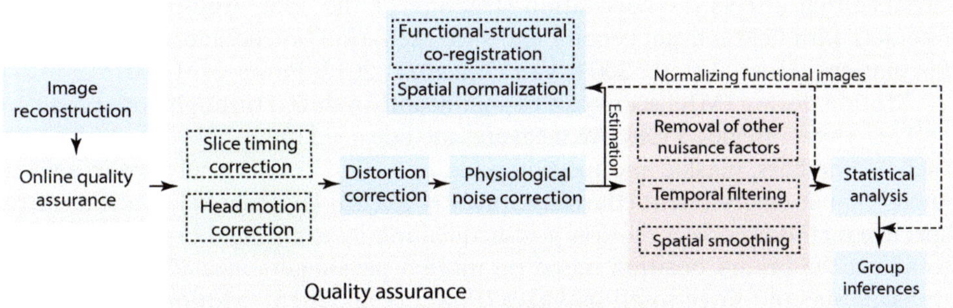

Fig. 4.13 fMRI preprocessing pipeline. (From Chen & Glover, 2015)

4.5.1 Quality Assurance

Quality assurance (QA) testing is essential to identify issues like scanner noise or signal drift that can corrupt data. Immediate post-scan reviews, such as checking motion parameters or viewing three-dimensional (3D) brain images, can catch and correct these issues early. Further, QA often continue throughout preprocessing, involving visual inspections and tests like analyzing slice intensity and calculating the functional signal-to-noise ratio (fSNR) to ensure data quality before further analysis (Murphy et al., 2009).

4.5.2 Slice Timing Correction

Most fMRI studies image brain slices sequentially, leading to variable acquisition times across slices within a single TR (relaxation time). If uncorrected, these slice-timing discrepancies can compromise studies requiring precise temporal resolution, such as those exploring causal relationships between brain regions or in rapid event-related designs. To address this, temporal interpolation is commonly used. This method aligns the signal amplitude of each slice to a consistent reference time by interpolating data from neighboring TRs. This technique is most effective when the slice sampling rate is much faster than the variability of the experimental signal (Huettel et al., 2004).

4.5.3 Head Motion Correction

Head motion is a significant issue in fMRI studies, especially those with long scan durations where subjects might become restless or drowsy, those involving tasks that require physical responses which can cause motion synchronized with stimuli, or studies involving specific groups such as children, the elderly, or individuals with diseases.

Subjects' head motion can degrade data quality by mixing signals from adjacent voxels, causing signal variability at cortical boundaries, and inducing spurious variance that affects the correlation structure of the data. Additionally, motion interacts with field inhomogeneity and slice excitation, introducing complex noise fluctuations (Huettel et al., 2004; Van Dijk et al., 2012; Power et al., 2015).

Motion during fMRI scans can be significantly reduced through both proactive and reactive strategies. Proactive measures include head immobilization techniques such as bite bars, masks, and fixation pads, as well as training subjects in a simulated scanner environment (Barnea-Goraly et al., 2014). Reactively, various retrospective methods correct motion post-acquisition by using rigid body realignment to adjust the brain's position based on motion parameters measured along three translational and three rotational axes (Power et al., 2015). Motion-related signal variance can also be mitigated by removing the motion parameters and their derivatives from the data or by realigning each brain volume to a fixed position through spatial interpolation (Friston et al., 1996; Satterthwaite et al., 2013; Yan et al., 2013). Additionally, problematic time points can be identified visually and excluded or adjusted using temporal interpolation.

Alternatively, prospective motion correction techniques adjust the slice plane in real-time during acquisition based on head position (Maclaren et al., 2013). These methods can be divided into those that use fMRI-derived motion information to adjust subsequent scans, suitable for slower movements, and more advanced approaches that employ external devices to track head orientation in real-time, offering improved correction but requiring extra setup.

4.5.4 Distortion Correction

fMRI signals can be adversely affected by geometric or intensity distortions arising from inhomogeneities in the static or excitation magnetic fields. These field variations can misshape and misplace tissue in the images, assuming a uniform linear relationship between signal frequency and spatial location. While hardware shimming in MR systems can partially compensate for these non-uniformities, additional techniques are employed to correct distortions. These methods include using a magnetic field map for additional data acquisition and applying image or k-space interpolation during the reconstruction process (Cusack et al., 2003; Sutton et al., 2003; Hutton et al., 2002; Jezzard, 2012).

4.5.5 Temporal Filtering

When the frequency distributions of signal and noise do not entirely overlap, temporal filtering can be effective in enhancing the functional signal-to-noise ratio (fSNR) of the data by removing noisy frequencies while preserving those of the signal. For example, in studies using a block-design paradigm where task-related signals are confined to narrow frequency bands, selectively suppressing non-task frequencies can significantly improve detection power. Additionally, high-pass filtering, often called detrending, is commonly used to eliminate slow scanner drift-induced fluctuations, routinely incorporated into most analysis software (Tanabe et al., 2002). This moderate temporal filtering not only boosts fSNR but also minimizes bias in subsequent statistical analyses by aligning the assumed models of the data more closely with their actual characteristics (Friston et al., 2000).

4.5.6 Spatial Smoothing

Spatial smoothing in fMRI data analysis offers three main benefits. First, it enhances the functional signal-to-noise ratio (fSNR) by suppressing noise among adjacent voxels, using a Gaussian kernel that matches the inherent spatial correlations of the data. This process improves both the temporal signal-to-noise ratio (tSNR) and the accuracy of statistical analyses by making the spatial structure of the data consistent with analytical models (Parrish et al., 2000; Tabelow et al., 2006).

Second, smoothing reduces anatomical or functional variations among subjects, enhancing group comparability. However, choosing the optimal kernel size is complex, influenced by the analysis goals: smaller kernels (around 4 mm) are recommended for single-subject studies, and larger kernels (6–8 mm) for group analyses. While larger kernels improve statistical robustness, they also reduce spatial resolution and may blur functional boundaries, hence the necessity to balance kernel size based on specific study requirements (Geissler et al., 2005; Sacchet & Knutson, 2013; White et al., 2001; Mikl et al., 2008; Alahmadi, 2021).

4.5.7 Physiological Noise Correction

BOLD contrast in fMRI is influenced by hemodynamic changes related to oxygenation, incorporating fluctuations from physiological processes like cardiac pulsatility and respiration (Birn, 2012). These physiological noises fall into two categories based on their spectral distributions:

4.5.7.1 Synchronized Fluctuations

Linked to cardiac (~0.8–1.3 Hz) and respiratory cycles (~0.1–0.3 Hz), these fluctuations arise from tissue movement and blood flow near large vessels for cardiac activity (Dagli et al., 1999), and chest movements affecting magnetic susceptibility for respiratory actions (Raj et al., 2001). While faster acquisition (e.g., TR <0.5 s) can mitigate these, longer TRs (≥2 s) used in many studies lead to aliasing of cardiac noise to lower frequencies. Various retrospective techniques model and eliminate this noise using both external recordings and direct data analysis (Hu et al., 1995; Glover et al., 2000; Pfeuffer et al., 2002; Verstynen & Deshpande, 2011).

4.5.7.2 Variability in Respiratory Volume and Heart Rate

Changes in breathing and heart rate affect arterial CO_2 levels, impacting blood flow and BOLD signal amplitude. These variations also contribute significant fluctuations in the BOLD signal during rest. Approaches to model these fluctuations include using external physiological data or directly from the fMRI data, similar to strategies for motion artifact removal (Wise et al., 2004; Birn et al., 2006, 2008a, b; Chang & Glover, 2009; Chang et al., 2009; Salimi-Khorshidi et al., 2014).

4.5.8 Functional-Structural Co-registration

The 3D stacks of anatomical and functional images often do not align due to differences in MR contrasts, slice orientations, voxel resolutions, and image distortions. This misalignment complicates the accurate mapping of functional activity (such as task activation maps) onto anatomical images. The process of aligning these functional and structural images is known as functional–structural co-registration. This technique typically involves resampling the anatomical data to match the spatial resolution of the functional data, followed by a rigid body transformation where a cost function (often mutual information) is minimized to achieve optimal alignment (Gholipour et al., 2007; Klein et al., 2009).

4.5.9 Spatial Normalization

In neuroscience studies involving multiple individuals, the variability in brain shapes and sizes necessitates standardizing each individual's brain to a common template. This standardization, known as spatial normalization, is crucial for aggregating brain activities accurately across subjects. Commonly used templates include the Talairach atlas (Talairach & Tournoux, 1988) and MNI templates, with studies highlighting differences and transformations between these coordinate systems (Brett et al., 2002; Lancaster et al., 2007). Spatial normalization methods vary, including intensity-based, landmark-based, or surface-based approaches (Gholipour et al., 2007; Klein et al., 2009). Typically, normalization is conducted by directly registering each individual's functional images to a functional template or through a two-step process: co-registering functional and structural images first, then registering the anatomical image to a high-resolution structural template. The direct method minimizes geometric distortions from differing imaging contrasts, while the two-step method is often more robust due to the higher resolution and quality of structural images. The choice of method depends on the specific scanning environment and imaging protocols. These steps are pictorially depicted in ◘ Fig. 4.14 (Xiong et al., 2023).

As we learned from the preprocessing of EEG data, the choice of a specific preprocessing pipeline for even fMRI data is influenced by various factors, including the type of stimulus, experimental hypotheses, and the acquisition environment (Strother, 2006; Huettel et al., 2004). For example, with interleaved slice acquisition, it's generally advisable to perform slice time correction before motion correction, whereas the order should be reversed for sequential acquisition. Furthermore, for processes that are linear, changing the order of operations won't affect the final results, such as those indicated in the pink rectangle. Additionally, there's a consid-

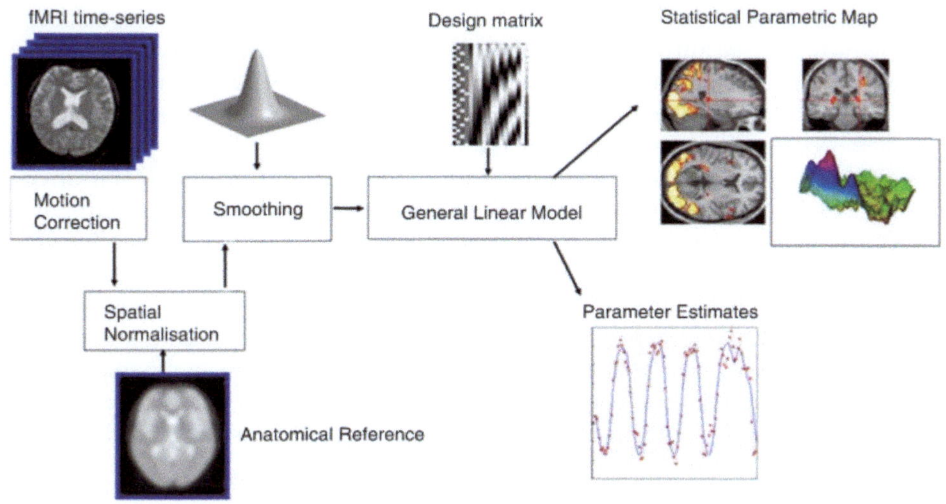

Fig. 4.14 fMRI signal processing pipeline. (From Xiong et al., 2023)

eration whether to normalize functional images before or after statistical analysis (Chen & Glover, 2015). Normalizing before analysis helps avoid additional smoothing and potential distortions from imperfect normalization, while normalizing after analysis facilitates statistical comparisons across subjects, supporting techniques like group independent component analysis and atlas-based graph analysis (Calhoun et al., 2001).

The neuroimaging community has access to a variety of tools that cover most preprocessing steps previously discussed. These tools are included in widely used fMRI software packages like AFNI (Cox & Hyde, 1997), ANTs (Avants et al., 2011), FreeSurfer (Fischl, 2012), FSL (Jenkinson et al., 2012), Nilearn (Abraham et al., 2014), and SPM (Penny et al., 2011), see ▪ Table 4.1 for details.

4.5.10 Preprocessing Effects

Preprocessing of fMRI data yields two main types of outputs (Esteban et al., 2019). The first type includes preprocessed time series, which are derived from the original data following retrospective signal corrections, spatiotemporal filtering, and resampling into a target space suited for analysis, such as a standardized anatomical reference. The second type comprises experimental confounds, which are additional time series like physiological recordings and estimated noise sources, utilized in the analysis (e.g., modeled as nuisance regressors). Commonly used confounds include motion parameters, framewise displacement, spatial standard deviation of the data after temporal differencing, and global signals. The preprocessing phase may also involve additional denoising steps and the estimation of confounds using dimensionality-reduction methods like principal component analysis or independent component analysis, as described in ▶ Sect. 4.4.2. Now, we will look at an example of a fMRI preprocessing pipeline.

4.5.10.1 **Example 1**

In this example, the authors present fMRIPrep (Esteban et al., 2019): a preprocessing pipeline for functional MRI. fMRIPrep utilizes sub-workflows that are dynamically configured based on the input data, incorporating tools from well-established, open-source neuroimaging packages (□ Table 4.1). The workflow engine Nipype (Neuroimaging in Python: Pipelines and Interfaces; ► http://nipy.org/nipype) (Gorgolewski et al., 2011) was employed to organize these workflows and handle execution details like resource management. The workflow was divided into two primary segments: anatomical MRI and fMRI processing streams, as shown in □ Fig. 4.15. Utilizing the brain imaging data structure (BIDS) (Gorgolewski et al., 2016), fMRIPrep automatically identifies the structure of input data and collects relevant metadata, such as imaging parameters, without requiring manual input. Additionally, according to the authors fMRIPrep adjusts to dataset inconsistencies, such as missing acquisitions or runs, using a series of heuristics (Esteban et al., 2019).

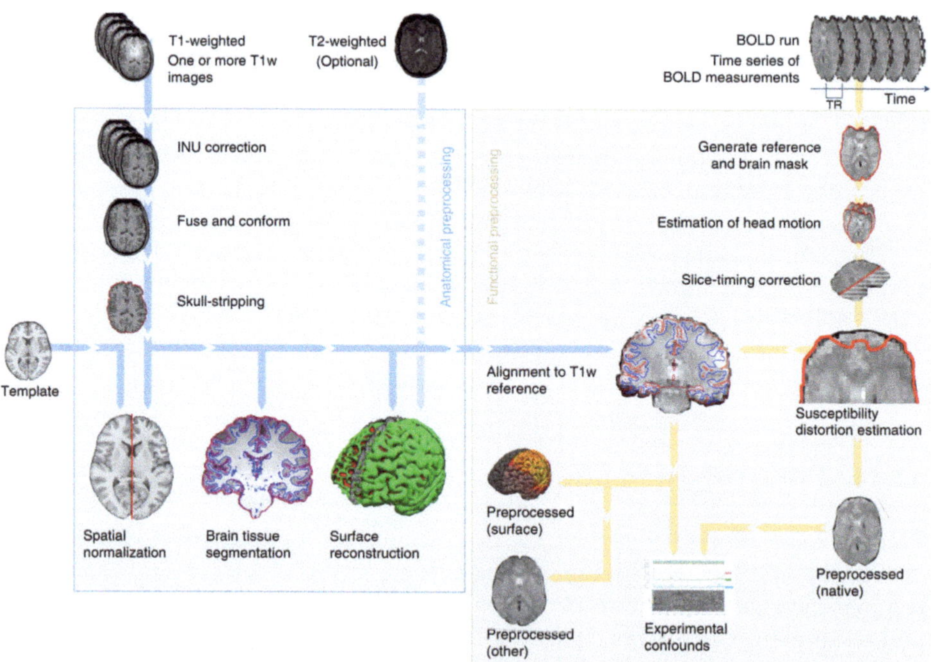

□ **Fig. 4.15** Using brain imaging data structure (BIDS) (Gorgolewski et al., 2016), the software automatically configures the optimal workflow for the provided dataset, eliminating the need for manual intervention. It autonomously locates necessary inputs such as one T1-weighted (T1w) image and one BOLD series, reads critical acquisition parameters like the repetition time (TR) and slice acquisition times, and identifies additional acquisitions like field maps that are essential for specific preprocessing steps, including the estimation of susceptibility distortion. (From Esteban et al., 2019)

4.5.10.2 **Example 2**

In this example (Fig. 4.16), we will visually explore the fMRI preprocessing pipeline and examine the various types of analysis performed post-preprocessing (Zhang et al., 2023). For any brain data, once preprocessing is complete, different analytical methods are applied to extract distinct features from the data. These

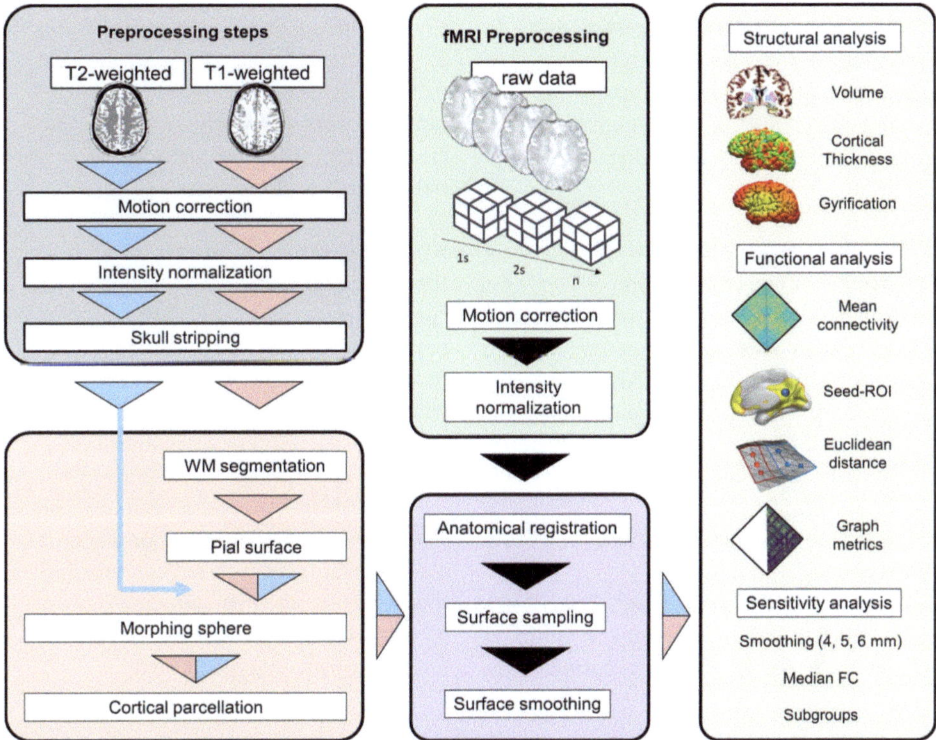

 Fig. 4.16 Analysis workflow: each participant underwent resting-state functional MRI, along with T1-weighted (T1w) and T2-weighted (T2w) MRI scans. Surface reconstruction was carried out using either a unimodal pipeline (UP; utilizing T1w signal alone) or a multimodal pipeline (MP; combining both T1w and T2w signals). In the UP workflow (indicated by a pink triangle), 3D T1w structural images were processed via the recon-all stream in Freesurfer, which included steps such as motion correction, intensity normalization, skull stripping, white matter segmentation, spherical morphing, and cortical parcellation. The MP workflow (represented by blue and pink triangles) incorporated T2w images (blue triangle) to enhance the Freesurfer workflow, adjusting spherical morphing and cortical parcellation to refine the pial surfaces further. Preprocessing of resting-state data involved correcting for head movement, normalizing intensity, registering anatomically, and smoothing the images. The rs-fMRI preprocessing pipeline was executed separately for both the UP and MP to ensure tailored processing. Outcomes in terms of structural (volume, cortical thickness, and gyrification index) and functional metrics (mean functional connectivity, seed-ROI maps, spatial topology, and graph analysis) were then compared to ascertain differences between the unimodal and multimodal pipelines. This comparative analysis aimed to evaluate the impact of each pipeline on the accuracy and reliability of both structural and functional outcomes. (From Zhang et al., 2023 originally published under CC-BY 4.0)

analytical approaches and their detailed applications will be discussed in ▶ Chap. 5, where we learn about the specific techniques used to analyze preprocessed brain data and the insights they provide.

Box 4.5 Essentials of Preprocessing fMRI Signals

Objective of fMRI Preprocessing: To enhance the clarity and reliability of functional magnetic resonance imaging data by reducing noise and correcting for artifacts and physiological fluctuations.

■ Key Preprocessing Steps:

Motion Correction: Compensates for patient movements during the scan, which can cause significant artifacts and misalignments in the fMRI data.

Slice Timing Correction: Adjusts the timing differences in image acquisition between different slices within a single brain volume, ensuring a consistent temporal representation across all brain regions.

Spatial Normalization: Transforms individual brain data into a common reference space to allow comparisons across subjects and groups in a standardized anatomical framework.

Smoothing: Applies a spatial filter to the data to increase signal-to-noise ratio by averaging the signal over neighboring voxels, which enhances the generalizability of the results across participants and sessions.

■ Importance for Analysis:

Proper preprocessing is essential to ensure the accuracy of fMRI studies, particularly those investigating brain function and connectivity. It is crucial for detecting subtle brain activations related to specific cognitive or behavioral tasks.

■ Outcome:

Preprocessed fMRI data provides a cleaner, more standardized basis for further analysis, such as statistical testing of brain activity patterns related to different experimental conditions.

4.6 Preprocessing of fNIRS Signal

During near-infrared spectroscopy (NIRS) of the brain, near-infrared (NIR) light is emitted onto the head's surface through optical fibers, known as optodes. After the light travels through the cortical regions, it becomes scattered and attenuated, and is then captured by a detector optode positioned at a specific distance from the source optode. The intensity of this attenuated light is analyzed to measure changes in oxygenated hemoglobin (HbO), deoxygenated hemoglobin (HbR), and/or total hemoglobin (HbT), which serve as indicators of brain activity (Ferrari & Quaresima, 2012). The model describing how light attenuates as it passes through the brain's highly scattering media is crucial for interpreting these changes accurately. There are two primary models for simulating light transport in the brain: the modified Beer–Lambert's law and the diffusion equation approach. The modified Beer–Lambert's law (Kocsis et al., 2006) is commonly employed to model light propagation within the brain because it supports real-time, online monitoring of how light travels through this medium.

4.6.1 Beer–Lambert Law

The transport of light in an absorbing media is given by the Beer–Lambert law which describes the attenuation of light at given a wavelength when it passes through a media containing chromophores. The Beer's law is described using the expression shown in Eq. (4.3).

$$A(t,\lambda) = \log\left(\frac{I(t,\lambda)}{I_0(t,\lambda)} \right) = \varepsilon(\lambda) \cdot C(\lambda) \cdot d(\lambda) \tag{4.3}$$

where in Eq. (4.3), $A(t,\lambda)$ is the attenuation (or optical density) of λ wavelength light when it passes through a given media at time t. $I(t,\lambda)$ is the intensity of λ wavelength light transmitted out of the given media at time t, and $I_0(t,\lambda)$ is the initial intensity of λ wavelength light incident on a media at time t. $\varepsilon(\lambda)$ is the molar extinction coefficient of a chromophore at λ of wavelength light expressed in $mM^{-1}\ cm^{-1}$. $C(\lambda)$ is the concentration of chromophore in the media at λ of wavelength light and $d(\lambda)$ is the direct path length of a photon from the emitting to the receiving optode placed on the media at λ of wavelength light, that is, the geometrical distance or the inter optode distance.

4.6.2 Modified Beer–Lambert Law

The Beer–Lambert law is traditionally used for purely absorbing media, but it does not hold up in media where scattering occurs, such as the brain. Therefore, to accommodate the complex nature of light transport through the brain, a modified version of the Beer–Lambert law is used. This modification accounts for both absorption by chromophores and the scattering effects within the brain. The modified Beer–Lambert law incorporates adjustments for the scattering-induced attenuation and the extended path length that photons travel due to scattering (Kocsis et al., 2006). In the brain, a highly scattering medium, the actual path length of photons exceeds the geometric distance between the optodes due to this scattering. Delpy et al. (1988) introduced a scaling factor known as the differential path-length factor (DPF) to correct for this increased path length. Consequently, the modified

Beer–Lambert law, which considers these scattering events, is expressed with the inclusion of the DPF to accurately describe NIR light attenuation in the brain as shown in Eq. (4.4).

$$A(t,\lambda) = \log\left(\frac{I(t,\lambda)}{I_0(t,\lambda)}\right) = \varepsilon(\lambda) \cdot C(\lambda) \cdot d(\lambda) \cdot \mathrm{DPF}(\lambda) + G(\lambda) \tag{4.4}$$

where $\mathrm{DPF}(\lambda)$ is the differential path length factor at λ of wavelength light as defined by Delpy et al. (1988) to correct for the path-length in scattering media.

The differential pathlength factor (DPF) or scaling factor is influenced by the number of scattering events within the medium. The DPF is influenced by several factors: it increases with an increase in the scattering coefficient and decreases with an increase in the absorption coefficient. It is also affected by the anisotropy factor (g) and the geometry of the medium. This factor is incorporated into the Beer–Lambert law to account for attenuation in scattering media (Matcher & Cooper, 1994). Typically, the DPF is considered relatively constant for a given tissue type because the variations in attenuation it measures are minor compared to the significant constant background attenuation inherent in the tissue. The term $G(\lambda)$ represents an unknown factor that combines the tissue's scattering coefficient with the geometry of the optodes, accounting for scattering losses. Consequently, due to these complexities, an absolute calculation of chromophore concentration directly from the Beer–Lambert equation is not feasible.

Equation (4.4) is useful primarily for measuring relative changes in chromophore concentrations by analyzing changes in optical density ($A(t, \lambda)$) at two different light wavelengths. This method helps negate the influence of the unknown G term by assuming a constant value of $G(\lambda)$ for all chromophores within the medium (Matcher & Cooper, 1994). Accurate quantification of chromophore concentration changes is achievable under the assumption that both the differential pathlength factor (DPF) and the distance between the optodes (d) remain constant throughout the measurement period. To calculate these changes, near-infrared spectroscopy (NIRS) measurements are taken at various wavelengths corresponding to specific chromophores. The modified Beer–Lambert law is then applied at these different wavelengths, and the equations are solved through matrix inversion to determine the concentration changes.

To quantify the change in concentration of HbO and HbR, 830 and 690 nm of light is used as these wavelengths of light are sensitive to changes in concentration of HbO and HbR, respectively. Then modified Beer–Lambert is written at two different wavelengths of light as shown in Eqs. (4.5) and (4.6), respectively.

$$\Delta \mathrm{OD}\lambda_1 = \varepsilon_{\mathrm{HbR}}^{\lambda 1} * L * [\mathrm{HbR}] + \varepsilon_{\mathrm{HbO}}^{\lambda 1} * L * [\mathrm{HbO}] \tag{4.5}$$

$$\Delta \mathrm{OD}\lambda_2 = \varepsilon_{\mathrm{HbR}}^{\lambda 2} * L * [\mathrm{HbR}] + \varepsilon_{\mathrm{HbO}}^{\lambda 2} * L * [\mathrm{HbO}] \tag{4.6}$$

The change in optical density calculated using Eqs. (4.5) and (4.6) is then used to calculate the changes in concentration of HbO and HbR as shown in Eqs. (4.7) and (4.8), respectively.

$$\Delta\left[\text{HbO}\right] = \frac{\Delta\text{OD}\lambda_1 \cdot \varepsilon_{\text{HbR}}^{\lambda 1} - \Delta\text{OD}\lambda_2 \cdot \varepsilon_{\text{HbR}}^{\lambda 2}}{\left(\varepsilon_{\text{HbO}}^{\lambda 1} \cdot \varepsilon_{\text{HbR}}^{\lambda 2} - \varepsilon_{\text{HbR}}^{\lambda 1} \cdot \varepsilon_{\text{HbO}}^{\lambda 2}\right) \cdot L} \tag{4.7}$$

$$\Delta\left[\text{HbR}\right] = \frac{\Delta\text{OD}\lambda_1 \cdot \varepsilon_{\text{HbO}}\left(\lambda_2\right) - \Delta\text{OD}\lambda_2 \cdot \varepsilon_{\text{HbO}}^{\lambda 1}}{\left(\varepsilon_{\text{HbR}}^{\lambda 1} \cdot \varepsilon_{\text{HbO}}^{\lambda 2} - \varepsilon_{\text{HbO}}^{\lambda 1} \cdot \varepsilon_{\text{HbR}}^{\lambda 2}\right) \cdot L} \tag{4.8}$$

HbT is obtained by adding the values of HbO and HbR.

$$\Delta\left[\text{HbT}\right] = \Delta\left[\text{HbO}\right] + \Delta\left[\text{HbR}\right] \tag{4.9}$$

where in Eqs. (4.5)–(4.9), $\Delta[\text{HbO}]$, $\Delta[\text{HbR}]$ and $\Delta[\text{HbT}]$ are changes in concentration of HbO (oxy-hemoglobin), HbR (de-oxy hemoglobin) and HbT (total hemoglobin), respectively.

An example preprocessing pipeline is shown in ◘ Fig. 4.17 (Piazza et al., 2020).

$\Delta\text{OD}\lambda_1$ and $\Delta\text{OD}\lambda_2$ are changes in optical density at wavelength 1 (830 nm) and wavelength 2 (690 nm), respectively. $\varepsilon_{\text{HbR}}^{\lambda 1}$ and $\varepsilon_{\text{HbR}}^{\lambda 2}$ is the absorption coefficient of HbR at wavelength 1 (830 nm) and wavelength 2 (690 nm), respectively; and $\varepsilon_{\text{HbO}}^{\lambda 1}$ and $\varepsilon_{\text{HbO}}^{\lambda 2}$ is absorption coefficient of HbO at wavelength 1 (830 nm) and wavelength 2 (690 nm), respectively.

L is the effective average path length of light through the tissue experiencing the absorption change. L is obtained from the product of the actual source-detector separation and differential pathlength factor (DPF). The source-detector separation is measurable along the surface of the head, and the extinction coefficient for a given chromophore can be looked up in tables (Strangman et al., 2002). The differential pathlength factor is either measured (with a time-domain or frequency domain instrument) or it is estimated as the distance between source and detector (for a continuous wave measurement) (Strangman et al., 2002).

The neuroscience community has access to a variety of tools that cover most preprocessing steps previously discussed. These tools are included in widely used fNIRS software packages like HOMER2 (Huppert et al., 2009), NIRS-SPM (Ye et al., 2009), fNIRS Toolbox (Santosa et al., 2018) and AtlasViewer (Aasted et al., 2015).

4.6.3 Preprocessing Effects

Preprocessing steps described in ▶ Sect. 4.3 can be applied to fNIRS signal either before or after conversion of signal to HbO (ΔHbO_2), HbR (ΔHHb), and HbT (ΔoxCCO). In the example shown in ◘ Fig. 4.18 (Pinti et al., 2021), preprocessing steps were applied after the conversion to HbO, HbR, and HbT (ΔoxCCO).

Head movement artifacts were corrected using a wavelet-based method (Molavi & Dumont, 2012) in the Homer2 software package, utilizing an interquartile range threshold of 1.5, proven effective in recovering the hemodynamic response. Homer2 is a fNIRS data processing software (Huppert et al., 2009), based on MATLAB (MathWorks, Natick, MA, USA).

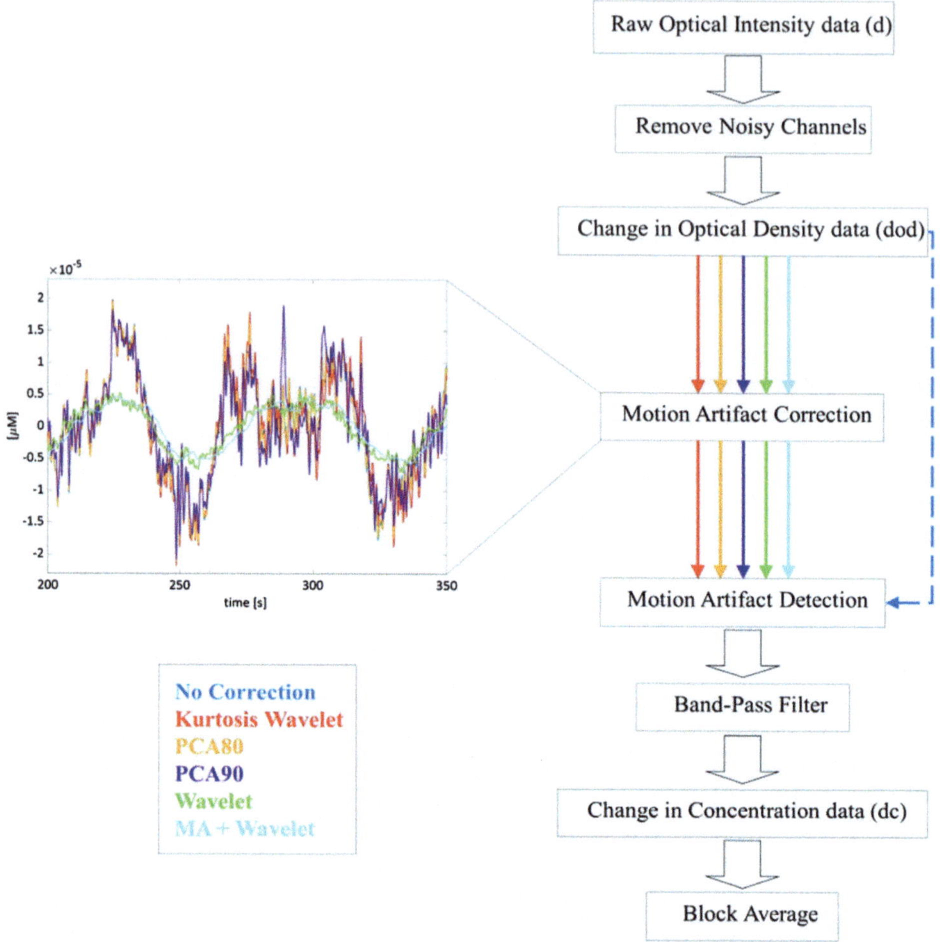

Fig. 4.17 Right panel: This displays the pre-processing analysis pipeline. The dotted arrow indicates the processing stream without any artifact correction techniques applied (No Correction). Left panel: This panel illustrates a comparison between various motion correction techniques and uncorrected data, as shown in a graph representing direct current (dc) data from a representative channel of one subject. (From Piazza et al., 2020)

A 150th order FIR band-pass filter ranging from 0.008 to 0.1 Hz was applied to the HbO (ΔHbO_2), HbR (ΔHHb), and HbT ($\Delta oxCCO$) signals to minimize very low and high frequency physiological noise. This filtering ensured inclusion of at least three harmonics of the stimulation frequencies for both right and left stimulations (~0.014 Hz) and the task presentation frequency (~0.09 Hz) within the filter's passband, maintaining filter stability. However, a cutoff frequency of 0.1 Hz does

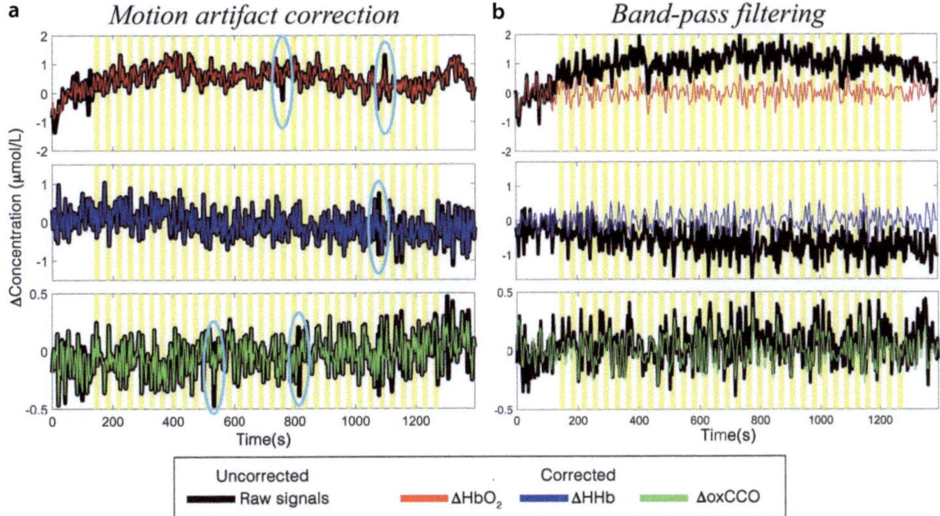

Fig. 4.18 Examples of applying preprocessing steps to the HbO (top), HbR (middle), and HbT (bottom) signals for one participant are illustrated here. Panel **a** displays the use of a motion artifact correction method with examples of artifacts highlighted in light blue circles. Panel **b** presents the results of FIR band-pass filtering. The raw, uncorrected HbO, HbR, and HbT (ΔoxCCO) signals are depicted in black, while the processed signals for HbO, HbR, and HbT (ΔoxCCO) are shown in red, blue, and green, respectively. The visual task blocks are indicated with yellow areas. (From Pinti et al., 2021 originally published under CC-BY 4.0)

not mitigate the impact of Mayer waves, a known issue in the fNIRS community (Pinti et al., 2021). Since band-pass filters cannot minimize Mayer waves due to their spectral overlap with hemodynamic changes, more sophisticated methods like short-separation channels or monitoring blood pressure changes are required to preserve the hemodynamic response effectively (Yücel et al., 2016).

To analyze brain activity, as shown in **Fig. 4.19**, authors used the general linear model (GLM) with finite impulse response (FIR) basis functions, avoiding assumptions about response shapes. This approach involves fitting shifted boxcar functions ($\Delta t = 2.88$ s) to the data, constructing a design matrix for each experimental condition (Pinti et al., 2021). Each experimental block lasted 36 s, with FIR bins spanning from 2 s before to 34 s after stimulus onset. The data for each participant and channel was processed separately for HbO, HbR, and HbT, obtaining averaged responses. For group-level analysis, authors baseline-corrected the responses and computed median β-values from the peak response period (8–29 s post-stimulus) (Pinti et al., 2021). The statistical significance was assessed using a one-sample t-test against zero, adjusted for multiple comparisons with the false discovery rate (FDR) correction, confirming normality with the Kolmogorov–Smirnov test.

4

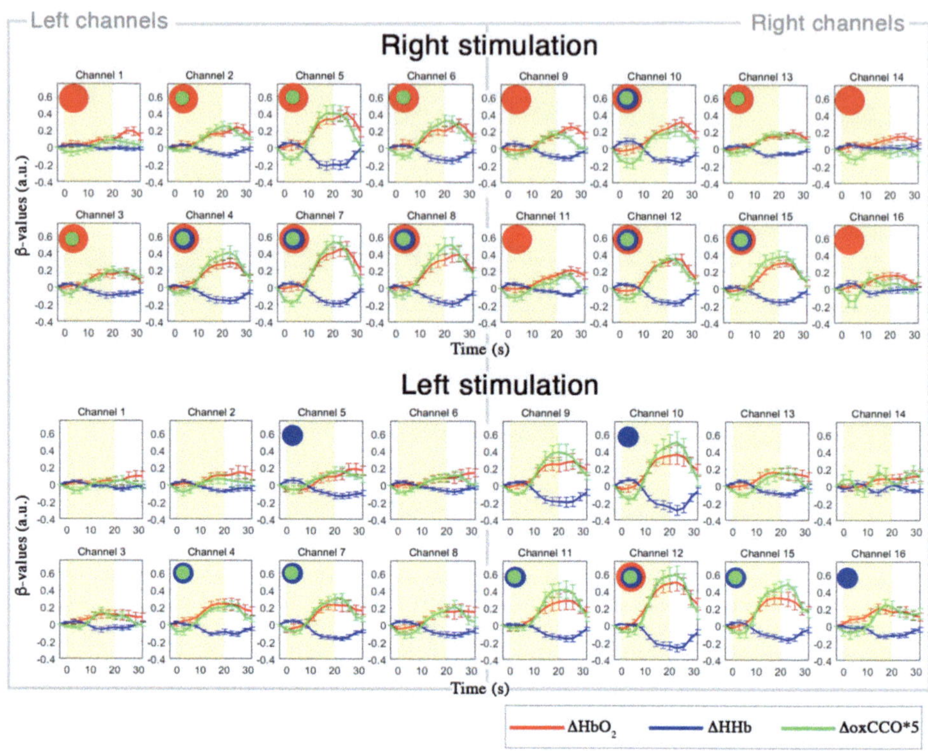

◘ Fig. 4.19 Grand-averaged responses (mean ± standard error) for both right (top) and left (bottom) visual hemifield conditions across each channel are shown. Responses from individual subjects, calculated using the FIR-GLM, are averaged across the group. The task block is highlighted in yellow. ΔHbO_2 is displayed in red, ΔHHb in blue, and $\Delta oxCCO$ (magnified by a factor of 5 for clarity) in green. Statistically significant changes are marked with red circles for ΔHbO_2, blue circles for ΔHHb, and green circles for $\Delta oxCCO$ ($p < 0.005$, FDR corrected for multiple comparisons). (From Pinti et al., 2021 originally published under CC-BY 4.0)

Box 4.6 Key Aspects of Preprocessing fNIRS Signals

Objective of fNIRS Preprocessing: To refine the functional near-infrared spectroscopy (fNIRS) data by minimizing noise and correcting for physiological and motion-related artifacts, thereby enhancing the quality and interpretability of the hemodynamic responses measured.

■ **Key Preprocessing Steps:**

Motion Correction: Addressing artifacts caused by subject movements, which can significantly distort the optical path and affect the measured light intensities. Techniques like spline interpolation, wavelet filtering, or correlation-based methods are often employed.

Physiological Noise Correction: Removing variations in the signal that are related to cardiac cycles, breathing, and blood pressure fluctuations. This is typically achieved using methods like adaptive filtering or independent component analysis to isolate these physiological components from the cerebral hemodynamic signals.

Detrending: Applying methods to remove slow drifts in the fNIRS signals that can occur due to changes in sensor contact, thermal effects, or other slow-varying extracerebral factors.

Low-Pass Filtering: Reducing high-frequency noise from the data to focus on the relevant hemodynamic response frequencies, which are generally lower than 0.5 Hz.

■ **Importance for Analysis:**
Effective preprocessing is crucial for ensuring that the fNIRS signals reflect true cerebral activity and not artifacts or external noise. This step is essential for accurate analysis, particularly in studies exploring cognitive functions, brain–computer interfaces, or clinical assessments of brain health.

■ **Outcome:**
Properly preprocessed fNIRS data results in more reliable and valid assessments of brain function, facilitating research into neural mechanisms and aiding in the development of therapeutic protocols.

4.7 Preprocessing of MEG Signal

Along with the artifacts mentioned in ▶ Sect. 4.1, raw MEG signals may contain system related artifacts which occurs due to issues such as superconducting quantum interference devices (SQUID) jumps, or sensors that are noisy, broken, or saturated (Gross et al., 2013). An example of MEG signal contaminated with cardiac artifacts is shown in ◘ Fig. 4.20 (Corsi, 2023).

The neuroscience community has access to a variety of tools that cover most preprocessing steps previously discussed. These tools are included in widely used MEG software packages like MNE (MEG/EEG neuroimaging environment) (Gramfort et al., 2013), FieldTrip (Oostenveld et al., 2011), brainstorm (Tadel et al., 2011), and SPM (Penny et al., 2011).

There are two primary strategies for managing artifacts in data analysis, as you have learned by now. The first approach involves identifying artifact-contaminated data segments either through visual inspection, automatic detection methods, or a combination of both. These contaminated segments are then excluded from further analysis (Gross et al., 2013). The second strategy involves using signal-processing techniques to reduce artifacts that do not originate from brain activity, while preserving the integrity of signals that do (Gross et al., 2013). This approach is similar to methods used in preprocessing EEG signals. An overview of these preprocessing and advanced signal-processing techniques applied to MEG signals is presented in ◘ Fig. 4.21 (Baillet, 2017). ▶ Chapter 5 looks deeper into these advanced signal-processing approaches, which are also applicable to other types of brain signals, as discussed further in the same chapter.

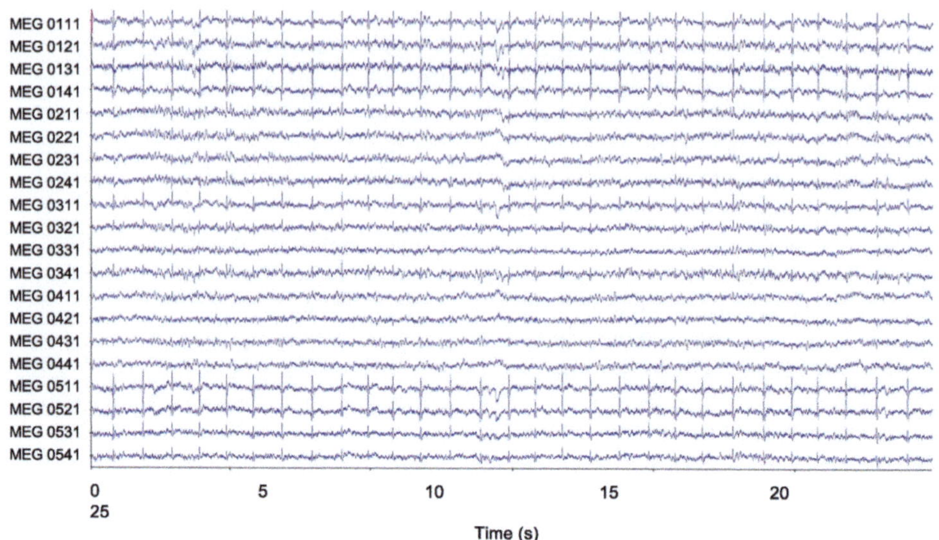

Fig. 4.20 Example of cardiac artifacts recorded with magnetometers. (From Corsi, 2023 originally published under CC-BY 4.0)

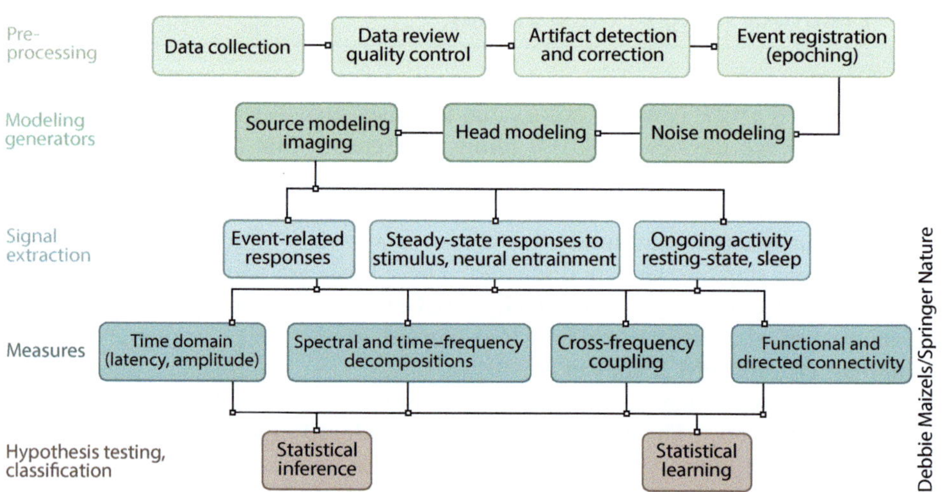

Fig. 4.21 MEG, like other brain imaging techniques, requires adherence to several key procedures during data analysis. First, preprocessing is essential to ensure optimal data quality before implementing advanced signal extraction techniques. This includes identifying and either removing or attenuating data segments contaminated by artifacts. Second, during the modeling phase of MEG, it's critical to carefully select parameters, such as choosing between a template or individual head shape, defining noise, and setting image reconstruction parameters. Third, signal extraction in MEG is highly adaptable and depends on the specific design of the experiment. Fourth, the potential measurements are extensive due to the multidimensional nature of the data (encompassing space, time, and frequency dimensions). Finally, the analysis concludes with statistical steps that may involve hypothesis testing, inference, or the application of statistical learning methods for signal classification and deriving other data insights. (From Baillet, 2017)

> **Box 4.7 Essentials of Preprocessing MEG Signals**
>
> **Objective of MEG Preprocessing**: To refine magnetoencephalography (MEG) data by reducing noise and correcting for artifacts, thus enhancing the quality and interpretability of the neural signals related to brain function.
>
> ■ **Key Preprocessing Steps:**
>
> **Artifact Rejection**: Identifying and removing segments of data contaminated by artifacts, such as those caused by eye movements, cardiac pulses, or external electromagnetic interference. Techniques often include independent component analysis (ICA) or signal-space projection.
>
> **Signal Space Separation (SSS)**: A technique specific to MEG that separates internal brain signals from external noise based on the spatial properties of the magnetic fields.
>
> **Temporal Filtering**: Applying filters to isolate the frequency bands relevant to the study's objectives while attenuating frequencies that may contain noise or are not of interest.
>
> **Head Movement Compensation**: Correcting for any head movements during the recording to ensure the accuracy of source localization in brain activity mapping.
>
> ■ **Importance for Analysis:**
>
> Proper preprocessing is crucial in MEG studies due to the method's high sensitivity to both internal and external magnetic fields. Efficient preprocessing ensures that the analyzed data accurately reflects true brain activity rather than artifacts or noise.
>
> ■ **Outcome:**
>
> Preprocessed MEG data offers a clearer, more accurate basis for subsequent analyses, such as functional connectivity analysis, source localization, and cognitive state assessment, making it vital for both research and clinical diagnostics.

4.8　Preprocessing of Invasive Brain Recordings

The quality of the brain signals acquired invasively during intracortical recordings are primarily influenced by the electrode itself and the cortical region from which it records. Both factors should be meticulously managed during the recording session to ensure optimal signal quality.

As you learned in ▶ Chap. 3 and as depicted in ▶ Fig. 3.7 that the low and high frequencies are often extracted from the raw signal by a high-pass and/or a low-pass filter. The high-frequency component of recorded neural signals captures rapid electrical changes, specifically neuronal spikes, while the low-frequency component is analyzed for slower electrical variations, such as local field potentials (LFPs), which arise from the activity of neuronal populations within a cortical area (Buzsáki, 2004). Filters to separate these components can be implemented via software or hardware, with no standardized frequency thresholds universally accepted for defining low and high frequencies (Pettersen & Einevoll, 2008). Filter parameters are customized based on the specific needs of the experiment and the characteristics of the neurons being studied. Typically, the low-frequency component is captured at frequencies below 200 Hz, and the high-frequency component at frequencies above 300 Hz. Common types of digital filters used in software include Bessel, Butterworth, and elliptic filters, each chosen based on the specific requirements of the analysis (Magri et al., 2009).

Spike sorting is the process of extracting and classifying spikes from the high-frequency component of cortical recordings (Lewicki, 1998), as shown in ■ Fig. 4.22 (Ahmadi et al., 2021). It can be performed online during the recording or offline afterward. Online spike sorting, designed to conserve computing resources and maintain real-time monitoring speed, typically involves a simple algorithm like setting a threshold to detect spikes. In contrast, offline spike sorting

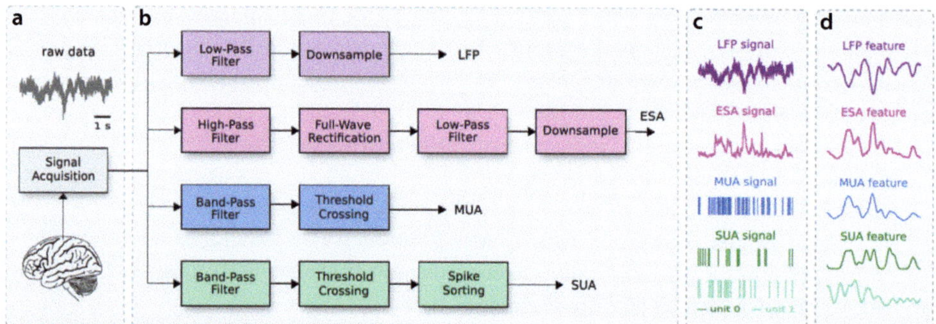

■ **Fig. 4.22**   Detailed schematic illustration of signal processing and feature extraction steps—**a Raw Neural Signal Acquisition**: The initial step involves the acquisition of raw neural signals from the motor cortex area of monkeys. This is achieved using a 96-channel intracortical Utah array, a high-density microelectrode array that enables precise recording of neural activity from multiple sites within the cortex simultaneously. **b Signal Processing Steps**: The acquired raw neural signals undergo various signal processing steps to isolate different types of neural signals. These steps include Preprocessing: Filtering to remove noise and artifacts. Segmentation: Dividing the continuous signal into segments or epochs. Normalization: Adjusting signal amplitudes for consistency. Specific processing techniques are applied to extract distinct neural signals: Local field potentials (LFP): Low-frequency components reflecting summed synaptic activity. Envelope of spiking activity (ESA): High-frequency envelope capturing overall spiking activity. Multi-unit activity (MUA): Spiking activity from a group of neurons. Single-unit activity (SUA): Isolated spikes from individual neurons. **c Processed Neural Signals**: Following the processing steps, the neural signals are categorized and identified as follows: LFP signals: Representing local field potentials with characteristic low-frequency oscillations. ESA signals: Depicting the envelope of spiking activity with high-frequency content. MUA signals: Illustrating multi-unit activity with multiple overlapping spikes. SUA signals: Showing single-unit activity with distinct, isolated spikes. **d Feature Extraction**: From each type of processed signal (LFP, ESA, MUA, and SUA), specific features are extracted for further analysis. These features may include amplitude and power: Measures of signal strength. Frequency components: Analysis of dominant frequencies and spectral content. Spike timing and rate: Temporal characteristics of spiking activity. Pattern recognition: Identification of specific neural patterns or motifs. These extracted features are crucial for understanding the neural coding mechanisms and for applications in brain–machine interfaces (BMIs) and neural prosthetics. This detailed schematic elucidates the comprehensive process from raw signal acquisition to the extraction of meaningful neural features, essential for advanced neurotechnological applications. (From Ahmadi et al., 2021 originally published under CC-BY 4.0)

allows for more detailed analysis using complex algorithms such as principal component analysis of spike waveforms (Rey et al., 2015; Chung et al., 2017).

Intracortical electrodes, placed extracellularly in the cortex, often detect signals from multiple nearby neurons. Online sorting preliminarily identifies spikes, including both single-unit activities (SUAs) and multi-unit activities (MUAs)—with SUAs representing signals from a single neuron and MUAs from multiple neurons (Rey et al., 2015). The electrode's small scale means that neuron signals vary in amplitude and waveform based on their proximity to the electrode. Offline sorting further distinguishes SUAs from MUAs, as depicted in ◘ Fig. 4.22b (Ahmadi et al., 2021), with SUAs marked in yellow and MUAs in gray. If online sorting is insufficient and the wideband signal is recorded, offline sorting can reapply threshold sorting from the wideband data. Given that spikes occur within milliseconds, spike sorting requires high temporal resolution, typically employing a sampling rate of 40 kHz to digitize the high-frequency signal (Rossant et al., 2016).

Box 4.8

Spike Sorting is a technique used in neuroscience to analyze and classify the electrical signals, or "spikes," generated by neurons during their activity. These spikes represent action potentials, which are rapid rises and falls in voltage across a neuron's membrane. Spike sorting is critical when recording electrical signals from multiple neurons simultaneously, as it helps to distinguish between the action potentials of different neurons based on the shapes and timing of their spikes

■ **The Process Involves Several Steps:**

Detection of spikes in the continuous voltage signal.

Alignment of these spikes to correct for timing variations.

Feature extraction where key characteristics of each spike's shape are quantified.

Clustering where spikes are grouped based on their similarities in these features, with each cluster presumably representing spikes from a single neuron.

Spike sorting allows researchers to study the firing patterns of individual neurons and how these patterns correlate with different behaviors, sensory inputs, or pathological states. This analysis is crucial for advancing our understanding of neural circuit functions and dysfunctions, and for developing neural prosthetic devices and other applications in brain–computer interfacing.

Further segmenting the LFP into different frequency bands can enhance signal detail (as shown in ◘ Fig. 4.23) and provide deeper insights into brain activities. These frequency bands typically include delta (1–4 Hz), theta (4–8 Hz), alpha (8–13 Hz), beta (13–30 Hz), and gamma (30–100 Hz). Despite extensive past research, there is ongoing exploration into new analytical methods to extract more valuable information from LFPs.

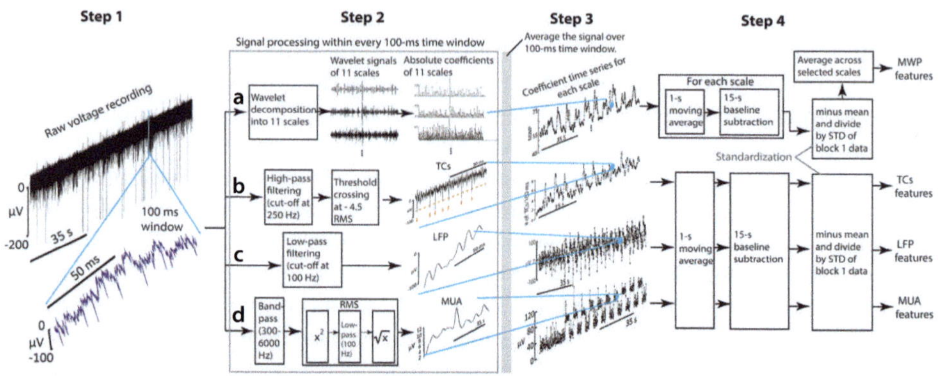

● **Fig. 4.23** Processing of a raw neural signal into distinct neural features involves several steps. In **Step 1**, a 100 ms section of the neural signal is selected from a larger raw voltage recording. **Step 2** involves multiple signal processing methods for this segment: Method (**a**): The raw signal is decomposed into 11 wavelet scales to extract the rectified wavelet coefficients for each scale. Method (**b**): A high-pass filter and a threshold set at −4.5 times the root mean square (RMS) value are used to detect the threshold crossings (TCs) within the 100 ms section. Method (**c**): A low-pass filter is applied to obtain the local field potential (LFP) from the raw signal. Method (**d**): A band-pass filter and customized RMS values are used to derive the multi-unit activity (MUA) from the raw signal. In **Step 3**, the signals processed through these methods are averaged over the 100 ms window to form a single data point that represents this segment in the averaged time series for the entire block. **Step 4** involves signal smoothing and standardization. To create the multi-waveform pattern (MWP) feature time series, a 1-s moving average and a 15-s wide mean subtraction are applied to the averaged time series of the block. This processed time series is then standardized and averaged across selected scales to produce a new time series for each channel. For other feature time series, the processed signal from Step 3 undergoes a 1-s moving average, followed by a 15-s mean subtraction and standardization, applied sequentially. From (Zhang et al., 2018) originally published under CC-BY 4.0

4.8.1 Preprocessing Effects

In the example shown in ● Fig. 4.23 (Zhang et al., 2018), authors applied wavelet decomposition to raw voltage data from a microelectrode array implanted in the motor cortex of a human with tetraplegia. The wavelet decomposition produced mean wavelet power (MWP) features, divided into three sub-frequency bands: low-frequency MWP (lf-MWP, 0–234 Hz), mid-frequency MWP (mf-MWP, 234–3.75 kHz), and high-frequency MWP (hf-MWP, >3.75 kHz) (Zhang et al., 2018).

ECoG signals, acquired from the surface of the brain, like all other brain signals, are susceptible to contamination by artifacts as detailed in ▶ Sect. 4.1. The preprocessing steps for ECoG signals to remove these artifacts are similar to those used for EEG signals and vary based on the type of artifacts encountered.

> **Box 4.9 Essentials of Preprocessing Invasive Brain Recordings**
>
> **Objective of Preprocessing Invasive Signals**: To enhance the quality and interpretability of data obtained from invasive brain recordings, such as those from electrocorticography (ECoG), and microelectrode arrays. This is crucial for both clinical diagnostics and neuroscientific research.
>
> ■ **Key Preprocessing Steps:**
>
> **Artifact Removal**: Vital for invasive recordings due to the close proximity of electrodes to sources of physiological noise such as muscle activity or heartbeats. Techniques like independent component analysis (ICA) or template subtraction are commonly used.
>
> **Signal Enhancement**: Employing filtering techniques to isolate neuronal signals from background noise. High-pass filters remove slow drifts, while low-pass filters curb high-frequency noise, sharpening the focus on relevant neuronal activity.
>
> **Normalization and Calibration**: Standardizing signal amplitudes to facilitate comparison across different recording sessions or different subjects. Calibration ensures consistency and accuracy in measurements, especially when longitudinal or comparative studies are involved.
>
> ■ **Challenges Specific to Invasive Recordings:**
>
> **Biocompatibility and Stability**: Ensuring that the signal remains stable over time despite potential issues with electrode displacement or tissue response.
>
> **Safety and Data Integrity**: Balancing the need for high-quality data with the imperative to minimize risk and discomfort for the patient.
>
> ■ **Importance for Analysis:**
>
> Proper preprocessing is essential to ensure that subsequent analyses, such as decoding neural patterns or mapping brain functions, are based on clean and reliable data. This is particularly critical in settings where these signals might be used to drive therapeutic devices or inform surgical decisions.
>
> ■ **Outcome:**
>
> Well-preprocessed data from invasive recordings provides a robust basis for advanced analyses, contributing to more accurate brain-mapping, improved clinical outcomes in surgical planning, and enhanced performance of neural prosthetics.

❯ Key Takeaways

1. **Enhancing Signal Quality**: The preprocessing steps detailed in this chapter, such as filtering, artifact removal, and signal normalization, are crucial for minimizing the effects of external noise and physiological artifacts. These steps ensure that the data reflects true neural activity as accurately as possible.

2. **Tailored Approaches**: Different brain signal recording techniques require specific preprocessing strategies, reflecting the unique characteristics and challenges associated with each method. For instance, motion correction is critical for fMRI data, while spike sorting is essential for analyzing data from invasive recordings.

3. **Impact on Research and Clinical Outcomes**: Effective preprocessing not only supports the integrity of basic neuroscience research but also enhances the clinical applicability of brain signal analysis. In clinical settings, these processes are crucial for accurate diagnosis and effective treatment planning, particularly in neurology and psychiatry.

4. **Advancements in Software and Techniques**: The ongoing development of specialized software and advanced algorithms continues to refine preprocessing techniques, making them more efficient and accessible. This evolution is crucial for keeping pace with the increasing complexity and volume of brain data being collected.

In conclusion, preprocessing is not merely a routine step in the analysis of brain signals; it is a cornerstone of neuroscientific analysis. It supports the integrity of research findings and the efficacy of clinical diagnostics by ensuring that the data underpinning them are as clean and accurate as possible. As we move forward, the innovations in preprocessing strategies will undoubtedly parallel advances in recording technologies, driving the neuroscience field towards ever more precise and insightful discoveries.

Conclusion

Throughout this chapter, we have explored the intricate process of preparing raw brain signals for analysis, an essential precursor to understanding cognitive processes and diagnosing neurological conditions. As we have seen, raw signals from brain recording techniques—whether from EEG, fMRI, fNIRS, MEG, intracortical recordings, or electrocorticography—are invariably contaminated by noise and artifacts from physiological and environmental sources. These unwanted signals can obscure the true neural signals, leading to potential misinterpretations and misleading results.

Our exploration detailed the typical sources of such noise and the techniques developed to combat their effects. Physiological artifacts, ranging from cardiac and muscular activities to eye movements, introduce significant challenges but are addressed through sophisticated preprocessing techniques that enhance the signal-to-noise ratio and prepare data for analysis. Environmental artifacts, caused by equipment and external electromagnetic fields, require equally rigorous strategies to mitigate their impact.

By learning about various preprocessing pipelines, we have seen how each brain recording technique necessitates specific approaches tailored to its unique challenges. These techniques, including artifact removal, signal enhancement, filtering, and normalization, are crucial for transforming raw data into usable information that can robustly support both research and clinical decisions.

This section offers practical exercises designed to help apply the material learned to concrete scenarios.

Artifact Identification and Removal in EEG

Objective: To practice identifying and removing common artifacts from EEG data.

Task: Students will use EEG processing software like EEGLAB to analyze sample EEG data containing artifacts such as eye blinks and muscle movements. They will apply techniques like independent component analysis (ICA) to isolate and remove these artifacts.

Motion Correction in fMRI Data

Objective: To understand the process and impact of motion correction in fMRI.

Task: Using fMRI analysis software such as SPM or FSL, students will perform motion correction on a dataset provided. They will compare the images before and after motion correction to assess the effectiveness of the correction algorithms.

Filtering MEG Data

Objective: To apply temporal filtering to MEG data to isolate relevant frequency bands.

Task: Students will use FieldTrip or MNE-Python to apply band-pass filters to MEG data, focusing on frequency bands typically associated with specific cognitive processes, and interpret the results.

Normalization of fNIRS Data

Objective: To practice normalizing fNIRS data to reduce inter-subject variability.

Task: Students will use an fNIRS data analysis tool to normalize a set of fNIRS data from multiple subjects performing the same task. They will assess how normalization impacts the comparability of data across subjects.

Preprocessing Pipeline for Invasive Recordings

Objective: To develop and implement a preprocessing pipeline for data from invasive brain recordings.

Task: Using synthetic or publicly available datasets of invasive recording data, students will create a preprocessing pipeline incorporating noise reduction, spike sorting, and alignment. The effectiveness of the pipeline will be evaluated based on the clarity and usability of the processed signals for further analysis.

Exploring Preprocessing Effects Using Simulation Software

Objective: To simulate the effects of different preprocessing techniques on the quality of brain signals.

Task: Students will use simulation software to generate brain signal data with adjustable noise levels and types. They will then apply various preprocessing techniques and analyze how each technique affects the signal-to-noise ratio and the detection of simulated neural events.

References

Aasted, C. M., Yücel, M. A., Cooper, R. J., Dubb, J., Tsuzuki, D., Becerra, L., et al. (2015). Anatomical guidance for functional near-infrared spectroscopy: AtlasViewer tutorial. *Neurophotonics, 2*(2), 020801.

Abraham, A., Pedregosa, F., Eickenberg, M., Gervais, P., Mueller, A., Kossaifi, J., et al. (2014). Machine learning for neuroimaging with scikit-learn. *Frontiers in Neuroinformatics, 8*, 14.

Ahmadi, N., Constandinou, T. G., & Bouganis, C. S. (2021). Inferring entire spiking activity from local field potentials. *Scientific Reports, 11*(1), 19045.

Alahmadi, A. A. (2021). Effects of different smoothing on global and regional resting functional connectivity. *Neuroradiology, 63*, 99–109.

Allen, P. J., Josephs, O., & Turner, R. (2000). A method for removing imaging artifact from continuous EEG recorded during functional MRI. *NeuroImage, 12*(2), 230–239.

Amin, U., Nascimento, F. A., Karakis, I., Schomer, D., & Benbadis, S. R. (2023). Normal variants and artifacts: Importance in EEG interpretation. *Epileptic Disorders, 25*(5), 591–648.

Anderer, P., Roberts, S., Schlögl, A., Gruber, G., Klösch, G., Herrmann, W., et al. (1999). Artifact processing in computerized analysis of sleep EEG—A review. *Neuropsychobiology, 40*(3), 150–157.

Aqil, M., Jbari, A., & Bourouhou, A. (2017). ECG signal denoising by discrete wavelet transform. *International Journal of Online Engineering, 13*(9), 51.

Ashburner, J., & Friston, K. J. (1999). Nonlinear spatial normalization using basis functions. *Human Brain Mapping, 7*(4), 254–266.

Avants, B. B., Tustison, N. J., Song, G., Cook, P. A., Klein, A., & Gee, J. C. (2011). A reproducible evaluation of ANTs similarity metric performance in brain image registration. *NeuroImage, 54*(3), 2033–2044.

Baillet, S. (2017). Magnetoencephalography for brain electrophysiology and imaging. *Nature Neuroscience, 20*(3), 327–339.

Barnea-Goraly, N., Weinzimer, S. A., Ruedy, K. J., Mauras, N., Beck, R. W., Marzelli, M. J., et al. (2014). High success rates of sedation-free brain MRI scanning in young children using simple subject preparation protocols with and without a commercial mock scanner—The Diabetes Research in Children Network (DirecNet) experience. *Pediatric Radiology, 44*, 181–186.

Birn, R. M. (2012). The role of physiological noise in resting-state functional connectivity. *NeuroImage, 62*(2), 864–870.

Birn, R. M., Diamond, J. B., Smith, M. A., & Bandettini, P. A. (2006). Separating respiratory-variation-related fluctuations from neuronal-activity-related fluctuations in fMRI. *NeuroImage, 31*(4), 1536–1548.

Birn, R. M., Murphy, K., & Bandettini, P. A. (2008a). The effect of respiration variations on independent component analysis results of resting state functional connectivity. *Human Brain Mapping, 29*(7), 740–750.

Birn, R. M., Smith, M. A., Jones, T. B., & Bandettini, P. A. (2008b). The respiration response function: The temporal dynamics of fMRI signal fluctuations related to changes in respiration. *NeuroImage, 40*(2), 644–654.

Brett, M., Johnsrude, I. S., & Owen, A. M. (2002). The problem of functional localization in the human brain. *Nature Reviews Neuroscience, 3*(3), 243–249.

Brienza, M., Davassi, C., & Mecarelli, O. (2019). Artifacts. In *Clinical electroencephalography* (pp. 109–130). Springer.

Bro, R., & Smilde, A. K. (2014). Principal component analysis. *Analytical Methods, 6*(9), 2812–2831.

Brunet, D., Murray, M. M., & Michel, C. M. (2011). Spatiotemporal analysis of multichannel EEG: CARTOOL. *Computational Intelligence and Neuroscience, 2011*, 1–15.

Buzsáki, G. (2004). Large-scale recording of neuronal ensembles. *Nature Neuroscience, 7*(5), 446–451.

Calhoun, V. D., Adali, T., Pearlson, G. D., & Pekar, J. J. (2001). A method for making group inferences from functional MRI data using independent component analysis. *Human Brain Mapping, 14*(3), 140–151.

Castellanos, N. P., & Makarov, V. A. (2006). Recovering EEG brain signals: Artifact suppression with wavelet enhanced independent component analysis. *Journal of Neuroscience Methods, 158*(2), 300–312.

Chaddad, A., Wu, Y., Kateb, R., & Bouridane, A. (2023). Electroencephalography signal processing: A comprehensive review and analysis of methods and techniques. *Sensors, 23*(14), 6434.

Chang, C., & Glover, G. H. (2009). Effects of model-based physiological noise correction on default mode network anti-correlations and correlations. *NeuroImage, 47*(4), 1448–1459.

Chang, C., Cunningham, J. P., & Glover, G. H. (2009). Influence of heart rate on the BOLD signal: The cardiac response function. *NeuroImage, 44*(3), 857–869.

Chen, J. E., & Glover, G. H. (2015). Functional magnetic resonance imaging methods. *Neuropsychology Review, 25*, 289–313.

Chung, J. E., Magland, J. F., Barnett, A. H., Tolosa, V. M., Tooker, A. C., Lee, K. Y., et al. (2017). A fully automated approach to spike sorting. *Neuron, 95*(6), 1381–1394.

Coffman, B. A., & Salisbury, D. F. (2020). MEG methods: A primer of basic MEG analysis. In *Neuroimaging in Schizophrenia* (pp. 191–210). Springer.

Cohen, M. X. (2014). *Analyzing neural time series data: Theory and practice.* MIT Press.

Corsi, M. C. (2023). Electroencephalography and magnetoencephalography. In *Machine learning for brain disorders* (pp. 285–312). Humana.

Cox, R. W., & Hyde, J. S. (1997). Software tools for analysis and visualization of fMRI data. *NMR in Biomedicine, 10*(4–5), 171–178.

Croft, R. J., & Barry, R. J. (2000). Removal of ocular artifact from the EEG: A review. *Neurophysiologie Clinique/Clinical Neurophysiology, 30*(1), 5–19.

Cusack, R., Brett, M., & Osswald, K. (2003). An evaluation of the use of magnetic field maps to undistort echo-planar images. *NeuroImage, 18*(1), 127–142.

Dagli, M. S., Ingeholm, J. E., & Haxby, J. V. (1999). Localization of cardiac-induced signal change in fMRI. *NeuroImage, 9*(4), 407–415.

De Clercq, W., Vergult, A., Vanrumste, B., Van Paesschen, W., & Van Huffel, S. (2006). Canonical correlation analysis applied to remove muscle artifacts from the electroencephalogram. *IEEE Transactions on Biomedical Engineering, 53*(12), 2583–2587.

Delorme, A., & Makeig, S. (2004). EEGLAB: An open source toolbox for analysis of single-trial EEG dynamics including independent component analysis. *Journal of Neuroscience Methods, 134*(1), 9–21.

Delpy, D. T., Cope, M., van der Zee, P., Arridge, S., Wray, S., & Wyatt, J. S. (1988). Estimation of optical pathlength through tissue from direct time of flight measurement. *Physics in Medicine & Biology, 33*(12), 1433.

Do, T. T., Nguyen, T. H., Le, T. A., Nguyen, S. A. T., Nguyen, Q. T. N., Tran, T. Q. V., et al. (2022). Automated EOG removal from EEG signal using independent component analysis and machine learning algorithms. In *8th International conference on the development of biomedical engineering in Vietnam: Proceedings of BME 8, 2020, Vietnam: Healthcare technology for smart city in low- and middle-income countries* (pp. 1001–1016). Springer International Publishing.

Esteban, O., Markiewicz, C. J., Blair, R. W., Moodie, C. A., Isik, A. I., Erramuzpe, A., et al. (2019). fMRIPrep: A robust preprocessing pipeline for functional MRI. *Nature Methods, 16*(1), 111–116.

Fatourechi, M., Bashashati, A., Ward, R. K., & Birch, G. E. (2007). EMG and EOG artifacts in brain computer interface systems: A survey. *Clinical Neurophysiology, 118*(3), 480–494.

Ferrari, M., & Quaresima, V. (2012). A brief review on the history of human functional near-infrared spectroscopy (fNIRS) development and fields of application. *NeuroImage, 63*(2), 921–935.

Fischl, B. (2012). FreeSurfer. *NeuroImage, 62*(2), 774–781.

Friston, K. J., Williams, S., Howard, R., Frackowiak, R. S., & Turner, R. (1996). Movement-related effects in fMRI time-series. *Magnetic Resonance in Medicine, 35*(3), 346–355.

Friston, K. J., Josephs, O., Zarahn, E., Holmes, A. P., Rouquette, S., & Poline, J. B. (2000). To smooth or not to smooth?: Bias and efficiency in FMRI time-series analysis. *NeuroImage, 12*(2), 196–208.

Geissler, A., Lanzenberger, R., Barth, M., Tahamtan, A. R., Milakara, D., Gartus, A., & Beisteiner, R. (2005). Influence of fMRI smoothing procedures on replicability of fine scale motor localization. *NeuroImage, 24*(2), 323–331.

Gholipour, A., Kehtarnavaz, N., Briggs, R., Devous, M., & Gopinath, K. (2007). Brain functional localization: A survey of image registration techniques. *IEEE Transactions on Medical Imaging, 26*(4), 427–451.

Glover, G. H., Li, T. Q., & Ress, D. (2000). Image-based method for retrospective correction of physiological motion effects in fMRI: RETROICOR. *Magnetic Resonance in Medicine, 44*(1), 162–167.

Goncharova, I. I., McFarland, D. J., Vaughan, T. M., & Wolpaw, J. R. (2003). EMG contamination of EEG: Spectral and topographical characteristics. *Clinical Neurophysiology, 114*(9), 1580–1593.

Gorgolewski, K., Burns, C. D., Madison, C., Clark, D., Halchenko, Y. O., Waskom, M. L., & Ghosh, S. S. (2011). Nipype: A flexible, lightweight and extensible neuroimaging data processing framework in python. *Frontiers in Neuroinformatics, 5*, 13.

Gorgolewski, K. J., Auer, T., Calhoun, V. D., Craddock, R. C., Das, S., Duff, E. P., et al. (2016). The brain imaging data structure, a format for organizing and describing outputs of neuroimaging experiments. *Scientific Data, 3*(1), 1–9.

Gramfort, A., Luessi, M., Larson, E., Engemann, D. A., Strohmeier, D., Brodbeck, C., et al. (2013). MEG and EEG data analysis with MNE-Python. *Frontiers in Neuroscience, 7*, 70133.

Greenacre, M., Groenen, P. J., Hastie, T., d'Enza, A. I., Markos, A., & Tuzhilina, E. (2022). Principal component analysis. *Nature Reviews Methods Primers, 2*(1), 100.

Gross, J., Baillet, S., Barnes, G. R., Henson, R. N., Hillebrand, A., Jensen, O., et al. (2013). Good practice for conducting and reporting MEG research. *NeuroImage, 65*, 349–363.

Han, J., Pei, J., & Tong, H. (2022). *Data mining: Concepts and techniques*. Morgan Kaufmann.

Hassan, M., Boudaoud, S., Terrien, J., Karlsson, B., & Marque, C. (2011). Combination of canonical correlation analysis and empirical mode decomposition applied to denoising the labor electrohysterogram. *IEEE Transactions on Biomedical Engineering, 58*(9), 2441–2447.

Haykin, S. (2002). *Adaptive filter theory* (4th ed.). Prentice Hall. ISBN: 0-13-090126-1.

Hu, X., Le, T. H., Parrish, T., & Erhard, P. (1995). Retrospective estimation and correction of physiological fluctuation in functional MRI. *Magnetic Resonance in Medicine, 34*(2), 201–212.

Huettel, S. A., Song, A. W., & McCarthy, G. (2004). *Functional magnetic resonance imaging*. Sinauer Associates.

Huppert, T. J., Diamond, S. G., Franceschini, M. A., & Boas, D. A. (2009). HomER: A review of time-series analysis methods for near-infrared spectroscopy of the brain. *Applied Optics, 48*(10), D280–D298.

Hutton, C., Bork, A., Josephs, O., Deichmann, R., Ashburner, J., & Turner, R. (2002). Image distortion correction in fMRI: A quantitative evaluation. *NeuroImage, 16*(1), 217–240.

Hyvärinen, A., & Oja, E. (2000). Independent component analysis: Algorithms and applications. *Neural Networks, 13*(4–5), 411–430.

Ille, N., Berg, P., & Scherg, M. (2002). Artifact correction of the ongoing EEG using spatial filters based on artifact and brain signal topographies. *Journal of Clinical Neurophysiology, 19*(2), 113–124.

Jenkinson, M., Beckmann, C. F., Behrens, T. E., Woolrich, M. W., & Smith, S. M. (2012). FSL. *NeuroImage, 62*(2), 782–790.

Jezzard, P. (2012). Correction of geometric distortion in fMRI data. *NeuroImage, 62*(2), 648–651.

Jiang, X., Bian, G. B., & Tian, Z. (2019). Removal of artifacts from EEG signals: A review. *Sensors, 19*(5), 987.

Khatun, S., Mahajan, R., & Morshed, B. I. (2016). Comparative study of wavelet-based unsupervised ocular artifact removal techniques for single-channel EEG data. *IEEE Journal of Translational Engineering in Health and Medicine, 4*, 1–8.

Kim, H. C., & Lee, J. H. (2022). Systematic evaluation of recursive approach of EEG-segment-based PCA for removal of helium-pump artefact from MRI. *Electronics Letters, 58*(15), 567–569.

Klein, A., Andersson, J., Ardekani, B. A., Ashburner, J., Avants, B., Chiang, M. C., Christensen, G. E., Collins, D. L., Gee, J., Hellier, P., Song, J. H., Jenkinson, M., Lepage, C., Rueckert, D., Thompson, P., Vercauteren, T., Woods, R. P., Mann, J. J., & Parsey, R. V. (2009). Evaluation of 14 nonlinear deformation algorithms applied to human brain MRI registration. *NeuroImage, 46*(3), 786–802.

Kocsis, L., Herman, P., & Eke, A. (2006). The modified Beer–Lambert law revisited. *Physics in Medicine & Biology, 51*(5), N91.

Laird, A. R., Robinson, J. L., McMillan, K. M., Tordesillas-Gutiérrez, D., Moran, S. T., Gonzales, S. M., et al. (2010). Comparison of the disparity between Talairach and MNI coordinates in functional neuroimaging data: Validation of the Lancaster transform. *NeuroImage, 51*(2), 677–683.

Lancaster, J. L., Tordesillas-Gutierrez, D., Martinez, M., Salinas, F., Evans, A., ZilleS, K., Mazziotta, J. C., & Fox, P. T. (2007). Bias between MNI and Talairach coordinates analyzed using the ICBM-152 brain template. *Human Brain Mapping, 28*(11), 1194–1205.

Lewicki, M. S. (1998). A review of methods for spike sorting: The detection and classification of neural action potentials. *Network: Computation in Neural Systems, 9*(4), R53.

Li, Y., Wang, P. T., Vaidya, M. P., Flint, R. D., Liu, C. Y., Slutzky, M. W., & Do, A. H. (2021). Electromyogram (EMG) removal by adding sources of EMG (ERASE)—A novel ICA-based algorithm for removing myoelectric artifacts from EEG. *Frontiers in Neuroscience, 14*, 597941.

Liu, T. T. (2016). Noise contributions to the fMRI signal: An overview. *NeuroImage, 143*, 141–151.

Maclaren, J., Herbst, M., Speck, O., & Zaitsev, M. (2013). Prospective motion correction in brain imaging: A review. *Magnetic Resonance in Medicine, 69*(3), 621–636.

Magri, C., Whittingstall, K., Singh, V., Logothetis, N. K., & Panzeri, S. (2009). A toolbox for the fast information analysis of multiple-site LFP, EEG and spike train recordings. *BMC Neuroscience, 10*, 1–24.

Matcher, S. J., & Cooper, C. E. (1994). Absolute quantification of deoxyhaemoglobin concentration in tissue near infrared spectroscopy. *Physics in Medicine & Biology, 39*(8), 1295.

McMenamin, B. W., Shackman, A. J., Greischar, L. L., & Davidson, R. J. (2011). Electromyogenic artifacts and electroencephalographic inferences revisited. *NeuroImage, 54*(1), 4–9.

Mert, A., & Akan, A. (2014). Detrended fluctuation thresholding for empirical mode decomposition based denoising. *Digital Signal Processing, 32*, 48–56.

Mijović, B., De Vos, M., Gligorijević, I., Taelman, J., & Van Huffel, S. (2010). Source separation from single-channel recordings by combining empirical-mode decomposition and independent component analysis. *IEEE Transactions on Biomedical Engineering, 57*(9), 2188–2196.

Mikl, M., Mareček, R., Hluštík, P., Pavlicová, M., Drastich, A., Chlebus, P., et al. (2008). Effects of spatial smoothing on fMRI group inferences. *Magnetic Resonance Imaging, 26*(4), 490–503.

Molavi, B., & Dumont, G. A. (2012). Wavelet-based motion artifact removal for functional near-infrared spectroscopy. *Physiological Measurement, 33*(2), 259.

Mowla, M. R., Ng, S. C., Zilany, M. S., & Paramesran, R. (2015). Artifacts-matched blind source separation and wavelet transform for multichannel EEG denoising. *Biomedical Signal Processing and Control, 22*, 111–118.

Murphy, K., Birn, R. M., Handwerker, D. A., Jones, T. B., & Bandettini, P. A. (2009). The impact of global signal regression on resting state correlations: Are anti-correlated networks introduced? *NeuroImage, 44*(3), 893–905.

Mutanen, T. P., Metsomaa, J., Liljander, S., & Ilmoniemi, R. J. (2018). Automatic and robust noise suppression in EEG and MEG: The SOUND algorithm. *NeuroImage, 166*, 135–151.

Muthukumaraswamy, S. D. (2013). High-frequency brain activity and muscle artifacts in MEG/EEG: A review and recommendations. *Frontiers in Human Neuroscience, 7*, 138.

Nakamura, M., & Shibasaki, H. (1987). Elimination of EKG artifacts from EEG records: A new method of non-cephalic referential EEG recording. *Electroencephalography and Clinical Neurophysiology, 66*(1), 89–92.

Niedermeyer, E., & da Silva, F. L. (Eds.). (2005). *Electroencephalography: Basic principles, clinical applications, and related fields*. Lippincott Williams & Wilkins.

Noorbasha, S. K., & Sudha, G. F. (2021). Removal of EOG artifacts and separation of different cerebral activity components from single channel EEG—An efficient approach combining SSA–ICA with wavelet thresholding for BCI applications. *Biomedical Signal Processing and Control, 63*, 102168.

Nunez, P. L., & Srinivasan, R. (2006). *Electric fields of the brain: The neurophysics of EEG*. Oxford University Press.

Onton, J., Westerfield, M., Townsend, J., & Makeig, S. (2006). Imaging human EEG dynamics using independent component analysis. *Neuroscience & Biobehavioral Reviews, 30*(6), 808–822.

Oostenveld, R., Fries, P., Maris, E., & Schoffelen, J. M. (2011). FieldTrip: Open source software for advanced analysis of MEG, EEG, and invasive electrophysiological data. *Computational Intelligence and Neuroscience, 2011*, 1–9.

Parrish, T. B., Gitelman, D. R., LaBar, K. S., & Mesulam, M. M. (2000). Impact of signal-to-noise on functional MRI. *Magnetic Resonance in Medicine, 44*(6), 925–932.

Pascual-Marqui, R. D. (2002). Standardized low-resolution brain electromagnetic tomography (sLORETA): Technical details. *Methods and Findings in Experimental and Clinical Pharmacology, 24*(Suppl D), 5–12.

Penny, W. D., Friston, K. J., Ashburner, J. T., Kiebel, S. J., & Nichols, T. E. (Eds.). (2011). *Statistical parametric mapping: The analysis of functional brain images*. Elsevier.

Pettersen, K. H., & Einevoll, G. T. (2008). Amplitude variability and extracellular low-pass filtering of neuronal spikes. *Biophysical Journal, 94*(3), 784–802.

Pfeuffer, J., Van De Moortele, P. F., Ugurbil, K., Hu, X., & Glover, G. H. (2002). Correction of physiologically induced global off-resonance effects in dynamic echo-planar and spiral functional imaging. *Magnetic Resonance in Medicine, 47*(2), 344–353.

Piazza, C., Bacchetta, A., Crippa, A., Mauri, M., Grazioli, S., Reni, G., et al. (2020). Preprocessing pipeline for fNIRS data in children. In *XV Mediterranean conference on medical and biological engineering and computing—MEDICON 2019: Proceedings of MEDICON 2019, September 26–28, 2019, Coimbra, Portugal* (pp. 235–244). Springer International Publishing.

Pinti, P., Siddiqui, M. F., Levy, A. D., Jones, E. J. H., & Tachtsidis, I. (2021). An analysis framework for the integration of broadband NIRS and EEG to assess neurovascular and neurometabolic coupling. *Scientific Reports, 11*(1), 3977.

Power, J. D., Schlaggar, B. L., & Petersen, S. E. (2015). Recent progress and outstanding issues in motion correction in resting state fMRI. *NeuroImage, 105*, 536–551.

Rahman, M. A., Khanam, F., Ahmad, M., & Uddin, M. S. (2020). Multiclass EEG signal classification utilizing Rényi min-entropy-based feature selection from wavelet packet transformation. *Brain Informatics, 7*(1), 7.

Raj, D., Anderson, A. W., & Gore, J. C. (2001). Respiratory effects in human functional magnetic resonance imaging due to bulk susceptibility changes. *Physics in Medicine and Biology, 46*(12), 3331–3340.

Ranjan, R., Sahana, B. C., & Bhandari, A. K. (2021). Ocular artifact elimination from electroencephalography signals: A systematic review. *Biocybernetics and Biomedical Engineering, 41*(3), 960–996.

Ranjan, R., Sahana, B. C., & Bhandari, A. K. (2022). Cardiac artifact noise removal from sleep EEG signals using hybrid denoising model. *IEEE Transactions on Instrumentation and Measurement, 71*, 1–10.

Rey, H. G., Pedreira, C., & Quiroga, R. Q. (2015). Past, present and future of spike sorting techniques. *Brain Research Bulletin, 119*, 106–117.

Rossant, C., Kadir, S. N., Goodman, D. F., Schulman, J., Hunter, M. L., Saleem, A. B., et al. (2016). Spike sorting for large, dense electrode arrays. *Nature Neuroscience, 19*(4), 634–641.

Sacchet, M. D., & Knutson, B. (2013). Spatial smoothing systematically biases the localization of reward-related brain activity. *NeuroImage, 66*, 270–277.

Safari, M., Shalbaf, R., Bagherzadeh, S., & Shalbaf, A. (2024). Classification of mental workload using brain connectivity and machine learning on electroencephalogram data. *Scientific Reports, 14*(1), 9153.

Salimi-Khorshidi, G., Douaud, G., Beckmann, C. F., Glasser, M. F., Griffanti, L., & Smith, S. M. (2014). Automatic denoising of functional MRI data: Combining independent component analysis and hierarchical fusion of classifiers. *NeuroImage, 90*, 449–468.

Santosa, H., Zhai, X., Fishburn, F., & Huppert, T. (2018). The NIRS brain AnalyzIR toolbox. *Algorithms, 11*(5), 73.

Satterthwaite, T. D., Elliott, M. A., Gerraty, R. T., Ruparel, K., Loughead, J., Calkins, M. E., et al. (2013). An improved framework for confound regression and filtering for control of motion artifact in the preprocessing of resting-state functional connectivity data. *NeuroImage, 64*, 240–256.

Sheoran, P., & Saini, J. S. (2020). A new method for automatic electrooculogram and eye blink artifacts correction of EEG signals using CCA and NAPCT. *Procedia Computer Science, 167*, 1761–1770.

Sheoran, M., Kumar, S., & Chawla, S. (2015). Methods of denoising of electroencephalogram signal: A review. *International Journal of Biomedical Engineering and Technology, 18*(4), 385–395.

Stachaczyk, M., Atashzar, S. F., & Farina, D. (2020). Adaptive spatial filtering of high-density EMG for reducing the influence of noise and artefacts in myoelectric control. *IEEE Transactions on Neural Systems and Rehabilitation Engineering, 28*(7), 1511–1517.

Stergiadis, C., Kostaridou, V. D., & Klados, M. A. (2022). Which BSS method separates better the EEG signals? A comparison of five different algorithms. *Biomedical Signal Processing and Control, 72*, 103292.

Strangman, G., Boas, D. A., & Sutton, J. P. (2002). Non-invasive neuroimaging using near-infrared light. *Biological Psychiatry, 52*(7), 679–693.

Strother, S. C. (2006). Evaluating fMRI preprocessing pipelines. *IEEE Engineering in Medicine and Biology Magazine, 25*(2), 27–41.

Sutton, B. P., Noll, D. C., & Fessler, J. A. (2003). Fast, iterative image reconstruction for MRI in the presence of field inhomogeneities. *IEEE Transactions on Medical Imaging, 22*(2), 178–188.

Sweeney, K. T., Ward, T. E., & McLoone, S. F. (2012). Artifact removal in physiological signals—Practices and possibilities. *IEEE Transactions on Information Technology in Biomedicine, 16*(3), 488–500.

Tabelow, K., Polzehl, J., Voss, H. U., & Spokoiny, V. (2006). Analyzing fMRI experiments with structural adaptive smoothing procedures. *NeuroImage, 33*(1), 55–62.

Tadel, F., Baillet, S., Mosher, J. C., Pantazis, D., & Leahy, R. M. (2011). Brainstorm: A user-friendly application for MEG/EEG analysis. *Computational Intelligence and Neuroscience, 2011*, 1–13.

Talairach, J., & Tournoux, P. (1988). *Co-planar stereotaxic atlas of the human brain.* Georg Theme Verlag.

Tanabe, J., Miller, D., Tregellas, J., Freedman, R., & Meyer, F. G. (2002). Comparison of detrending methods for optimal fMRI preprocessing. *NeuroImage, 15*(4), 902–907.

Tatum, W. O., Dworetzky, B. A., & Schomer, D. L. (2011). Artifact and recording concepts in EEG. *Journal of Clinical Neurophysiology, 28*(3), 252–263.

Urigüen, J. A., & Garcia-Zapirain, B. (2015). EEG artifact removal—State-of-the-art and guidelines. *Journal of Neural Engineering, 12*(3), 031001.

Van Dijk, K. R., Sabuncu, M. R., & Buckner, R. L. (2012). The influence of head motion on intrinsic functional connectivity MRI. *NeuroImage, 59*(1), 431–438.

Verstynen, T. D., & Deshpande, V. (2011). Using pulse oximetry to account for high and low frequency physiological artifacts in the BOLD signal. *NeuroImage, 55*(4), 1633–1644.

White, T., O'Leary, D., Magnotta, V., Arndt, S., Flaum, M., & Andreasen, N. C. (2001). Anatomic and functional variability: The effects of filter size in group fMRI data analysis. *NeuroImage, 13*(4), 577–588.

Wise, R. G., Ide, K., Poulin, M. J., & Tracey, I. (2004). Resting fluctuations in arterial carbon dioxide induce significant low frequency variations in BOLD signal. *NeuroImage, 21*(4), 1652–1664.

Xiong, F., Pan, Y., & Bai, L. (2023). Research applications of functional magnetic resonance imaging (fMRI) in neuroscience. In *PET/MR: Functional and molecular imaging of neurological diseases and neurosciences* (pp. 47–78). Springer.

Yan, C. G., Cheung, B., Kelly, C., Colcombe, S., Craddock, R. C., Di Martino, A., et al. (2013). A comprehensive assessment of regional variation in the impact of head micromovements on functional connectomics. *NeuroImage, 76*, 183–201.

Yan, X., Boudrias, M. H., & Mitsis, G. D. (2022). Removal of transcranial alternating current stimulation eeg artifacts using blind source separation and wavelets. *IEEE Transactions on Biomedical Engineering, 69*(10), 3183–3192.

Ye, J. C., Tak, S., Jang, K. E., Jung, J., & Jang, J. (2009). NIRS-SPM: Statistical parametric mapping for near-infrared spectroscopy. *NeuroImage, 44*(2), 428–447.

Yücel, M. A., Selb, J., Aasted, C. M., Lin, P. Y., Borsook, D., Becerra, L., & Boas, D. A. (2016). Mayer waves reduce the accuracy of estimated hemodynamic response functions in functional near-infrared spectroscopy. *Biomedical Optics Express, 7*(8), 3078–3088.

Zhang, M., Schwemmer, M. A., Ting, J. E., Majstorovic, C. E., Friedenberg, D. A., Bockbrader, M. A., et al. (2018). Extracting wavelet based neural features from human intracortical recordings for neuroprosthetics applications. *Bioelectronic Medicine, 4*, 1–14.

Zhang, L., Pini, L., & Corbetta, M. (2023). Different MRI structural processing methods do not impact functional connectivity computation. *Scientific Reports, 13*(1), 8589.

Machine Learning with Brain Data

Feature Engineering and Analysis Methods

Contents

© The Author(s), under exclusive license to Springer Nature Switzerland AG 2025
U. Chaudhary, *Expanding Senses using Neurotechnology*, https://doi.org/10.1007/978-3-031-76081-5_5

This chapter talks about brain signal analysis, focusing on the crucial steps of feature extraction and selection, followed by the application of machine learning techniques. It begins by detailing the inherent complexity of brain signals and the necessity for robust feature engineering to transform these signals into a more interpretable and manageable form. Various types of features—time-domain, frequency-domain, time-frequency domain, and spectral domain—are explored, each providing unique insights into brain function. The feature extraction process is vital for reducing data dimensionality while preserving essential information, which is critical for subsequent analyses. The chapter also discusses different machine learning methodologies applied to the refined data, including supervised, semi-supervised, unsupervised, and reinforcement learning, each offering distinct advantages for specific applications in brain–computer interfaces (BCIs) and other neurotechnological applications. Practical examples of how these methods are applied to real-world data, such as EEG, fMRI, and MEG signals, illustrate the theoretical concepts discussed. The groundwork laid here sets the stage for further detailed discussions in subsequent volumes, focusing on the expansive applications of these technologies in neurology, rehabilitation, and beyond.

Learning Objectives

1. Understand the Complexity of Brain Signals: Recognize the inherent complexities in brain signals and the necessity of robust feature engineering to make these signals interpretable and manageable for further analysis. Exploration of feature extraction and feature selection techniques.
2. Explore Machine Learning Applications: Investigate different machine learning methodologies such as supervised, semi-supervised, unsupervised, reinforcement learning, and deep learning.
3. Apply Machine Learning to Real-World Data: Through practical examples, learn how to apply machine learning techniques to real-world brain signal data such as EEG, fMRI, and MEG, emphasizing the transformative potential of these methods in enhancing the functionality of BCIs and improving the quality of life for individuals with disabilities.
4. Link Theoretical Concepts to Practical Applications: Integrate the theoretical concepts of feature engineering and machine learning with practical demonstrations, illustrating their utility in actual neurotechnology applications and setting the stage for advanced studies.

Brain signals are inherently complex and laden with vast informational content. To manage this complexity, feature engineering is employed to transform processed brain signals into more expressive, interpretable features that capture the essence of the underlying neural processes. This transformation is crucial for reducing the dimensionality of the data while preserving the vital information contained within the brain signals.

A feature can be defined as any measurable property or characteristic of a phenomenon being observed, and in the context of brain signals, features may be binary, categorical, or continuous (Guyon & Elisseeff, 2006). The diversity in feature definition allows for flexibility in how data are summarized and interpreted.

Feature extraction, also known as feature engineering, involves isolating important signal characteristics—attributes that differentiate one form of brain activity from another—from irrelevant noise (Krusienski et al., 2012). This process refines the data into a format that is both compact and meaningful, making it suitable for subsequent analysis either by human experts or computational models (Ren et al., 2014).

The feature extraction process follows the initial stages of signal acquisition and noise removal. It involves identifying and collecting key attributes from the brain signals, which are then used to address specific problems or hypotheses in neuroscience research. This step is pivotal as it simplifies the data and enhances the efficiency and effectiveness of the analytical models applied to understand and interpret brain function.

In this chapter, you will explore the various features of brain signals and the diverse machine-learning techniques applied to these features to study brain processes and develop interfaces between brain signals and external devices. This intersection of neuroscience and technology is encapsulated in brain–computer or brain–machine interfaces (BCI/BMI). These interfaces represent a groundbreaking field that extends motor and sensory capabilities, offering significant enhancements for individuals with disabilities. The foundational concepts introduced here lay the groundwork for a more detailed examination in Volume 2, where the potential of BCI/BMI technology will be discussed extensively. Volume 2 also explores how these technologies can transform accessibility and improve quality of life, providing new dimensions of interaction for those with impairments.

This chapter will briefly introduce the names of various methods used in the different steps of analysis of brain data. Detailed discussions on the application and intricacies of these methods are beyond the scope of this text, as they are extensively covered in numerous specialized books. For those interested in a deeper understanding of the exact algorithms and their implementation, I recommend consulting these comprehensive resources (Kotsiantis et al., 2006, 2007; Ayodele, 2010; Lemm et al., 2011; Mwangi et al., 2014; Shalev-Shwartz & Ben-David, 2014; Krishnan & Athavale, 2018; Hu & Zhang, 2019; Carleo et al., 2019; Hosseini et al., 2020; Badillo et al., 2020; Greener et al., 2022; Sharifani & Amini, 2023). Finally, you will also see several examples illustrating the complete signal processing workflow, from pre-processing to machine learning techniques applied to EEG, fMRI, fNIRS, MEG, and invasive recordings.

5.1 Brain Signals' Feature Extraction and Selection

Brain signal feature extraction is essential in analyzing neural data, bridging raw neurophysiological recordings and interpretable outcomes in various applications, including diagnostic tools and brain–computer interfaces (BCI). This section elucidates the nature of brain signal features and the methodologies employed in feature extraction, which are crucial for enhancing the effectiveness of machine learning algorithms applied to brain data.

5.1.1 Brain Signal Feature Extraction

Brain signal features are quantifiable attributes or characteristics derived from neurophysiological data that effectively represent the underlying brain activity in a simplified form. These features can be broadly categorized based on their nature into time-domain, frequency-domain, time-frequency domain, and spectral domain features, each offering unique insights into the brain's functioning.

5.1.1.1 Time-Domain Features

These are the simplest forms of features and include statistical measures such as mean, median, variance, skewness, and kurtosis of the signal amplitude over time. Time-domain features are directly derived from the amplitude variations in the neural recordings and are particularly useful for capturing the raw temporal dynamics of the neural activity (Cohen, 2014; Riaz et al., 2015; Vidaurre et al., 2009).

5.1.1.2 Frequency-Domain Features

These features are derived from the analysis of the spectrum of brain signals and include power spectral density, spectral entropy, and band powers within specific frequency bands (delta, theta, alpha, beta, and gamma). Frequency-domain features are crucial for understanding the oscillatory dynamics of the brain and have been extensively used in both research and clinical applications (Niedermeyer & da Silva, 2005).

5.1.1.3 Time-Frequency Domain Features

These features provide a combined view of how the power of different frequency bands evolves over time and include measures derived from wavelet transforms or short-time Fourier transforms. Time-frequency features are valuable for analyzing non-stationary signals like EEG, where the spectral properties change over time (Cohen, 2014).

5.1.1.4 Spectral Domain Features

Spectral-domain features, extracted from the frequency analysis of brain signals, help understand the oscillatory activity of the brain and are extensively utilized in various neuroscientific research and clinical applications. These features provide a detailed representation of how power distribution across different frequency bands correlates with cognitive states, emotional responses, and pathological conditions (Herman et al., 2008).

◻ Figure 5.1 presents an overview of various features categorized across four distinct domains. ◻ Table 5.1 provides an overview of some of feature extraction approaches (Yadav et al., 2020).

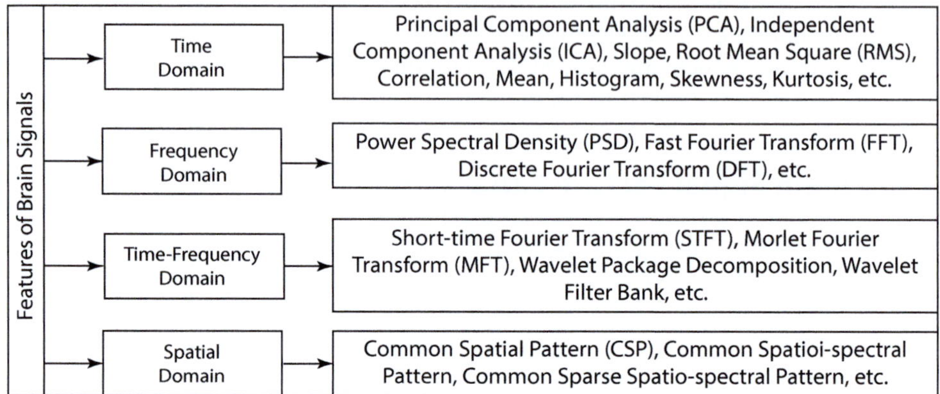

◻ **Fig. 5.1** An overview of different feature extraction methods—feature extraction is a pivotal process in the analysis and interpretation of data. This process involves reducing the amount of resources required to describe a large set of data accurately. When performing feature extraction, essential information is extracted from raw data and transformed into a reduced set of features, which can then be efficiently processed while maintaining the integrity of the original dataset. Some of the widely used methods are **Principal Component Analysis (PCA)**: PCA is a statistical technique used to emphasize variation and bring out strong patterns in a dataset. The method transforms the original variables into a new set of variables, which are linear combinations of the original variables. These new variables, or principal components, are orthogonal and ranked according to the variance of data along them, with the first principal component having the largest possible variance. This method is widely used for dimensionality reduction in data preprocessing, noise reduction, and exploratory data analysis. **Linear Discriminant Analysis (LDA)**: LDA is used to find the linear combination of features that best separates two or more classes of objects or events. The resulting combination is used as a linear classifier or, more commonly, for dimensionality reduction before classification. The goal is to maximize the ratio of between-class variance to the within-class variance in any particular dataset, thereby ensuring maximum separability. **Independent Component Analysis (ICA)**: ICA is a computational method for separating a multivariate signal into additive subcomponents. This technique is based on the assumption that the subcomponents are non-Gaussian signals and statistically independent from each other. ICA is pivotal for isolating independent sources from multichannel brain recordings, such as EEG. This method is fundamental in identifying artifacts or separate functional components within the brain signals, helping to clarify underlying neural mechanisms or identifying sources of noise such as eye blinks or heartbeats. **Wavelet Transforms**: Utilized extensively in EEG signal processing, wavelet transforms allow for the decomposition of brain signals into components at various frequencies and resolutions. This capability is crucial for analyzing non-stationary signals, like EEG, which vary in frequency and amplitude over time, thereby enabling detailed examination of transient, high-frequency events alongside more stable, low-frequency trends. **Autoencoders**: In deep learning applications related to brain signal analysis, autoencoders help in reducing data dimensionality and extracting meaningful features from large-scale neural data. By learning to compress and decompress the input data (e.g., from high-dimensional neural recordings), autoencoders facilitate the identification of inherent data structures and patterns, which is essential for tasks such as anomaly detection or neural decoding in BCIs. Each of these feature extraction methods is adapted to the specific characteristics and challenges of brain signal analysis, aiding in the precise interpretation of neural activities and enhancing the performance of computational models used in neuroscience and clinical diagnostics

> **Box 5.1   Feature Extraction**
> The process of transforming raw data into a set of measurable characteristics or features that capture the essential information, making the data more amenable to modeling or further analysis. For more details see the legend of ◘ Fig. 5.1.

> **Box 5.2   Time-Domain Features**
> Attributes derived from signal measurements taken directly in the time domain, such as amplitude, latency, rise time, and duration, which are used to describe signal characteristics without transforming the data into another domain.

Depending on the specific brain signal and its intended application, different features outlined in ◘ Fig. 5.1 may be utilized. Interestingly, the method for extracting these features shares the same name as the feature itself. This transformation process, known as feature extraction in brain signal analysis, involves condensing processed brain data into a manageable set of key features that preserve the necessary information for various applications. This step not only streamlines the analytical process but also substantially improves the effectiveness of machine learning models, as detailed in ▸ Sect. 5.2, by decreasing dimensionality and eliminating superfluous or redundant data.

◘ **Table 5.1**   Overview of various feature extraction approaches

Classifying property	Approach	Characteristics
Dimension reduction	Principal component analysis (PCA)	– **Linear transformation**: Based on linear transformation concepts – **Uncorrelated variables**: Transforms correlated observations into a set of uncorrelated variables – **Optimal representation**: Achieves optimal data representation through minimal mean squared error (MSE) – **Classification limitation**: No guarantee of producing useful classification results – **Noise reduction**: Effective for noise reduction and dimensionality reduction – **Artifact independence**: Artifacts do not correlate with EEG signals
	Independent component analysis (ICA)	– **Signal segregation**: Segregates mixed signals into their original sources – **Statistical independence**: Assumes mutual statistical independence of underlying sources – **Artifact removal**: Robust approach for artifact removal – **Artifact independence**: Artifacts do not correlate with EEG signals. – **Power spectrum corruption**: Prone to corrupting the power spectrum of the signals

(continued)

5

■ **Table 5.1** (continued)

Classifying property	Approach	Characteristics
Space	Common spatial pattern (CSP)	– **Spatial filtering**: Based on spatial filtering for 2-class problems with multiclass extensions – **Spatial resolution dependency**: Performance is highly dependent on spatial resolution – **Feature extraction**: Extracts features that maximize variance between two classes
Time-frequency	AutoRegressive components (AR)	– **Spectrum model**: Typically used as a spectrum model for EEG signal analysis – **Frequency resolution**: Offers high frequency resolution for short time segments – **Non-stationary signals**: Generally unsuitable for non-stationary signals – **Adaptive variant**: Includes an adaptive variant known as Multivariate Adaptive AutoRegressive (MVAAR)
	Matched filtering (MF)	– **Pattern detection**: Detects specific patterns based on their matches with predetermined known signals or templates – **Temporal features**: Suitable for detecting waveforms with consistent temporal features – **Signal-to-noise ratio**: Enhances the signal-to-noise ratio for specific patterns
	Continuous wavelet transform (CWT)	– **Frequency and temporal information**: Provides both frequency and temporal information of signals – **Non-stationary signals**: Appropriate for analyzing non-stationary signals – **Resolution trade-off**: Offers a trade-off between time and frequency resolution.
	Discrete wavelet transform (DWT)	– **Frequency and temporal information**: Provides both frequency and temporal information – **Non-stationary signals**: Suitable for analyzing non-stationary signals – **Reduced redundancy**: Offers reduced redundancy and complexity compared to CWT – **Multi-resolution analysis**: Enables multi-resolution analysis by decomposing signals into different frequency components

> **Box 5.3   Frequency-Domain Features**
> Features extracted by analyzing the frequency components of signals, such as power spectral density or specific frequency band powers (e.g., delta, theta, alpha, beta, gamma waves in EEG analysis).

> **Box 5.4   Time-Frequency Domain Features**
> Features that provide information about both the frequency content of the signal and how it changes over time, often extracted using methods like short-time Fourier transform (STFT) or wavelet transform.

> **Box 5.5   Spectral Domain Features**
> Characteristics obtained from the signal spectrum, providing insights into the distribution of power across different frequencies within a signal.

5.1.2   Brain Signal Feature Selection

Once feature extraction has effectively condensed the processed brain signals into a more manageable and informative set of features, the next imperative step is to determine the most relevant features that contribute significantly to the predictive modeling task. Feature selection methods play a pivotal role here, aiming to remove redundant or irrelevant features, thereby reducing the dimensionality of the data further. This reduction not only simplifies the computational demands but also improves the accuracy and generalizability of subsequent machine-learning models (Dash & Liu, 1997; Guyon & Elisseeff, 2003; Chandrashekar & Sahin, 2014; Li et al., 2017). By focusing on the most influential features, researchers can ensure that the computational resources are allocated efficiently, leading to more robust and interpretable models that are better suited for practical applications in brain–computer interfaces and other areas of neuroscience (McFarland & Wolpaw, 2005).

Thus, feature selection in brain signal analysis is a crucial step aimed at improving the efficiency and effectiveness of machine learning models by identifying and retaining the most relevant features from the data. This process reduces computational complexity, enhances model interpretability, and potentially improves prediction performance by removing irrelevant or redundant information that could lead to overfitting (Dash & Liu, 1997; Guyon & Elisseeff, 2003; Chandrashekar & Sahin, 2014; Li et al., 2017). We will now have an overview of some of the common feature selection methods used in brain signal analysis.

5.1.2.1 Filter Methods

Filter methods are the most straightforward feature selection techniques used in brain signal analysis. They evaluate the importance of features based on univariate statistics (e.g., correlation with the target variable) before the machine learning model is run. These methods are computationally efficient and have the advantage of being independent of any learning algorithm. Some of the common statistical measures used include:

- **Correlation Coefficient**: This involves measuring the correlation between each feature and the target variable. Features with low correlation are considered irrelevant and are removed from the dataset (Guyon & Elisseeff, 2003).
- **Mutual Information**: This measures the amount of information one can obtain about one random variable by observing another. It is particularly useful in feature selection for it assesses the dependency between the features and the target variable without assuming any linear relationship (Kraskov et al., 2004).
- **Chi-Squared Test**: This test is used primarily for categorical features to determine the independence between each feature and the target.
- **ANOVA F-Test**: Analyzes variance among group means in a sample, typically for continuous variables.

5.1.2.2 Wrapper Methods

Wrapper methods evaluate subsets of variables, allowing, unlike filter approaches, to find the best subset for the model's performance. This method involves selecting a subset of features, using a model to train them, and calculating performance to gauge the effectiveness of that subset (Dash & Liu, 1997; Guyon & Elisseeff, 2003). The process is repeated until an optimal subset is selected. Techniques include:

- **Sequential Feature Selection**: This can be either forward selection, where features are iteratively added to the model, or backward elimination, where features are removed from a full model until an optimal subset is achieved.
- **Recursive Feature Elimination**: This is an aggressive elimination strategy that recursively removes attributes and builds a model on those remaining attributes. It uses model accuracy to identify which attributes (or combination of attributes) contribute the most to predicting the target attribute.

5.1.2.3 Embedded Methods

Embedded methods integrate feature selection as part of the model training process and are algorithm-specific (Lal et al., 2006). They combine the qualities of filter and wrapper methods, being more accurate than filter methods while being less computationally intensive than wrapper methods. Examples include:

- **Lasso (L1 Regularization)**: Lasso is a regression analysis method that performs both variable selection and regularization in order to enhance the prediction accuracy and interpretability of the statistical model it produces. Lasso includes a penalty term that constrains the size of the coefficient estimates, effectively reducing some coefficients to zero and thus performing feature selection (Muthukrishnan & Rohini, 2016).
- **Decision Trees**: Trees are built by splitting a node into two or more sub-nodes based on the feature that results in the best split (maximizes the separation in the data). Features used at the top (near the root) of the trees are typically more important than features used at the leaves (Kingsford & Salzberg, 2008).

5.1.2.4　Dimensionality Reduction

Although not strictly a feature selection method, dimensionality reduction is often used in brain signal analysis to reduce the number of random variables under consideration by obtaining a set of principal variables. Techniques such as principal component analysis (PCA) and linear discriminant analysis (LDA) transform features into a lower-dimensional space where the axes represent the directions of maximum variance (in PCA) or maximum separation between classes (in LDA) (Martinez & Kak, 2001).

5.1.2.5　Hybrid Methods

Hybrid methods combine the strengths of more than one of the abovementioned methods to improve model performance and reduce the likelihood of selecting redundant features. A typical hybrid method could start with a filter method to reduce the number of features and then apply a wrapper or embedded method to refine the feature set further. One example is the genetic algorithm, which is used as part of a hybrid method by using a filter-based method to reduce the number of features and then applying a genetic algorithm to explore the space of feature subsets (Oh et al., 2004).

Figure 5.2 presents an overview of various feature selection methods. Table 5.2 provides an overview of some of feature selection approaches (Yadav et al., 2020).

> **Box 5.6　Feature Selection**
> A technique used to select a subset of relevant features for use in model construction. This helps improve model performance by eliminating irrelevant or redundant data, reducing overfitting, and enhancing generalization. For more details see the legend of Fig. 5.2.

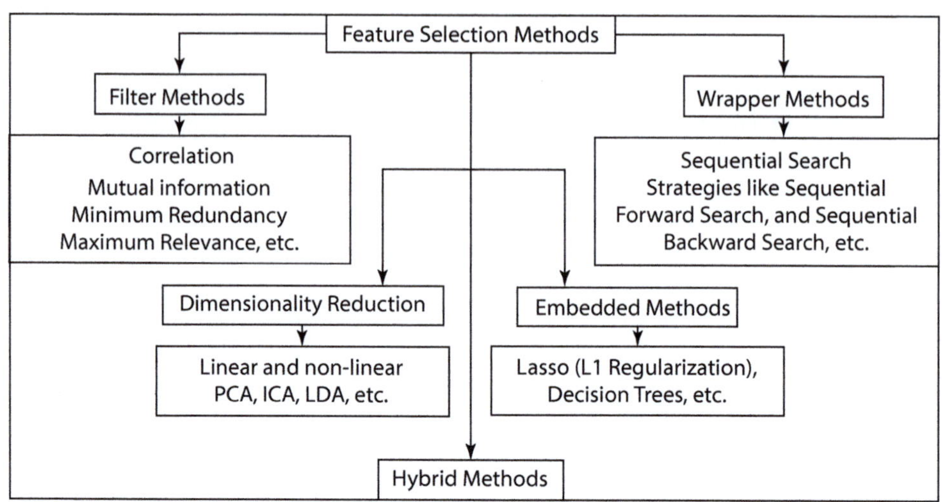

Fig. 5.2 Feature selection is a critical step in the analysis of brain signals, where the goal is to identify the most relevant features from a vast dataset, such as those obtained from EEG (electroencephalography), MEG (magnetoencephalography), or fMRI (functional magnetic resonance imaging). Effective feature selection can significantly enhance the performance of machine learning models by eliminating redundant or irrelevant features, thus improving both the accuracy and efficiency of subsequent analyses. **Filter Methods**: These methods apply a statistical measure to assign a scoring to each feature based on their correlation with the outcome variable. Features are selected independently of any machine learning algorithms. In brain signal analysis, filter methods might involve correlation coefficients, chi-squared test, or mutual information scores to evaluate the relationship between signal features and neural states or responses. These methods are computationally efficient, making them suitable for high-dimensional datasets like those from neuroimaging studies. **Wrapper Methods**: Wrapper methods consider the selection of a subset of features as a search problem, where different combinations are prepared, evaluated, and compared with each other. A predictive model is used to score each combination of features based on their predictive power. Common techniques include recursive feature elimination, which is often used with SVM (support vector machines) to find the best subset of EEG features for classifying different cognitive states in a BCI application. **Embedded Methods**: These methods perform feature selection during the model training process and are specific to given learning algorithms. Examples include Lasso and Ridge regression, which introduce a penalty term for complexity (number of features) to the model fitting process. In brain signal analysis, embedded methods are particularly useful in scenarios like decoding fMRI patterns, where the selection of voxels (3D pixels in brain scans) is inherently linked to the regression model used to predict different brain states or activities. **Hybrid Methods**: Hybrid methods combine filter and wrapper methods to try and exploit the strengths of both. An initial filter might be used to reduce the dimensionality before a wrapper method is applied to search for an optimal subset of features. This approach is valuable in brain signal data, which is typically complex and high-dimensional, as it balances between the thoroughness of wrapper methods and the speed of filter methods. Each of these methods has its merits and limitations and can be chosen based on the specific requirements of the brain signal analysis task, such as the nature of the data, the computational resources available, and the specific objectives of the study. Effective feature selection not only improves the interpretability of the models but also enhances their predictive performance, which is crucial for advancing neurotechnology applications like BCIs and precision medicine in neurology

■ **Table 5.2** An overview of some of feature selection approaches

Approach	Characteristics
Genetic algorithm (GA)	– **High-resource consumption**: Requires substantial computational resources due to complex operations and large search spaces – **Premature convergence**: Possibility of converging on suboptimal solutions prematurely, leading to less than optimal feature selection outcomes – **Evolutionary technique**: Uses mechanisms inspired by biological evolution, such as selection, crossover, and mutation, to optimize feature subsets
Sequential forward selection (SFS)/ sequential backward selection (SBS)	– **Suboptimal approaches**: Typically result in suboptimal feature subsets due to their greedy nature, which only considers local optimality – **Stepwise selection**: SFS adds features sequentially, while SBS removes features sequentially, based on their contribution to the model's performance – **Simple implementation**: Easy to implement and understand, but may not find the best subset of features
Sequential forward floating search (SFFS)/sequential backward floating search (SBFS)	– **Variants of SFS/SBS**: Modified versions of SFS and SBS to enhance performance – **Plus l-take away r approach**: Incorporates a strategy where features can be added and removed dynamically, allowing more flexibility compared to standard SFS/SBS – **Improved performance**: Partially overcomes the limitations of SFS and SBS by avoiding local optima more effectively, leading to better feature subsets – **Dynamic adjustment**: Adjusts the selection process dynamically to potentially re-evaluate previous decisions, improving the overall feature selection process

5.2 Machine Learning with Brain Data: Why?

Moving from feature extraction to feature selection and machine learning creates a streamlined and powerful framework for analyzing brain signals. Initially, feature extraction transforms the raw brain data into a set of manageable and interpretable attributes, capturing the essential characteristics of the neural activity. Following this, feature selection refines the dataset further by isolating the most significant features, effectively reducing noise, complexity, and computational burden. This process is critical for enhancing the efficacy of the machine learning models that follow, as it ensures that only the most relevant features are included, thereby improving both the speed and accuracy of the learning algorithms. Machine learning is essential for analyzing brain data due to the complexity and vast volume of information that the brain generates. Some of the reasons why machine learning is particularly suited to this task are:

- **Complex Pattern Recognition**: The human brain is an intricate network that produces highly complex and dynamic signals. Machine learning algorithms excel at identifying patterns in data that are too subtle or complex for traditional statistical methods to detect. This capability is crucial for interpreting brain signals, which often contain nuanced information that can indicate cognitive states, neural disorders, or responses to stimuli (Mitchell et al., 2004; Raghavendra et al., 2020).
- **Handling High-Dimensional Data**: Brain data, whether derived from EEG, fMRI, or other neuroimaging techniques, typically involve thousands of measurements that create high-dimensional datasets. Machine learning can effectively manage high-dimensional data, extracting useful insights without becoming bogged down by the curse of dimensionality—a problem where performance worsens as the dimensionality of the data increases (Haufe et al., 2014; Sadiq et al., 2021).
- **Predictive Modeling**: Machine learning provides powerful predictive capabilities that are essential for applications such as brain–computer interfaces (BCIs) or for predicting the onset of neurological events like seizures. These models can learn from past data to forecast future brain activities, enabling proactive interventions or responsive technologies that adapt to a user's mental state (Siddiqui et al., 2020).
- **Adaptability**: Machine learning models can adapt to the unique neurological patterns of individual users, improving over time with more data. This adaptability is vital for personalized medicine, where treatments and monitoring need to be tailored to individual physiological and neurological responses (Bonkhoff & Grefkes, 2022).
- **Automation and Efficiency**: Analyzing brain data manually is time-consuming and prone to error, especially given the volume and velocity of data produced by modern neuroimaging technologies. Machine learning automates the data analysis process, significantly increasing efficiency and allowing researchers to focus on higher-level interpretation and decision-making (Lemm et al., 2011).
- **Decoding and Mapping Brain Activity**: Machine learning techniques are instrumental in decoding the information represented by brain activity, aiding in mapping functional areas of the brain. This is crucial for both scientific understanding and clinical applications, such as surgical planning for epilepsy treatment or rehabilitation strategies for stroke victims (Mitchell et al., 2004; Glaser et al., 2020; Bonkhoff & Grefkes, 2022).
- **Development of Diagnostic Tools**: Machine learning algorithms can help develop new diagnostic tools that identify patterns associated with neurological diseases and disorders. By learning from a broad range of cases, these models can help clinicians diagnose conditions earlier and with greater accuracy than traditional methods.

Thus, the application of machine learning to brain data is not just beneficial; it is essential for advancing our understanding of the brain and improving clinical neurology and psychology. This technology enables us to process complex data efficiently, uncover patterns invisible to the naked eye, and translate these findings into practical applications that enhance human health and cognitive capabilities. In the next section, you will learn about different machine-learning algorithms that are typically applied to brain data.

5.3 Machine Learning with Brain Data: How?

Machine learning algorithms can be broadly classified into four categories: supervised, semi-supervised, unsupervised, and reinforcement learning, as shown in ◘ Fig. 5.3 (Mishra et al., 2023).

5.3.1 Supervised Learning

In supervised learning, the process involves training a machine using a labeled dataset provided by the operator, which contains both the inputs and the desired outputs. The task of the machine learning algorithm is to develop a model that can map these inputs to their correct outputs. Although the operator already knows the solutions to these problems, the algorithm works by recognizing patterns within the data, learning from these observations, and then making predictions based on this learning. Throughout this training process, the algorithm's predictions are evaluated and corrected by the operator, and this iterative correction continues until the algorithm reaches an optimal level of accuracy and performance (Kotsiantis et al., 2007).

For instance, consider an ambient lighting system in a home controlled by brain signals, interfacing the user's brain activity with a digital lighting system. To toggle the light ON (1) or OFF (0), the system must distinguish between two distinct brain signal patterns corresponding to these outputs. Specifically, an increase in the alpha band power (if the system is EEG signal-based) or an increase in oxyhemoglobin concentration (if the system is fNIRS signal-based) triggers the light to turn ON. At the same time, a decrease leads to it turning OFF. In this setup, each brain signal input is clearly labeled with a corresponding desired output (ON or OFF), thus illustrating the fundamental concept of supervised learning, where the model is trained to associate specific inputs with predefined outputs.

Thus, as shown in ◘ Fig. 5.3, supervised learning algorithms leverage predefined inputs and known outputs to develop a model based on data points with established outcomes. Once trained, the system can make predictions or decisions about new data based on this prior training.

Supervised learning is typically categorized into two types: regression and classification. Regression algorithms, including linear regression, nonlinear regression, Gaussian process regression, and regression trees, are primarily used to predict continuous variables. On the other hand, classification algorithms such as support vector machines (SVM), logistic regression, decision trees, naïve Bayes, discriminant analysis, and k-nearest neighbor (kNN) are designed to categorize data into discrete classes, typically binary outcomes like yes-no, male-female, true-false, etc. (Lemm et al., 2011; Badillo et al., 2020; Greener et al., 2022).

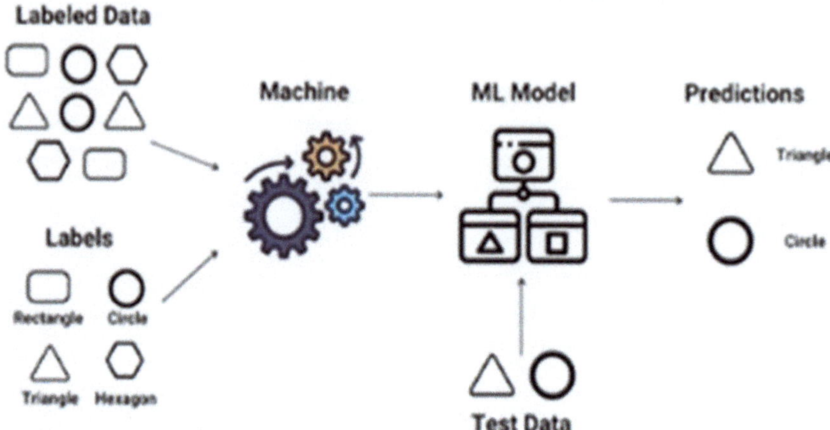

Fig. 5.3 An overview of supervised machine learning—supervised machine learning is a cornerstone of computational neuroscience and brain signal analysis, playing a crucial role in interpreting complex data from brain imaging and electrophysiological recordings. In supervised learning, models are trained on a labeled dataset containing inputs paired with the correct outputs, enabling the algorithm to learn the mapping between the two. Once trained, these models can predict outcomes for new, unseen data. This methodology is particularly useful in applications ranging from decoding mental states from EEG signals to predicting neurological outcomes based on fMRI data. **Key Concepts of Supervised Machine Learning are—Training and Testing**: The process involves dividing the available data into training and testing sets. The training set is used to teach the model to recognize patterns that correlate inputs with outputs, while the testing set is used to evaluate the model's performance on unseen data. This separation helps in assessing the generalizability and robustness of the model. **Model Types**: **Linear Models** (e.g., Linear Regression, Logistic Regression): These models are often used for their simplicity and interpretability, especially in tasks like fMRI-based activity prediction where understanding the contribution of each brain region is crucial. **Support Vector Machines (SVM)**: SVMs are powerful for classification tasks, such as distinguishing between different cognitive states in EEG data, due to their ability to handle high-dimensional space and their robustness in maximizing the margin between different classes. **Decision Trees and Random Forests**: These models are useful for their ability to handle nonlinear relationships and their intrinsic feature selection capabilities, making them suitable for complex brain signal classification where multiple signal features contribute to the outcomes. **Neural Networks**: Particularly deep learning models are increasingly used for their ability to model complex, hierarchical relationships in data, such as temporal dynamics in EEG or spatial patterns in fMRI, which are critical for tasks like seizure prediction or brain activity mapping. **Performance Metrics**: The choice of metrics for evaluating a supervised learning model depends on the specific task. Common metrics include accuracy, precision, recall, F1 score for classification tasks, and mean squared error for regression tasks. In brain signal analysis, it is also important to consider clinical relevance, such as the ability of a model to improve diagnostic accuracy or predict therapeutic outcomes. **Challenges in Brain Signal Analysis—High Dimensionality**: Brain data typically involves thousands of features (e.g., voxels, electrode channels), which can lead to overfitting. Techniques such as dimensionality reduction or regularization are often employed to address this. **Noise and Variability**: Brain signals are inherently noisy and subject to variability both within and across individuals, posing challenges in creating robust models. **Interpretability**: In clinical and neuroscience contexts, models must not only be accurate but also interpretable, as understanding the

underlying biological or cognitive processes is crucial. **Applications—Disease Diagnosis**: Using patterns recognized in brain imaging to identify markers of neurological diseases such as Alzheimer's or schizophrenia. **BCI**: Interpreting EEG signals to translate into commands in a brain–computer interface, enabling communication for individuals with severe motor disabilities. **Cognitive and Emotional State Recognition**: Analyzing patterns in brain signals to infer the subject's mental state, which can be used in neuromarketing or psychological research. (From Mishra et al., 2023)

> **Box 5.7 Supervised Learning**
> A type of machine learning that involves training a model on a labeled dataset, where the input data are paired with correct outputs (labels), to learn the mapping from inputs to outputs. For more details see the legend of ▪ Fig. 5.3.

5.3.2 Semi-Supervised Learning

Semi-supervised learning is akin to supervised learning, yet it incorporates both labeled and unlabeled data. Labeled data come with informative tags that clarify the content for the algorithm, facilitating understanding and processing. In contrast, unlabeled data do not have these descriptive tags. By utilizing a mix of both labeled and unlabeled data, machine learning algorithms are equipped to infer labels for the unlabeled data, effectively extending their learning capacity beyond the explicitly provided examples (Kotsiantis et al., 2006; Lemm et al., 2011; Badillo et al., 2020; Greener et al., 2022).

Let us revisit our ambient lighting system concept; a commercially available system relying solely on supervised learning may face limitations, as not all users may have the ability to modulate their alpha band power (in EEG signals) or oxy-hemoglobin concentration (for systems based on fNIRS signals). Implementing a semi-supervised learning approach could potentially broaden the product's appeal. This system would retain predefined labels from the supervised learning setup but also possess the capability to adapt by processing new, unlabeled data, thereby enhancing its versatility and user-friendliness.

Thus, as shown in ▪ Fig. 5.4, semi-supervised learning algorithms leverage both labeled and unlabeled input data to build a model. This approach is typically divided into clustering and classification.

> **Box 5.8 Semi-Supervised Learning**
> A learning technique that combines a small amount of labeled data with a large amount of unlabeled data during training, which is useful when acquiring labeled data is expensive or labor-intensive. For more details see the legend of ▪ Fig. 5.4.

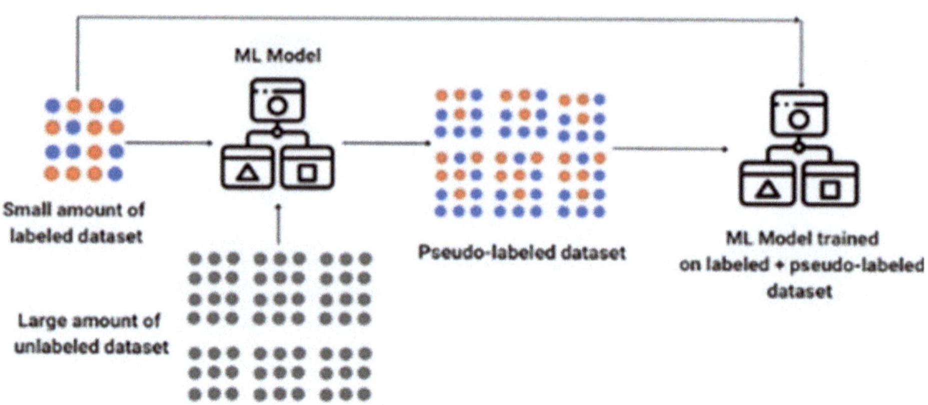

Fig. 5.4 An overview of semi-supervised machine learning—semi-supervised machine learning is an increasingly important method in the field of brain signal analysis, particularly beneficial where acquiring labeled data can be time-consuming or costly, such as in clinical diagnostics and neuroscience research. Semi-supervised learning sits between supervised and unsupervised learning, leveraging both labeled and unlabeled data to build more robust models. This approach is especially relevant in brain signal analysis where the richness of unlabeled data (e.g., continuous EEG recordings) can be harnessed to enhance learning accuracy and efficiency. **Fundamental Concepts of Semi-Supervised Machine Learning are—Combining Labeled and Unlabeled Data**: Semi-supervised learning algorithms utilize a small amount of labeled data alongside a larger set of unlabeled data. The idea is that the unlabeled data, when used correctly, can provide a wealth of information about the structure of the feature space that can help guide the learning process and improve model performance. **Approaches Used: Self-Training**: A supervised model is first trained with the available labeled data. It is then used to predict labels for the unlabeled data, incorporating the most confident predictions back into the training set in subsequent iterations. This method is useful in situations like artifact detection in EEG signals, where labeled examples are limited. **Co-training**: Two models are simultaneously trained on separate views of the data (for instance, spatial vs. temporal features of brain signals). Each model then labels unlabeled examples for the other to train on, iteratively improving each other. **Graph-Based Methods**: These methods construct a graph where nodes represent samples (labeled and unlabeled) and edges represent similarity between samples. Label information is propagated from labeled to unlabeled nodes based on their similarity, aiding in tasks like clustering similar neural patterns across different brain regions or states. **Generative Models**: These models attempt to learn the underlying distribution of both labeled and unlabeled data. Learning this joint distribution helps in better understanding and classifying brain signal data, which is crucial for complex decision-making tasks in neurological studies. **Challenges in Brain Signal Analysis—Data Quality and Complexity**: Brain signals are complex and often contaminated with noise (e.g., movement artifacts), making the distinction between relevant and irrelevant features difficult in both labeled and unlabeled datasets. **Variability in Data**: Variability between subjects (inter-subject variability) and within subjects (intra-subject variability) in brain data can significantly affect the reliability of semi-supervised learning models, necessitating sophisticated normalization, and standardization techniques. **Scalability**: Managing and processing large datasets with both labeled and unlabeled data demand significant computational resources, especially when dealing with high-dimensional data such as multi-channel EEG or whole-brain fMRI scans. **Applications—Neurological Disorder Diagnosis**: Enhancing the diagnostic accuracy by incorporating unlabeled patient data to refine the models that identify patterns associated with disorders like epilepsy or autism. **Sleep Stage Classification**: Leveraging unlabeled segments of sleep recordings to improve the classification models trained with smaller labeled datasets. **Functional Connectivity Analysis**: Using unlabeled data to explore and map the functional connections within the brain, supporting research into cognitive processes and brain disorders.

Ethical Considerations: In clinical settings, the reliability of semi-supervised methods must be thoroughly validated to prevent misdiagnoses. Furthermore, ethical considerations regarding the use of patient data, especially unlabeled data, must be addressed to ensure privacy and consent compliance. (From Mishra et al., 2023)

5.3.3 Reinforcement Learning

Reinforcement learning centers on structured learning strategies, where the machine learning algorithm is equipped with a series of actions, parameters, and goals. The algorithm experiments with various strategies within the defined framework, assessing each outcome to identify the most effective approach. This method embodies the trial-and-error learning paradigm, where the machine refines its actions based on previous outcomes and continuously adjusts its strategy to optimize performance and achieve desired results (Lemm et al., 2011; Badillo et al., 2020; Greener et al., 2022).

Reconsidering our ambient lighting system example, imagine a scenario where the system learns to control the lighting based on the user's individual strategies for turning the lights ON and OFF, without any predefined requirement to alter specific brain signal features or perform designated tasks. The users employ their unique methods to interact with the system, which in turn adapts to these personal strategies. This adaptive capability of the system exemplifies the principles of reinforcement learning.

Thus, as shown in ◘ Fig. 5.5, in reinforcement learning, algorithms learn through direct interaction with their environment, striving to achieve specific goals. This process involves agents—programs designed to perform certain tasks within a given environment, which can be either real or virtual. The agent executes actions, which are changes that affect the environment's state. Decisions about which actions to take are governed by policies, the set of rules the agent follows. The outcomes of these actions are assessed by rewards, which can be positive or negative, providing feedback to the agent.

Reinforcement learning systems are typically categorized into model-free and model-based approaches. Model-free algorithms do not attempt to build a detailed model of the environment. Instead, they learn the optimal policy through direct interaction, akin to a trial-and-error method. These algorithms are subdivided into value-based methods, like Q-learning, deep Q neural networks (DQN), and state-action-reward-state-action (SARSA), which aim to ascertain the best strategy by precisely evaluating the value function for each state. On the other hand, policy-based methods, such as policy gradient techniques, determine the optimal policy directly without calculating a value function (Lemm et al., 2011; Badillo et al., 2020; Greener et al., 2022).

Conversely, model-based algorithms develop an explicit model of the environment, which the agent uses to predict the outcomes of actions, including the expected rewards and subsequent states. This method allows the agent to systematically explore and learn about the environment, optimizing its policy based on the model's predictions about future states and rewards.

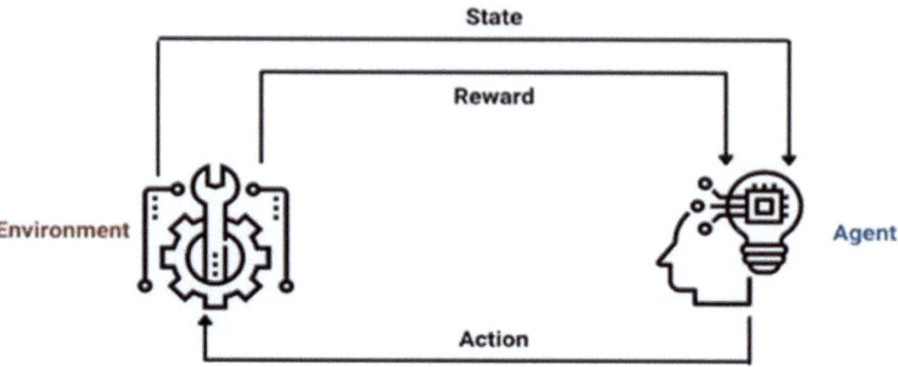

Fig. 5.5 An overview of reinforcement machine learning—reinforcement learning (RL) is a type of machine learning that is distinctly suited for making sequential decisions, learning optimal actions through trial-and-error interactions with a dynamic environment. This methodology is increasingly being applied within the context of brain signal analysis, where it holds significant potential for developing adaptive brain–computer interfaces (BCIs) and personalized neurotherapeutic strategies. In RL, an agent learns to achieve a goal in an uncertain, potentially complex environment by performing actions and receiving feedback in the form of rewards or punishments. **Core Principles of Reinforcement Learning—Agent and Environment**: The RL framework consists of an agent that interacts with its environment. In brain signal analysis, the agent could be a software interfacing with neural activity data, where it attempts to optimize a therapeutic or communicative outcome based on the user's brain signals. **State, Action, Reward**: The agent makes observations about the state of the environment (e.g., current brain signal patterns) and based on these observations, takes actions (e.g., adjusting the parameters of a neurostimulator) aimed at maximizing some notion of cumulative reward (e.g., symptom relief, communication effectiveness). **Policy**: The policy is a strategy used by the agent to decide the next action based on the current state. It can be deterministic or stochastic. In brain signal applications, developing robust policies that can adapt to the variability in individual neural patterns is crucial. **Value Function**: This function estimates what the future rewards might be, given a state and an action. It helps in evaluating which states are beneficial in the long run. In neurological contexts, this could involve predicting the long-term benefits of different treatment protocols in neurorehabilitation. **Model of the Environment**: Some RL approaches require a model of the environment, which is a prediction of what the next state and reward will be, given a state and action. In brain signal analysis, this might involve modeling how neural responses to various stimuli or interventions will result in changes in mental state or cognitive function. **Challenges in Applying RL to Brain Signal Analysis—High Dimensionality**: Brain data features high-dimensional input spaces, which can make the state space intractably large. **Noise and Variability**: Neural signals are notoriously noisy and exhibit significant variability, complicating the process of reliably mapping states to rewards. **Real-Time Processing**: Many brain signal applications, such as BCIs, require real-time processing and decision-making, presenting computational challenges. **Applications of RL in Brain Signal Analysis—Adaptive BCIs**: RL can be used to continuously refine the interface's responses based on ongoing user feedback, enhancing the efficacy and user satisfaction of BCIs. **Neurotherapy**: For conditions like Parkinson's disease or epilepsy, RL algorithms can help in dynamically adjusting treatment plans based on patient-specific responses observed through neuroimaging or electrophysiological data. **Cognitive Rehabilitation**: RL methods can be applied to develop adaptive cognitive training programs, where the difficulty and type of tasks are adjusted in real-time based on the user's performance and brain activity. **Ethical and Practical Considerations—Safety and Reliability**: Ensuring the safety and reliability of RL-driven systems in clinical settings is paramount, as inappropriate actions could have significant adverse effects. **Transparency and Control**: Maintaining transparency in how decisions are made and allowing for human override where necessary are important to ensure trust and control in automated systems. (From Mishra et al., 2023)

> **Box 5.9 Reinforcement Learning**
> A type of machine learning where an agent learns to make decisions by performing actions in an environment to maximize some notion of cumulative reward. For more details see the legend of ▫ Fig. 5.5.

5.3.4 Unsupervised Learning

In unsupervised learning, the algorithm independently explores data to discern underlying patterns and relationships, operating without predefined answers or direct human guidance. The focus is on discovering correlations and insights by processing the available data. The machine learning algorithm undertakes the task of interpreting and structuring large datasets, often by clustering similar data points or organizing them into a coherent format. Over time, as it processes more information, the algorithm's capacity for decision-making and data interpretation enhances, allowing it to offer increasingly sophisticated analyses and insights (Lemm et al., 2011; Badillo et al., 2020; Greener et al., 2022).

Unsupervised learning is not directly applicable to our ambient lighting system scenario, as this type of learning does not use predefined outputs. Nevertheless, unsupervised learning can be indirectly beneficial in such systems. It can process brain data to develop a model that enhances a lighting system that incorporates semi-supervised or reinforcement learning methods. This indirect application allows the unsupervised learning algorithm to contribute to understanding and modeling the data, which can then improve the system's overall functionality.

Thus, as shown in ▫ Fig. 5.6, unsupervised learning algorithms work solely with input datasets and do not require predefined outputs, enabling the machine to infer patterns without external guidance.

This approach is typically divided into clustering and dimensionality reduction. Clustering algorithms—such as K-means, k-medoids, hierarchical clustering, self-organizing maps, fuzzy c-means, and Gaussian mixtures—group data based on shared similarities, organizing them into clusters. Dimensionality reduction techniques simplify datasets by reducing the number of input variables, aiming to streamline classification and enhance model fitting with minimal information loss. Examples include principal component analysis, factor analysis, independent component analysis, and random projection (Ayodele, 2010; Lemm et al., 2011; Carleo et al., 2019; Badillo et al., 2020; Greener et al., 2022). While unsupervised learning can handle more complex tasks than supervised learning, it poses greater challenges in terms of implementation.

Finally, it is important to emphasize a particular group of machine learning algorithms known as deep learning. These algorithms are notable for their distinctive and powerful characteristics.

Unsupervised Learning

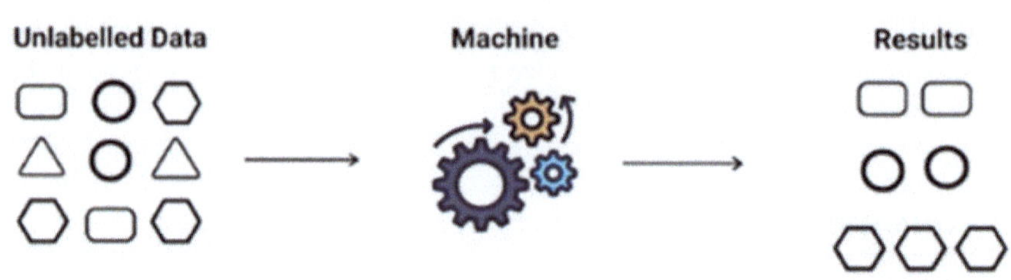

Fig. 5.6 An overview of unsupervised machine learning—unsupervised machine learning is a powerful analytical technique used extensively in brain signal analysis, particularly valuable for uncovering hidden patterns and intrinsic structures in data without prior labeling. This form of machine learning is crucial in neuroscientific studies where the complexity and volume of data, such as those derived from EEG (electroencephalography), MEG (magnetoencephalography), or fMRI (functional magnetic resonance imaging), can be immense and not always accompanied by explicit labels. **Core Concepts of Unsupervised Machine Learning—Data Exploration**: Unsupervised learning algorithms explore data to find natural clustering, groupings, or representations that explain significant variations in the data. This is especially useful in brain signal analysis for identifying subtypes of neurological conditions based on signal patterns or discovering stages of brain development without predefined categories. **Common Algorithms—Clustering**: Methods like k-means clustering, hierarchical clustering, and DBSCAN (density-based spatial clustering of applications with noise) are used to group data points that exhibit similar characteristics. In brain signal analysis, clustering can be applied to segment brain regions that respond similarly to certain stimuli or to categorize patient populations by similarity in brain activity patterns. **Association**: This involves finding rules or patterns that describe large portions of the data, such as frequent co-occurrences of brain signal patterns across different regions during specific cognitive tasks. **Dimensionality Reduction**: Techniques such as PCA (principal component analysis) and t-SNE (t-distributed stochastic neighbor embedding) reduce the number of random variables under consideration, by extracting a smaller number of essential features from complex data sets. These techniques are particularly important in neuroimaging data analysis, where reducing data dimensions can help in visualizing and processing high-dimensional datasets more effectively. **Anomaly Detection**: Identifying unusual data points that do not fit well with the general pattern of the data. This is crucial in clinical diagnostics to detect abnormal brain activities indicative of disorders like seizures or tumors. **Challenges in Brain Signal Analysis—Complexity and Variability**: Neural data are inherently complex and variable not only across different individuals but also within the same individual under different conditions. This variability makes unsupervised learning particularly challenging as it must uncover meaningful patterns without overfitting to random noise. **Scalability**: Brain data are typically large in scale; thus, unsupervised learning algorithms need to be efficient enough to handle large datasets without compromising the speed and accuracy of the analysis. **Interpretability**: While unsupervised learning can reveal new insights, the interpretations of these findings must be cautiously approached, especially in a clinical or neuroscience context, to avoid erroneous conclusions. **Applications of Unsupervised Learning in Brain Signal Analysis—Functional Brain Network Identification**: Mapping functional connectivity patterns among different brain regions to understand underlying neural mechanisms or to study brain network alterations in disorders like schizophrenia or Alzheimer's disease. **Sleep Stage Classification**: Analyzing EEG signals to identify and categorize different stages of sleep automatically, based on the intrinsic patterns and transitions observed in the brain signals. **Genotype-Phenotype Mapping**: Relating genetic data with phenotypic brain imaging data to identify genetic markers that influence brain structure and function, often performed without predefined hypotheses about the relationships. **Ethical Considerations—Bias and Generalization**: Care must be taken to ensure that the algorithms do not perpetuate biases present in the data, and that findings are generalizable across different populations. **Data Privacy**: Handling brain data involves sensitive information, requiring stringent measures to ensure privacy and confidentiality. (From Mishra et al., 2023)

> **Box 5.10 Unsupervised Learning**
> Machine learning techniques used to identify patterns in data without the need for human-labeled responses or outputs, often used for clustering or anomaly detection. For more details see the legend of ▫ Fig. 5.6.

5.4 Deep Learning with Brain Data

Deep learning relies on multi-layered neural networks, consisting of input, hidden, and output layers, to process extensive datasets by mimicking the human brain's structure and functions. The input layer receives data features, while multiple hidden layers—ranging from a few to several depending on the problem's complexity—process these inputs through weighted connections and generate outputs. The final layer, the output layer, delivers the results for the provided inputs. Deep learning algorithms progressively enhance result accuracy through iterative computation and prediction at each layer (LeCun et al., 2015). Artificial neural networks (ANNs), convolutional neural networks (CNNs), and recurrent neural networks (RNNs). ANNs, also known as simulated neural networks (SNNs), form the foundation of deep learning. They are designed to emulate biological neural networks, capable of learning complex nonlinear relationships through activation functions, making them potent universal function approximators (Zhang et al., 2021a; Chaki & Woźniak, 2023; Ranjan et al., 2024).

Deep learning (DL) models are broadly classified into three categories based on their learning functionalities: discriminative, generative, and hybrid models. Discriminative models, such as multilayer perceptrons (MLPs) (Murtagh, 1991), convolutional neural networks (CNNs) (Gu et al., 2018), recurrent neural networks (RNNs) (Connor et al., 1994; Schuster & Paliwal, 1997), and their variants, focus on distinguishing between different types of data inputs. CNNs are specialized for processing data with inherent correlations (e.g., spatial or temporal), using filters or kernels to extract significant features through convolution operations. These networks feature several layers that detect and interpret patterns at increasing levels of complexity, from simple to intricate (Shrestha & Mahmood, 2019). RNNs incorporate feedback connections in their architecture, enabling them to handle sequential data effectively. This makes them particularly suitable for time-series analysis. Popular RNN variants include long short-term memory (LSTM) networks, side-output residual networks (SRNs), and gated recurrent units (GRUs), each tailored for specific types of sequential data processing (de Bardeci et al., 2021; Supakar et al., 2022).

Generative models like generative adversarial networks (GANs) (Goodfellow et al., 2020), restricted Boltzmann machines (RBMs) (Hinton, 2012), self-organizing maps (SOMs) (Kohonen, 1990), deep belief networks (DBNs) (Hinton, 2009), and autoencoders (AEs), along with their derivatives, are designed to generate new data instances that mimic the training data (Bank et al., 2023).

Hybrid models combine the characteristics of both discriminative and generative models, often integrating deep transfer learning (DTL) and deep reinforce-

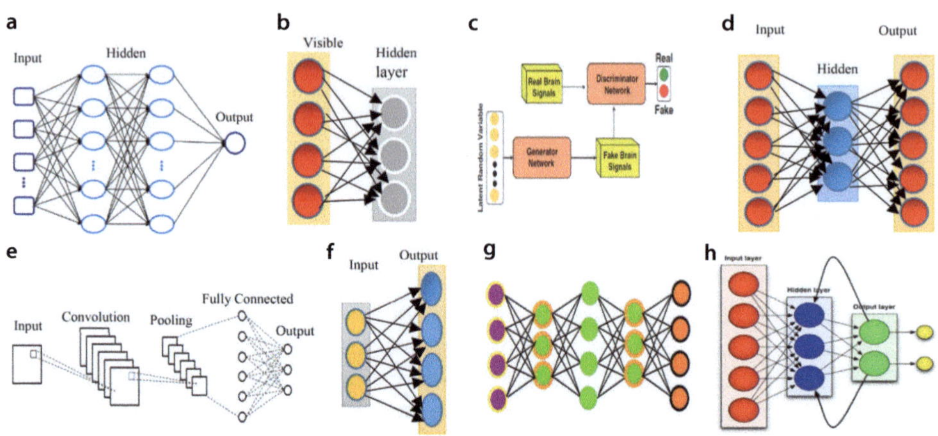

◘ Fig. 5.7 Illustrations of diverse deep learning architectures include: **a** Multilayer perceptron (MLP), which utilizes multiple layers to process input data sequentially through each layer; **b** Restricted Boltzmann machine (RBM), a stochastic neural network that efficiently learns the probability distribution over its set of inputs; **c** Generative adversarial network (GAN), where two neural networks contest with each other to generate new, synthetic instances of data that can pass for real data; **d** Autoencoder (AE), which is designed to learn efficient codings of the input data; **e** Convolutional neural network (CNN), primarily used for processing data that has a grid-like topology, such as images; **f** Self-organizing map (SOM), which produces a low-dimensional, discretized representation of the input space of the training samples; **g** Deep belief network (DBN), a generative model with multiple layers of latent variables; and **h** Recurrent neural network (RNN), which processes sequences by maintaining a state that can represent information from an arbitrary length of sequence. (From Ranjan et al., 2024)

ment learning (DRL) to enhance learning efficiency and adaptability across diverse scenarios. These models leverage the strengths of both approaches to improve performance and learning capabilities. ◘ Figure 5.7 provides a visual representation of these deep-learning models (Ranjan et al., 2024).

The fundamental distinction between deep learning and traditional machine learning lies in their data processing capabilities and learning approaches. Traditional machine learning algorithms typically require structured and labeled data to train effectively, relying on predefined tags to classify and identify objects within the data. In contrast, deep learning thrives on its ability to handle unstructured and unlabeled data. Deep learning methods automatically extract features and patterns directly from the data by utilizing a hierarchical structure of multilayered neural networks. This capability enables the system to enhance its accuracy in identifying and classifying data through progressive learning and adaptation without manual labeling. Deep learning methods can be both supervised and unsupervised. Finally, deep learning models can also engage in reinforcement learning, a sophisticated form of learning where the model enhances its accuracy by receiving positive feedback from prior actions. While distinct from unsupervised learning, this advanced method allows the model to adapt and improve autonomously by evaluating its previous outcomes and modifying future actions accordingly to

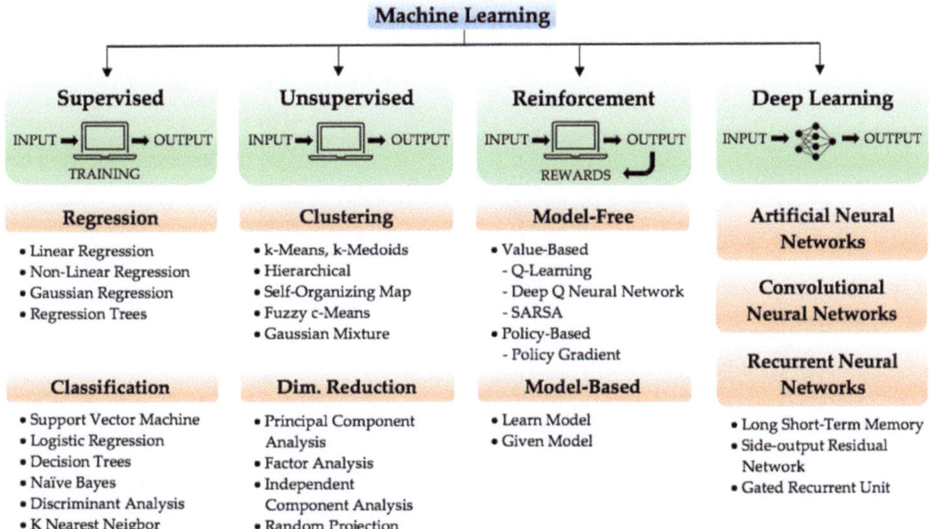

Fig. 5.8 Machine learning algorithms categorized into four primary types: supervised, unsupervised, reinforcement, and deep learning methods. (From Luján et al., 2021 originally published under CC-BY 4.0)

achieve better results (Janiesch et al., 2021). Figure 5.8 presents a selection of machine-learning algorithms (Luján et al., 2021). Note that the methods depicted in Fig. 5.8 are not comprehensive.

Having navigated through the intricate processes of feature extraction and selection, and the diverse array of machine and deep learning methods, we now transition to a practical demonstration of how these theoretical frameworks are applied to actual brain signal data. In the forthcoming section, we will explore specific examples that illustrate the application of these advanced techniques. This practical application not only exemplifies the theoretical concepts discussed but also highlights these methods' real-world utility and transformative potential in neurotechnology. We will see how these methodologies are employed across various types of brain signals, such as EEG, fMRI, fNIRS, MEG, and invasive providing a clear view of their effectiveness in enhancing our understanding and interaction with brain functions.

Table 5.3 provides an overview of some of classification approaches (Yadav et al., 2020).

Box 5.11 Neural Networks

Computational models composed of interconnected groups of artificial neurons, where each connection can transmit a signal from one neuron to another, designed to approximate complex functions and capable of learning from data.

Table 5.3 An overview of some classification approaches

Classifying property	Approach	Characteristics
Generative model	Bayesian analysis	– **Class belongingness**: Observed feature vector is associated with the class with maximum belongingness. – **Decision boundaries**: Produces nonlinear decision boundaries. – **BCI systems**: Usually unpopular in BCI systems due to complexity and computational demands.
Linear	Linear discriminant analysis (LDA)	– **Ease of use**: Simple classifier with adequate accuracy. – **Computational efficiency**: Relatively low computational requirements, making it suitable for real-time applications. – **Noise sensitivity**: Poor performance in the presence of noise or outliers, requiring careful preprocessing of data. – **Regularization**: Requires regularization to handle multicollinearity and overfitting. – **Class variants**: Typically based on two classes; multiclass variants available for more complex tasks. – **Enhanced versions**: Better-quality versions include BLDA (Bayesian LDA) and FLDA (Fisher's LDA).
	Support vector machine (SVM)	– **Flexibility**: Offers both linear and nonlinear (Gaussian) approaches, adaptable to various data structures. – **Class options**: Supports both binary and multiclass classification, enhancing versatility. – **Optimization**: Maximizes the margin between classes by maximizing the distance between the nearest training samples and the decision hyperplane. – **Noise sensitivity**: Performance can degrade in the presence of noise or outliers, necessitating robust preprocessing. – **Regularization**: Regularization techniques are essential to prevent overfitting and handle noisy data. – **Performance**: High-speed performance, suitable for large datasets and real-time applications.
Nonlinear	k-nearest neighbor classifier (k-NNC)	– **Multiclass approach**: Naturally supports multiclass classification, making it suitable for a wide range of applications. – **Distance metric**: Utilizes metric distances (e.g., Euclidean) between the test feature and its neighbors for classification. – **Feature vector dimension**: Works best with low-dimensional feature vectors, as high dimensionality can reduce performance due to the curse of dimensionality. – **Dimensionality sensitivity**: Highly sensitive to the dimensionality of feature vectors, which can impact accuracy and computation time.
	Artificial neural network (ANN)	– **Multiclass approach**: Supports multiclass classification, handling complex patterns in data. – **Flexibility**: Exhibits excellent flexibility, capable of modeling nonlinear relationships. – **Architectures**: Numerous architectures exist, including probabilistic neural networks (PNN), fuzzy ARTMAP ANN, finite impulse response neural network (FIRNN), and probability estimating guarded neural classifier (PeGNC). – **Learning capabilities**: Can learn from large datasets and improve performance over time. – **Complexity**: Computationally intensive, requiring significant resources and careful tuning of hyperparameters.

5.5 Feature Engineering and Machine Learning with EEG

5.5.1 Example 1

In this example, the authors (Samal & Hashmi, 2024) provide a detailed overview of utilizing EEG technology to detect the emotional states of individuals, illustrating the specific methodologies employed, as shown in ▪ Fig. 5.9 (Samal & Hashmi, 2024).

Once the EEG signal has been acquired from the region under investigation, different pre-processing steps, as discussed in ▶ Chap. 4, are applied to prevent noise contamination that can impact further classification. To recap from ▶ Chap. 4 and merge with the feature extraction and selection and machine learning methods described in this chapter, the steps involved in analyzing EEG signals for emotion detection are:

- **Artifact Removal**: Various biological signals or external interferences can introduce artifacts into EEG data. These unwanted signals may arise from activities such as blinking (EOG), eye or muscle movements (EMG), heartbeats (ECG), or external sources. Due to their similar amplitudes, distinguishing these artifacts from true EEG data can be challenging. To ensure the EEG signal is free from these artifacts and suitable for accurate feature extraction, noise removal or attenuation is essential (Garcés & Orosco, 2008).

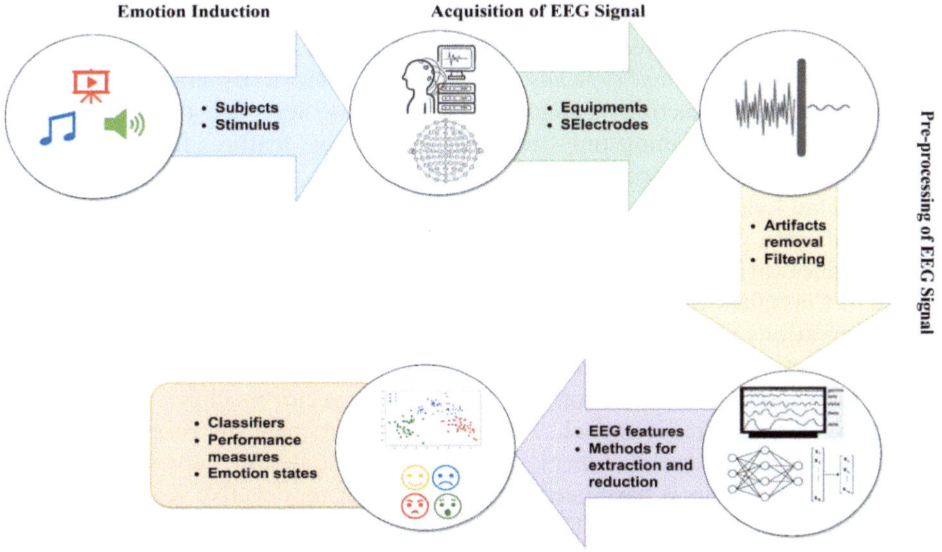

▪ **Fig. 5.9** A summary of the signal processing techniques used on EEG data for emotion detection. (From Samal & Hashmi, 2024 originally published under CC-BY 4.0)

Popular preprocessing methods like independent component analysis (ICA), principal component analysis (PCA), common spatial patterns (CSP), and common average reference (CAR) are widely used, with ICA being particularly effective for artifact removal.

- **Filtering**: Frequency domain filters help eliminate artifacts in EEG signals by narrowing the bandwidth under analysis without altering or distorting the signals. Common filters include notch, high-pass, and low-pass filters. High-pass filters are typically set with a cutoff frequency below 0.5 Hz to remove very low-frequency components like breathing, while low-pass filters with a cutoff around 50–60 Hz reduce high-frequency noise. Notch filters are designed to specifically block a single frequency, such as the 50 Hz power supply frequency, ensuring its complete exclusion.

- **Baseline Correction and Removal**: For many emotion identification applications, such as those discussed by Jiménez-Guarneros and Alejo-Eleuterio (2022), it is crucial to correct or remove the baseline or pre-stimulus signal during preprocessing. This allows for effective comparison with the post-stimulus signal. The sliding window technique is commonly employed for this purpose, with window sizes and methods tailored to the specific requirements of the study.

- **Feature Extraction**: The next step is feature extraction after preprocessing and noise reduction. For emotion recognition using EEG data, extracting features that accurately represent a person's emotional state is essential. Classification algorithms then use these features to identify emotions. The effectiveness of emotion recognition largely depends on the quality of the features extracted. Commonly, EEG features are analyzed in the time, frequency, and time-frequency domains, as shown in ▶ Sect. 5.1.

- **Feature Selection** or reduction is crucial in EEG-based emotion recognition to manage BCI systems' typically large feature vectors. By employing feature selection and reduction strategies, as shown in ▶ Sect. 5.1, the complexity is minimized by retaining only the most informative features for the classifier. This approach enhances the efficiency and accuracy of model training.

- **Machine Learning**: Emotion recognition primarily involves categorizing input signals into predefined classes. Selecting an effective classifier that accurately identifies various emotions is crucial for a successful emotion classification system. Classifiers use mathematical functions to predict categories in unknown observations within a validation dataset, as described in ▶ Sect. 5.3. Various classification methods are used in affective computing to analyze EEG data. Basic machine learning algorithms like support vector machines (Zhang et al., 2016), decision trees (Li et al., 2022a, b, c, d), and linear discriminant analysis, alongside advanced classifiers such as recurrent neural networks and long short-term memory, have been employed. Other models like K-nearest neighbor (KNN) (Mehmood & Lee, 2016) and Random Forest (RF) (Zhang et al., 2021b) are also popular for emotion recognition.

These steps are standard across various brain signal analyses. However, the specific methods employed within each step can vary based on the application and the characteristics of the signal, as demonstrated in the subsequent examples in this chapter.

5.5.2 **Example 2**

In this example, authors (Ranjan et al., 2024) have used deep learning methods for diagnosing Schizophrenia with EEG signals. The process, as usual, includes several stages: signal acquisition, pre-processing, signal transformation (if necessary), feature extraction, classification through deep learning models, and finally, signal post-processing. After acquiring the signals, they are integrated into the deep learning model in three distinct ways: (1) EEG signals are directly fed into the deep learning model after basic pre-processing; (2) After pre-processing, key features are extracted from the signals and then input into the deep learning model; (3) In the final method, EEG signals undergo preprocessing and are then transformed into images, such as time-frequency representations (TFR), which are subsequently used in deep learning models for analysis. ◘ Figure 5.10 pictorially summarizes the methods used (Ranjan et al., 2024).

As you might expect by now, these methods can be adapted to various brain signals depending on the specific application.

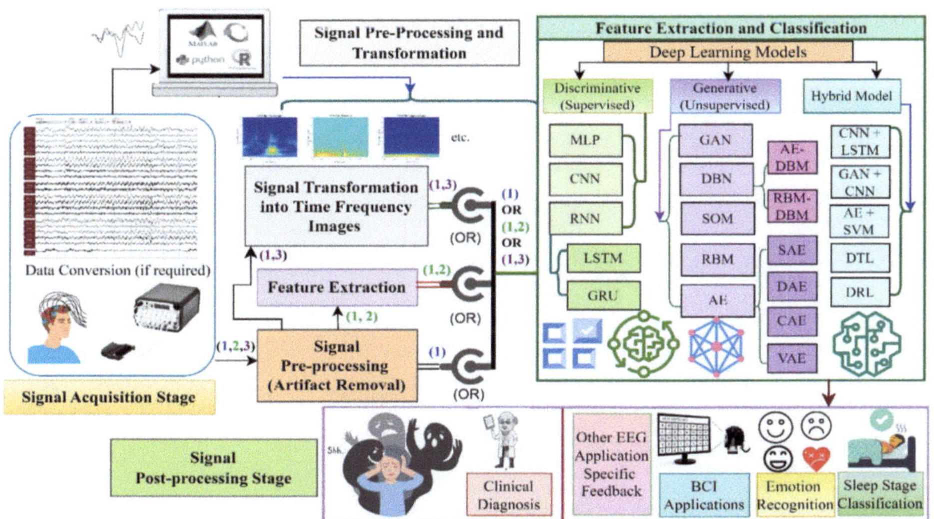

◘ **Fig. 5.10** Overview of the deep learning (DL)-based EEG signal analysis framework. The process includes signal acquisition, pre-processing, optional signal transformation, feature extraction, classification using deep learning models, and final signal post-processing. The abbreviations in the figure highlight different DL methods. *MLP* multilayer perceptron, *CNN* convolutional neural networks, *RNN* recurrent neural networks, *LSTM* long short-term memory, *GRU* gated recurrent unit, *GAN* generative adversarial network, *DBN* deep belief network, *SOM* self-organizing map, *RBM* restricted Boltzmann machine, *AE* autoencoder, *SAE* autoencoders such as sparse, *DAE* denoising, *CAE* contractive autoencoders, *VAE* variational autoencoders, *DTL* deep transfer learning, *DRL* deep reinforcement learning. (From Ranjan et al., 2024)

5.6   Feature Engineering and Machine Learning with fMRI

5.6.1   Example 1

In the example, shown in ◘ Fig. 5.11, the authors (Özmen et al., 2024) used a machine learning approach to classify depression disorders and improve the classification accuracy using weighted-3D-discrete wavelet transform (DWT). The steps employed are described below (Özmen et al., 2024).

— **Preprocessing**: The preprocessing of task-related fMRI images was conducted using the MATLAB-based SPM software package, which included steps such as realignment, slice timing correction, co-registration, spatial normalization, and spatial smoothing with a Gaussian Kernel of 6 mm full-width at half minimum (FWHM) (Chen & Glover, 2015). fMRI scans were affected by various artifacts, including head movements, noise from the scanning system, and thermal noise from diverse sources. Additionally, variations in the head shapes and sizes of the subjects resulted in differing numbers of voxels across the fMRI data (Rondina et al., 2013).

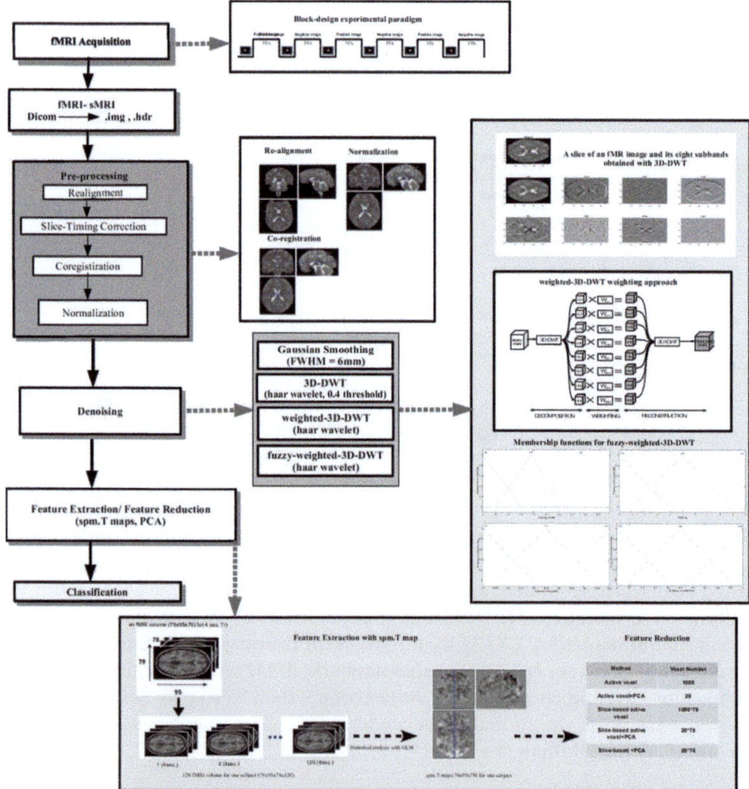

◘ **Fig. 5.11**   The figure illustrates the key steps undertaken in the study, from preprocessing to the application of machine learning methods using fMRI data to classify patients with depression. (From Özmen et al., 2024)

The initial preprocessing step involved motion correction of fMRI images, where consecutive volumes were realigned to a reference volume using a six-parameter rigid body transformation. Post-realignment, the aligned images were averaged to create a mean image for co-registration.

Spatial normalization involved registering structural images to standard templates like Talairach or MNI using affine transformations (Khullar et al., 2011). The spatial smoothing stage employed a Gaussian filter to smooth fMRI images, which helped suppress noise but also blurred image details and altered intensity variations. Various denoising techniques, including Gaussian smoothing, weighted-3D-DWT, fuzzy-weighted-3D-DWT, and classical 3D-DWT, were explored to assess their impact on classifying patients with depression (Özmen et al., 2024).

- **Statistical analysis**: In the statistical analysis of fMRI, signal changes during task-based experimental paradigms were compared to control conditions. Voxel intensity values in brain regions were calculated using statistical parameter mapping (SPM), a voxel-based, univariate approach that utilizes the general linear model (GLM) (Myers & Montgomery, 1997). This method enables the examination of each voxel individually for transient hemodynamic changes.
- **Feature extraction**: Activation maps were used to characterize brain functions for each subject, with SPM-based analysis generating voxel values distributed according to a known probability density function for a visual stimuli-related task. Two feature selection methods were applied to these spm.T maps: voxel-based and transform-based. Transform-based feature selection, such as principal component analysis (PCA), reduces dimensionality by transforming data into a new coordinate system, as read in ▶ Chap. 4. Voxel-based feature selection retains data in the original system but ranks voxels by intensity, selecting the top "n" voxels with the highest values, known as the active voxel method (Mitchell et al., 2004).
- **Machine learning**: This study employed support vector machine (SVM), k-nearest neighbor (k-NN) (Misaki et al., 2010), Naive Bayes (Kuncheva & Rodríguez, 2010), and random forest classifiers (Tripoliti et al., 2010)—commonly used in fMRI studies—to classify patients with depression (Pereira et al., 2009).

5.6.2 Example 2

In this example, the authors (Nenning & Langs, 2022) applied machine learning to structural neuroimaging. Structural neuroimaging is fundamental in clinical diagnostic neuroradiology. While it primarily reveals the overt structural characteristics of the brain, it significantly aids in diagnosing and making treatment decisions for various diseases. It enables the quantification of brain structures and their deviations related to disease, serving as potential markers for clinical outcomes. With the widespread availability of structural neuroimaging, machine learning techniques are increasingly employed to leverage the growing dataset of images to develop robust models for segmentation, classification, and prediction tasks, as shown in ◼ Fig. 5.12.

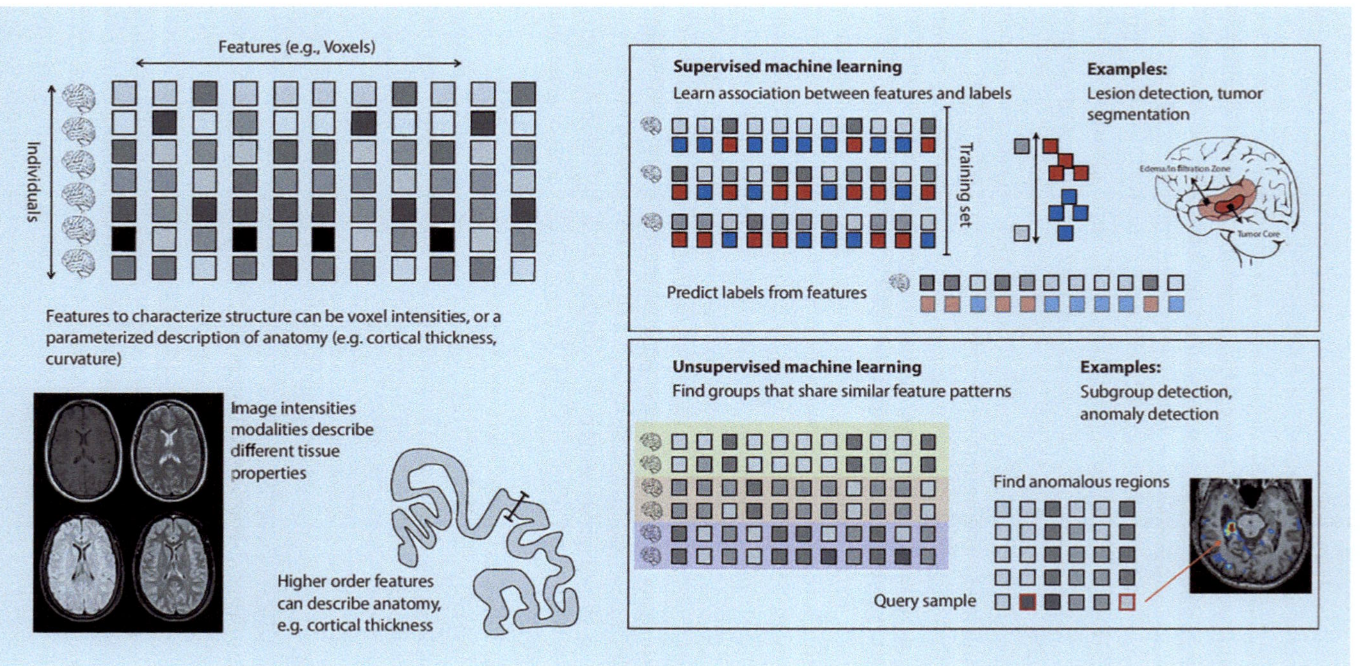

Fig. 5.12 Machine learning is adept at detecting, segmenting, and quantifying characteristics of anatomical structures, as well as identifying abnormalities associated with various diseases. Supervised machine learning methods train on paired examples that include imaging data and corresponding annotations, such as labeled tumors, to learn how to identify and classify similar cases in new datasets. Unsupervised machine learning, on the other hand, analyzes extensive collections of observations to identify inherent structures, characterize normal variability within anatomical features, or detect clusters of observations that share similar characteristics. (From Nenning & Langs, 2022 originally published under CC-BY 4.0)

5.7 Feature Engineering and Machine Learning with fNIRS

Figure 5.13 (Shamsi & Najafizadeh, 2021) illustrates the various stages involved in analyzing fNIRS signals, beginning with acquisition and culminating in applying machine learning techniques. The outcomes of this process may include the classification of patients with different disorders or the generation of outputs to control external devices, similar to other forms of brain signal analysis. It should be noted that the specific methods employed at each stage vary depending on the application and signal characteristics.

5.7.1 Example 1

In this study, the authors (Kumar et al., 2023) proposed a one-dimensional (1D) convolutional neural network (CNN) to classify mental workload utilizing fNIRS data. The 1D CNN analyzes the temporal dynamics of brain activity, extracting features from the time-series fNIRS data to classify different levels of mental workload, as shown in Fig. 5.14 (Kumar et al., 2023).

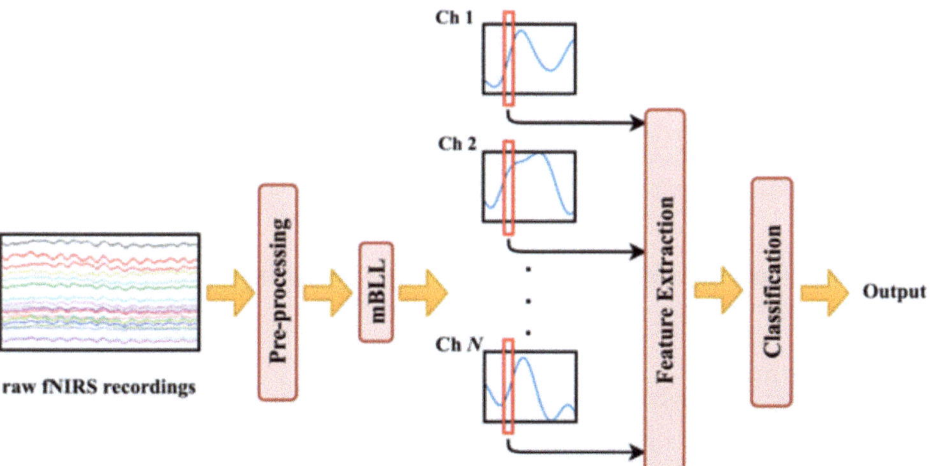

Fig. 5.13 Overview of fNIRS signal analysis. *mBLL* modified Beer-Lambert Law. (From Shamsi & Najafizadeh, 2021)

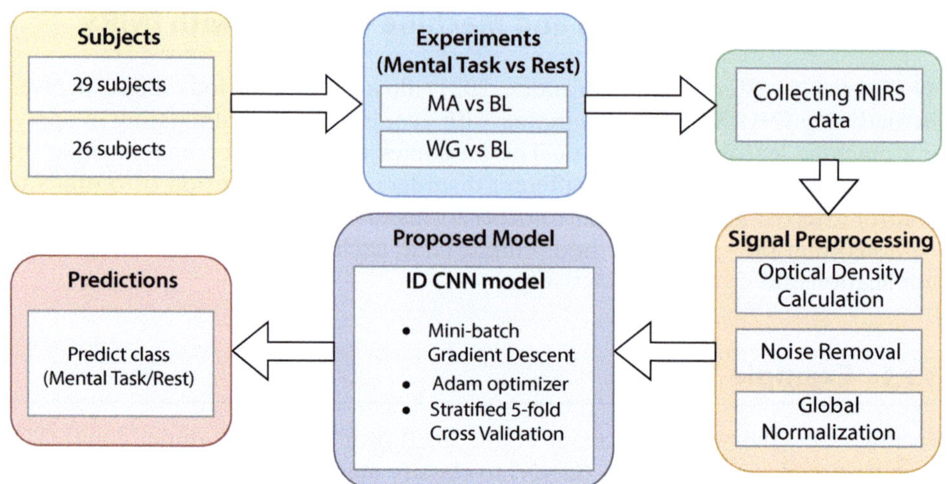

Fig. 5.14 Proposed methodology illustrations. Two datasets were utilized in the studies: **Dataset 1** involved 29 subjects (15 females, 14 males, average age 28.5 ± 3.7 years) performing mental arithmetic (MA) and baseline (BL) tasks. The experiment employed 36 fNIRS channels on each subject's scalp. During MA tasks, participants solved subtraction problems presented on a screen and were instructed to repeat the calculations mentally. During BL, subjects rested, viewing a black crosshair to maintain low cognitive load. **Dataset 2** comprised 26 participants (17 females, 9 males, average age 26.1 ± 3.5 years) engaged in word generation (WG) and rest (BL) tasks. Each participant completed three sessions, each containing 10 trials of WG and BL. For WG, subjects created words beginning with a specified letter without repeating words. Trials were spaced randomly over time. The BL conditions were identical to those in Dataset 1. The data underwent pre-processing as outlined in ▶ Chap. 3, and a 1D CNN model was applied, detailed in the "Proposed Model" section of the figure. This model determined whether the signal corresponded to a mental task or a rest period. (From Kumar et al., 2023)

5.8   Feature Engineering and Machine Learning with MEG

5.8.1   Example 1

In this example, the authors (Cetin & Temurtas, 2021) employed two types of visual stimuli (face and scrambled face) across 9414 trials to examine neuronal electrical activity during visual decoding tasks involving face perception. The raw data were preprocessed, as explained in ▶ Chap. 4, to extract useful data from the inherently noisy MEG signals, after which feature extraction and classification strategies were employed, as shown in ▣ Fig. 5.15 (Cetin & Temurtas, 2021). Two neural network models, a probabilistic neural network (PNN) (Specht, 1990) and a multilayer neural network (MLNN) (Alpaydin, 2020), were developed to classify the MEG signals. The classification outcomes were with accuracies of 82.36% for the PNN and 77.78% for MLNN. Notably, the PNN, which operates without the back-propagation algorithm and does not require training with the entire dataset, offered a quicker classification process, presenting a viable alternative to traditional methods (Abadi et al., 2015; Cetin & Temurtas, 2021).

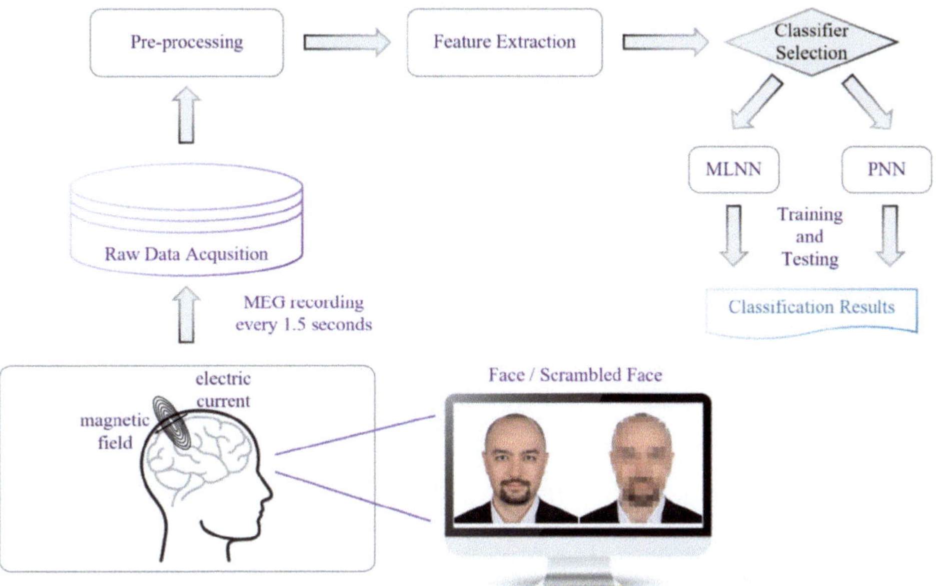

Fig. 5.15 The block diagram illustrates the comprehensive workflow employed in the study, from the recording of MEG signals to their classification. The process involves several key steps: (1) pre-processing of signals using a bandpass filter, (2) extraction of source signals through spatial filtering, (3) derivation of feature vectors from the source signals, and (4) classification based on these feature vectors. (From Cetin & Temurtas, 2021)

5.8.2 Example 2

In this example, for detecting early indicators of Alzheimer's disease (AD), the authors (Giovannetti et al., 2021) introduced an innovative Deep-MEG method, which utilizes image-based representations of magnetoencephalography (MEG) data and deep convolutional neural network-based ensemble classifiers. Authors employed functional connectivity (FC) measures, capturing interactions between brain bio-magnetic signals from different brain regions as MEG data representations. These FC measures across various frequency bands were converted into image formats, and a deep transfer learning model was applied to extract comprehensive feature sets and enhance classification outcomes, as shown in Fig. 5.16 (Giovannetti et al., 2021). In a longitudinal study, the authors evaluated the Deep-MEG architecture using resting-state MEG data and magnetic resonance imaging scans from 87 participants. The approach achieved 89% and 87% accuracy rates for early AD prediction among 54 mild cognitive impairment subjects and a broader group of 87 individuals, including 33 healthy controls. These findings demonstrate the efficacy of the Deep-MEG method in identifying early spectral-temporal connectivity alterations and their spatial dynamics, marking it as an effective tool for early AD detection.

5

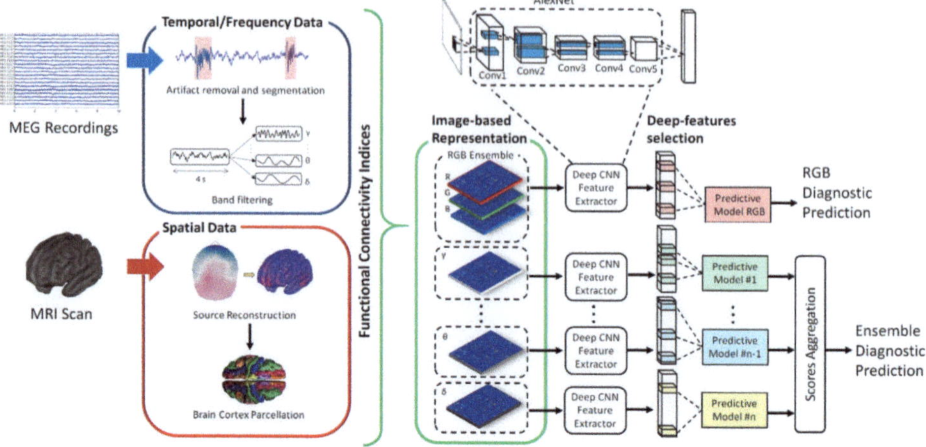

Fig. 5.16 Schematic layout of the MEG data analysis pipeline—spatial, temporal, and frequency-filtered data from MEG recordings and MRI scans were analyzed to assess brain information processing. Functional connectivity (FC) measures like phase locking value (PLV) and magnitude coefficient (MC) were calculated for signal pairs across 90 brain regions. AlexNet, a deep neural network with 60 million parameters, was utilized for its robust feature extraction capabilities without retraining. This network includes five convolutional layers and three fully connected layers, ending with a softmax classification layer. The FC representations for each patient were resized to 227×227 pixels using bicubic interpolation and processed through the pre-trained AlexNet to extract features. For classification, linear discriminant analysis (LDA) and support vector machines (SVMs) were employed to categorize MEG recordings into healthy control, stable, or progressive mild cognitive impairment categories. (From Giovannetti et al., 2021 originally published under CC-BY 4.0)

5.9 Feature Engineering and Machine Learning with Invasive Brain Signals

5.9.1 Example 1

In the example shown in ■ Fig. 5.17, authors (Thengone et al., 2020) employed the spike signals from an individual with spinal cord injury with two 96-channel Blackrock microelectrode arrays implanted in the precentral gyrus. The participant's goal was to move the cursor toward a target auditory tone. The signal processing and the experimental setup are shown in ■ Fig. 5.17.

More detailed examples of these systems and their implementations with individuals with disabilities are thoroughly explored in Volume 2. Volume 2 provides an in-depth analysis of the transformative impact these technologies have on enhancing the daily lives of individuals with disabilities, providing in-depth analyses and case studies that illustrate the practical benefits and improvements facilitated by such advanced systems.

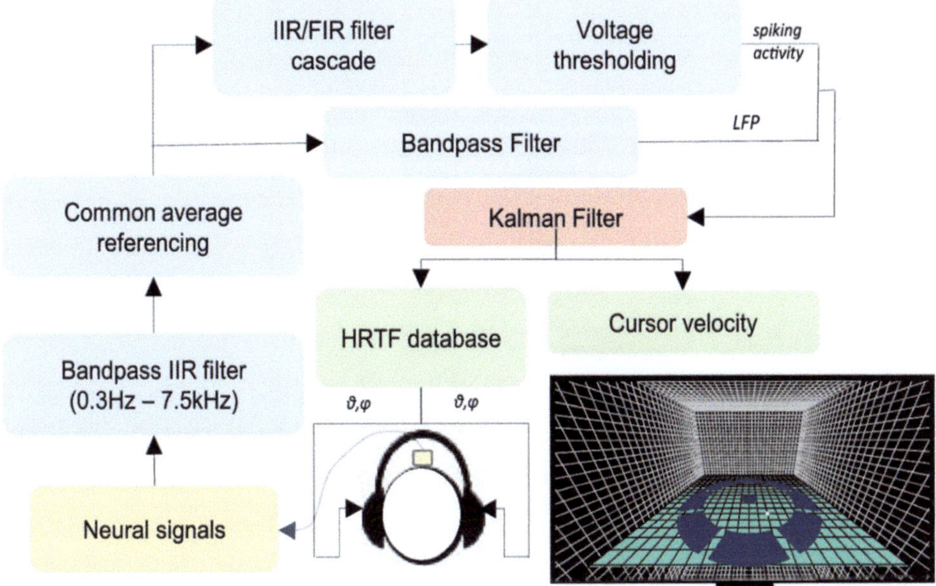

Fig. 5.17 The acquired signals were initially filtered analogically using a Butterworth bandpass filter and subsequently digitized at a rate of 30,000 samples per second by a 96-channel NeuroPort neural signal processor. Additional digital high-pass filtering was applied to refine signal quality for neural spike analysis. Decoder calibration for the intracortical brain–computer interface (iBCI) was conducted during an open-loop cursor movement task, using a steady-state Kalman decoder. The implemented auditory iBCI control task was the Radial-8 task, where the central target and eight peripheral targets were displayed in an oval layout on the screen. The auditory signals created an illusion of a virtual 3D room with targets distributed uniformly around the viewer in the front-back plane. Each trial was initiated with a sound played through earphones corresponding to the target location without any visual cues. HRFT—head-response transfer function was employed to deliver real-time, task-relevant auditory feedback to the user through headphones. (From Thengone et al., 2020)

5.9.2 Example 2

Although **Fig. 5.18** (Adama & Bogdan, 2023) illustrates the use of EEG signals, the flowchart presents methods that are also applicable to ECoG signals. It should be clear by now that the overall signal-processing steps remain consistent across different types of brain signals. However, the specific methods employed within each step can vary depending on the nature of the signals and their particular applications.

5

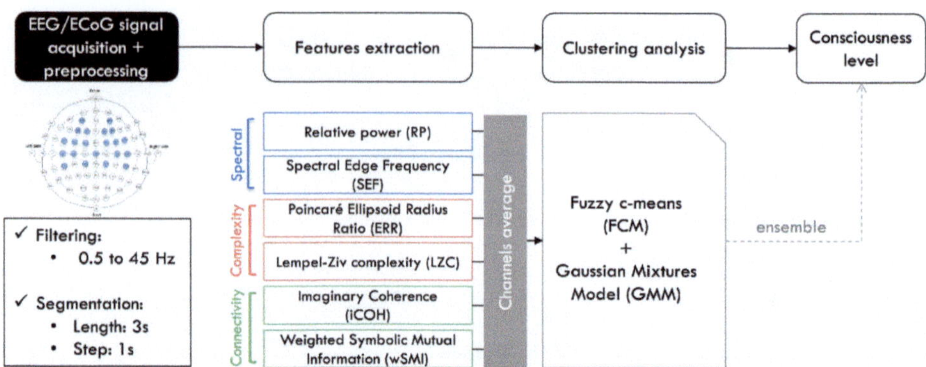

Fig. 5.18 The signal processing and analysis pipeline detailed here involves multiple steps to ensure precise analysis, as you have seen in all the previous examples. Initially, the recorded signal undergoes filtering to remove unwanted noise and is then segmented into manageable parts for detailed examination. Subsequently, various features are extracted from these segments. Each extracted feature is averaged across a specifically selected group of channels to ensure consistency and reduce variability. Following this, clustering analysis is performed to categorize the data effectively. Finally, the probability of patient consciousness is determined by applying a decision rule to the clustered results, providing a systematic approach to assess the patient's condition. (From Adama & Bogdan, 2023 originally published under CC-BY 4.0)

> **Key Takeaways**

1. **The Importance of Feature Engineering in Brain Signal Analysis**: This chapter emphasizes the significance of feature extraction and selection in simplifying brain signals and making them interpretable. By categorizing features into time-domain, frequency-domain, time-frequency domain, and spectral domain, the chapter illustrates how these techniques help reduce data complexity while preserving critical information, making brain signals more manageable for machine learning analysis.

2. **Diverse Machine Learning Applications in Neurotechnology**: The chapter explores various machine learning techniques—supervised, semi-supervised, unsupervised, and reinforcement learning—highlighting their transformative role in analyzing neurophysiological data. These methods enhance brain–computer interface (BCI) applications and are particularly effective in tasks such as emotion detection, neurological disorder diagnosis, and cognitive state prediction.

3. **Real-World Applications of Machine Learning in Brain Data**: Practical examples show how machine learning methodologies are applied to different types of brain signals, such as EEG, fMRI, and MEG. These applications demonstrate how advanced analytical techniques, including deep learning models like CNNs and RNNs, are integral to developing neuroprosthetics, BCIs, and diagnostic tools for conditions like schizophrenia and Alzheimer's disease.

4. **Future Implications of Advanced Machine Learning in Neuroscience**: The chapter sets the stage for future explorations into deep learning, hybrid models, and reinforcement learning, which promise further advancements in neurotechnology. These emerging methods will likely enable more personalized and adaptive brain–computer interfaces, contributing to clinical diagnostics, rehabilitation, and enhancing the quality of life for individuals with disabilities.

As we look forward to the next volume, which will look deeper into the practical applications of BCIs, it is clear that the groundwork laid by understanding and processing brain signals is fundamental. The continuous advancements in feature engineering and machine learning will undoubtedly propel the field of neurotechnology forward, enhancing our ability to interpret and interact with the human brain. This endeavor not only promises substantial improvements in medical diagnostics and treatments but also paves the way for significant enhancements in the quality of life for those with physical limitations, offering them new avenues for interaction and communication.

Conclusion

In this chapter, we have explored the intricate process of extracting and selecting features from brain signals and the application of various machine-learning techniques to enhance our understanding and interfacing with neural data. Brain signals are inherently complex, containing rich information that, when effectively harnessed, can provide profound insights into neural processes and enable the development of innovative brain–computer interfaces (BCIs). The methods discussed, from feature extraction to machine learning, underscore the critical role of sophisticated analytical techniques in simplifying and interpreting brain data.

Feature engineering plays a pivotal role in the signal processing pipeline, transforming raw neurophysiological data into a manageable and informative format. This process facilitates easier data analysis and enhances machine learning models' performance by highlighting the most relevant features without extraneous noise. The selection of features is a finely tuned process that seeks to maintain the integrity and significance of the brain signal while reducing dimensionality and improving model accuracy.

The use of machine learning, from supervised to unsupervised learning models, in analyzing brain data is indispensable. These models manage high-dimensional data and uncover not immediately apparent patterns, proving essential in applications such as predictive modeling and real-time decision-making in BCIs. Machine learning models' adaptability and predictive power enable them to play a crucial role in developing technologies that improve life quality, particularly for individuals with disabilities.

Moreover, the transition from theoretical concepts to practical applications demonstrates the real-world utility of these computational techniques. The examples provided have illustrated the application of these methods with common steps across various brain signal types, as shown in ◨ Fig. 5.19, including EEG, fMRI, and more, showcasing their effectiveness in real-life scenarios.

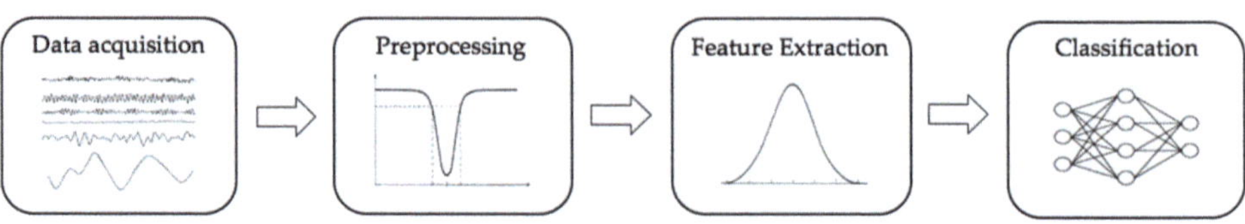

■ **Fig. 5.19** Brain signal analysis involves several critical signal processing steps fundamental across various brain signals, such as EEG, fMRI, fNIRS, MEG, and invasive brain signals. These steps are essential for enhancing the quality of the signals, facilitating effective analysis, and ensuring that subsequent interpretations are reliable and accurate. The following is a detailed breakdown of these common processing steps applied in the context of neurotechnology. **Data Acquisition**: The initial step involves using appropriate sensors and equipment to capture brain signals. The quality of signal acquisition is pivotal, as it impacts all subsequent processing stages. For instance, EEG signals are typically collected using scalp electrodes, whereas MEG signals are captured using highly sensitive magnetometers. **Preprocessing**: This stage is critical for reducing noise and improving the signal-to-noise ratio (SNR). Pre-processing techniques include (1) Filtering: Applying high-pass, low-pass, and bandpass filters to remove frequency-based noise components. For example, high-frequency noise can be removed from EEG signals using a low-pass filter, whereas line noise at 50/60 Hz is commonly removed using a notch filter. (2) Artifact Rejection: Identifying and removing artifacts that are not of cerebral origin, such as eye blinks, cardiac signals, or muscle noise. Techniques such as independent component analysis (ICA) are frequently used in EEG to isolate and remove these artifacts. (3) Downsampling: Reducing the signal's sampling rate to decrease data size and processing requirements, particularly when high-resolution data does not contribute additional useful information. **Feature Extraction**: Extracting meaningful features from the processed signals is crucial for further analysis. Common features include (1) Time-Domain Features: Such as amplitude, variance, and peak measurements, which provide insights into the power and distribution of the signal over time. (2) Frequency-Domain Features: Spectral power densities or band power within specific frequency bands (e.g., alpha, beta, gamma) that are linked to different cognitive states or brain activities. (3) Time-Frequency Features: Combining time and frequency information through methods like wavelet transforms, which effectively analyze non-stationary signals typical of brain activity. **Machine Learning**: Applying statistical techniques and machine learning methods to the extracted features to identify patterns, make predictions, or infer the cognitive or clinical state of the individual. Each step is tailored to the specific characteristics and requirements of the brain signal being analyzed. These processing steps ensure that the data are optimally prepared for high-level analysis, such as decoding brain states or diagnosing neurological conditions. This rigorous processing framework is essential for maximizing the reliability and efficacy of neurotechnology applications, from basic research to clinical diagnostics and brain–computer interfaces. (From Luján et al., 2021 originally published under CC-BY 4.0)

This section offers practical exercises designed to help apply the material learned to concrete scenarios.

Feature Extraction from EEG Data

Objective: To practice extracting both time-domain and frequency-domain features from EEG datasets.

Task: Students will use EEG processing software like EEGLAB or FieldTrip to extract specific features such as peak amplitudes, average power in specific frequency bands, and spectral entropy from a given EEG dataset.

Classification of Brain States Using Supervised Learning

Objective: To apply machine learning algorithms to classify different brain states from EEG features.

Task: Using a labeled dataset where brain states (e.g., alert, relaxed, and sleep stages) are pre-defined, students will train a classifier such as SVM or neural network to distinguish between these states based on the extracted features. Performance will be evaluated based on accuracy, precision, and recall.

Clustering of fMRI Data with Unsupervised Learning

Objective: To explore pattern recognition in brain activity without pre-labeled data.

Task: Students will apply clustering algorithms like k-means or hierarchical clustering to fMRI data to identify patterns or regions of similar activity during different cognitive tasks. The clusters will be analyzed to infer functional connectivity within the brain.

Developing a Reinforcement Learning Model for a BCI

Objective: To create a simple brain–computer interface using reinforcement learning.

Task: Students will simulate or use a real BCI setup to control a virtual or physical object. The system will use reinforcement learning to improve its predictions based on continuous feedback received from the user's brain signals.

Feature Selection Challenge

Objective: To understand the impact of feature selection on machine learning model performance.

Task: Students will use a dataset with a large number of features (high dimensionality) and apply different feature selection techniques to reduce the number of features. They will compare the performance of machine learning models trained with the full set of features and the reduced set to evaluate the effectiveness of the feature selection methods.

Normalization Techniques Comparison

Objective: To compare the effects of different normalization techniques on the same dataset.

Task: Students will apply various normalization techniques such as min-max scaling, z-score normalization, and normalization by decimal scaling on EEG or MEG data. They will then visualize the data and evaluate which normalization method provides the best clarity and separability for subsequent analyses.

References

Abadi, M. K., Subramanian, R., Kia, S. M., Avesani, P., Patras, I., & Sebe, N. (2015). DECAF: MEG-based multimodal database for decoding affective physiological responses. *IEEE Transactions on Affective Computing, 6*(3), 209–222.

Adama, S., & Bogdan, M. (2023). Assessing consciousness in patients with disorders of consciousness using soft-clustering. *Brain Informatics, 10*(1), 16.

Alpaydin, E. (2020). *Introduction to machine learning.* MIT Press.

Ayodele, T. O. (2010). Types of machine learning algorithms. *New Advances in Machine Learning, 3*(19–48), 5–1.

Badillo, S., Banfai, B., Birzele, F., Davydov, I. I., Hutchinson, L., Kam-Thong, T., et al. (2020). An introduction to machine learning. *Clinical Pharmacology & Therapeutics, 107*(4), 871–885.

Bank, D., Koenigstein, N., & Giryes, R. (2023). Autoencoders. In *Machine learning for data science handbook: Data mining and knowledge discovery handbook* (pp. 353–374). Springer.

Bonkhoff, A. K., & Grefkes, C. (2022). Precision medicine in stroke: Towards personalized outcome predictions using artificial intelligence. *Brain, 145*(2), 457–475.

Carleo, G., Cirac, I., Cranmer, K., Daudet, L., Schuld, M., Tishby, N., et al. (2019). Machine learning and the physical sciences. *Reviews of Modern Physics, 91*(4), 045002.

Cetin, O., & Temurtas, F. (2021). A comparative study on classification of magnetoencephalography signals using probabilistic neural network and multilayer neural network. *Soft Computing, 25*(3), 2267–2275.

Chaki, J., & Woźniak, M. (2023). Deep learning for neurodegenerative disorder (2016 to 2022): A systematic review. *Biomedical Signal Processing and Control, 80*, 104223.

Chandrashekar, G., & Sahin, F. (2014). A survey on feature selection methods. *Computers & Electrical Engineering, 40*(1), 16–28.

Chen, J. E., & Glover, G. H. (2015). Functional magnetic resonance imaging methods. *Neuropsychology Review, 25*, 289–313.

Cohen, M. X. (2014). *Analyzing neural time series data: Theory and practice.* MIT Press.

Connor, J. T., Martin, R. D., & Atlas, L. E. (1994). Recurrent neural networks and robust time series prediction. *IEEE Transactions on Neural Networks, 5*(2), 240–254.

Dash, M., & Liu, H. (1997). Feature selection for classification. *Intelligent Data Analysis, 1*(1–4), 131–156.

de Bardeci, M., Ip, C. T., & Olbrich, S. (2021). Deep learning applied to electroencephalogram data in mental disorders: A systematic review. *Biological Psychology, 162*, 108117.

Garcés, M. A., & Orosco, L. L. (2008). EEG signal processing in brain–computer interface. In *Smart wheelchairs and brain-computer interfaces* (pp. 95–110). Academic Press.

Giovannetti, A., Susi, G., Casti, P., Mencattini, A., Pusil, S., López, M. E., et al. (2021). Deep-MEG: Spatiotemporal CNN features and multiband ensemble classification for predicting the early signs of Alzheimer's disease with magnetoencephalography. *Neural Computing and Applications, 33*(21), 14651–14667.

Glaser, J. I., Benjamin, A. S., Chowdhury, R. H., Perich, M. G., Miller, L. E., & Kording, K. P. (2020). Machine learning for neural decoding. *eNeuro, 7*(4). https://doi.org/10.1523/ENEURO.0506-19.2020

Goodfellow, I., Pouget-Abadie, J., Mirza, M., Xu, B., Warde-Farley, D., Ozair, S., et al. (2020). Generative adversarial networks. *Communications of the ACM, 63*(11), 139–144.

Greener, J. G., Kandathil, S. M., Moffat, L., & Jones, D. T. (2022). A guide to machine learning for biologists. *Nature Reviews Molecular Cell Biology, 23*(1), 40–55.

Gu, J., Wang, Z., Kuen, J., Ma, L., Shahroudy, A., Shuai, B., et al. (2018). Recent advances in convolutional neural networks. *Pattern Recognition, 77*, 354–377.

Guyon, I., & Elisseeff, A. (2003). An introduction to variable and feature selection. *Journal of Machine Learning Research, 3*, 1157–1182.

Guyon, I., & Elisseeff, A. (2006). An introduction to feature extraction. In *Feature extraction: Foundations and applications* (pp. 1–25). Springer.

Haufe, S., Dähne, S., & Nikulin, V. V. (2014). Dimensionality reduction for the analysis of brain oscillations. *NeuroImage, 101*, 583–597.

Herman, P., Prasad, G., McGinnity, T. M., & Coyle, D. (2008). Comparative analysis of spectral approaches to feature extraction for EEG-based motor imagery classification. *IEEE Transactions on Neural Systems and Rehabilitation Engineering, 16*(4), 317–326.

Hinton, G. E. (2009). Deep belief networks. *Scholarpedia, 4*(5), 5947.

Hinton, G. E. (2012). A practical guide to training restricted Boltzmann machines. In *Neural networks: Tricks of the trade* (2nd ed., pp. 599–619). Springer.

Hosseini, M. P., Hosseini, A., & Ahi, K. (2020). A review on machine learning for EEG signal processing in bioengineering. *IEEE Reviews in Biomedical Engineering, 14*, 204–218.

Hu, L., & Zhang, Z. (Eds.). (2019). *EEG signal processing and feature extraction*. Springer.

Janiesch, C., Zschech, P., & Heinrich, K. (2021). Machine learning and deep learning. *Electronic Markets, 31*(3), 685–695.

Jiménez-Guarneros, M., & Alejo-Eleuterio, R. (2022). A class-incremental learning method based on preserving the learned feature space for EEG-based emotion recognition. *Mathematics, 10*(4), 598.

Khullar, S., Michael, A., Correa, N., Adali, T., Baum, S. A., & Calhoun, V. D. (2011). Wavelet-based fMRI analysis: 3-D denoising, signal separation, and validation metrics. *NeuroImage, 54*(4), 2867–2884.

Kingsford, C., & Salzberg, S. L. (2008). What are decision trees? *Nature Biotechnology, 26*(9), 1011–1013.

Kohonen, T. (1990). The self-organizing map. *Proceedings of the IEEE, 78*(9), 1464–1480.

Kotsiantis, S. B., Zaharakis, I. D., & Pintelas, P. E. (2006). Machine learning: A review of classification and combining techniques. *Artificial Intelligence Review, 26*, 159–190.

Kotsiantis, S. B., Zaharakis, I., & Pintelas, P. (2007). Supervised machine learning: A review of classification techniques. *Frontiers in Artificial Intelligence and Applications, 160*(1), 3–24.

Kraskov, A., Stögbauer, H., & Grassberger, P. (2004). Estimating mutual information. *Physical Review E, 69*(6), 066138.

Krishnan, S., & Athavale, Y. (2018). Trends in biomedical signal feature extraction. *Biomedical Signal Processing and Control, 43*, 41–63.

Krusienski, D. J., McFarland, D. J., Principe, J. C., & Wolpaw, E. (2012). BCI signal processing: Feature extraction. In J. R. Wolpaw & E. W. Wolpaw (Eds.), *Brain-computer interfaces: Principles and practice* (pp. 123–146). Oxford University Press.

Kumar, A., Karmakar, S., Agarwal, I., & Pal, T. (2023). Mental workload classification with one-dimensional CNN using fNIRS signal. In *International conference on pattern recognition and machine intelligence, December 2023* (pp. 746–755). Springer Nature.

Kuncheva, L. I., & Rodríguez, J. J. (2010). Classifier ensembles for fMRI data analysis: An experiment. *Magnetic Resonance Imaging, 28*(4), 583–593.

Lal, T. N., Chapelle, O., Weston, J., & Elisseeff, A. (2006). Embedded methods. In *Feature extraction: Foundations and applications* (pp. 137–165). Springer.

LeCun, Y., Bengio, Y., & Hinton, G. (2015). Deep learning. *Nature, 521*(7553), 436–444.

Lemm, S., Blankertz, B., Dickhaus, T., & Müller, K. R. (2011). Introduction to machine learning for brain imaging. *NeuroImage, 56*(2), 387–399.

Li, J., Cheng, K., Wang, S., Morstatter, F., Trevino, R. P., Tang, J., & Liu, H. (2017). Feature selection: A data perspective. *ACM Computing Surveys (CSUR), 50*(6), 1–45.

Li, D., Yang, Z., Hou, F., Kang, Q., Liu, S., Song, Y., et al. (2022a). EEG-based emotion recognition with haptic vibration by a feature fusion method. *IEEE Transactions on Instrumentation and Measurement, 71*, 1–11.

Li, D., Xie, L., Chai, B., & Wang, Z. (2022b). A feature-based on potential and differential entropy information for electroencephalogram emotion recognition. *Electronics Letters, 58*(4), 174–177.

Li, J., Hua, H., Xu, Z., Shu, L., Xu, X., Kuang, F., & Wu, S. (2022c). Cross-subject EEG emotion recognition combined with connectivity features and meta-transfer learning. *Computers in Biology and Medicine, 145*, 105519.

Li, Q., Zhang, T., Chen, C. P., Yi, K., & Chen, L. (2022d). Residual GCB-Net: Residual graph convolutional broad network on emotion recognition. *IEEE Transactions on Cognitive and Developmental Systems., 15*, 1673.

Luján, M. Á., Jimeno, M. V., Mateo Sotos, J., Ricarte, J. J., & Borja, A. L. (2021). A survey on EEG signal processing techniques and machine learning: Applications to the neurofeedback of autobiographical memory deficits in schizophrenia. *Electronics, 10*(23), 3037.

Martinez, A. M., & Kak, A. C. (2001). PCA versus LDA. *IEEE Transactions on Pattern Analysis and Machine Intelligence, 23*(2), 228–233.

McFarland, D. J., & Wolpaw, J. R. (2005). Sensorimotor rhythm-based brain-computer interface (BCI): Feature selection by regression improves performance. *IEEE Transactions on Neural Systems and Rehabilitation Engineering, 13*(3), 372–379.

Mehmood, R. M., & Lee, H. J. (2016). A novel feature extraction method based on late positive potential for emotion recognition in human brain signal patterns. *Computers & Electrical Engineering, 53*, 444–457.

Misaki, M., Kim, Y., Bandettini, P. A., & Kriegeskorte, N. (2010). Comparison of multivariate classifiers and response normalizations for pattern-information fMRI. *NeuroImage, 53*(1), 103–118.

Mishra, A., Yadav, P., & Kim, S. (2023). Artificial intelligence accelerators. In *Artificial intelligence and hardware accelerators* (pp. 1–52). Springer International Publishing.

Mitchell, T. M., Hutchinson, R., Niculescu, R. S., Pereira, F., Wang, X., Just, M., & Newman, S. (2004). Learning to decode cognitive states from brain images. *Machine Learning, 57*(1), 145–175.

Murtagh, F. (1991). Multilayer perceptrons for classification and regression. *Neurocomputing, 2*(5–6), 183–197.

Muthukrishnan, R., & Rohini, R. (2016, October). LASSO: A feature selection technique in predictive modeling for machine learning. In *2016 IEEE international conference on advances in computer applications (ICACA)* (pp. 18–20). IEEE.

Mwangi, B., Tian, T. S., & Soares, J. C. (2014). A review of feature reduction techniques in neuroimaging. *Neuroinformatics, 12*, 229–244.

Myers, R. H., & Montgomery, D. C. (1997). A tutorial on generalized linear models. *Journal of Quality Technology, 29*(3), 274–291.

Nenning, K. H., & Langs, G. (2022). Machine learning in neuroimaging: From research to clinical practice. *Die Radiologie, 62*(Suppl 1), 1–10.

Niedermeyer, E., & da Silva, F. L. (Eds.). (2005). *Electroencephalography: Basic principles, clinical applications, and related fields.* Lippincott Williams & Wilkins.

Oh, I. S., Lee, J. S., & Moon, B. R. (2004). Hybrid genetic algorithms for feature selection. *IEEE Transactions on Pattern Analysis and Machine Intelligence, 26*(11), 1424–1437.

Özmen, G., Özşen, S., Paksoy, Y., Güler, Ö., & Tekdemir, R. (2024). Machine learning based detection of depression from task-based fMRI using weighted-3D-DWT denoising method. *Multimedia Tools and Applications, 83*(4), 11805–11829.

Pereira, F., Mitchell, T., & Botvinick, M. (2009). Machine learning classifiers and fMRI: A tutorial overview. *NeuroImage, 45*(1), S199–S209.

Raghavendra, U., Acharya, U. R., & Adeli, H. (2020). Artificial intelligence techniques for automated diagnosis of neurological disorders. *European Neurology, 82*(1–3), 41–64.

Ranjan, R., Sahana, B. C., & Bhandari, A. K. (2024). Deep learning models for diagnosis of schizophrenia using EEG signals: Emerging trends, challenges, and prospects. *Archives of Computational Methods in Engineering, 31*, 1–40.

Ren, J., Zabalza, J., Marshall, S., & Zheng, J. (2014). Effective feature extraction and data reduction with hyperspectral imaging in remote sensing. *IEEE Signal Processing Magazine, 31*(4), 149–154.

Riaz, F., Hassan, A., Rehman, S., Niazi, I. K., & Dremstrup, K. (2015). EMD-based temporal and spectral features for the classification of EEG signals using supervised learning. *IEEE Transactions on Neural Systems and Rehabilitation Engineering, 24*(1), 28–35.

Rondina, J. M., Hahn, T., de Oliveira, L., Marquand, A. F., Dresler, T., Leitner, T., et al. (2013). SCoRS—A method based on stability for feature selection and mapping in neuroimaging. *IEEE Transactions on Medical Imaging, 33*(1), 85–98.

Sadiq, M. T., Yu, X., & Yuan, Z. (2021). Exploiting dimensionality reduction and neural network techniques for the development of expert brain–computer interfaces. *Expert Systems with Applications, 164*, 114031.

Samal, P., & Hashmi, M. F. (2024). Role of machine learning and deep learning techniques in EEG-based BCI emotion recognition system: A review. *Artificial Intelligence Review, 57*(3), 1–66.

Schuster, M., & Paliwal, K. K. (1997). Bidirectional recurrent neural networks. *IEEE Transactions on Signal Processing, 45*(11), 2673–2681.

Shalev-Shwartz, S., & Ben-David, S. (2014). *Understanding machine learning: From theory to algorithms*. Cambridge University Press.

Shamsi, F., & Najafizadeh, L. (2021). Multi-class fNIRS classification of motor execution tasks with application to brain-computer interfaces. In *Biomedical signal processing: Innovation and applications* (pp. 1–32). Springer.

Sharifani, K., & Amini, M. (2023). Machine learning and deep learning: A review of methods and applications. *World Information Technology and Engineering Journal, 10*(7), 3897–3904.

Shrestha, A., & Mahmood, A. (2019). Review of deep learning algorithms and architectures. *IEEE Access, 7*, 53040–53065.

Siddiqui, M. K., Morales-Menendez, R., Huang, X., & Hussain, N. (2020). A review of epileptic seizure detection using machine learning classifiers. *Brain Informatics, 7*(1), 5.

Specht, D. F. (1990). Probabilistic neural networks. *Neural Networks, 3*(1), 109–118.

Supakar, R., Satvaya, P., & Chakrabarti, P. (2022). A deep learning based model using RNN-LSTM for the detection of schizophrenia from EEG data. *Computers in Biology and Medicine, 151*, 106225.

Thengone, D. J., Hosman, T., Simeral, J. D., & Hochberg, L. R. (2020). Auditory-reliant intracortical brain computer interfaces for effector control by a person with tetraplegia. In *HCI international 2020–late breaking posters: 22nd international conference, HCII 2020, Copenhagen, Denmark, July 19–24, 2020, proceedings, part II* (pp. 102–109). Springer International Publishing.

Tripoliti, E. E., Fotiadis, D. I., Argyropoulou, M., & Manis, G. (2010). A six stage approach for the diagnosis of the Alzheimer's disease based on fMRI data. *Journal of Biomedical Informatics, 43*(2), 307–320.

Vidaurre, C., Krämer, N., Blankertz, B., & Schlögl, A. (2009). Time domain parameters as a feature for EEG-based brain–computer interfaces. *Neural Networks, 22*(9), 1313–1319.

Yadav, D., Yadav, S., & Veer, K. (2020). A comprehensive assessment of brain computer interfaces: Recent trends and challenges. *Journal of Neuroscience Methods, 346*, 108918.

Zhang, Y., Ji, X., & Zhang, S. (2016). An approach to EEG-based emotion recognition using combined feature extraction method. *Neuroscience Letters, 633*, 152–157.

Zhang, X., Yao, L., Wang, X., Monaghan, J., Mcalpine, D., & Zhang, Y. (2021a). A survey on deep learning-based non-invasive brain signals: Recent advances and new frontiers. *Journal of Neural Engineering, 18*(3), 031002.

Zhang, S., Hu, B., Bian, J., Zhang, M., & Zheng, X. (2021b). A novel emotion recognition method incorporating MST-based brain network and FVMD-GAMPE. In *2021 IEEE international conference on bioinformatics and biomedicine (BIBM), December* (pp. 1153–1158). IEEE.

Neuromodulation in Neural Circuits

Neurotransmitters and Synaptic Plasticity

Contents

Test your learning and check your understanding of this book's contents: use the "Springer Nature Flashcards" app to access questions using ▶ https://sn.pub/kmb-jyz. To use the app, please follow the instructions in ▶ Chap. 1.

This chapter introduces the fundamental concepts of neuromodulation within the nervous system, focusing on the intricate roles of neurotransmitters, synaptic plasticity, and neural circuits in modulating neural functions. It elucidates the mechanisms by which neurotransmitters influence neuronal excitability and interaction and explores how these chemical messengers contribute to the dynamic regulation of neural circuits. Synaptic plasticity is discussed as a critical process for learning and memory, facilitated by modifications in synaptic strength through mechanisms such as long-term potentiation (LTP) and long-term depression (LTD). This chapter sets the stage for understanding how these biological processes underpin not only basic neural activity but also complex behaviors and cognitive functions. The insights gained underscore the significance of neuromodulation in neuroscience research and its potential therapeutic applications for neurological and psychiatric conditions. As the foundational entry into a series on brain stimulation techniques, this discussion prepares the reader for deeper explorations into how these neuroscientific principles can be applied in clinical and therapeutic settings.

Learning Objectives

1. Understand Neuromodulation: Recognize neuromodulation as a critical mechanism in the nervous system that adjusts neuronal activity in response to various neurotransmitters and neural activities.
2. Explore Neurotransmitter Functions: Differentiate between excitatory and inhibitory neurotransmitters and understand how they influence neural excitability and synaptic plasticity.
3. Grasp Synaptic Plasticity Mechanisms: Explain the processes and significance of synaptic plasticity, including long-term potentiation (LTP) and long-term depression (LTD), and their roles in learning and memory.
4. Examine Neural Circuits: Describe the structure and function of neural circuits, including feedforward and feedback loops, and their implications for information processing and cognitive functions.

In this book's initial segment, we explored various techniques for acquiring signals produced by the brain. Building on this foundation, the subsequent chapters will shift our focus toward methodologies designed to influence brain activity through various stimulation techniques. This transition marks a critical pivot from passive observation to active modulation of neural functions, which holds significant implications for both research and therapeutic applications.

Before we learn about the specifics of these stimulation techniques in the ensuing chapters, it is crucial first to establish an understanding of neural communication within the brain. This chapter explores the underlying mechanisms of neuromodulation, setting the stage for a deeper appreciation of how brain stimulation can be effectively employed.

This chapter is designed to deepen our understanding of the complex interactions among neurotransmitters, synaptic plasticity, and neural circuits and how these elements collectively shape the modulation of neural functions. Such modulation is foundational to both basic neural activities and cognitive processes, underscoring the pivotal role of neuromodulation in neuroscience. Current research is increasingly unraveling the intricacies of neuromodulatory systems and their extensive effects on diverse cognitive functions and behavioral outcomes. This expanding body of knowledge offers critical insights into the fundamental mechanisms of brain function and highlights potential therapeutic avenues for addressing a range of neurological and psychiatric disorders (Marder, 2012; Tritsch & Sabatini, 2012; Dayan, 2012). Through exploring these dynamics, this chapter aims to illuminate the essential contributions of neuromodulation to both core and complex brain functions, enhancing our grasp of its significance in broader neuroscientific investigations.

6.1 Neuromodulation

Neuromodulation is a pivotal concept in neuroscientific research, elucidating the complex neurobiological mechanisms that govern the regulation of neural activity within the brain. This exploration begins at the cellular level, where neurons, the brain's primary structural and functional units, are interconnected through complex networks of synapses, as illustrated in ◘ Fig. 6.1a (Mirza et al., 2019). Neuronal communication unfolds through an intricate interplay of electrical and chemical signals, a process elaborately depicted in ◘ Fig. 6.1b (Mirza et al., 2019). Central to this communication are neurotransmitters, the biochemical messengers that bind to receptors on neuronal surfaces, subtly modulating their excitability and, consequently, influencing overall neural activity (Marder, 2012).

The nervous system, a sophisticated network of neurons and synaptic connections, orchestrates various physiological functions and behaviors. At the core of this system's functionality is the modulation of neural activity, a dynamic process that incorporates various factors, including neurotransmitter dynamics, synaptic plasticity, and the broader neural circuits that structure brain function. This multifaceted modulation is crucial for the nervous system's ability to adapt and respond to complex environmental stimuli and internal changes (Berridge, 2007; Marder, 2012).

Neuromodulation is crucial in enabling the brain to adjust its neural connections in response to both external and internal stimuli, enhancing the functionality and adaptability of neural networks (Lee & Dan, 2012). This dynamic process involves certain neurons that employ one or multiple neurotransmitters to exert control over various neuronal groups (Rolls & Treves, 1997), markedly different from the traditional synaptic transmission that typically involves a presynaptic neuron directly impacting a single postsynaptic target (Eccles, 1973).

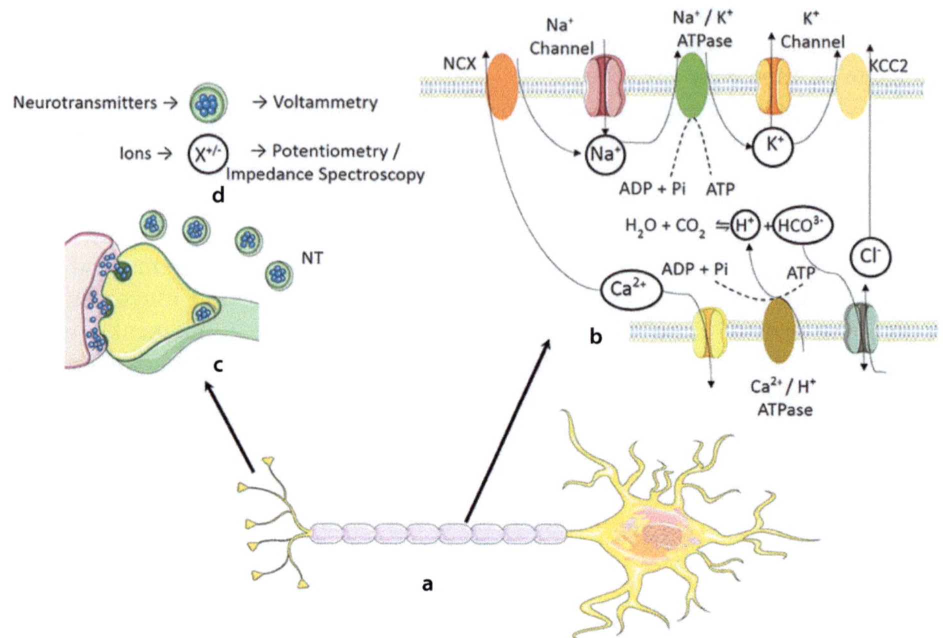

Fig. 6.1 **a** In a typical neuron, neural activity induces transients in both ions and neurotransmitters. **b** The propagation of an action potential along the axon generates ionic shifts. This process involves the activation of ion pumps such as Na$^+$/ATPase and Ca^{2+}/ATPase, which cause extracellular acidification and alkalinization. **c** As the neural response propagates across neurons, neurotransmitters are released into the synaptic cleft. **d** The detection of the two primary classes of neurochemicals—neurotransmitters and ions—is facilitated by electrochemical techniques. Specifically, voltammetry can measure neurotransmitters, while ions are typically assessed through potentiometry. (From Mirza et al., 2019 originally published under CC-BY)

Neuromodulators, released by a specialized group of neurons, diffuse across broad areas of the nervous system, affecting numerous neurons simultaneously. This extensive influence is particularly significant in synaptic plasticity—the brain's ability to modify the strength of synapses according to the level of neuronal activity (Cesca et al., 2010; Choquet & Triller, 2013; Südhof, 2018). Such changes in synaptic strength are instrumental in remodeling neural circuits, thereby affecting critical cognitive functions like attention and learning (Angela & Dayan, 2005; Selemon, 2013; Doya, 2002). This broader impact of neuromodulation underscores its vital role in the brain's continuous adaptation and optimization processes.

Box 6.1 Neuromodulation

The process by which neural activity is regulated by neurotransmitters and neuromodulators, which adjust the properties of neurons and synapses, influencing the overall dynamics of neural circuits.

6.2 Neurotransmitters and Neuromodulation

6.2.1 Fundamentals of Neurotransmitters

Neurotransmitters, the essential chemical messengers of the nervous system, play a critical role in transmitting signals across synapses—the specialized connections that facilitate communication between neurons, as illustrated in ▣ Fig. 6.1c (Mirza et al., 2019). These molecules are pivotal in regulating neural activity by influencing the excitability of neurons. Functionally, neurotransmitters can be classified into two main categories: excitatory and inhibitory. Excitatory neurotransmitters enhance the probability that the receiving neuron will fire an action potential, thereby promoting neural activity. In contrast, inhibitory neurotransmitters reduce the likelihood of action potential generation in the receiving neuron, thus dampening neural activity (Brock & Cunnane, 1987; Agnati et al., 1995).

Among the array of neurotransmitters, classic examples include glutamate, the principal excitatory neurotransmitter that plays a critical role in neural activation and synaptic plasticity, and GABA (gamma-aminobutyric acid), the main inhibitory neurotransmitter in the brain, pivotal for reducing neuronal excitability and thus modulating neural circuitry dynamics (Petroff, 2002; Cooper et al., 2003). The balance between these excitatory and inhibitory influences is crucial for the proper functioning of neural networks and the overall stability of neural systems.

Glutamate's role extends beyond mere excitation; it is fundamental in learning and memory by reinforcing synaptic connections, a process known as long-term potentiation (LTP). Conversely, GABA's inhibitory actions are not merely suppressive but are essential for creating the rhythmic oscillations that underlie many cognitive processes, including attention and memory consolidation (Yizhar et al., 2011; Mody & Maguire, 2012).

Recent advancements have expanded our understanding of the nuanced roles neurotransmitters play within the brain. Studies have highlighted the existence of neuromodulatory neurotransmitters, such as dopamine, serotonin, and acetylcholine, which modulate the strength of synaptic transmission and influence a wide range of brain functions, from mood regulation to cognitive processing (Tritsch & Sabatini, 2012; Sara, 2009). These neurotransmitters do not simply toggle neurons on or off but rather adjust the neural circuits' sensitivity, influencing behavioral outcomes and mental states over varying timescales and contexts.

6.2.2 Modulatory Effects of Neurotransmitters

The functional role of neurotransmitters transcends the basic classification into excitatory and inhibitory types, revealing a complex landscape of subtle modulatory effects on neural circuits. Notably, dopamine and serotonin, which are primarily recognized as neuromodulators, play pivotal roles in a wide array of brain functions. These functions range from regulating mood and reward systems to enhancing cognitive flexibility (Bromberg-Martin et al., 2010; Cools & D'Esposito, 2011; Schmitt et al., 2006). Unlike traditional neurotransmitters that directly activate or inhibit synaptic responses, dopamine and serotonin subtly adjust the strength and efficacy of synaptic transmissions, exerting a broad, regulatory influence on neuronal interactions.

Dopamine, for instance, plays a pivotal role in the brain's reward system, influencing how rewards are anticipated and experienced. It adjusts the neural circuits' responsiveness to stimuli, thereby shaping learning, decision-making, and behavioral responses based on reward anticipation and outcomes (Schultz, 2016). Serotonin, on the other hand, contributes to a broad spectrum of psychological and physiological functions, including mood regulation, anxiety control, and the modulation of sleep and appetite. Its influence permeates the brain, affecting both emotional states and executive functions, thereby modulating cognitive processes and behavioral flexibility (Dayan & Huys, 2009).

The modulatory actions of these neurotransmitters are mediated through a complex interplay with specific receptor subtypes distributed across different brain regions, allowing for a finely tuned regulation of neural circuit dynamics. This receptor-specific action enables neurotransmitters like dopamine (Missale et al., 1998) and serotonin (Nichols & Nichols, 2008) to exert widespread effects on brain function, influencing everything from high-level cognitive processes to fundamental physiological responses.

The nuanced modulatory roles of neurotransmitters highlight the sophistication of neural communication and the complexity of brain function. Understanding these roles offers profound insights into the neural basis of behavior, cognition, and emotion, providing a foundational framework for developing therapeutic strategies aimed at addressing psychiatric and neurological disorders by targeting specific neurotransmitter systems (Lüscher & Malenka, 2011; Russo & Nestler, 2013).

> **Box 6.2 Neurotransmitters**
> Chemical messengers that transmit signals across a synapse from one neuron to another, primarily responsible for increasing (excitatory) or decreasing (inhibitory) neuronal excitability.

6.3   Synaptic Plasticity: The Basis for Learning and Memory

6.3.1   Mechanisms of Synaptic Plasticity

Synaptic plasticity encompasses the adaptive capability of synapses to modulate their strength over time, serving as a cornerstone for the neural mechanisms of learning and memory. This dynamic property enables the nervous system to encode experiences and adapt to new information, with long-term potentiation (LTP) and long-term depression (LTD) being pivotal processes that facilitate such synaptic modulation. LTP, characterized by an enhancement of synaptic efficacy, is typically induced by the activation of NMDA (N-methyl-D-aspartate) receptors, leading to an influx of calcium ions that trigger intracellular pathways supporting the strengthening of synaptic connections. On the other hand, LTD involves a reduction in synaptic efficacy, often mediated through processes such as the endocytosis of AMPA (α-amino-3-hydroxy-5-methyl-4-isoxazolepropionic acid) receptors, resulting in a diminished postsynaptic response (Malenka & Bear, 2004; Citri & Malenka, 2008). This dynamic is shown in ◘ Fig. 6.2 (Citri & Malenka, 2008).

These mechanisms of synaptic plasticity—LTP and LTD—are not merely binary opposites but represent a continuum of synaptic adjustments that underlie the brain's capacity to store and process information. The interplay between these processes enables the fine-tuning of synaptic transmission, crucial for the encoding and retrieval of memories, as well as the adaptation of behavior in response to changing environments (Citri & Malenka, 2008).

Recent research has further elaborated on the molecular and cellular pathways that govern LTP and LTD, highlighting the roles of various signaling molecules, synaptic proteins, and gene expression changes that sustain these long-lasting forms of synaptic modulation. For instance, the role of BDNF (Brain-Derived Neurotrophic Factor) (Lu et al., 2008) in facilitating LTP through the receptor

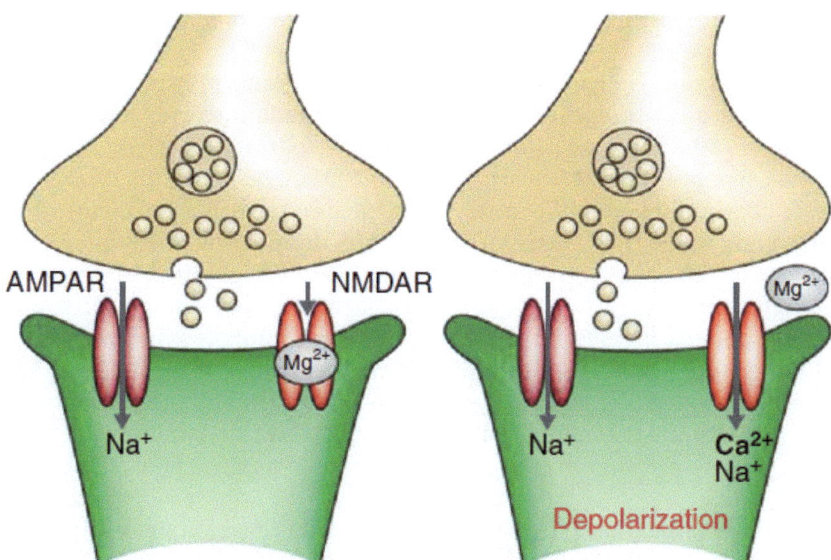

◘ **Fig. 6.2** Model of synaptic transmission at excitatory synapses. During basal synaptic transmission, synaptically released glutamate binds both the N-methyl-D-aspartate (NMDA) and α-amino-3-hydroxy-5-methyl-4-isoxazole propionic acid receptors (AMPARs). Na⁺ flows through the AMPAR channel but not through the N-methyl-D-aspartate receptors NMDAR channel because of the Mg²⁺ block of this channel. Depolarization of the postsynaptic cell (right) relieves the Mg²⁺ block of the NMDAR channel and allows both Na⁺ and Ca²⁺ to flow into the dendritic spine. The resultant increase in Ca²⁺ in the dendritic spine is necessary for triggering the subsequent events that drive synaptic plasticity. (Adapted from Citri & Malenka, 2008)

tyrosine kinase (TrkB) receptors (Minichiello, 2009) underscores the intricate link between neuromodulation and synaptic plasticity, illustrating how synaptic strength adjustments are embedded within broader neurobiological contexts.

Moreover, emerging studies have identified the impact of external factors, such as stress and exercise, on synaptic plasticity, suggesting that LTP and LTD are modulated by a wide range of physiological and environmental conditions, thereby influencing learning and memory processes in a context-dependent manner (Duman & Aghajanian, 2012; Zovkic & Sweatt, 2013).

> **Box 6.3   Synaptic Plasticity**
> The ability of synapses to strengthen or weaken over time, based on increases or decreases in their activity; a fundamental property that underlies learning and memory.

6.3.1.1 **Long-Term Potentiation (LTP)**

Long-term potentiation (LTP) is an enduring enhancement in synaptic efficiency following repeated stimulations, pivotal in the synaptic basis of learning and memory. This phenomenon, crucial for our understanding of memory formation, involves a series of well-coordinated events beginning with synaptic activation and culminating in the lasting augmentation of synaptic transmission (Citri & Malenka, 2008).

- **Induction Phase**: The initiation of LTP typically occurs through high-frequency stimulation of the presynaptic neuron. This stimulation leads to the release of glutamate, which then activates N-methyl-D-aspartate (NMDA) receptors on the postsynaptic neuron (Furukawa et al., 2005). The activation of these receptors is conditional upon the postsynaptic membrane being sufficiently depolarized to relieve the Mg^{2+} block of the NMDA receptor channels.
- **Calcium Influx and Signaling Cascades**: The critical entry of calcium ions through the activated NMDA receptors into the postsynaptic neuron marks a pivotal moment in LTP induction. This calcium influx is instrumental, setting off a series of intracellular signaling pathways that orchestrate the molecular alterations underlying LTP. Key players in these signaling cascades include calcium/calmodulin-dependent protein kinase II (CaMKII) and protein kinase A (PKA), which are vital for the amplification of synaptic signals (Lisman et al., 2002).
- **Synaptic Modifications**: These signaling cascades precipitate several adaptive changes in synaptic structure and function. One of the primary outcomes is the increased insertion and phosphorylation of α-amino-3-hydroxy-5-methyl-4-isoxazolepropionic acid (AMPA) receptors into the postsynaptic membrane, which boosts the postsynaptic cell's responsiveness to glutamate. Concurrently, alterations in the presynaptic neuron may augment neurotransmitter release, further enhancing synaptic efficacy (Nicoll & Roche, 2013). These changes collectively contribute to the potentiation of synaptic transmission, as visually represented in ◘ Fig. 6.3.

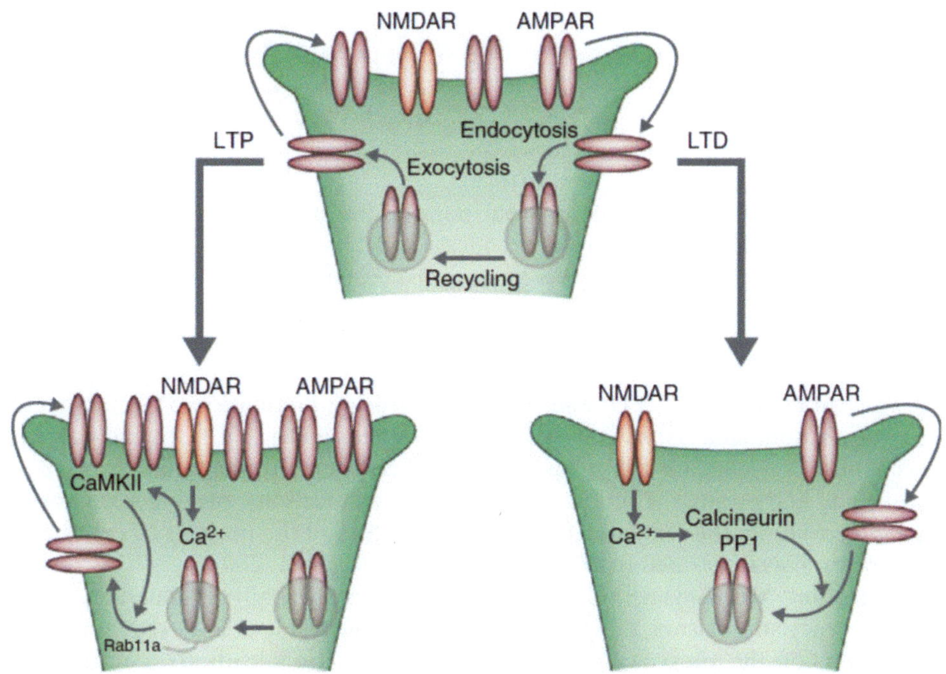

Fig. 6.3 Model of AMPAR Trafficking in synaptic plasticity: the diagram illustrates AMPAR (α-amino-3-hydroxy-5-methyl-4-isoxazolepropionic acid receptor) trafficking during long-term potentiation (LTP) and long-term depression (LTD). In the basal state, AMPARs oscillate between the postsynaptic membrane and internal compartments, shifting laterally outside the synaptic zone to endocytic zones for clathrin- and dynamin-dependent endocytosis into early endosomes. Normally, these receptors move to recycling endosomes and re-enter the plasma membrane via exocytosis, migrating laterally back into the synapse where MAGUK proteins anchor them. Upon LTP induction, calcium-driven processes enhance receptor exocytosis and synaptic stabilization, involving CAMKII and Rab11a-mediated fusion of recycling endosomes. Conversely, LTD induction leads to increased extrasynaptic receptor endocytosis, driven by calcium and primarily mediated by calcineurin and protein phosphatase 1 (PP1). Whereas basal receptor cycling maintains a balance between endocytosis and recycling, after LTD, receptors may be retained intracellularly and potentially degraded. (Adapted from Citri & Malenka, 2008)

- **Maintenance and Expression**: The enduring nature of LTP reflects modifications at both pre- and postsynaptic sites, resulting in a sustained increase in synaptic strength. This persistent strengthening facilitates an enhanced response to subsequent stimulations, embodying the synaptic basis for learning and memory.
- **Role in Learning and Memory**: LTP is acclaimed as a primary cellular mechanism underlying learning and memory. The process of strengthening synaptic connections through LTP is believed to embody the neural substrate for storing and consolidating information, playing a foundational role in memory formation and retention (Bliss & Collingridge, 1993; Takeuchi et al., 2014). The relationship between LTP and cognitive processes underscores the significance of synaptic plasticity in the neural encoding of memories, with extensive

research supporting LTP's central role in various forms of learning and memory across different brain regions (Bliss & Collingridge, 2013; Huganir & Nicoll, 2013).

> **Box 6.4   Long-Term Potentiation (LTP)**
> A long-lasting enhancement in signal transmission between two neurons that results from stimulating them synchronously; widely considered one of the major cellular mechanisms that underlies learning and memory.

6.3.1.2 Long-Term Depression (LTD)

Long-term depression (LTD) embodies the enduring diminution of synaptic efficacy, a phenomenon as pivotal as long-term potentiation (LTP) for the nuanced mechanisms of learning and memory. This process is instrumental in the selective weakening and eventual elimination of synaptic connections, playing a key role in synaptic pruning, the refinement of neural circuits, and the purging of extraneous information, as illustrated in ☐ Fig. 6.3 (Citri & Malenka, 2008).

- **Induction Mechanisms**: Unlike LTP, which is typically initiated by high-frequency stimulation, LTD induction is often the result of low-frequency stimulation of the presynaptic neuron. This stimulation leads to a modest yet sustained increase in postsynaptic calcium levels, contrasting with the sharp and transient calcium influx characterizing LTP induction. The nuanced differences in calcium dynamics between LTP and LTD underscore the diversity of neural signaling pathways and their roles in synaptic modulation (Mulkey & Malenka, 1992; Dudek & Bear, 1992).
- **Calcium Dynamics and Intracellular Signaling**: The gradual rise in calcium concentration within the postsynaptic neuron during LTD induction activates a distinct set of intracellular signaling mechanisms, prominently involving phosphatases such as calcineurin. This contrasts with the kinase-driven pathways predominantly engaged during LTP, highlighting the complexity of synaptic plasticity mechanisms (Mulkey et al., 1994; Wang et al., 2005).
- **Synaptic Modifications**: The activation of these LTD-specific signaling pathways culminates in the removal of AMPA receptors from the postsynaptic membrane, diminishing the neuron's sensitivity to glutamatergic signaling. Concurrently, alterations in presynaptic function may reduce neurotransmitters' release, further decreasing synaptic strength. These modifications collectively contribute to a reduced synaptic efficacy, effectively dampening the response to future stimulations (Lüscher & Malenka, 2012).
- **Maintenance, Expression, and Functional Implications**: LTD is critical for the adaptive remodeling of neural networks, facilitating the removal of outdated or infrequently used synaptic connections. This synaptic weakening and elimination process is essential for maintaining synaptic plasticity and homeostasis, enabling the nervous system to refine memory representations and discard irrelevant information, thereby optimizing cognitive efficiency (Bear & Malenka, 1994; Linden & Connor, 1995).

LTD and LTP together constitute the yin and yang of synaptic plasticity, with their interplay ensuring the dynamic regulation of synaptic strength essential for learning, memory, and cognitive flexibility (Malenka & Bear, 2004). The equilibrium between LTD and LTP is vital for sustaining healthy cognitive functions, preserving the integrity of neural networks, and fostering the brain's overall adaptability (Nabavi et al., 2014). Insights into the mechanisms of LTD and LTP enrich our understanding of the physiological underpinnings of learning and memory, offering potential avenues for therapeutic interventions in various neurological conditions (Dhuriya & Sharma, 2020).

> **Box 6.5   Long-Term Depression (LTD)**
> A long-lasting decrease in synaptic strength between two neurons, believed to play a role in the pruning of old connections and in learning and memory by removing less-used synaptic pathways.

6.3.1.3   Role in Neural Circuit Function

Synaptic plasticity extends its influence beyond the confines of individual synapses, playing a pivotal role in shaping the architecture and function of broader neural networks. The process of synaptic strengthening, for instance, does not occur in isolation but significantly impacts neural pathways. When synapses within a specific circuit undergo potentiation, this pathway becomes more likely to be activated in response to relevant stimuli, a phenomenon crucial for memory consolidation mechanisms. Such preferential activation ensures that neural representations associated with significant or frequently encountered stimuli are more readily accessible, thereby facilitating the retention and recall of important information (Martin et al., 2000; Kandel, 2001; Takeuchi et al., 2014).

Conversely, the process of synaptic weakening plays a complementary role in the dynamic modulation of neural circuits. Long-term depression (LTD) or other forms of synaptic weakening contribute to the reconfiguration of neural networks, enabling the brain to adapt to new information or experiences. This synaptic downscaling is instrumental in the selective pruning of less-used or irrelevant synaptic connections, which, in turn, facilitates the forgetting of outdated or no longer relevant information. Such synaptic and circuitry adjustments are critical for maintaining the efficiency and flexibility of neural processing, preventing the overload of neural networks with superfluous information and ensuring that cognitive resources are allocated to more pertinent or recently acquired knowledge (Herry et al., 2008).

The dynamic interplay between synaptic strengthening and weakening underscores the brain's remarkable capacity for plasticity, enabling continuous learning and adaptation throughout an individual's life. By modulating the strength and efficacy of synaptic connections within and across neural circuits, the brain orchestrates a delicate balance between the retention of meaningful experiences and the removal of unnecessary information, optimizing cognitive functions and behavioral responses to the environment (Zatorre et al., 2012; Tonegawa et al., 2015).

These synaptic and circuitry plasticity processes contribute fundamentally to our understanding of learning and memory and offer valuable insights into the physiological basis for a wide range of cognitive phenomena, from decision-making and problem-solving to the emotional responses that underpin human experience. Furthermore, elucidating the mechanisms underlying synaptic plasticity and its effects on neural circuitry has profound implications for developing therapeutic strategies for neurological and psychiatric disorders characterized by dysregulated synaptic function and plasticity (Malenka & Bear, 2004; Di Filippo et al., 2008; Mayford et al., 2012; Wondolowski & Dickman, 2013).

> **Box 6.6   Excitatory Neurotransmitters**
> Neurotransmitters that promote the generation of electrical signals in the receiving neuron, typically by causing depolarization of the neuron; glutamate is the primary excitatory neurotransmitter in the brain.

> **Box 6.7   Inhibitory Neurotransmitters**
> Neurotransmitters that prevent the generation of electrical signals in the receiving neuron, typically by causing hyperpolarization of the neuron; gamma-aminobutyric acid (GABA) is the primary inhibitory neurotransmitter in the brain.

6.3.2 Neural Circuits and Network Dynamics

6.3.2.1 Structure and Function of Neural Circuits

Neural circuits, the complex networks of interconnected neurons, form the essential infrastructure for processing and transmitting information within the nervous system. These circuits embody various architectural and functional paradigms, such as feedforward or feedback loops, which are crucial in controlling neural signal direction, integration, and modulation. These configurations are vital for numerous neural processes, directing the flow of information from sensory reception to motor execution. They also facilitate complex cognitive functions by enabling dynamic interactions and communication across different brain regions. This organizational complexity allows the nervous system to perform its wide range of functions efficiently and adaptively (Douglas & Martin, 2004; Harris & Shepherd, 2015).

Feedforward circuits are characterized by a unidirectional flow of information, where input from one neuron or neural ensemble moves to another without reciprocal interaction. This linear pathway ensures the efficient transmission of signals across different processing stages, crucial for tasks requiring rapid and streamlined responses. Conversely, feedback loops involve recursive pathways where outputs of a circuit loop back to influence its own input or the inputs of upstream neurons. This arrangement allows for the regulation of signal flow, refining sensory percep-

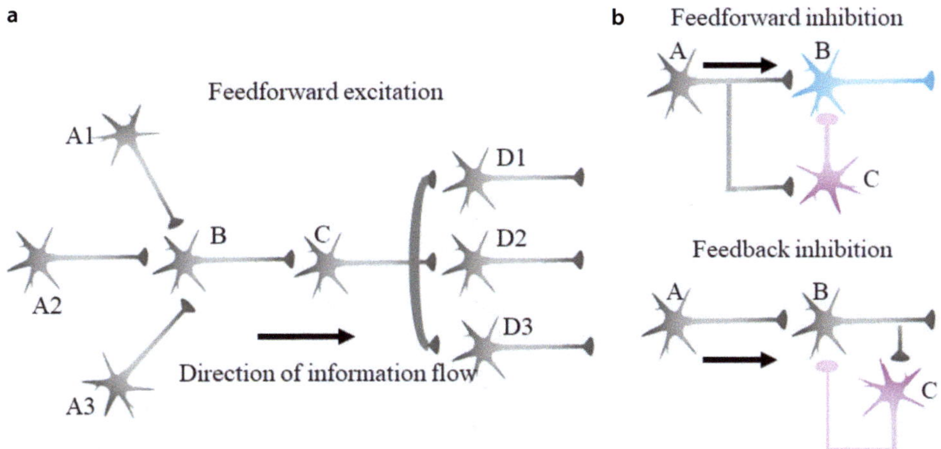

Fig. 6.4   **a** Feedforward excitation involves the transmission of information through a sequence of excitatory neurons, labeled A to D. Here, multiple A neurons (three distinct neurons) converge to synapse onto neuron B, demonstrating convergent excitation. Neuron C, in turn, diverges to synapse onto three separate D neurons, exemplifying divergent excitation. **b** Feedforward inhibition is characterized by an inhibitory neuron C that receives input from a presynaptic excitatory neuron A and sends inhibitory outputs to a postsynaptic neuron B. Conversely, feedback inhibition features the same inhibitory neuron C receiving input from and directing inhibitory outputs back to a postsynaptic excitatory neuron B, creating a regulatory feedback loop. (Adapted from Kim & Zhen, 2009 and based on Luo, 2021)

tions, stabilizing network activity, and facilitating learning and memory by reinforcing or diminishing specific neural pathways based on feedback (Salinas & Sejnowski, 2001; Marder, 2012). **Figure 6.4 depicts the feedforward and feedback circuits.

The functional characteristics of neural circuits are not static but exhibit plasticity, allowing the nervous system to adapt to changes in environmental stimuli, learn from experiences, and recover from injury. Synaptic plasticity, including mechanisms such as long-term potentiation (LTP) and long-term depression (LTD), underlies the adaptability of neural circuits, adjusting synaptic strength in response to activity levels. This capacity for modification ensures that neural circuits can reorganize and optimize their function in response to internal and external demands, playing a critical role in the nervous system's development, learning, and overall adaptability (Buonomano & Merzenich, 1998; Feldman, 2012).

Understanding the structural and functional dynamics of neural circuits and their capacity for plasticity offers profound insights into the principles of neural computation and the basis of behavior. It also provides a framework for exploring the pathophysiology of neurological and psychiatric disorders, where aberrations in circuit function and plasticity are often implicated. This knowledge not only deepens our comprehension of the brain's intricate workings but also guides the development of therapeutic interventions aimed at restoring or compensating for dysfunctional circuitry in various neurological conditions (Denève & Machens, 2016; Aitchison & Lengyel, 2017).

> ### Box 6.8   Neural Circuits
> Networks of neurons interconnected by synapses, responsible for processing and transmitting information through electrical and chemical signals.

> ### Box 6.9   Feedforward and Feedback Loops
> Structural motifs in neural circuits where feedforward refers to the unidirectional flow of information from input to output, and feedback denotes the process where outputs of a system are routed back as inputs.

6.3.2.2 Modulation Through Circuit Dynamics

The dynamics of neural circuits, encapsulated in phenomena such as oscillatory behavior and synchronization, play a pivotal role in the modulation of neural activity. Oscillations in brain activity, exemplified by theta (4–8 Hz) and gamma (30–100 Hz) rhythms, are integral to the coordination of information transfer and processing across disparate brain regions. These rhythmic activities are thought to underpin critical aspects of neural coding and inter-regional communication, facilitating a wide range of cognitive processes (Buzsaki & Draguhn, 2004). Moreover, the synchronization of neuronal firing within circuits serves to bolster signal transmission and integration, significantly influencing cognitive functions, including attention, perception, and memory consolidation (Fries, 2005; Jensen et al., 2007).

The regulation of neural activity encompasses a complex network of interactions among neurotransmitters, synaptic plasticity mechanisms, and the structural configurations of neural circuits, as described in the section above. Understanding these interactions offers profound insights into the operational principles of the nervous system, shedding light on the neurobiological substrates of behavior, cognition, and the pathophysiology of neurological disorders (Feldman, 2012; Dayan & Abbott, 2005).

As introduced earlier in this chapter, a cornerstone of neuromodulation is the principle of synaptic plasticity—the capacity of synapses to undergo long-term strengthening (LTP) or weakening (LTD) in response to changes in activity. This synaptic dynamism is fundamental to learning and the encoding of memories. Complementing synaptic adjustments, the broader concept of neuroplasticity—entailing the brain's ability to restructure and forge new neural connections—plays a pivotal role in the brain's adaptability. Such plastic changes enable the brain to adapt to new learning experiences, navigate changes in the environment, and recover from injuries, highlighting the brain's remarkable capacity for reorganization and adaptation (Malenka & Bear, 2004; Pascual-Leone et al., 2005). ■ Figure 6.5 illustrates the analogy between the progression from neurons to neural circuits and finally to brain function and the building of language from letters to words to sentences, forming a complete article (Luo, 2021). It starts with individual neurons (letters), progresses to core circuit motifs (words), expands to larger-scale architectural plans (sentences), and culminates in the brain's complete

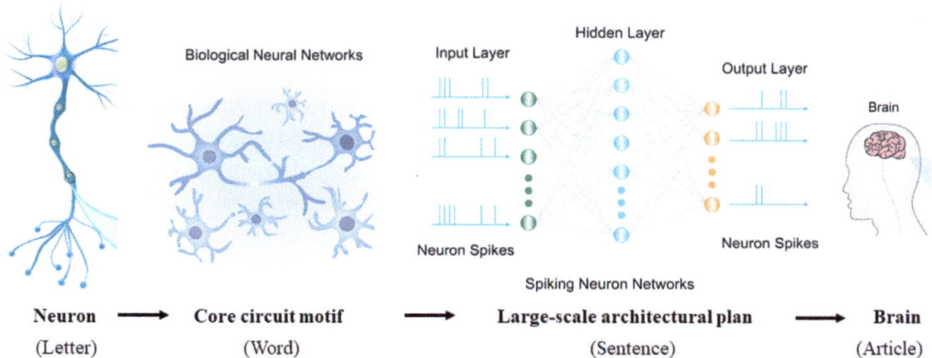

Fig. 6.5 This figure presents an analogy between the structural development and functional evolution of neural circuits and the composition of language, moving from letters to words to sentences, and culminating in a complete article representing the brain. On the left side, labeled "Structure," there is an illustration of a neuron, with its dendrites in blue and axon in red, representing the basic building block of neural circuits, akin to a letter forming part of a word. Progressing to the right, these "letters" form a "Core circuit motif," depicted as an interconnected group of neurons with green stars symbolizing neuronal bodies and red dots indicating synapses, analogous to words that form a sentence. This core circuit motif is further integrated into a "Larger-scale architectural plan," resembling a neural pathway with multiple neurons, each represented by a colored star connected by multi-colored lines, symbolizing various pathways or sentences forming a paragraph. The figure implies that just as letters combine into words and sentences to create a comprehensive article, neurons form circuits that then integrate into larger networks, ultimately giving rise to the complex functionalities of the brain. (Adapted from Xiang et al., 2021; Yuan et al., 2022 originally published under CC-BY 4.0 and based on Luo, 2021)

functionality (an article). The development and evolution of the brain are represented as a continuum from basic structural units to complex functional networks (Luo, 2021).

> ### Box 6.10   Dopamine and Serotonin
> Neuromodulators known for their role in regulating mood, emotion, and cognition, with broad influences over multiple neural processes and synaptic modulation.

Building on the foundational understanding of neural operation, brain stimulation techniques such as deep brain stimulation (DBS), transcranial magnetic stimulation (TMS), and transcranial electrical stimulation (tES) exploit these principles of neural activity modulation. By directly influencing neural circuits and pathways, these techniques offer therapeutic avenues for correcting dysregulated neural activity associated with various neurological and psychiatric conditions. This transition from theoretical understanding to clinical application marks a significant advancement in neuroscience, bridging the gap between basic research and therapeutic intervention (Lozano et al., 2019; George & Aston-Jones, 2010).

▶ Chapters 7–9 will explore the diverse spectrum of neuromodulation techniques that have transformed our approach to diagnosing, understanding, and treating neurological disorders. Brain stimulation, fundamentally, involves altering neural activity through the precise delivery of stimuli—electrical, magnetic, or chemical—to targeted brain areas. The clinical ramifications of these techniques extend well beyond academic interest, providing tangible benefits in the treatment of various challenging conditions, from movement disorders to depression, underscoring the therapeutic potential of neuromodulation in contemporary neuroscience and medicine.

❯ **Key Takeaways**

1. **Neuromodulation as a Core Mechanism in Neural Function**: The chapter highlights neuromodulation as a fundamental process in neural circuits, wherein neurotransmitters such as dopamine, serotonin, and acetylcholine influence the overall dynamics of neuronal excitability and synaptic plasticity. These neuromodulators play pivotal roles in regulating behaviors, cognitive functions, and emotional responses, making neuromodulation crucial to both basic and complex neural processes.

2. **Synaptic Plasticity and Its Role in Learning and Memory**: Synaptic plasticity, particularly long-term potentiation (LTP) and long-term depression (LTD), is identified as a key mechanism through which the brain encodes, stores, and adapts information. The chapter discusses how these processes adjust synaptic strength, enabling learning and memory, and outlines the molecular pathways, such as calcium signaling and receptor modulation, that sustain these changes.

3. **Neural Circuits and Their Dynamic Plasticity**: The structure and functionality of neural circuits, including feedforward and feedback loops, are explored as essential for processing sensory information and regulating cognitive functions. The chapter emphasizes that neural circuits are not static but exhibit plasticity, allowing the brain to reorganize itself in response to environmental stimuli and learning experiences.

4. **Clinical Implications and Therapeutic Potential**: The principles of neuromodulation and synaptic plasticity provide the foundation for developing neuromodulatory therapies such as deep brain stimulation (DBS) and transcranial magnetic stimulation (TMS). These techniques, which influence neural activity by directly modulating circuits, have shown potential in treating neurological and psychiatric disorders, underscoring the therapeutic value of understanding these neural mechanisms.

As we progress through the book, the focus will shift from understanding these basic mechanisms to applying this knowledge through various brain stimulation techniques, discussed in the upcoming chapters. These techniques, which range from non-invasive to invasive approaches, leverage the principles outlined here to modulate brain activity directly, offering promising new therapies for a range of conditions.

The foundational concepts introduced in this chapter set the stage for a deeper exploration of how we can influence the brain's activity through targeted interventions. By bridging the gap between theoretical neuroscientific concepts and practical therapeutic applications, the subsequent chapters aim to enrich our approach to neuroscience and neurology, enhancing both our understanding and our ability to intervene effectively in brain-based disorders.

Conclusion

In this chapter we deep-dived into the complexities of neural communication, highlighting the pivotal role of neuromodulation in the brain's adaptability and functionality. As we have explored, the intricate dance between neurotransmitters, synaptic plasticity, and neural circuits is foundational not only to basic neural operations but also to complex cognitive processes. The discussion on neurotransmitters reveals their critical roles as chemical messengers that enhance or inhibit neural signals across synaptic pathways, while the sections on synaptic plasticity provide a window into the brain's capacity to learn and adapt through the strengthening or weakening of synaptic connections.

Neuromodulation, we have learned, is not a mere background function but a dynamic and essential process of brain activity that enables the nervous system to adapt to new information and changing environments. This adaptive capability is seen in the modulation of neural circuits, which reconfigure and optimize brain functionality in response to both internal changes and external stimuli. The emerging research in this field not only advances our understanding of fundamental brain functions but also opens new avenues for treating neurological and psychiatric disorders.

This section offers practical exercises designed to help apply the material learned to concrete scenarios.

Modeling Neural Circuits

Objective: To construct models of neural circuits using feedforward and feedback loops.

Task: Students will use software like NEURON or MATLAB to create simple neural circuit models that demonstrate how feedforward and feedback mechanisms affect signal processing within the circuit.

Simulating Neurotransmitter Effects

Objective: To simulate the effects of different neurotransmitters on neuron excitability.

Task: Using a computational tool like Brian 2, students will simulate how varying levels of excitatory and inhibitory neurotransmitters (e.g., glutamate and GABA) alter the firing patterns of neurons.

Analyzing Neuromodulation Effects

Objective: To analyze the role of neuromodulators like dopamine and serotonin on cognitive processes.

Task: Students will review case studies or research articles where the modulation of dopamine and serotonin was manipulated, discussing the resultant effects on behavior and neural circuit functions.

Neural Circuit Mapping

Objective: To map the structure and function of specific neural circuits.

Task: Using data from brain imaging studies, students will identify neural circuits involved in specific cognitive functions or behaviors, and present their findings on how these circuits are influenced by synaptic plasticity.

Role-Playing Clinical Scenarios

Objective: To apply the knowledge of neuromodulation in therapeutic contexts.

Task: Students will role-play as neurologists or psychiatrists, developing treatment plans for hypothetical patients with disorders influenced by neuromodulation, such as depression or Parkinson's disease, incorporating strategies like pharmacotherapy or deep brain stimulation.

References

Agnati, L. F., Zoli, M., Strömberg, I., & Fuxe, K. (1995). Intercellular communication in the brain: Wiring versus volume transmission. *Neuroscience, 69*(3), 711–726.

Aitchison, L., & Lengyel, M. (2017). With or without you: Predictive coding and Bayesian inference in the brain. *Current Opinion in Neurobiology, 46*, 219–227.

Angela, J. Y., & Dayan, P. (2005). Uncertainty, neuromodulation, and attention. *Neuron, 46*(4), 681–692.

Bear, M. F., & Malenka, R. C. (1994). Synaptic plasticity: LTP and LTD. *Current Opinion in Neurobiology, 4*(3), 389–399.

Berridge, K. C. (2007). The debate over dopamine's role in reward: The case for incentive salience. *Psychopharmacology, 191*, 391–431.

Bliss, T. V., & Collingridge, G. L. (1993). A synaptic model of memory: Long-term potentiation in the hippocampus. *Nature, 361*(6407), 31–39.

Bliss, T. V., & Collingridge, G. L. (2013). Expression of NMDA receptor-dependent LTP in the hippocampus: Bridging the divide. *Molecular Brain, 6*, 1–14.

Brock, J. A., & Cunnane, T. C. (1987). Relationship between the nerve action potential and transmitter release from sympathetic postganglionic nerve terminals. *Nature, 326*(6113), 605–607.

Bromberg-Martin, E. S., Matsumoto, M., & Hikosaka, O. (2010). Dopamine in motivational control: Rewarding, aversive, and alerting. *Neuron, 68*(5), 815–834.

Buonomano, D. V., & Merzenich, M. M. (1998). Cortical plasticity: From synapses to maps. *Annual Review of Neuroscience, 21*(1), 149–186.

Buzsaki, G., & Draguhn, A. (2004). Neuronal oscillations in cortical networks. *Science, 304*(5679), 1926–1929.

Cesca, F., Baldelli, P., Valtorta, F., & Benfenati, F. (2010). The synapsins: Key actors of synapse function and plasticity. *Progress in Neurobiology, 91*(4), 313–348.

Choquet, D., & Triller, A. (2013). The dynamic synapse. *Neuron, 80*(3), 691–703.

Citri, A., & Malenka, R. C. (2008). Synaptic plasticity: Multiple forms, functions, and mechanisms. *Neuropsychopharmacology, 33*(1), 18–41.

Cools, R., & D'Esposito, M. (2011). Inverted-U–shaped dopamine actions on human working memory and cognitive control. *Biological Psychiatry, 69*(12), e113–e125.

Cooper, J. R., Bloom, F. E., & Roth, R. H. (2003). *The biochemical basis of neuropharmacology.* Oxford University Press.

Dayan, P. (2012). Twenty-five lessons from computational neuromodulation. *Neuron, 76*(1), 240–256.

Dayan, P., & Abbott, L. F. (2005). *Theoretical neuroscience: Computational and mathematical modeling of neural systems.* MIT Press.

Dayan, P., & Huys, Q. J. (2009). Serotonin in affective control. *Annual Review of Neuroscience, 32*, 95–126.

Denève, S., & Machens, C. K. (2016). Efficient codes and balanced networks. *Nature Neuroscience, 19*(3), 375–382.

Dhuriya, Y. K., & Sharma, D. (2020). Neuronal plasticity: Neuronal organization is associated with neurological disorders. *Journal of Molecular Neuroscience, 70*(11), 1684–1701.

Di Filippo, M., Sarchielli, P., Picconi, B., & Calabresi, P. (2008). Neuroinflammation and synaptic plasticity: Theoretical basis for a novel, immune-centred, therapeutic approach to neurological disorders. *Trends in Pharmacological Sciences, 29*(8), 402–412.

Douglas, R. J., & Martin, K. A. (2004). Neuronal circuits of the neocortex. *Annual Review of Neuroscience, 27*, 419–451.

Doya, K. (2002). Metalearning and neuromodulation. *Neural Networks, 15*(4–6), 495–506.

Dudek, S. M., & Bear, M. F. (1992). Homosynaptic long-term depression in area CA1 of hippocampus and effects of N-methyl-D-aspartate receptor blockade. *Proceedings of the National Academy of Sciences, 89*(10), 4363–4367.

Duman, R. S., & Aghajanian, G. K. (2012). Synaptic dysfunction in depression: Potential therapeutic targets. *Science, 338*(6103), 68–72.

Eccles, J. C. (1973). *The understanding of the brain.* McGraw-Hill.

Feldman, D. E. (2012). The spike-timing dependence of plasticity. *Neuron, 75*(4), 556–571.

Fries, P. (2005). A mechanism for cognitive dynamics: Neuronal communication through neuronal coherence. *Trends in Cognitive Sciences, 9*(10), 474–480.

Furukawa, H., Singh, S. K., Mancusso, R., & Gouaux, E. (2005). Subunit arrangement and function in NMDA receptors. *Nature, 438*(7065), 185–192.

George, M. S., & Aston-Jones, G. (2010). Non-invasive techniques for probing neurocircuitry and treating illness: Vagus nerve stimulation (VNS), transcranial magnetic stimulation (TMS) and transcranial direct current stimulation (tDCS). *Neuropsychopharmacology, 35*(1), 301–316.

Harris, K. D., & Shepherd, G. M. (2015). The neocortical circuit: Themes and variations. *Nature Neuroscience, 18*(2), 170–181.

Herry, C., Ciocchi, S., Senn, V., Demmou, L., Müller, C., & Lüthi, A. (2008). Switching on and off fear by distinct neuronal circuits. *Nature, 454*(7204), 600–606.

Huganir, R. L., & Nicoll, R. A. (2013). AMPARs and synaptic plasticity: The last 25 years. *Neuron, 80*(3), 704–717.

Jensen, O., Kaiser, J., & Lachaux, J. P. (2007). Human gamma-frequency oscillations associated with attention and memory. *Trends in Neurosciences, 30*(7), 317–324.

Kandel, E. R. (2001). The molecular biology of memory storage: A dialogue between genes and synapses. *Science, 294*(5544), 1030–1038.

Kim, J. S., & Zhen, M. (2009). Neuronal polarity. In M. D. Binder, N. Hirokawa, & U. Windhorst (Eds.), *Encyclopedia of neuroscience*. Springer.

Lee, S. H., & Dan, Y. (2012). Neuromodulation of brain states. *Neuron, 76*(1), 209–222.

Linden, D. J., & Connor, J. A. (1995). Long-term synaptic depression. *Annual Review of Neuroscience, 18*(1), 319–357.

Lisman, J., Schulman, H., & Cline, H. (2002). The molecular basis of CaMKII function in synaptic and behavioural memory. *Nature Reviews Neuroscience, 3*(3), 175–190.

Lozano, A. M., Lipsman, N., Bergman, H., Brown, P., Chabardes, S., Chang, J. W., et al. (2019). Deep brain stimulation: Current challenges and future directions. *Nature Reviews Neurology, 15*(3), 148–160.

Lu, Y., Christian, K., & Lu, B. (2008). BDNF: A key regulator for protein synthesis-dependent LTP and long-term memory? *Neurobiology of Learning and Memory, 89*(3), 312–323.

Luo, L. (2021). Architectures of neuronal circuits. *Science, 373*(6559), eabg7285.

Lüscher, C., & Malenka, R. C. (2011). Drug-evoked synaptic plasticity in addiction: From molecular changes to circuit remodeling. *Neuron, 69*(4), 650–663.

Lüscher, C., & Malenka, R. C. (2012). NMDA receptor-dependent long-term potentiation and long-term depression (LTP/LTD). *Cold Spring Harbor Perspectives in Biology, 4*(6), a005710.

Malenka, R. C., & Bear, M. F. (2004). LTP and LTD: An embarrassment of riches. *Neuron, 44*(1), 5–21.

Marder, E. (2012). Neuromodulation of neuronal circuits: Back to the future. *Neuron, 76*(1), 1–11.

Martin, S. J., Grimwood, P. D., & Morris, R. G. (2000). Synaptic plasticity and memory: An evaluation of the hypothesis. *Annual Review of Neuroscience, 23*(1), 649–711.

Mayford, M., Siegelbaum, S. A., & Kandel, E. R. (2012). Synapses and memory storage. *Cold Spring Harbor Perspectives in Biology, 4*(6), a005751.

Minichiello, L. (2009). TrkB signalling pathways in LTP and learning. *Nature Reviews Neuroscience, 10*(12), 850–860.

Mirza, K. B., Golden, C. T., Nikolic, K., & Toumazou, C. (2019). Closed-loop implantable therapeutic neuromodulation systems based on neurochemical monitoring. *Frontiers in Neuroscience, 13*, 808.

Missale, C., Nash, S. R., Robinson, S. W., Jaber, M., & Caron, M. G. (1998). Dopamine receptors: From structure to function. *Physiological Reviews, 78*(1), 189–225.

Mody, I., & Maguire, J. (2012). The reciprocal regulation of stress hormones and GABAA receptors. *Frontiers in Cellular Neuroscience, 6*, 4.

Mulkey, R. M., & Malenka, R. C. (1992). Mechanisms underlying induction of homosynaptic long-term depression in area CA1 of the hippocampus. *Neuron, 9*(5), 967–975.

Mulkey, R. M., Endo, S., Shenolikar, S., & Malenka, R. C. (1994). Involvement of a calcineurin/inhibitor-1 phosphatase cascade in hippocampal long-term depression. *Nature, 369*(6480), 486–488.

Nabavi, S., Fox, R., Proulx, C. D., Lin, J. Y., Tsien, R. Y., & Malinow, R. (2014). Engineering a memory with LTD and LTP. *Nature, 511*(7509), 348–352.

Nichols, D. E., & Nichols, C. D. (2008). Serotonin receptors. *Chemical Reviews, 108*(5), 1614–1641.

Nicoll, R. A., & Roche, K. W. (2013). Long-term potentiation: Peeling the onion. *Neuropharmacology, 74*, 18–22.

Pascual-Leone, A., Amedi, A., Fregni, F., & Merabet, L. B. (2005). The plastic human brain cortex. *Annual Review of Neuroscience, 28*, 377–401.

Petroff, O. A. (2002). Book review: GABA and glutamate in the human brain. *The Neuroscientist, 8*(6), 562–573.

Rolls, E., & Treves, A. (1997). *Neural networks and brain function*. Oxford University Press.

Russo, S. J., & Nestler, E. J. (2013). The brain reward circuitry in mood disorders. *Nature Reviews Neuroscience, 14*(9), 609–625.

Salinas, E., & Sejnowski, T. J. (2001). Correlated neuronal activity and the flow of neural information. *Nature Reviews Neuroscience, 2*(8), 539–550.

Sara, S. J. (2009). The locus coeruleus and noradrenergic modulation of cognition. *Nature Reviews Neuroscience, 10*(3), 211–223.

Schmitt, J. A. J., Wingen, M., Ramaekers, J. G., Evers, E. A. T., & Riedel, W. J. (2006). Serotonin and human cognitive performance. *Current Pharmaceutical Design, 12*(20), 2473–2486.

Schultz, W. (2016). Dopamine reward prediction-error signalling: A two-component response. *Nature Reviews Neuroscience, 17*(3), 183–195.

Selemon, L. D. (2013). A role for synaptic plasticity in the adolescent development of executive function. *Translational Psychiatry, 3*(3), e238.

Südhof, T. C. (2018). Towards an understanding of synapse formation. *Neuron, 100*(2), 276–293.

Takeuchi, T., Duszkiewicz, A. J., & Morris, R. G. (2014). The synaptic plasticity and memory hypothesis: Encoding, storage and persistence. *Philosophical Transactions of the Royal Society B: Biological Sciences, 369*(1633), 20130288.

Tonegawa, S., Liu, X., Ramirez, S., & Redondo, R. (2015). Memory engram cells have come of age. *Neuron, 87*(5), 918–931.

Tritsch, N. X., & Sabatini, B. L. (2012). Dopaminergic modulation of synaptic transmission in cortex and striatum. *Neuron, 76*(1), 33–50.

Wang, H. X., Gerkin, R. C., Nauen, D. W., & Bi, G. Q. (2005). Coactivation and timing-dependent integration of synaptic potentiation and depression. *Nature Neuroscience, 8*(2), 187–193.

Wondolowski, J., & Dickman, D. (2013). Emerging links between homeostatic synaptic plasticity and neurological disease. *Frontiers in Cellular Neuroscience, 7*, 223.

Xiang, Z., Tang, C., Chang, C., & Liu, G. (2021). A new viewpoint and model of neural signal generation and transmission: Signal transmission on unmyelinated neurons. *Nano Research, 14*, 590–600.

Yizhar, O., Fenno, L. E., Prigge, M., Schneider, F., Davidson, T. J., O'shea, D. J., et al. (2011). Neocortical excitation/inhibition balance in information processing and social dysfunction. *Nature, 477*(7363), 171–178.

Yuan, R., Duan, Q., Tiw, P. J., Li, G., Xiao, Z., Jing, Z., et al. (2022). A calibratable sensory neuron based on epitaxial VO2 for spike-based neuromorphic multisensory system. *Nature Communications, 13*(1), 3973.

Zatorre, R. J., Fields, R. D., & Johansen-Berg, H. (2012). Plasticity in gray and white: Neuroimaging changes in brain structure during learning. *Nature Neuroscience, 15*(4), 528–536.

Zovkic, I. B., & Sweatt, J. D. (2013). Epigenetic mechanisms in learned fear: Implications for PTSD. *Neuropsychopharmacology, 38*(1), 77–93.

Invasive Brain Stimulation Techniques

DBS for Neurological and Psychiatric Disorders

Contents

© The Author(s), under exclusive license to Springer Nature Switzerland AG 2025
U. Chaudhary, *Expanding Senses using Neurotechnology*, https://doi.org/10.1007/978-3-031-76081-5_7

Test your learning and check your understanding of this book's contents: use the "Springer Nature Flashcards" app to access questions using ▶ https://sn.pub/kmb-jyz. To use the app, please follow the instructions in ▶ Chap. 1.

This chapter explores the specialized domain of invasive brain stimulation, with a particular emphasis on deep brain stimulation (DBS) as a groundbreaking technique in neuromodulation. It provides a detailed exploration of the mechanisms, applications, and potential of surgically implanted devices in treating severe neurological conditions unresponsive to less invasive methods. Historical perspectives trace DBS from its inception in the nineteenth century with the pioneering work of Fritsch and Hitzig, through its evolution driven by technological advances in neuroimaging and computational modeling. This chapter discusses the application of DBS to a variety of neurological and psychiatric disorders, emphasizing its role in Parkinson's disease, and explores its expanding use in conditions like Alzheimer's disease and major depression. Future prospects of DBS, including the development of closed-loop systems and the integration of machine learning for optimizing stimulation parameters, are also examined. This overview highlights the sophisticated integration of science, technology, and medicine in DBS and its significant impact on enhancing patient outcomes.

Learning Objectives

1. Understand the Fundamentals of DBS: Gain knowledge of Deep Brain Stimulation (DBS), including its components, mechanism of action, and how it modifies neural activity in specific brain regions to treat disorders.
2. Explore the Historical Development of DBS: Trace the evolution of DBS from early electrical stimulation experiments to its current status as a sophisticated treatment for movement and affective disorders.
3. Recognize DBS Clinical Applications: Identify the neurological and psychiatric conditions treated with DBS, such as Parkinson's disease, essential tremor, dystonia, and treatment-resistant depression, and understand the criteria for selecting candidates for DBS.
4. Analyze Technological Advances in DBS: Examine the advancements in DBS technology, including the shift from open-loop to closed-loop systems, and the role of computational modeling and imaging in enhancing the precision and efficacy of DBS.

In ▶ Chap. 6, we explored the broader landscape of neuromodulation and its pivotal role in understanding and manipulating neural function, we now transition to a more focused examination of invasive brain stimulation techniques. In this chapter, we take a closer look at surgically implanted devices—an important aspect of neuromodulation that allows for precise, targeted interventions in the brain's intricate networks. We will uncover the mechanisms, applications, and transformative potential of invasive techniques in treating severe neurological conditions that are less responsive to external interventions, setting the stage for a deeper understanding of their scientific and clinical implications.

7.1 Brain Stimulation

Brain stimulation, a rapidly advancing domain within the field of neuroscience, comprises a diverse assortment of strategies aimed at modulating neuronal activity, employing both invasive and non-invasive methods.

The foundational history of brain stimulation traces back to the nineteenth century, initiated by seminal experiments by Fritsch and Hitzig. Their critical research demonstrated that electrical currents could elicit motor responses when applied to specified areas of the canine cerebral cortex, establishing a fundamental understanding of the electrical properties of neural functions and the specific localization of motor control (Fritsch & Hitzig, 1870). Moving into the twentieth century, the development of electroconvulsive therapy (ECT) in the 1930s marked a significant advancement, transforming the treatment landscape for psychiatric disorders and heralding a new era in therapeutic brain stimulation (Cerletti & Bini, 1938).

The development of brain stimulation technologies has been significantly shaped by advances in neuroimaging, particularly through methods like fMRI and PET scans that enable the in vivo visualization of brain regions under stimulation. While PET scan discussions are beyond the scope of this text, other imaging modalities play a crucial role. Electrophysiological techniques, such as EEG and MEG, which you read in ▶ Chaps. 2 and 4, provide immediate feedback on the brain's response to electrical stimulation. Additionally, computational models are instrumental in forecasting how these stimulations might modify neural circuits and affect brain functionality (George & Aston-Jones, 2010; Pascual-Leone et al., 2005).

Central to the theoretical framework of brain stimulation is the modulation of neuronal excitability, a critical aspect of neurophysiology that allows for the alteration of neural activity. The modulation of neuronal excitability is achieved through various sophisticated methods:

1. **Electrical Stimulation**: This modality applies targeted electrical currents directly to neurons, thereby altering their membrane potential and, consequently, their activity. The response elicited—neuronal excitation or inhibition—depends on the specific parameters of the stimulation used. Deep brain stimulation (DBS) exemplifies this approach, where surgically implanted electrodes deliver precise electrical pulses to deep-seated brain structures to manage movement disorders such as Parkinson's disease, showing significant symptom improvement (Benabid et al., 1987).
2. **Magnetic Stimulation**: Techniques like Transcranial Magnetic Stimulation (TMS) utilize magnetic fields to induce electrical currents within the brain through electromagnetic induction. This non-invasive approach can modulate

neural activity by transiently depolarizing or hyperpolarizing neurons in specified regions, thereby influencing cortical excitability and inter-regional neural communication (Barker et al., 1985; Rossi et al., 2009).

3. **Optogenetic Stimulation**: A cutting-edge advancement in neurotechnology, optogenetics merges genetic engineering with optical stimulation to manipulate neuronal activity with high temporal and spatial precision. By sensitizing specific neuron populations to light, optogenetics allows for an unprecedented level of control over neural circuits, offering profound insights into the neural substrates of behavior and brain function (Deisseroth, 2011). Optogenetic stimulation is out of the scope of this book.

These methodologies can be divided into invasive and non-invasive techniques:

1. **Invasive Techniques**: Procedures like DBS are invasive, necessitating surgical procedures to implant stimulation devices within the brain. These approaches are often reserved for more severe neurological conditions that are refractory to conventional treatments and have shown effectiveness in reducing symptoms and improving life quality (Kringelbach et al., 2007; Mayberg et al., 2005).

2. **Non-Invasive Techniques**: Modalities such as TMS and transcranial Direct Current Stimulation (tDCS) are non-invasive, applying stimulation from outside the skull. These techniques have gained traction for their potential in treating psychiatric disorders, aiding in neurorehabilitation, and enhancing cognitive functions, with ongoing research into optimizing their efficacy and application scope (George & Aston-Jones, 2010; Nitsche et al., 2008).

As brain stimulation technologies advance, they not only provide powerful tools for treating neurological and psychiatric disorders but also serve as essential windows into the complex workings of the brain, enabling the refinement of theoretical models of brain function and the development of novel therapeutic avenues.

Brain stimulation constitutes a sophisticated medical and scientific strategy to modify neural activity through the precise delivery of stimuli—be it electrical, magnetic, or chemical—to designated areas of the nervous system. It is grounded in several theoretical frameworks that provide insight into its mechanisms and potential effects:

1. **Network Modulation Theory**: This theory posits that the impacts of brain stimulation are network-wide rather than solely at the stimulation site. In the case of deep brain stimulation (DBS), the electrical impulses delivered to a specific brain nucleus are understood to affect not just that nucleus but also the interconnected network of neurons and pathways. Consequently, DBS has the potential to influence the overall dynamics of neural networks, possibly explaining its effectiveness in conditions such as Parkinson's disease and depression, where network-wide dysregulation is evident (Lozano & Lipsman, 2013; Kringelbach et al., 2007).

2. **Neuroplasticity Theory**: This perspective maintains that brain stimulation can induce enduring alterations in neural structures and connections. It builds on the concept of neuroplasticity—the brain's intrinsic capacity to rewire and adapt structurally and functionally in response to various stimuli, including

targeted brain stimulation. Such changes can solidify with repeated stimulation, potentially leading to long-term improvements in neurological functions and recovery from deficits. This principle is crucial in rehabilitation contexts, where techniques like repetitive TMS (rTMS) and transcranial Direct Current Stimulation (tDCS) are employed to foster recovery from stroke and other neurological injuries by promoting adaptive plasticity (Cirillo et al., 2017; Fregni & Pascual-Leone, 2007).

In the forthcoming section of this chapter and ▶ Chap. 8, we will embark on an exhaustive examination of several brain stimulation methodologies, shown in ◘ Fig. 7.1 (Belmaker & Lichtenberg, 2023), except ECT.

1. **Deep Brain Stimulation (DBS)**: This invasive technique involves implanting electrodes within certain brain areas to deliver electrical impulses. These impulses can adjust abnormal neural activity associated with various disorders,

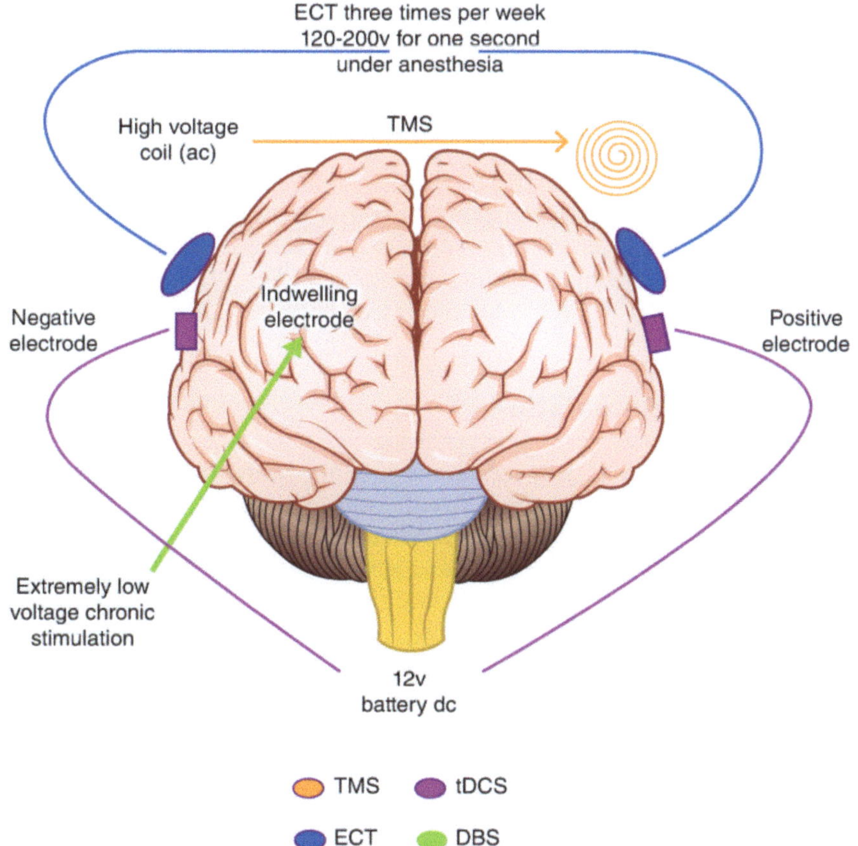

◘ **Fig. 7.1** Pictorial depiction of different brain stimulation techniques. (From Belmaker & Lichtenberg, 2023)

notably Parkinson's disease and treatment-resistant depression (Mayberg et al., 2005; Kringelbach et al., 2007).

2. **Transcranial Magnetic Stimulation (TMS)**: TMS is a non-invasive method that induces electrical activity in the brain via electromagnetic induction, influencing cortical excitability and functional connectivity, and is utilized in the treatment of depression and as a tool for neuroscientific research (Rossi et al., 2009; Chervyakov et al., 2015).

3. **Transcranial Direct Current Stimulation (tDCS)**: tDCS modulates neuronal activity by delivering a constant, low direct current through electrodes placed on the scalp, with applications ranging from cognitive enhancement to stroke rehabilitation (Fregni & Pascual-Leone, 2007; Brunoni et al., 2012).

4. **Transcranial Alternating Current Stimulation (tACS)**: This technique applies oscillating electrical currents at various frequencies to entrain brain activity and modulate cognitive processes, such as learning and memory (Vossen et al., 2015; Herrmann et al., 2013).

Each technique's section will provide its history, technical framework, and clinical utility, along with discussions on the most recent research advancements. The objective is to provide a theoretical background, ensuring you acquire an understanding of the contemporary landscape and the prospective trajectory of brain stimulation within neuroscience.

> **Box 7.1 Deep Brain Stimulation (DBS)**
>
> A surgical treatment involving the implantation of electrodes within certain areas of the brain. These electrodes deliver electrical impulses that regulate abnormal impulses associated with various neurological and psychiatric disorders.

7.2 Deep Brain Stimulation

Deep brain stimulation (DBS) represents a sophisticated and invasive form of brain stimulation that necessitates surgical intervention to place electrodes within targeted brain areas. Over the years, DBS has become an integral component of treatment protocols in both neurology and psychiatry due to its potential to alleviate symptoms in various neurological conditions, particularly those unresponsive to conventional therapies.

The genesis of DBS can be traced back to the early experiments of the twentieth century, which explored electrical stimulation of the brain. The modern era of DBS began in the 1980s when Alim Benabid and Pierre Pollak discovered its beneficial effects on motor symptoms in Parkinson's disease, leading to significant clinical interest and technological advancement in the subsequent years (Benabid et al., 1987, 1994).

The DBS system consists of three critical components: the implanted electrodes, a pulse generator commonly referred to as a "brain pacemaker," and a wire that connects the two. The electrodes deliver electrical impulses that can modulate the abnormal neuronal activity underlying certain disorders. The pulse generator, often placed under the skin near the collarbone, can be programmed externally, allowing for adjustments in stimulation parameters to achieve optimal therapeutic effects (Kringelbach et al., 2007; Lozano et al., 2019).

Clinically, DBS has shown efficacy in treating a spectrum of movement disorders, including Parkinson's disease, essential tremor, and dystonia. Its application has also extended into the realm of psychiatry, showing promise in managing treatment-resistant depression, obsessive-compulsive disorder, and Tourette syndrome (Mayberg et al., 2005; Schuurman et al., 2000).

Recent trends in DBS research involve optimizing electrode placement using advanced imaging techniques, individualizing stimulation parameters through closed-loop systems that adjust stimulation in real time based on neural signals and exploring the potential for treating additional psychiatric and neurological disorders (Widge et al., 2018; Ramirez-Zamora et al., 2018).

This chapter will provide an extensive review of DBS, offering an in-depth narrative of its historical evolution, describing the engineering intricacies of DBS devices, articulating its numerous clinical indications, and critically analyzing emergent research directions. This discourse aims to enrich the reader's comprehension of DBS as a transformative tool in modern medicine and its potential trajectory in the landscape of therapeutic brain stimulation.

7.2.1 Historical Development

Deep brain stimulation (DBS) finds its genesis within the realm of functional neurosurgery, a field that emerged in the nascent years of the twentieth century. The earliest functional neurosurgeons undertook lesioning procedures, creating deliberate areas of damage within the brain to address neurological and psychiatric ailments including Parkinson's disease, epilepsy, and obsessive-compulsive disorder. While occasionally effective, these procedures were inherently irreversible and frequently associated with profound side effects.

The inception of brain electrical stimulation as a therapeutic measure can be traced to the seminal work by Walter Rudolf Hess during the 1920s. Hess's experiments in stimulating the brains of cats led to monumental insights into the functioning of the autonomic nervous system and laid the groundwork for neuromodulation (Hess, 1928).

The advent of modern DBS occurred during the closing decades of the twentieth century, propelled by the technological advancements and conceptual parallels drawn from the cardiac pacemaker's success. It was during this time that significant progress was made in developing implantable devices capable of modulating brain activity via precise electrical impulses. A pivotal moment was reached with the work of Dr. Alim-Louis Benabid and colleagues in 1987, who demonstrated that high-frequency stimulation of the thalamus could effectively alleviate tremors in patients with Parkinson's disease, a discovery that heralded the potential of DBS for movement disorders (Benabid et al., 1987).

The technological maturation and clinical application of DBS accelerated rapidly thereafter. By the late 1990s, the U.S. Food and Drug Administration (FDA) had approved DBS for the treatment of essential tremor and subsequently for Parkinson's disease, endorsing the technique as a mainstream medical intervention (FDA, 1997). Since its regulatory endorsement, DBS has undergone continuous refinement and diversification, with its application now extending to a spectrum of other movement disorders and ongoing investigations into its efficacy for various psychiatric conditions.

A visualization of this historical and technological evolution is depicted in ◘ Fig. 7.2 (Krauss et al., 2021), which illustrates a timeline chronicling the major milestones and developments in DBS technology. From its conceptualization to contemporary practice, DBS has transformed from a novel experimental therapy to a well-established treatment, expanding the frontiers of functional neurosurgery and offering hope to patients with previously intractable conditions.

7.2.2 Technological Aspects

Deep brain stimulation (DBS) represents an advanced, invasive neurostimulation technique that entails the surgical placement of electrodes within specific deep brain structures targeted for therapeutic intervention, as shown in ◘ Fig. 7.3 (Kricheldorff et al., 2022).

The configuration of a state-of-the-art DBS system comprises three main components as explained below and shown in ◘ Fig. 7.4 (Rosa & Lisanby, 2012).

1. **Electrode Lead**: The electrode lead is a thin, insulated wire that ends in multiple contacts (usually four to eight) made of biocompatible materials such as platinum-iridium. These contacts deliver electrical pulses directly to targeted brain areas. The design of the electrode, including the number and arrangement of contacts, is crucial for optimizing stimulation efficacy and minimizing side effects (Butson & McIntyre, 2008).
2. **Neurostimulator (Pulse Generator)**: The neurostimulator is a battery-powered device that generates the electrical pulses sent through the electrode lead. Implanted under the skin, typically in the chest or abdomen, the neurostimulator's settings can be adjusted externally by clinicians to customize therapy for individual patients.
3. **Extension Cable**: An extension cable connects the electrode leading to the neurostimulator. This cable is routed subcutaneously from the scalp, down the neck, to the site of the neurostimulator implant, ensuring a continuous pathway for electrical signals.

The surgical procedure for DBS implantation unfolds in a series of carefully orchestrated steps, leveraging state-of-the-art neuroimaging and stereotactic techniques to ensure precision and efficacy. The process begins with the utilization of advanced neuroimaging modalities, such as magnetic resonance imaging (MRI) or computed tomography (CT), to accurately pinpoint the exact location within the

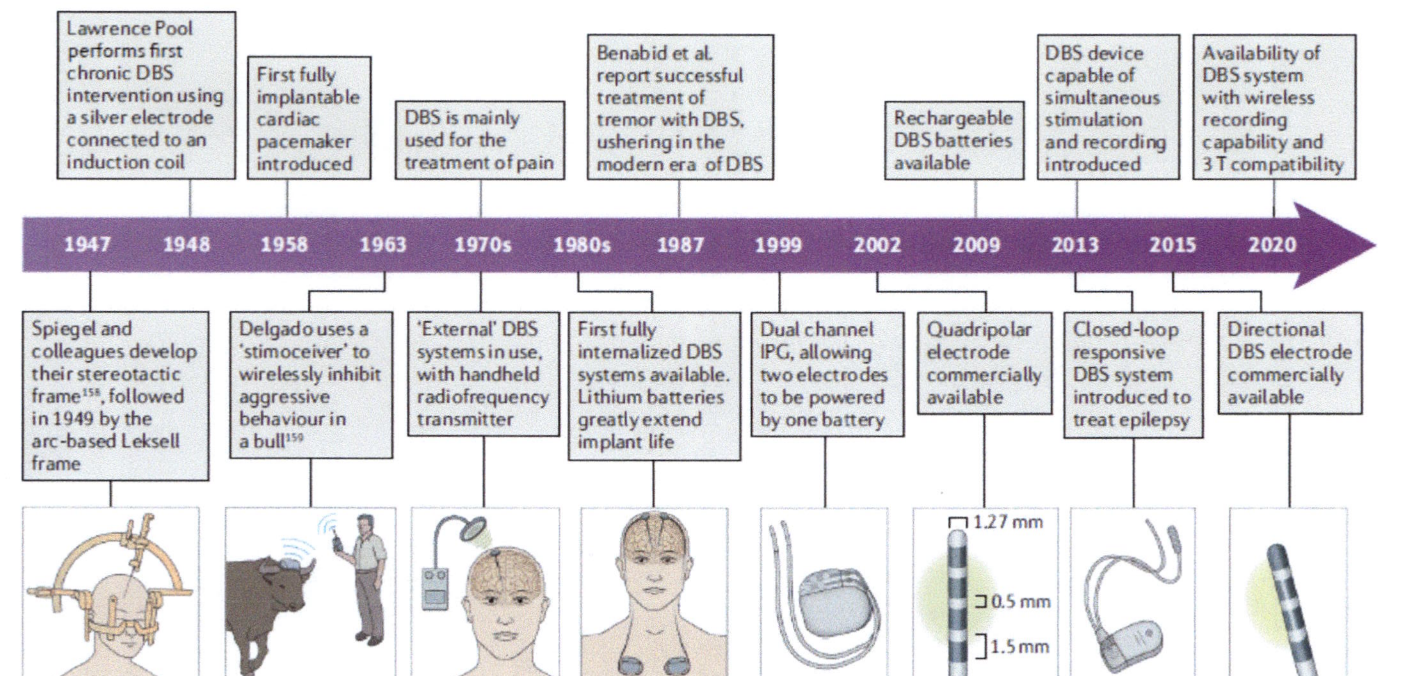

Fig. 7.2 Timeline of technology development for DBS. *DBS* deep brain stimulation, *IPG* implantable pulse generator. (From Krauss et al., 2021. The references in the figure are: 158—Spiegel et al., 1947; 159—Delgado, 1969)

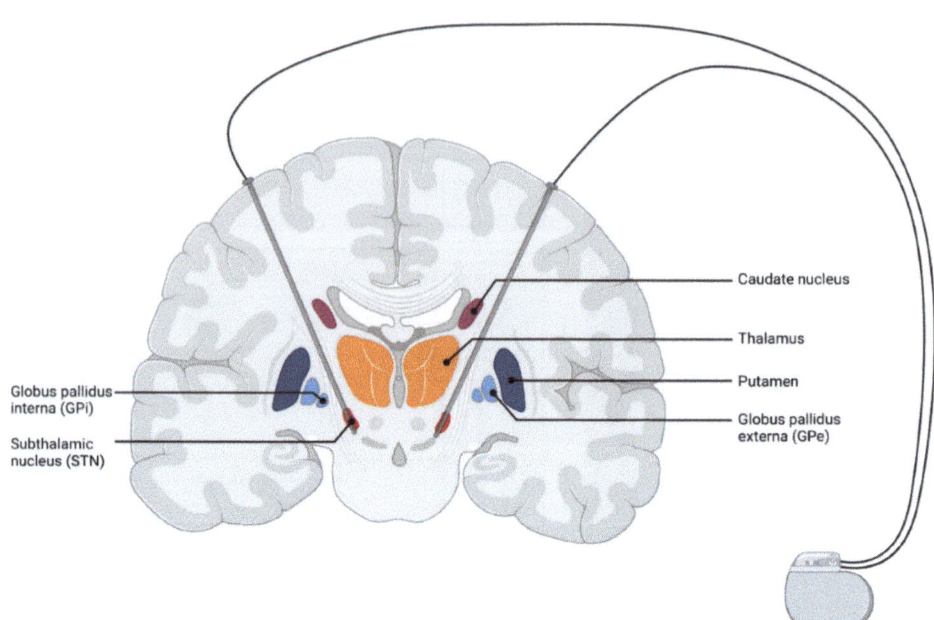

◘ Fig. 7.3 Detailed coronal section of the brain, showcasing the anatomical features of the basal ganglia (BG), with an intricate diagram of a deep brain stimulation (DBS) apparatus aimed at the subthalamic nucleus. The system includes an implanted pulse generator, typically located subcutaneously in the chest, connected via leads that extend intracranially. These leads deliver precise electrical currents to the subthalamic nucleus, influencing neuronal activity within this critical area of the brain involved in motor control. (From Kricheldorff et al., 2022 originally published under CC-BY 4.0)

◘ Fig. 7.4 Deep brain stimulation (DBS). (From Rosa & Lisanby, 2012)

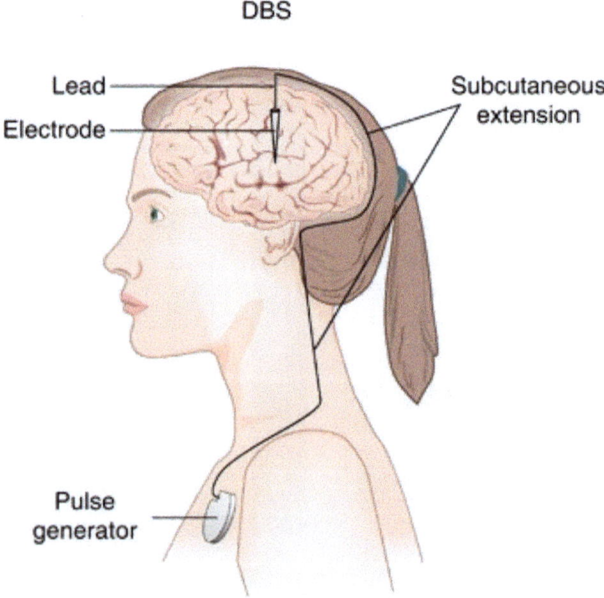

brain that will serve as the target for electrode placement (Benabid et al., 1987, 1994; Krauss et al., 2021).

This initial step is critical for optimizing the therapeutic outcomes of the DBS treatment. Following the neuroimaging-guided targeting, the patient undergoes the surgical procedure under either local or general anesthesia, depending on the specific requirements of the case and the patient's condition. During this phase, a neurosurgeon implants the electrodes into the targeted brain region with the aid of a stereotactic apparatus, ensuring the electrodes are positioned with utmost precision. The final phase involves connecting the implanted electrodes to the neurostimulator. After implantation, a clinician or a neurologist programs the neurostimulator to deliver electrical stimulation at parameters meticulously tailored to meet the patient's individual needs and medical condition.

> **Box 7.2   Neurostimulator (Pulse Generator)**
> A device, similar to a pacemaker, that generates the electrical pulses needed for DBS. It is typically implanted under the skin in the upper chest area and connected to the brain electrodes by a subcutaneous wire.

> **Box 7.3   Electrode Lead**
> A flexible wire with multiple electrodes at its end that is implanted in the brain. It delivers electrical pulses directly to specific brain areas to modulate neural activity.

7.2.2.1 Common DBS Electrode Configuration

The common DBS electrode configuration is intricately designed to cater to the complex requirements of neuromodulation, providing targeted stimulation to specific brain regions while minimizing unwanted side effects. A typical DBS electrode consists of a lead that houses multiple electrode contacts. These contacts are the points through which electrical impulses are delivered to the brain tissue. The leads are thin, insulated wires that protect the integrity of the transmitted electrical signals. Modern DBS electrodes commonly feature four to eight contacts, arranged linearly along the distal end of the lead. Each contact is made from biocompatible materials, such as platinum-iridium, to ensure safety and durability within the brain's environment., as shown in ◘ Fig. 7.5a (Krauss et al., 2021). The spacing between the contacts on a DBS electrode can vary, affecting the distribution and focus of the electrical field generated by stimulation. Some electrodes are designed with contacts spaced closely together, typically around 0.5–1.5 mm apart, allowing for more precise targeting of neural tissue. Others may have wider spacing, up to 3 mm or more, to create a broader stimulation field that can influence larger brain areas.

The number and shape of contacts also contribute to the stimulation's characteristics. While cylindrical contacts are standard, allowing for omnidirectional

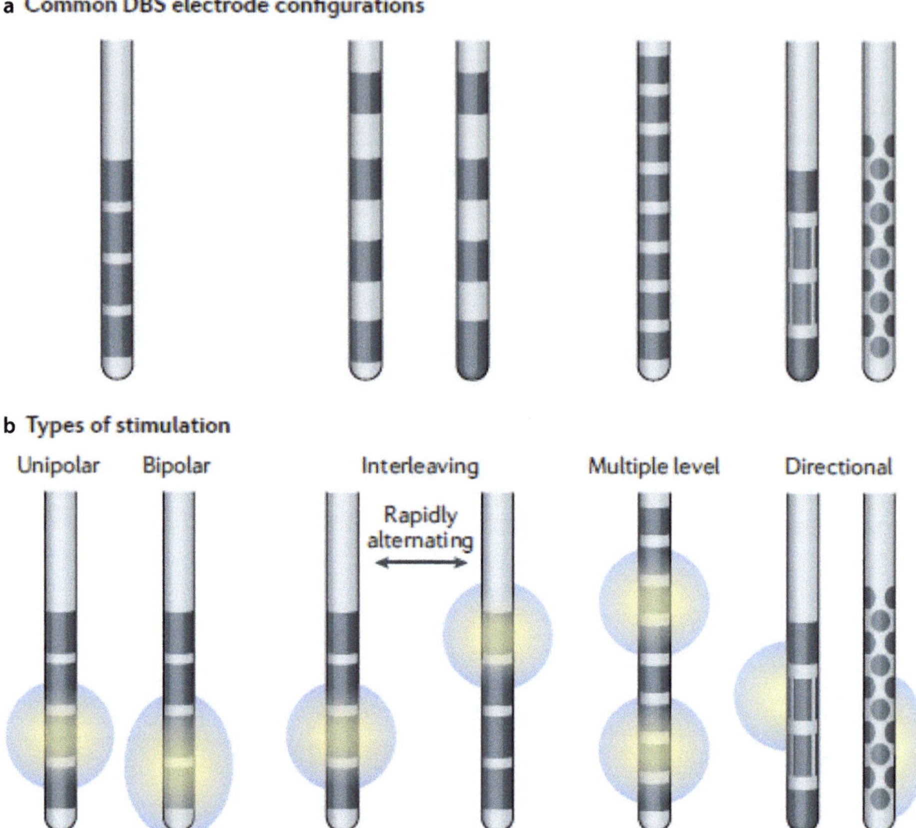

Fig. 7.5 DBS electrode configurations. Configurations of DBS electrodes. **a** Typical electrode arrangements employed in deep brain stimulation (DBS) are depicted, with dark gray areas representing the electrode contacts capable of current delivery. The design of these electrodes varies, especially in terms of contact spacing, as well as the contact number and configuration. Wider spacing between contacts allows for a broader range of neural targets to be addressed, while tighter spacing between contacts aids in achieving finer control over the stimulation delivered. **b** Stimulation methodologies are contingent on the specific DBS system utilized. Unipolar stimulation involves the flow of current either from the power source to a contact or in the reverse direction. Bipolar stimulation is characterized by the movement of current between electrode contacts, where at least one acts as an anode and another as a cathode. Interleaving stimulation encompasses the sequential application of varying stimulation parameters. Multiple-level stimulation offers the possibility of activating several neural targets simultaneously, assuming these targets are aligned with the electrode's path. Directional stimulation allows for the targeted redirection of current, tailored according to the local anatomical structure or to address specific clinical symptoms. (From Krauss et al., 2021)

stimulation, recent advancements have introduced directional (segmented) contacts. Directional electrodes can steer the electrical field toward specific areas while avoiding others, offering enhanced control over the stimulation's effects and potentially reducing side effects by sparing non-targeted tissue.

Each electrode can be independently programmed to emit electrical pulses, thereby enabling clinicians to fine-tune the stimulation to the patient's needs by selecting which electrodes to activate and determining their intensity, frequency, and pulse width (Butson & McIntyre, 2008). This flexibility is vital for optimizing therapeutic outcomes and minimizing side effects, as it allows for the modulation of neural activity in a highly targeted manner.

7.2.2.2 Electrical Configuration

DBS can employ various stimulation designs, each with distinct neurophysiological implications, as shown in ◘ Fig. 7.5b (Krauss et al., 2021). The most common stimulation paradigms include:

1. **Monopolar Stimulation**: In this configuration, one electrode serves as the active source of stimulation while the pulse generator casing acts as the return path. Monopolar stimulation provides a broad and diffuse electric field, useful for influencing a wide area of neural tissue but with an increased risk of affecting unintended regions (Volkmann et al., 2001).
2. **Bipolar Stimulation**: Involves using two electrodes on the lead for both the active stimulation and the return path, creating a more focused electric field between them. This design can help target specific neural populations more precisely, reducing the likelihood of side effects related to stimulating adjacent areas (Rizzone et al., 2001).
3. **Interleaving Stimulation Mode**: It represents an advanced DBS programming strategy where two different sets of stimulation parameters are alternated within a single electrode array. This approach allows clinicians to target multiple neuronal populations or address complex symptoms by varying the location (contacts used), frequency, pulse width, or amplitude of the stimulation in a rapid sequence. Interleaving stimulation is particularly beneficial for patients who require distinct parameter settings to optimally control different symptoms that cannot be simultaneously managed with a single set of parameters.
4. **Multiple-Level Stimulation Mode**: Multiple-level stimulation, also known as multi-target stimulation, involves stimulating more than one brain region or neural pathway simultaneously or sequentially using either a single electrode with multiple contact points or multiple electrodes. This mode is predicated on the understanding that some neurological and psychiatric disorders involve dysfunction in multiple neural networks. By targeting several areas, multiple-level stimulation aims to modulate complex network dynamics more effectively, offering a broader therapeutic impact. This approach is being explored in conditions like Parkinson's disease, where both the subthalamic nucleus and the globus pallidus internus may be targeted to optimize symptom control (Bronstein et al., 2011).
5. **Directional Stimulation**: Recent advancements in electrode design have introduced directional leads that can steer the electric field in specific directions. This allows for even more precise targeting of brain regions and pathways, potentially enhancing therapeutic effects and reducing adverse outcomes (Steigerwald et al., 2016).

The surgical procedure for DBS, complemented by the intricate design of the DBS system and its programmable stimulation paradigms, underscores the technique's potential as an effective treatment modality. As research progresses, understanding the optimal electrode configurations and stimulation designs for specific conditions will be paramount in enhancing DBS's therapeutic efficacy and broadening its clinical applications.

> **Box 7.4   Directional Electrodes**
> Advanced DBS electrodes designed to steer electrical currents towards specific brain areas, enhancing the precision of stimulation and reducing side effects.

7.2.3 DBS Mechanism

The neurophysiological mechanisms by which DBS achieves its therapeutic effects are rooted in complex interactions at both the neuronal and neural network levels. Despite ongoing research, the precise neurophysiological mechanisms of DBS remain partially understood. It is hypothesized that DBS modulates neural circuit activity through direct and indirect pathways, capable of both exciting and inhibiting neuronal firing depending on the stimulation parameters set by the clinician (McIntyre et al., 2004). The mechanisms underlying the therapeutic effects of DBS involve intricate interactions between electrical stimulation, neurotransmitter dynamics, synaptic plasticity, and changes in neural network activity.

(a) **Neurotransmitter Release and Synaptic Plasticity**: One of the primary mechanisms by which DBS exerts its effects is through the modulation of neurotransmitter release, as shown in ◘ Fig. 7.6a (Lozano et al., 2019). Upon electrical stimulation, there is an immediate response involving the release of various neurotransmitters (Gradinaru et al., 2009). This neurotransmitter release can trigger calcium waves within the neural tissue, leading to the subsequent release of gliotransmitters. These molecules play a crucial role in modulating synaptic activity and plasticity, the brain's ability to change and adapt in response to new information or environmental changes.

> **Box 7.5 Subthalamic Nucleus (STN)**
> A small group of cells located in the midbrain that plays a critical role in regulating movement. It is often a target for DBS in Parkinson's disease treatment.

> **Box 7.6 Globus Pallidus Internus (GPi)**
> A part of the basal ganglia involved in the regulation of voluntary movement, commonly targeted by DBS to treat dystonia and Parkinson's disease.

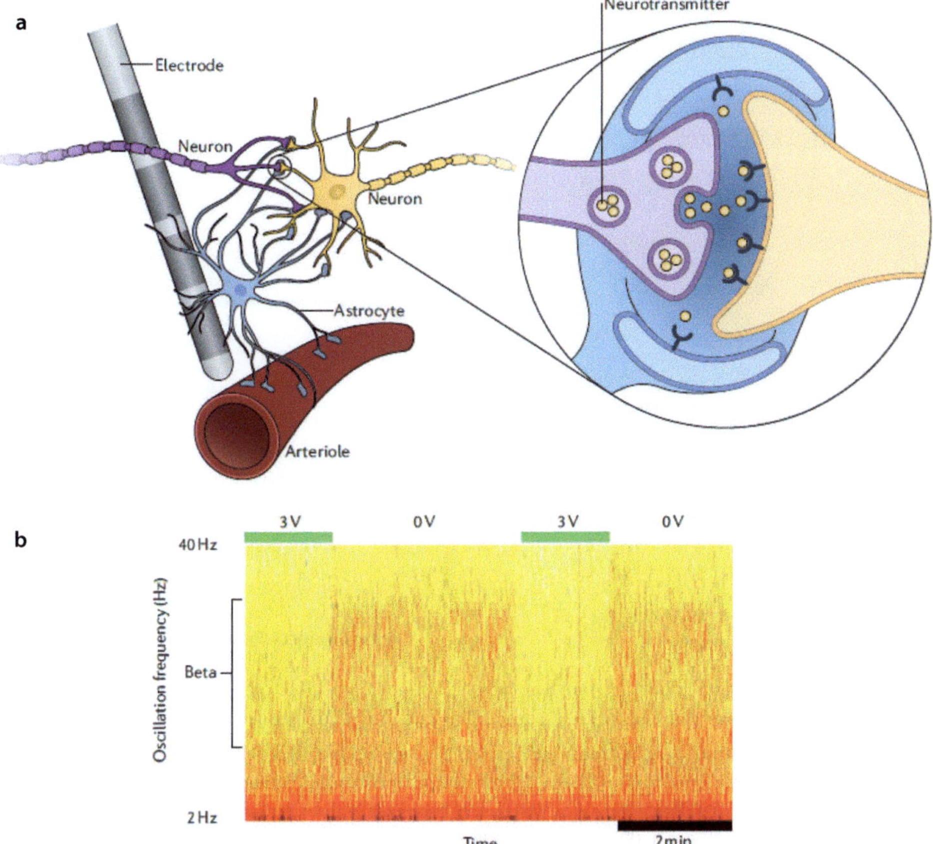

Fig. 7.6 Mechanisms underlying deep brain stimulation (DBS). **a** The initiation of DBS triggers the release of neurotransmitters, which in turn initiates a cascade of calcium signaling within neural tissue. This signaling is crucial for the subsequent release of gliotransmitters, molecules that play a key role in modulating the plasticity of synapses. Such changes in synaptic plasticity can lead to the dilation of arterioles, enhancing blood flow in the targeted brain region. **b** DBS notably impacts the local field potentials within the subthalamic nucleus, specifically affecting neural oscillations. When DBS is applied at a strength of 3 V, there is a swift reduction in the activity within the beta frequency band, a change that is closely associated with symptom relief. This reduction in beta activity is transient, however, with levels returning to baseline once the stimulation is discontinued. (From Lozano et al., 2019)

The alterations in synaptic plasticity can result in physiological changes such as arteriole dilation, which in turn enhances regional blood flow, thereby improving the delivery of oxygen and nutrients to the stimulated area (Tawfik et al., 2010).

(b) **Local Field Potential Changes and Neural Network Activity**: Another significant aspect of DBS's mechanism involves its impact on local field potentials within targeted brain structures, such as the subthalamic nucleus (STN) in Parkinson's disease, as shown in **Fig. 7.7** (Spiliotis et al., 2022). DBS has

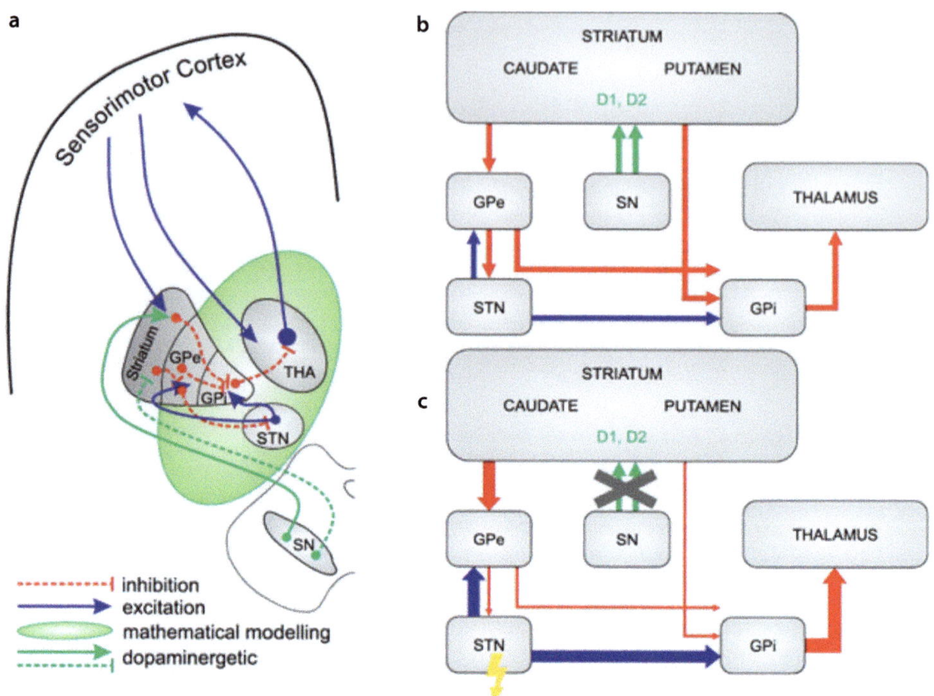

Fig. 7.7 The illustration portrays the **a** basal ganglia and thalamus network, highlighting their connections and anatomical positions. **b** Shows the network in a healthy state, and **c** represents the network under Parkinsonian conditions where the crossed-out green arrows indicate diminished dopaminergic input from the substantia nigra. The striatum is divided into two functional areas, one predominantly expressing D1 receptors, and the other, D2 receptors. The sensorimotor cortex connects to both the thalamus and the striatum, the latter serving as the input region for the basal ganglia network. This network includes the striatum, the globus pallidus with its external and internal parts (GPe and GPi, respectively), the subthalamic nucleus (STN), and the substantia nigra, focusing here only on the pars compacta (SN). The green area in diagram **a** denotes the portion of the network that has been subject to mathematical modeling. Blue lines with arrows signify excitatory connections, while red lines represent inhibitory ones. Green arrows illustrate the dopaminergic projections from the SN that activate the striatum via D1 receptors in the direct pathway and inhibit it via D2 receptors in the indirect pathway. In diagrams **b** and **c**, bold lines indicate strengthened synaptic projections, and thin lines denote reductions. The thunderbolt arrow marks the target area for deep brain stimulation (DBS) in the STN. It is hypothesized that under DBS, the increased excitation (bold blue projection from STN to GPi) will be mitigated, and the disinhibition (thin red projection from GPe to GPi) will be normalized. (From Spiliotis et al., 2022 originally published under CC-BY 4.0)

been observed to induce rapid changes in the pattern of local field potentials, notably reducing activity within the beta frequency band when stimulation is applied at specific intensities (e.g., 3 V), as shown in **Fig. 7.6b** (Lozano et al., 2019). This reduction in beta-band activity is associated with symptomatic relief in Parkinson's disease, as excessive beta-band synchronization in the STN is linked to motor symptoms. Interestingly, once the stimulation is ceased, the beta-band activity resumes, which correlates with the return of symptoms

(Kühn et al., 2008). This observation underscores the direct and immediate influence of DBS on neural network activity and its potential to disrupt pathological patterns of synchronization within neural circuits.

7.2.4 Current Deep Brain Stimulation (DBS) Design

Current deep brain stimulation (DBS) technologies have evolved to include sophisticated designs aimed at improving therapeutic outcomes and reducing side effects. These designs can be broadly categorized into open-loop and closed-loop systems, as shown in ■ Fig. 7.8 (Parastarfeizabadi & Kouzani, 2017), each with distinct mechanisms of action, as shown in ■ Fig. 7.9 (Bouthour et al., 2019).

> **Box 7.7 Beta Oscillations**
> Brain wave patterns in the beta frequency range (13–30 Hz) often associated with motor control and active concentration; modulation of these oscillations is a target of DBS in Parkinson's disease.

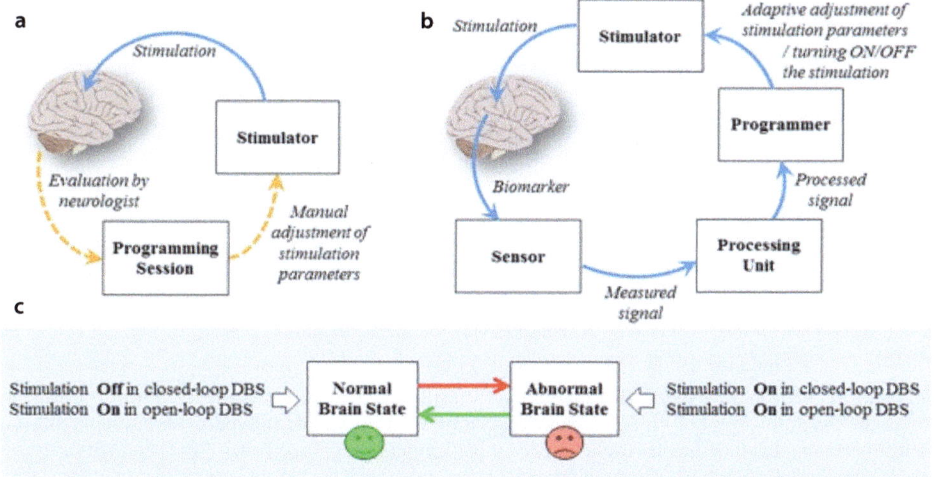

■ **Fig. 7.8** Comparison of open-loop versus closed-loop deep brain stimulation (DBS). **a** In open-loop DBS, the stimulation settings are manually adjusted by a neurologist every 3–12 months following the implantation of the DBS device, with no automatic adjustments based on the patient's neural activity. **b** Conversely, closed-loop DBS dynamically adjusts the stimulation parameters automatically in response to real-time brain activity data, using specific biomarkers to guide adjustments. **c** The illustration also shows how these two systems respond to changes in brain states. In closed-loop DBS, stimulation is halted when the brain returns to a normal state, thus optimizing energy use and potentially reducing side effects. In contrast, open-loop DBS continues to deliver constant stimulation, irrespective of the brain's current state, which can lead to inefficiencies and unnecessary stimulation when not needed. (From Parastarfeizabadi & Kouzani, 2017 originally published under CC-BY 4.0)

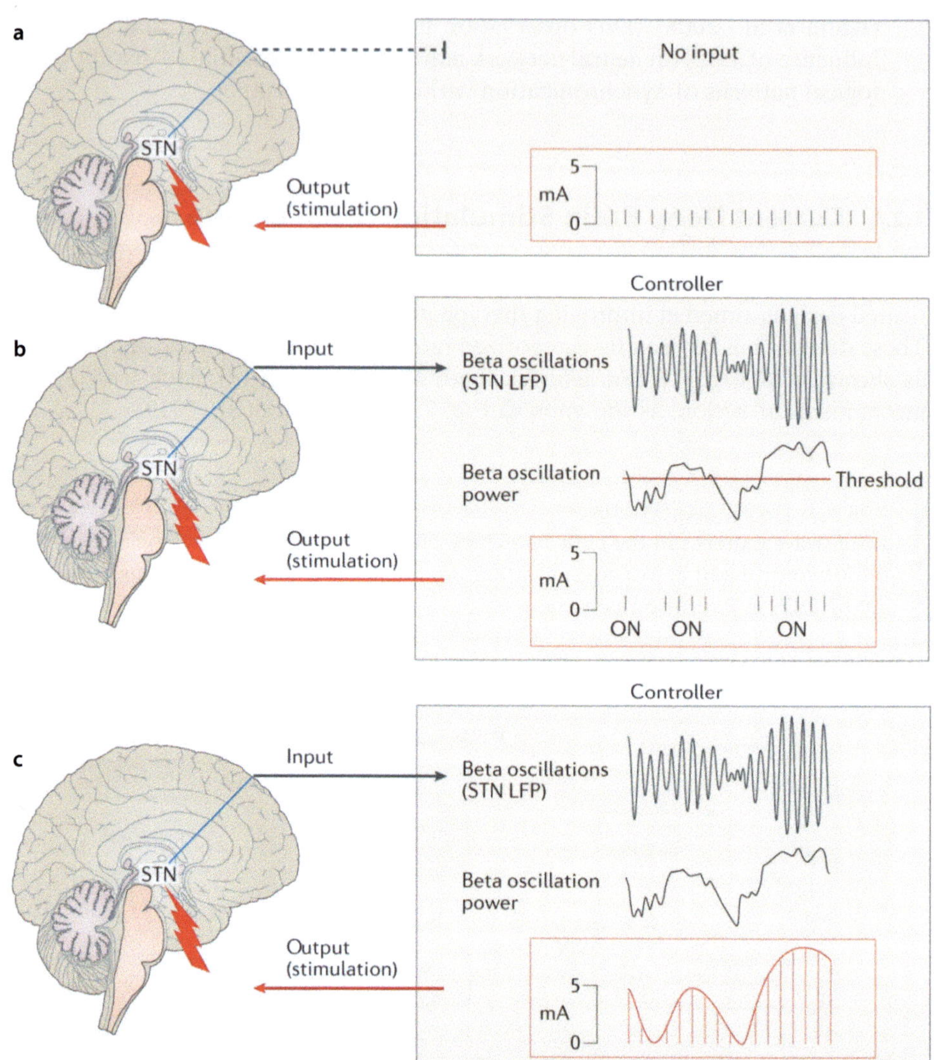

☐ **Fig. 7.9** Overview of deep brain stimulation (DBS) configurations. **a** Open-loop DBS provides continuous stimulation at a fixed amplitude without a direct relationship between input signals and output. Here, clinicians pre-set the stimulation parameters, which the controller then uses to deliver consistent stimulation. **b** Closed-loop responsive (or on-demand) DBS adjusts the controller's output based solely on one variable: the power of the input signal, typically the local field potential (LFP) recorded at the electrode tip within the subthalamic nucleus (STN). Stimulation with a fixed amplitude is initiated when beta oscillation power (13–30 Hz) crosses a specific threshold, operating on an all-or-nothing principle. **c** Closed-loop adaptive DBS allows multiple input variables (such as power, duration, or shape of oscillations) to influence several output variables, thereby permitting input changes to dictate stimulation adjustments. This sophisticated system continually alters stimulation parameters to align with fluctuating input variables, utilizing algorithms that may be enhanced by machine learning for dynamic input-output calibration. In this model, the LFP recorded at the STN electrode tip acts as the input, with stimulation amplitude being dynamically regulated to correspond with variations in beta power. (From Bouthour et al., 2019)

7.2.4.1 Open-Loop DBS Design

Open-loop DBS systems deliver continuous stimulation at a fixed amplitude, operating independently of the patient's neural activity. In this configuration, clinicians manually set the stimulation parameters, such as frequency, pulse width, and amplitude, based on clinical observations and patient feedback. The controller in the system then consistently delivers stimulation according to these predefined settings, without dynamically adjusting to real-time changes in the brain's activity (Lozano et al., 2019).

> **Box 7.8 Open-Loop System**
> A type of DBS system where stimulation parameters are set by a clinician and do not change in response to changes in the patient's neurological condition.

7.2.4.2 Closed-Loop DBS Designs

Closed-loop DBS, in contrast, introduces an interactive element to stimulation, where the system's output is directly influenced by its input signals, typically derived from recorded brain activity, as shown in ◘ Fig. 7.10 (Parastarfeizabadi & Kouzani, 2017). This design aims to create a more responsive and tailored therapeutic approach.

(a) Closed-Loop Responsive (on-demand) DBS: This version of closed-loop DBS is characterized by its response to a single variable, such as the power of a specific frequency band within the local field potential (LFP) recorded at the electrode tip. For example, if the power of beta oscillations (13–30 Hz) within the subthalamic nucleus (STN) exceeds a predefined threshold, the system delivers fixed amplitude stimulation. This approach operates on an all-or-nothing principle, activating stimulation only when necessary (Rosin et al., 2011).

(b) Closed-Loop Adaptive DBS: Adaptive DBS systems offer a more nuanced modulation of stimulation parameters, adjusting in response to several input variables like power, duration, or waveform of neural oscillations. The controller detects these variables and modifies the stimulation's amplitude, frequency, or pulse width to match the changing neural activity, as shown in ◘ Fig. 7.8. This version can employ algorithms, potentially refined by machine learning, to optimize the input-output relationship continuously. In some implementations, the LFP recorded in the STN serves as the input, with the system dynamically adjusting stimulation amplitude in accordance with beta power fluctuations (Little et al., 2013).

These advancements in DBS design, particularly in closed-loop systems, represent a significant leap toward more personalized and efficient neuromodulation therapies. By adapting stimulation parameters in real-time based on the patient's neural state, closed-loop DBS holds the promise for enhancing therapeutic efficacy while

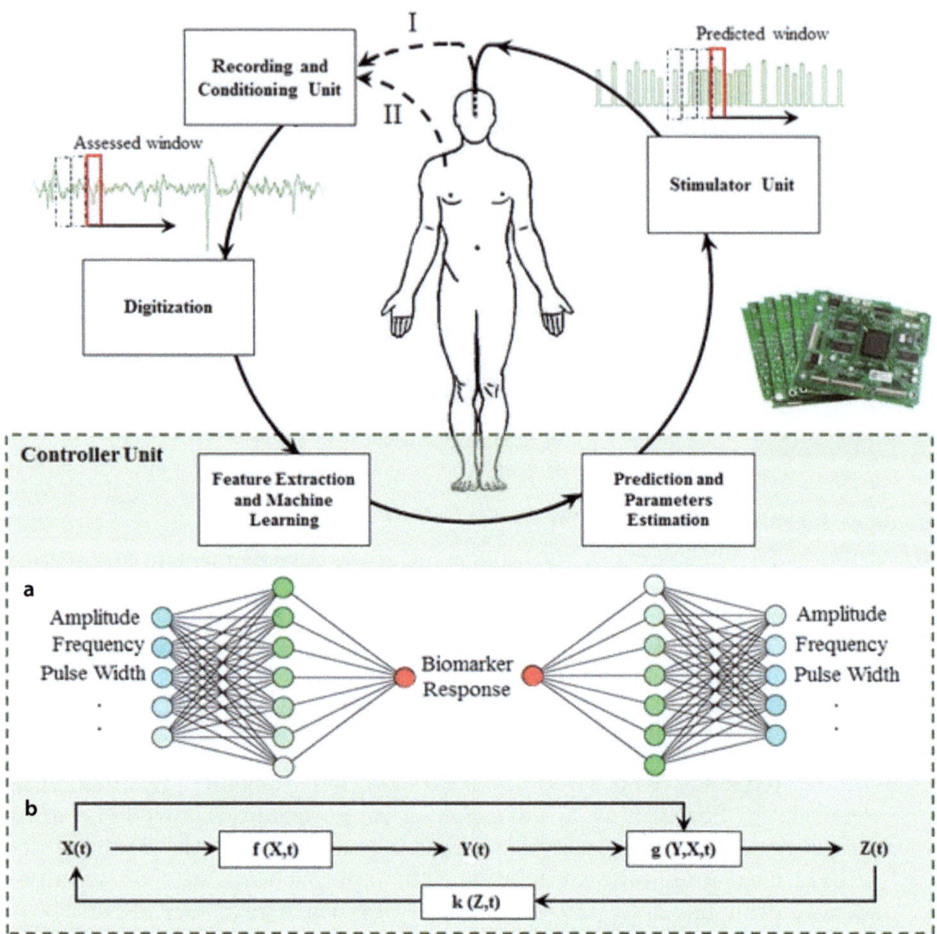

Fig. 7.10 The real-time closed-loop deep brain stimulation (DBS) programming process involves several critical steps to adjust stimulation parameters dynamically based on the patient's current neural state. Initially, a recording unit captures the biomarker signal through an electrode placed either inside (I) or outside (II) the brain, depending on the biomarker's nature. This signal undergoes conditioning, including amplification and filtration, before being digitized and forwarded to the controller unit. Within the controller unit, **a** computational model assesses the biomarker signal across various dimensions (such as amplitude, frequency, and pulse width) to determine a response signal. This signal facilitates the prediction of optimized stimulation parameters. **b** Outlines the controller's structure, where $X(t)$ represents the input vector, $Y(t)$ denotes neural circuit states, and $Z(t)$ signifies the biomarker response. The functions mapping the input to neural state and neural state to biomarker response are represented by $f(X, t)$ and $g(X, Y, t)$, respectively. The function $k(Z, t)$ embodies the controller mechanism that analyzes the biomarker response and adjusts the stimulation parameters accordingly. These estimated parameters are then modified in the stimulation unit to generate stimulation pulses directed to the stimulation electrodes. This procedure utilizes a brief time-window of the recorded signal to predict stimulation parameters, continuously advancing the biomarker signal's time-window and concurrently executing calculations to anticipate and adjust the forthcoming stimulation-window, ensuring the process's real-time nature. (From Parastarfeizabadi & Kouzani, 2017 originally published under CC-BY 4.0)

minimizing adverse effects, paving the way for more targeted and effective treatment strategies for neurological disorders.

These mechanisms highlight the multifaceted nature of DBS and its capacity to induce beneficial neurophysiological changes through direct electrical stimulation and indirect modulation of neurochemical and neurovascular processes. This modulation of neural activity is key to normalizing the dysfunctional patterns of brain activity associated with various neurological and psychiatric disorders, thus providing symptomatic relief and improving patients' quality of life.

> **Box 7.9   Closed-Loop System**
> An advanced DBS system that adjusts the stimulation parameters in real-time based on feedback from the patient's brain activity, providing dynamic and personalized treatment.

7.3 Clinical Applications

Deep brain stimulation (DBS) has established itself as a pivotal intervention in the realm of neurological and psychiatric treatments, particularly for patients grappling with Parkinson's disease (PD) who do not respond adequately to medication, as shown in ◘ Fig. 7.11 (Senevirathne et al., 2023). This advanced therapeutic technique has been demonstrated to markedly alleviate symptoms such as tremors, bradykinesia, and rigidity, which are characteristic of PD, leading to substantial enhancements in patients' quality of life (Deuschl et al., 2006). The versatility of DBS extends beyond Parkinson's disease, encompassing a range of other movement disorders including dystonia and essential tremor, where its application has similarly yielded positive outcomes (Krauss et al., 1999; Schuurman et al., 2000).

The utility of DBS further spans into the treatment of epilepsy, showcasing significant promise in reducing seizure frequency and severity, thereby offering a new lease on life for patients with this challenging condition (Fisher et al., 2010). Moreover, the exploration of DBS in neurodegenerative diseases such as Alzheimer's disease (AD), as shown in ◘ Fig. 7.11 (Senevirathne et al., 2023), has opened new avenues for potentially mitigating cognitive decline and improving patient outcomes (Laxton et al., 2010).

In the psychiatric domain, DBS has been investigated for its efficacy in managing treatment-resistant forms of obsessive-compulsive disorder and major depression, as shown in ◘ Fig. 7.12 (Johnson et al., 2024). Seminal studies by Mayberg et al. (2005) have illuminated the capacity of DBS to induce remarkable clinical improvements in patients suffering from severe depression, signifying a breakthrough in the treatment of psychiatric disorders that are resistant to conventional therapies.

The U.S. Food and Drug Administration (FDA) has recognized the therapeutic value of DBS, approving its use for the treatment of Parkinson's disease, essential tremor, dystonia, and obsessive-compulsive disorder, thereby endorsing its safety

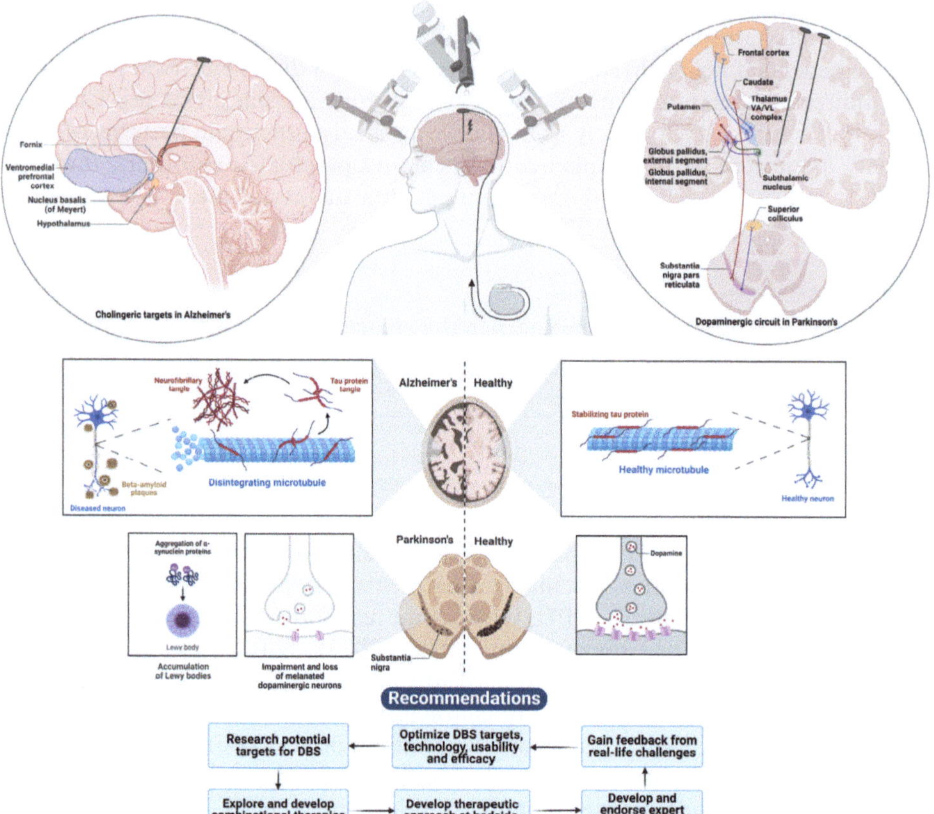

◘ Fig. 7.11 This illustration provides an overview of neurobiology of Parkinson's disease (PD) and Alzheimer's disease (AD). While both disorders involve the degeneration of brain cells, the pathologies and affected brain circuits differ significantly, influencing how DBS is applied clinically. In PD, DBS is an FDA-approved treatment and is primarily used to alleviate motor symptoms associated with the disease. This involves targeting areas like the subthalamic nucleus or the globus pallidus internus, which are critical in the regulation of movement. Neurobiological studies suggest that PD is characterized by the degeneration of dopaminergic neurons in the substantia nigra, leading to a disruption of the neuronal circuits that control movement. DBS helps by modulating the electrical signaling pathways in these circuits, thereby improving motor function and reducing symptoms such as tremor, rigidity, and bradykinesia. Conversely, AD involves widespread neuronal loss and the accumulation of amyloid plaques and tau tangles, primarily affecting memory and cognitive functions. The application of DBS in AD is still experimental, with research focusing on areas like the fornix and nucleus basalis of Meynert, which are linked to memory and cognitive processes. Initial studies have shown that DBS may enhance neural activity and potentially slow disease progression in AD, but optimal stimulation parameters and target regions are still under investigation. The decision to use a low-frequency system (LFS) or high-frequency system (HFS) should be tailored based on the specific symptoms and disease characteristics of each patient, for both PD and AD, to maximize therapeutic outcomes while minimizing side effects. There is an urgent need for foundational neurobiological research to better understand the differential effects of DBS in PD and AD. This calls for extended clinical trials to refine DBS protocols, particularly to determine the most effective stimulation frequencies and locations for AD. (From Senevirathne et al., 2023 originally published under CC-BY 4.0)

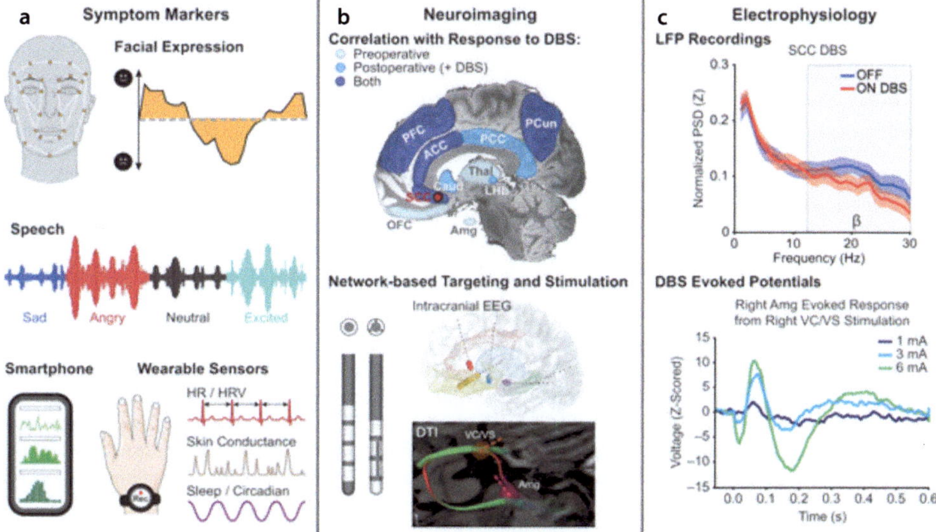

◘ Fig. 7.12 **a** This section presents examples of potential biomarkers for assessing depression symptom severity. These include facial expressions analyzed via video, speech patterns evaluated through audio recordings, smartphone interaction metrics, and data collected from wearable sensors. **b** Top: Neuroimaging studies have identified changes in regional brain volumes, neurotransmitter concentrations, and activity (including glucose metabolism, cerebral blood flow, and BOLD signals) both before and after surgery, which correlate with symptom improvement in major depressive disorder (MDD) following deep brain stimulation (DBS). This representation simplifies the anatomical regions involved. Bottom: In a clinical trial scenario, intracranial EEG and Diffusion Tensor Imaging (DTI) using fiber tractography pinpointed a therapeutic network involving the amygdala (as a recording site to gauge depressive symptoms) and the ventral capsule/ventral striatum (VC/VS) (as a stimulation target) for DBS treatment in MDD (Scangos et al., 2021). Directionally capable DBS electrodes may enhance targeting of therapeutic neural pathways. **c** Top: Local field potential (LFP) recordings are employed to assess the neurophysiological impacts of DBS and to identify effective stimulation markers. In one study by Sendi et al. (2021) DBS at the left subcallosal cingulate (SCC) diminished beta power, which, when deactivated, correlated with a more significant reduction in symptoms 1-week post-surgery. Bottom: Stimulus-evoked potentials, induced by VC/VS stimulation and recorded via intracranial EEG from the amygdala, aid in discerning functional connectivity within the DBS therapeutic network (Scangos et al., 2021). These findings guide the selection of stimulation parameters to optimize engagement of the network. *ACC* anterior cingulate cortex, *Amg* amygdala, *Caud* caudate, *LHb* lateral habenula, *OFC* orbitofrontal cortex, *PCC* posterior cingulate cortex, *PCun* precuneus, *PFC* prefrontal cortex, *SCC* subcallosal cingulate, *Thal* thalamus. (From Johnson et al., 2024)

and efficacy for these conditions. This regulatory approval not only underscores the significant clinical benefits associated with DBS but also highlights its role as a transformative treatment modality within the medical field. DBS target for different disorders is shown in ◘ Fig. 7.13 (Wiseman et al., 2024).

As DBS continues to evolve, ongoing research and technological advancements are poised to expand its indications and refine its application, promising to enhance the lives of individuals affected by a wide spectrum of neurological and psychiatric conditions.

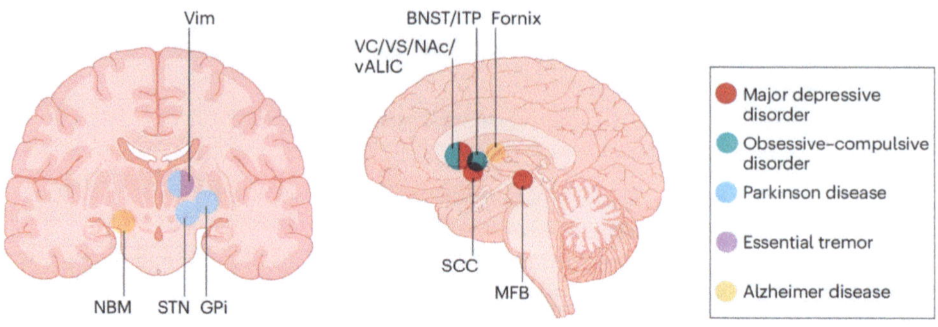

Fig. 7.13 Coronal (left) and sagittal (right) views illustrate the brain areas targeted in deep brain stimulation for various disorders: major depressive disorder (marked in red), obsessive–compulsive disorder (green), Parkinson's disease (blue), essential tremor (purple), and Alzheimer's disease (yellow). The specific regions include the bed nucleus of the stria terminalis (BNST), globus pallidus interna (GPi), inferior thalamic peduncle (ITP), medial forebrain bundle (MFB), nucleus accumbens (NAc), nucleus basalis of Meynert (NBM), subcallosal cingulate (SCC), subthalamic nucleus (STN), ventral anterior limb of the internal capsule (vALIC), ventral capsule/ventral striatum (VC/VS), and the ventral intermediate nucleus of the thalamus (Vim). (From Wiseman et al., 2024)

7.4 Latest Research Trends

Recent progress in deep brain stimulation (DBS) technology has been directed toward enhancing the precision and adaptability of stimulation, notably through the introduction of directional electrodes and network-focused stimulation strategies. Directional electrodes are designed to allow focused stimulation of specific brain areas or neural pathways, minimizing the stimulation of adjacent non-targeted regions. This advancement facilitates more accurate targeting, potentially improving therapeutic outcomes and reducing side effects (Steigerwald et al., 2016).

Moreover, the development and implementation of closed-loop DBS systems and the implementation of machine learning mark a pivotal advancement in neuromodulation therapy. These innovative systems are capable of dynamically adjusting stimulation parameters in response to real-time changes in brain activity, thereby offering a more responsive and individualized treatment approach. By monitoring specific biomarkers of neural activity, such as local field potentials, closed-loop systems can modulate the intensity, frequency, and duration of stimulation based on the patient's immediate neurological state, as shown in Fig. 7.14 (Oliveira et al., 2023).

This represents a significant evolution from traditional open-loop DBS systems, which operate on fixed settings regardless of ongoing neural changes (Little et al., 2013), as shown in Fig. 7.15 (Bouthour et al., 2019).

The integration of these advanced technologies into DBS devices not only underscores the rapid progress in the field but also highlights the move towards more personalized and effective treatment modalities for a range of neurological and psychiatric disorders.

The application of deep brain stimulation (DBS) is being vigorously explored beyond its established uses, venturing into areas such as Alzheimer's disease, chronic pain, and addiction. These research endeavors significantly broaden the

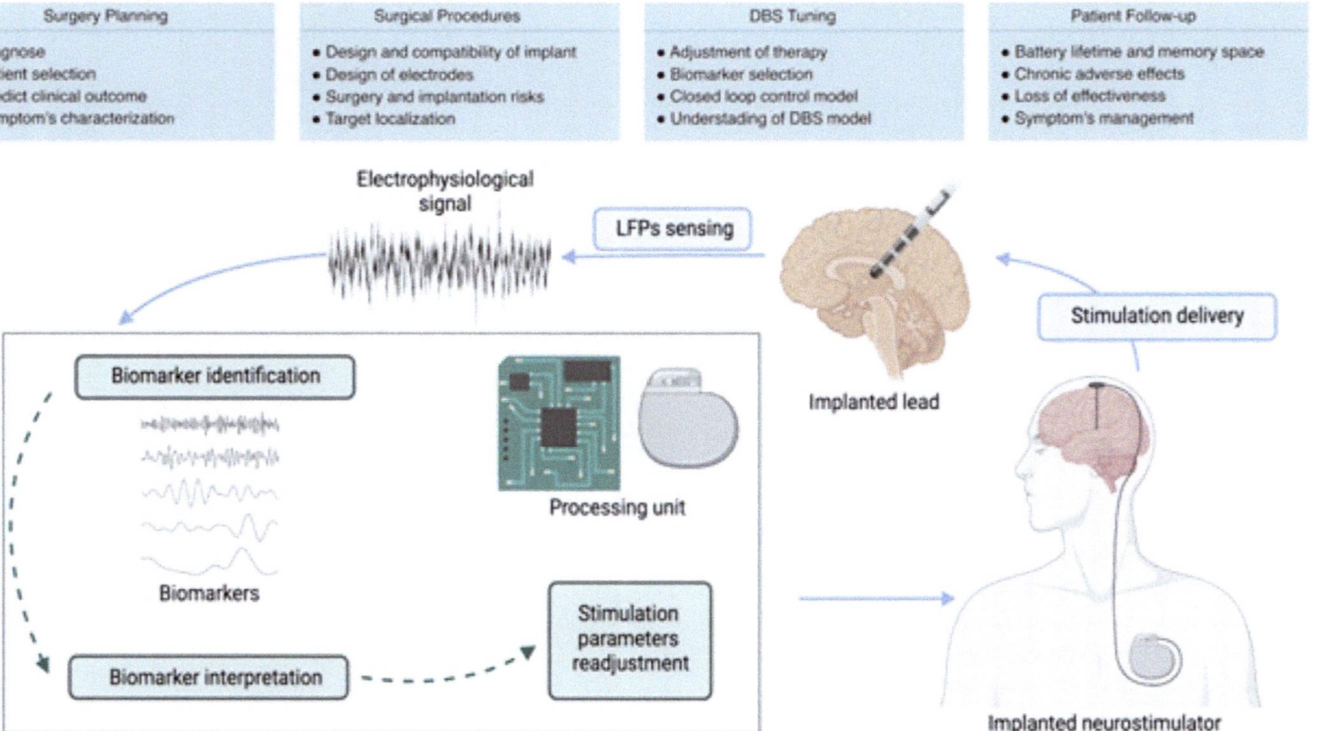

Fig. 7.14 Closed-loop DBS—the upper panel delineates a comprehensive list of challenges encountered during various phases of DBS therapy implementation, where machine learning (ML) techniques are crucial. These stages include initial patient assessment, precise electrode placement, real-time monitoring of neural activity, and long-term management of therapy adjustments. The diagram illustrates a sophisticated closed-loop feedback system designed for aDBS. This system is centered around the sensing of local field potentials (LFPs) and the identification and interpretation of electrophysiological biomarkers. Machine learning algorithms analyze these biomarkers to optimize stimulation parameters dynamically, responding to fluctuating neural states to enhance therapeutic outcomes. This integration of ML enables the aDBS system to adapt to the patient's changing physiological conditions, potentially improving the efficacy and precision of the therapy. (From Oliveira et al., 2023 originally published under CC-BY 4.0)

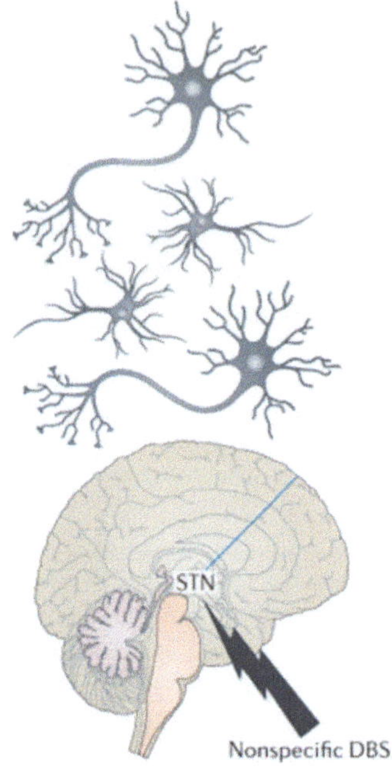

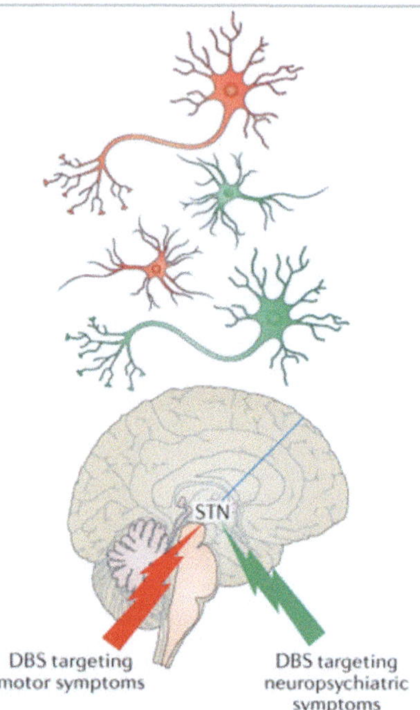

Fig. 7.15   **a** Current advancements in deep brain stimulation (DBS) for Parkinson's disease (PD) focus on the targeted stimulation of specific neural populations that influence critical brain circuits, thereby alleviating motor symptoms. Presently, these systems predominantly administer high-frequency stimulation to the subthalamic nucleus (STN), impacting both motor and affective cortical and subcortical neuronal networks. **b** Looking ahead, the scope of DBS could expand, with future stimulation protocols for PD potentially customized to address various symptoms within the same patient. The ongoing use of chronically implanted sensors and stimulators might unveil causal relationships between motor symptoms and co-occurring psychiatric conditions, potentially broadening DBS applications to include psychiatric symptoms in neuropsychiatric diseases and specific psychiatric circuit disorders. Concurrent circuit analysis in animal studies may reveal neuronal biomarkers that correlate with the underlying mechanisms of symptoms in both motor (red neurons) and affective (green neurons) networks in PD patients. Such insights could lead to the development of innovative stimulation paradigms designed to meet the unique needs of individual patients and minimize unintended stimulation of non-target networks. (From Bouthour et al., 2019)

prospective therapeutic scope of DBS, introducing potential treatments for conditions previously thought intractable (Fisher et al., 2010; Laxton et al., 2010). A pivotal avenue of current research is aimed at unraveling the intricate mechanisms through which DBS achieves its therapeutic effects. This includes in-depth studies into how DBS influences neurotransmitter systems and the dynamics of neural networks, areas that are crucial for customizing DBS therapies to individual patient needs and conditions (Lozano et al., 2019; Gratwicke et al., 2015).

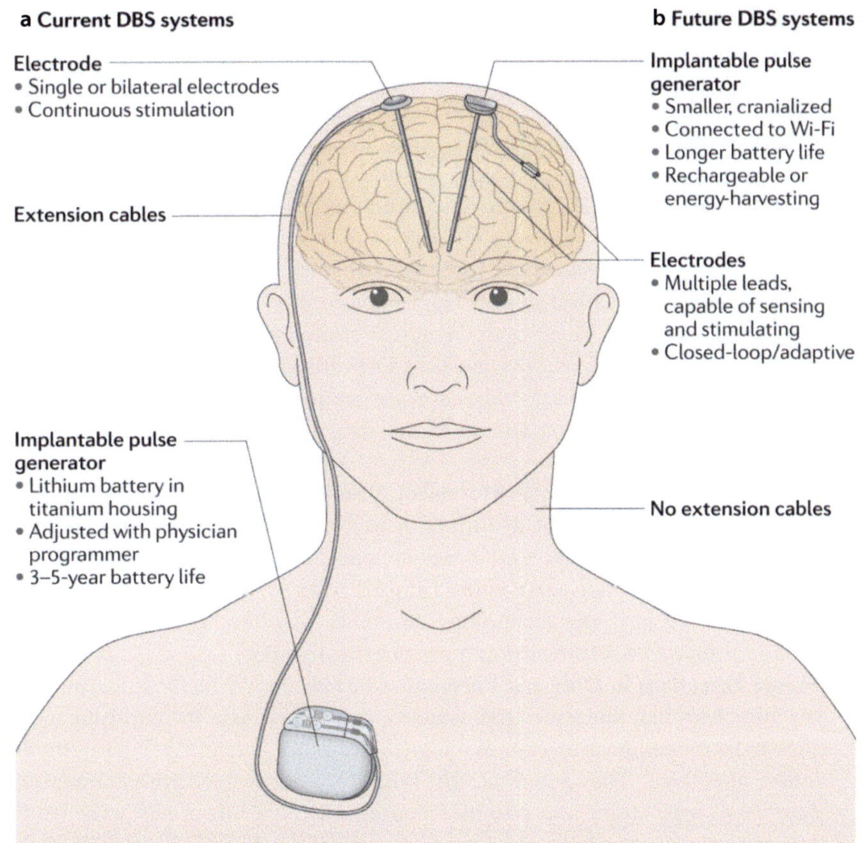

◘ Fig. 7.16 **a** A typical current deep brain stimulation (DBS) setup consists of surgically implanted electrodes in specific brain regions, connected to a pulse generator typically placed in the chest area. This configuration utilizes fixed stimulation parameters that are manually adjusted by clinicians during follow-up visits based on patient responses and symptomatology. **b** A predicted future DBS configuration envisages a more advanced, adaptive system. This next-generation DBS setup is expected to feature enhanced electrodes with the capability for real-time sensing and adjustment. Integrated with sophisticated machine learning algorithms, these systems will automatically optimize stimulation parameters in response to continuous feedback from the patient's neural activity. This evolution aims to provide more precise, personalized therapy, minimizing side effects and maximizing therapeutic efficacy for neurological disorders. (From Krauss et al., 2021)

DBS thus remains at the forefront of neuromodulation technologies, representing a beacon of hope for patients afflicted with a range of neurological and psychiatric disorders. With ongoing advancements in our comprehension of brain function and continuous innovations in DBS technology, the horizon for DBS applications and its therapeutic efficacy is expected to broaden. This evolution signals the dawn of a new epoch in neurotherapeutic interventions, one where DBS plays a central role in providing relief and improving the quality of life for patients worldwide. The future vision of DBS is shown in ◘ Fig. 7.16 (Krauss et al., 2021).

> **Key Takeaways**
> 1. **Deep Brain Stimulation (DBS) as a Transformative Therapy**: DBS has evolved from early experimental brain stimulation techniques into a sophisticated and widely used clinical treatment for movement disorders such as Parkinson's disease, essential tremor, and dystonia. This chapter highlights DBS's therapeutic potential in managing neurological conditions that are resistant to medication, providing significant symptom relief, and improving the quality of life for patients.
> 2. **Technological Advancements in DBS**: This chapter discusses how the technology behind DBS has progressed from open-loop systems to more advanced closed-loop designs. These systems adjust stimulation in real time based on patient-specific neural activity, making DBS more precise, personalized, and effective. Technological innovations like directional electrodes and computational models have further enhanced the accuracy and outcomes of the therapy.
> 3. **Clinical Applications Beyond Movement Disorders**: DBS's role has expanded beyond treating movement disorders to psychiatric conditions such as treatment-resistant depression and obsessive-compulsive disorder. This chapter also explores ongoing research into using DBS for Alzheimer's disease and epilepsy, showing how the technology continues to offer hope in managing a broader range of neurological and psychiatric disorders.
> 4. **Future Directions in DBS and Personalized Medicine**: With the integration of machine learning, adaptive DBS systems, and biomarkers for real-time monitoring, future applications are moving toward more personalized neuromodulation therapies. The potential for DBS to influence cognitive functions, emotional regulation, and complex neural networks opens new avenues for treating conditions previously deemed untreatable.

In essence, DBS stands as a testament to the remarkable progress in neuroscience and neurotechnology. It encapsulates a journey from concept to clinical practice, highlighting the blend of innovation, precision, and potential that continues to drive the field forward. As researchers continue to explore and refine these invasive brain stimulation techniques, researchers remain committed to unlocking even more profound insights into the brain's workings and improving therapeutic outcomes for patients across the globe.

Conclusion

As we conclude this chapter on invasive brain stimulation with a focus on deep brain stimulation (DBS), it is essential to reflect on the transformational journey and the sophisticated advancements that DBS has undergone. Starting from its roots in basic neuroscience research to becoming a critical therapeutic tool, DBS has drastically enhanced our ability to treat severe neurological conditions that were once considered beyond the reach of traditional medical interventions.

Building on the comprehensive exploration of neuromodulation discussed in ▶ Chap. 6, this chapter provided an overview of the specialized techniques that allow for direct intervention within the brain's complex network. We have seen how DBS leverages precise electrical impulses to modulate specific brain regions, offering relief and functional improvements to patients suffering from debilitating disorders such as Parkinson's disease, essential tremor, and dystonia.

The historical overview provided has shown us the evolution from the early experimental days to the current sophisticated implementations that benefit from the integration of real-time imaging and computational modeling. These enhancements have not only improved the precision of DBS but have also broadened its applicability, making it a cornerstone in the field of therapeutic neurostimulation.

Looking forward, the potential expansion of DBS applications continues to be supported by ongoing research and technological innovation. Studies exploring the use of DBS for conditions such as Alzheimer's disease, depression, and even rare neuropsychiatric disorders promise to widen the scope of this technology significantly. Each step forward in this field not only enhances our understanding of neural function but also improves the quality of life for patients around the world.

This section offers practical exercises designed to help apply the material learned to concrete scenarios.

DBS System Components Assembly

Objective: Familiarize students with the hardware components of a DBS system.

Task: Students will use a model or virtual simulation tool to assemble a DBS system, identifying and connecting components such as the neurostimulator, electrode leads, and extension cables.

Case Study Analysis: Parkinson's Disease Treatment with DBS

Objective: Understand the clinical decision-making process for treating Parkinson's disease with DBS.

Task: Students will analyze a series of case studies of Parkinson's patients treated with DBS, discussing the rationale for electrode placement, stimulation parameters, and evaluating the outcomes.

Programming a Closed-Loop DBS System

Objective: Explore the functionality and programming of closed-loop DBS systems.

Task: Using a computer simulation, students will program a closed-loop DBS system to respond to simulated changes in neuronal activity. They will adjust parameters like frequency, pulse width, and amplitude based on the feedback provided by the system.

Ethical Debate: The Use of DBS in Psychiatric Disorders

Objective: Discuss the ethical implications of using DBS in psychiatric treatments.

Task: Students will participate in a debate on the ethical considerations of implementing DBS for conditions such as depression and obsessive-compulsive disorder, focusing on patient consent, potential risks, and benefits.

DBS Electrode Placement and Surgical Planning

Objective: Learn about the surgical planning and challenges involved in DBS electrode placement.

Task: Students will use 3D brain mapping software to plan the electrode placement for a hypothetical DBS procedure. They will choose target areas based on the disorder being treated and discuss how to avoid critical brain areas to minimize side effects.

Design a Research Proposal for a New DBS Application

Objective: Develop a proposal for extending the use of DBS to a new neurological condition.

Task: Students will identify a neurological condition not currently treated by DBS and develop a research proposal that outlines the rationale, proposed methodology, expected challenges, and potential impact of using DBS as a treatment.

References

Barker, A. T., Jalinous, R., & Freeston, I. L. (1985). Non-invasive magnetic stimulation of human motor cortex. *The Lancet, 325*(8437), 1106–1107.

Belmaker, R. H., & Lichtenberg, P. (2023). Electroconvulsive therapy, transcranial magnetic stimulation, deep brain stimulation and tDCS. In *Psychopharmacology reconsidered: A concise guide exploring the limits of diagnosis and treatment* (pp. 115–121). Springer International Publishing.

Benabid, A. L., Pollak, P., Louveau, A., Henry, S., & De Rougemont, J. (1987). Combined (thalamotomy and stimulation) stereotactic surgery of the VIM thalamic nucleus for bilateral Parkinson disease. *Stereotactic and Functional Neurosurgery, 50*(1–6), 344–346.

Benabid, A. L., Pollak, P., Gross, C., Hoffmann, D., Benazzouz, A., Gao, D. M., et al. (1994). Acute and long-term effects of subthalamic nucleus stimulation in Parkinson's disease. *Stereotactic and Functional Neurosurgery, 62*(1–4), 76–84.

Bouthour, W., Mégevand, P., Donoghue, J., Lüscher, C., Birbaumer, N., & Krack, P. (2019). Biomarkers for closed-loop deep brain stimulation in Parkinson disease and beyond. *Nature Reviews Neurology, 15*(6), 343–352.

Bronstein, J. M., Tagliati, M., Alterman, R. L., Lozano, A. M., Volkmann, J., Stefani, A., et al. (2011). Deep brain stimulation for Parkinson disease: An expert consensus and review of key issues. *Archives of Neurology, 68*(2), 165–165.

Brunoni, A. R., Nitsche, M. A., Bolognini, N., Bikson, M., Wagner, T., Merabet, L., et al. (2012). Clinical research with transcranial direct current stimulation (tDCS): Challenges and future directions. *Brain Stimulation, 5*(3), 175–195.

Butson, C. R., & McIntyre, C. C. (2008). Current steering to control the volume of tissue activated during deep brain stimulation. *Brain Stimulation, 1*(1), 7–15.

Cerletti, U., & Bini, L. (1938). Un nuovo metodo di shockterapie: L'elettro-shock. *Bollettino dell'Accademia Medica di Roma, 64*, 136–138.

Chervyakov, A. V., Chernyavsky, A. Y., Sinitsyn, D. O., & Piradov, M. A. (2015). Possible mechanisms underlying the therapeutic effects of transcranial magnetic stimulation. *Frontiers in Human Neuroscience, 9*, 303.

Cirillo, G. D. P. G., Di Pino, G., Capone, F., Ranieri, F., Florio, L., Todisco, V., et al. (2017). Neurobiological after-effects of non-invasive brain stimulation. *Brain Stimulation, 10*(1), 1–18.

Deisseroth, K. (2011). Optogenetics. *Nature Methods, 8*(1), 26–29.

Delgado, J. M. R. (1969). *Physical control of the mind: Toward a psychocivilized society* (Vol. 41). World Bank Publications.

Deuschl, G., Schade-Brittinger, C., Krack, P., Volkmann, J., Schäfer, H., Bötzel, K., Daniels, C., Deutschländer, A., Dillmann, U., Eisner, W., Gruber, D., Hamel, W., Herzog, J., Hilker, R., Klebe, S., Kloß, M., Koy, J., Krause, M., Kupsch, A., …, Voges, J. (2006). A randomized trial of deep-brain stimulation for Parkinson's disease. The New England Journal of Medicine, 355(9), 896–908.

Fisher, R., Salanova, V., Witt, T., Worth, R., Henry, T., Gross, R., Oommen, K., Osorio, I., Nazzaro, J., Labar, D., Kaplitt, M., Sperling, M., Sandok, E., Neal, J., Handforth, A., Stern, J., DeSalles, A., Chung, S., Shetter, A., …, Graves, N. (2010). Electrical stimulation of the anterior nucleus of thalamus for treatment of refractory epilepsy. Epilepsia, 51(5), 899–908.

Fregni, F., & Pascual-Leone, A. (2007). Technology insight: Noninvasive brain stimulation in neurology—Perspectives on the therapeutic potential of rTMS and tDCS. *Nature Clinical Practice Neurology, 3*(7), 383–393.

Fritsch, G., & Hitzig, E. (1870). Über die elektrische Erregbarkeit des Grosshirns. *Archiv für Anatomie, Physiologie und Wissenschaftliche Medicin, 37*, 300–332.

George, M. S., & Aston-Jones, G. (2010). Noninvasive techniques for probing neurocircuitry and treating illness: Vagus nerve stimulation (VNS), transcranial magnetic stimulation (TMS) and transcranial direct current stimulation (tDCS). *Neuropsychopharmacology, 35*(1), 301–316.

Gradinaru, V., Mogri, M., Thompson, K. R., Henderson, J. M., & Deisseroth, K. (2009). Optical deconstruction of parkinsonian neural circuitry. *Science, 324*(5925), 354–359.

Gratwicke, J., Jahanshahi, M., & Foltynie, T. (2015). Parkinson's disease dementia: A neural networks perspective. *Brain, 138*(6), 1454–1476.

Herrmann, C. S., Rach, S., Neuling, T., & Strüber, D. (2013). Transcranial alternating current stimulation: A review of the underlying mechanisms and modulation of cognitive processes. *Frontiers in Human Neuroscience, 7*, 279.

Hess, W. R. (1928). Stammganglien-Reizversuche. *Berichte der gesamten Physiologie, 42*, 554–555.

Johnson, K. A., Okun, M. S., Scangos, K. W., Mayberg, H. S., & de Hemptinne, C. (2024). Deep brain stimulation for refractory major depressive disorder: A comprehensive review. *Molecular Psychiatry, 29*, 1–13.

Krauss, J. K., Pohle, T., Weber, S., Ozdoba, C., & Burgunder, J. M. (1999). Bilateral stimulation of globus pallidus internus for treatment of cervical dystonia. *The Lancet, 354*(9181), 837–838.

Krauss, J. K., Lipsman, N., Aziz, T., Boutet, A., Brown, P., Chang, J. W., et al. (2021). Technology of deep brain stimulation: Current status and future directions. *Nature Reviews Neurology, 17*(2), 75–87.

Kricheldorff, J., Göke, K., Kiebs, M., Kasten, F. H., Herrmann, C. S., Witt, K., & Hurlemann, R. (2022). Evidence of neuroplastic changes after transcranial magnetic, electric, and deep brain stimulation. *Brain Sciences, 12*(7), 929.

Kringelbach, M. L., Jenkinson, N., Owen, S. L., & Aziz, T. Z. (2007). Translational principles of deep brain stimulation. *Nature Reviews Neuroscience, 8*(8), 623–635.

Kühn, A. A., Kempf, F., Brücke, C., Doyle, L. G., Martinez-Torres, I., Pogosyan, A., et al. (2008). High-frequency stimulation of the subthalamic nucleus suppresses oscillatory β activity in patients with Parkinson's disease in parallel with improvement in motor performance. *Journal of Neuroscience, 28*(24), 6165–6173.

Laxton, A. W., Tang-Wai, D. F., McAndrews, M. P., Zumsteg, D., Wennberg, R., Keren, R., Wherrett, J., Naglie, G., Hamani, C., Smith, G. S., & Lozano, A. M. (2010). A phase I trial of deep brain stimulation of memory circuits in Alzheimer's disease. *Annals of Neurology, 68*(4), 521–534.

Little, S., Pogosyan, A., Neal, S., Zavala, B., Zrinzo, L., Hariz, M., et al. (2013). Adaptive deep brain stimulation in advanced Parkinson disease. *Annals of Neurology, 74*(3), 449–457.

Lozano, A. M., & Lipsman, N. (2013). Probing and regulating dysfunctional circuits using deep brain stimulation. *Neuron, 77*(3), 406–424.

Lozano, A. M., Lipsman, N., Bergman, H., Brown, P., Chabardes, S., Chang, J. W., et al. (2019). Deep brain stimulation: Current challenges and future directions. *Nature Reviews Neurology, 15*(3), 148–160.

Mayberg, H. S., Lozano, A. M., Voon, V., McNeely, H. E., Seminowicz, D., Hamani, C., et al. (2005). Deep brain stimulation for treatment-resistant depression. *Neuron, 45*(5), 651–660.

McIntyre, C. C., Savasta, M., Kerkerian-Le Goff, L., & Vitek, J. L. (2004). Uncovering the mechanism (s) of action of deep brain stimulation: Activation, inhibition, or both. *Clinical Neurophysiology, 115*(6), 1239–1248.

Nitsche, M. A., Cohen, L. G., Wassermann, E. M., Priori, A., Lang, N., Antal, A., et al. (2008). Transcranial direct current stimulation: State of the art 2008. *Brain Stimulation, 1*(3), 206–223.

Oliveira, A. M., Coelho, L., Carvalho, E., Ferreira-Pinto, M. J., Vaz, R., & Aguiar, P. (2023). Machine learning for adaptive deep brain stimulation in Parkinson's disease: Closing the loop. *Journal of Neurology, 270*(11), 5313–5326.

Parastarfeizabadi, M., & Kouzani, A. Z. (2017). Advances in closed-loop deep brain stimulation devices. *Journal of Neuroengineering and Rehabilitation, 14*, 1–20.

Pascual-Leone, A., Amedi, A., Fregni, F., & Merabet, L. B. (2005). The plastic human brain cortex. *Annual Review of Neuroscience, 28*, 377–401.

Ramirez-Zamora, A., Giordano, J. J., Gunduz, A., Brown, P., Sanchez, J. C., Foote, K. D., et al. (2018). Evolving applications, technological challenges and future opportunities in neuromodulation: Proceedings of the fifth annual deep brain stimulation think tank. *Frontiers in Neuroscience, 11*, 734.

Rizzone, M., Lanotte, M., Bergamasco, B., Tavella, A., Torre, E., Faccani, G., et al. (2001). Deep brain stimulation of the subthalamic nucleus in Parkinson's disease: Effects of variation in stimulation parameters. *Journal of Neurology, Neurosurgery & Psychiatry, 71*(2), 215–219.

Rosa, M. A., & Lisanby, S. H. (2012). Somatic treatments for mood disorders. *Neuropsychopharmacology, 37*(1), 102–116.

Rosin, B., Slovik, M., Mitelman, R., Rivlin-Etzion, M., Haber, S. N., Israel, Z., et al. (2011). Closed-loop deep brain stimulation is superior in ameliorating parkinsonism. *Neuron, 72*(2), 370–384.

Rossi, S., Hallett, M., Rossini, P. M., Pascual-Leone, A., & Safety of TMS Consensus Group. (2009). Safety, ethical considerations, and application guidelines for the use of transcranial magnetic stimulation in clinical practice and research. *Clinical Neurophysiology, 120*(12), 2008–2039.

Scangos, K. W., Khambhati, A. N., Daly, P. M., Makhoul, G. S., Sugrue, L. P., Zamanian, H., et al. (2021). Closed-loop neuromodulation in an individual with treatment-resistant depression. *Nature Medicine, 27*(10), 1696–1700.

Schuurman, P. R., Bosch, D. A., Bossuyt, P. M., Bonsel, G. J., Van Someren, E. J., De Bie, R. M., et al. (2000). A comparison of continuous thalamic stimulation and thalamotomy for suppression of severe tremor. *New England Journal of Medicine, 342*(7), 461–468.

Sendi, M. S., Waters, A. C., Tiruvadi, V., Riva-Posse, P., Crowell, A., Isbaine, F., et al. (2021). Intraoperative neural signals predict rapid antidepressant effects of deep brain stimulation. *Translational Psychiatry, 11*(1), 551.

Senevirathne, D. K. L., Mahboob, A., Zhai, K., Paul, P., Kammen, A., Lee, D. J., et al. (2023). Deep brain stimulation beyond the clinic: Navigating the future of Parkinson's and Alzheimer's disease therapy. *Cells, 12*(11), 1478.

Spiegel, E. A., Wycis, H. T., Marks, M., & Lee, A. J. (1947). Stereotaxic apparatus for operations on the human brain. *Science, 106*(2754), 349–350.

Spiliotis, K., Starke, J., Franz, D., Richter, A., & Köhling, R. (2022). Deep brain stimulation for movement disorder treatment: Exploring frequency-dependent efficacy in a computational network model. *Biological Cybernetics, 116*(1), 93–116.

Steigerwald, F., Müller, L., Johannes, S., Matthies, C., & Volkmann, J. (2016). Directional deep brain stimulation of the subthalamic nucleus: A pilot study using a novel neurostimulation device. *Movement Disorders, 31*(8), 1240–1243.

Tawfik, V. L., Chang, S. Y., Hitti, F. L., Roberts, D. W., Leiter, J. C., Jovanovic, S., & Lee, K. H. (2010). Deep brain stimulation results in local glutamate and adenosine release: Investigation into the role of astrocytes. *Neurosurgery, 67*(2), 367–375.

U.S. Food and Drug Administration (FDA). (1997). Recently-approved devices—Activa tremor control therapy—P960009. Retrieved from https://www.fda.gov/medical-devices/recently-approved-devices/activa-tremor-control-therapy-p960009

Volkmann, J., Allert, N., Voges, J., Weiss, P. H., Freund, H. J., & Sturm, V. (2001). Safety and efficacy of pallidal or subthalamic nucleus stimulation in advanced PD. *Neurology, 56*(4), 548–551.

Vossen, A., Gross, J., & Thut, G. (2015). Alpha power increase after transcranial alternating current stimulation at alpha frequency (α-tACS) reflects plastic changes rather than entrainment. *Brain Stimulation, 8*(3), 499–508.

Widge, A. S., Malone, D. A., Jr., & Dougherty, D. D. (2018). Closing the loop on deep brain stimulation for treatment-resistant depression. *Frontiers in Neuroscience, 12*, 175.

Wiseman, M., Sewell, I. J., Nestor, S. M., Giacobbe, P., Hamani, C., Lipsman, N., & Rabin, J. S. (2024). Cognitive effects of focal neuromodulation in neurological and psychiatric disorders. *Nature Reviews Psychology, 3*, 1–19.

Non-invasive Brain Stimulation Techniques

TMS, tDCS, and tACS Methods

Contents

U. Chaudhary, *Expanding Senses using Neurotechnology*, https://doi.org/10.1007/978-3-031-76081-5_8

Test your learning and check your understanding of this book's contents: use the "Springer Nature Flashcards" app to access questions using ▶ https://sn.pub/kmb-jyz. To use the app, please follow the instructions in ▶ Chap. 1.

This chapter explores the growing field of non-invasive brain stimulation (NIBS) techniques, highlighting their potential as effective therapeutic options that avoid the need for surgery. These techniques, which include transcranial magnetic stimulation (TMS), transcranial direct current stimulation (tDCS) and transcranial alternating current stimulation (tACS), are explored for their unique mechanisms and applications in modulating neuronal activity. Through a detailed examination of each method, this chapter highlights their potential to influence brain oscillations, enhance cognitive functions, and treat a variety of neuropsychiatric and neurological conditions. Special attention is given to the scientific principles underlying each technique, their clinical implications, and the potential for inducing long-term changes in the brain. The discussion is anchored in the latest research findings and enriched with illustrative diagrams and case studies to provide a comprehensive overview of current capabilities and future directions in NIBS. The overarching goal of this chapter is to underscore the transformative potential of these technologies in advancing brain research and therapy, setting the stage for their expanded use in clinical settings.

Learning Objectives

1. Understand Non-Invasive Brain Stimulation (NIBS) Methods: Gain a fundamental understanding of various NIBS methods such as Transcranial Magnetic Stimulation (TMS), Transcranial Direct Current Stimulation (tDCS), and Transcranial Alternating Current Stimulation (tACS), and their underlying mechanisms.
2. Explore the Effects of NIBS on Neuronal Activity: Examine how these techniques influence brain oscillations, neuronal excitability, and synaptic plasticity, contributing to their therapeutic effects.
3. Analyze Clinical Applications of NIBS: Identify the neuropsychiatric and neurological conditions that can be treated with NIBS, focusing on the efficacy and potential of these techniques in clinical settings.
4. Assess the Technological Advancements in NIBS: Discuss recent technological innovations in NIBS, including the development of more precise stimulation methods and the integration of NIBS with neuroimaging.

Non-invasive brain stimulation (NIBS) encompasses a range of techniques designed to modulate neuronal activity without the need for surgical intervention. These methods, including transcranial magnetic stimulation (TMS) and transcranial current stimulation (tES), have shown promise in modulating brain oscillations and neural synchronization, which are crucial for various cognitive and neurological functions (Rossi et al., 2009; Giordano et al., 2017). The therapeutic potential of NIBS extends to a wide range of neurological and neuropsychiatric disorders. For example, anodal tDCS, a type of tES, has been investigated for its efficacy in improving cognitive deficits following traumatic brain injuries, showing promise in enhancing working memory and modulating neural connectivity (Weber et al., 2014; Hill et al., 2016). Furthermore, NIBS can induce functional and structural plasticity in the brain, with evidence suggesting its role in enhancing synaptic efficacy and promoting neurogenesis (Di Pino et al., 2014; Polanía et al., 2018), which is a topic of ▶ Chap. 9.

In this chapter, we will explore various types of NIBS techniques, as shown in ▣ Fig. 8.1 (Wang et al., 2023). I will provide a brief overview of TUS and tNIR before diving deeper into TMS, tDCS, and tACS.

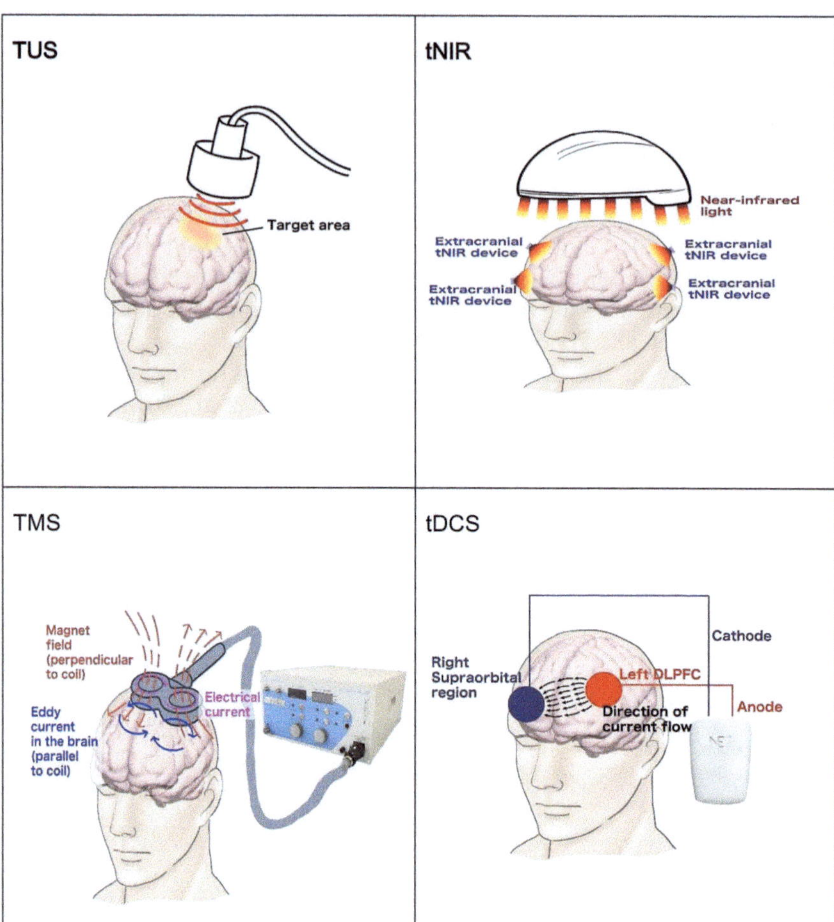

Fig. 8.1 The schematic diagram represents various types of non-invasive brain stimulation techniques, each targeting specific areas of the brain to modulate neurological functions without surgical intervention. These include **Transcranial Ultrasound Stimulation (TUS)**: Utilizes focused ultrasound waves to modulate neural activity. This technique can target deep brain structures with high spatial precision, influencing neuronal excitability and potentially treating conditions like depression and Parkinson's disease. **Transcranial Near-Infrared Light Therapy (tNIR)**: Employs near-infrared light to penetrate the skull and stimulate neural activity. tNIR is thought to enhance mitochondrial function and increase neuroplasticity, which can be beneficial in treating brain injuries and neurodegenerative diseases. **Transcranial Magnetic Stimulation (TMS)**: Uses magnetic fields to induce electrical currents in specific brain regions, such as the dorsolateral prefrontal cortex (DLPFC), which is often targeted for its role in mood regulation and cognitive functions. TMS is widely used for treating depression and has potential applications in cognitive enhancement. **Transcranial Direct Current Stimulation (tDCS)**: Applies a constant, low-intensity electrical current to the brain via electrodes placed on the scalp, influencing the dorsolateral prefrontal cortex (DLPFC) among other regions. tDCS is explored for its ability to enhance cognitive performance, treat psychiatric conditions, and aid in stroke recovery by modulating cortical excitability. Each of these techniques offers unique mechanisms and potential therapeutic benefits for a range of neurological and psychological conditions, highlighting the diverse applications of non-invasive brain stimulation in medical and research settings. From Wang et al. (2023) originally published under CC-BY 4.0

8.1 Transcranial Ultrasound Stimulation (TUS)

Ultrasound is a mechanical wave with frequencies exceeding the human auditory range, spanning from 20 kHz to 1 GHz. This wave propagates through tissues by alternating phases of compression and rarefaction, effectively transmitting energy via molecular motion. Ultrasound technology is categorized based on intensity; high-intensity ultrasounds utilize intensities >3 W/cm^2, while low-intensity ultrasounds (LIU) operate with intensities <3 W/cm^2 (Chávez-Martínez et al., 2020). High-intensity focused ultrasound (HIFU) is specifically employed for therapeutic ablations by causing significant tissue heating, which can destroy targeted tissue areas effectively. In contrast, LIU is known for its gentle mechanical effects on tissues, which do not induce heating or damage, making it suitable for diagnostic purposes and therapeutic applications that require subtle influence over cellular functions.

Beyond its direct mechanical impacts, LIU has demonstrated potential in modulating the activity of neurotrophic factors, which in turn may influence neural activity and neuroplasticity (Tufail et al., 2010). This modulation is crucial for therapies aimed at enhancing brain function or recovery after injury. In the context of transcranial ultrasound stimulation (TUS), specifically transcranial focused ultrasound (tFUS), LIU is applied non-invasively to penetrate the skull and target deep brain regions effectively. This technique allows for precise targeting of neural tissues, offering potential therapeutic benefits in neuromodulation and the treatment of neurological disorders (Meng et al., 2021).

8.2 Transcranial Near-Infrared (tNIR)

Transcranial near-infrared (tNIR) stimulation is a non-invasive brain stimulation technique that employs near-infrared light-to-modulate brain function. This method utilizes wavelengths typically ranging from 650 to 1350 nm, which can penetrate the skull and reach cortical brain tissue (Tedford et al., 2015; Pitzschke et al., 2015). tNIR is particularly appealing due to its ability to influence neural activity without the invasiveness of surgical procedures or the discomfort associated with electrical methods.

The primary mechanism through which tNIR is believed to affect brain activity is through photobiomodulation (Salehpour et al., 2018). This process involves the absorption of infrared light by cytochrome c oxidase, a component of the mitochondrial electron transport chain. The absorption of light enhances mitochondrial energy production, leading to an increase in adenosine triphosphate (ATP), the cellular energy currency. This boost in ATP can enhance cellular metabolism and support various cellular functions, including those critical for neural health and activity (Poyton & Ball, 2011). tNIR has been explored for several applications in neurotherapy and cognitive enhancement (Berman et al., 2017; Nizamutdinov et al., 2021; Liebert et al., 2021).

While promising, tNIR technology still faces several challenges such as, (1) effective treatment requires precise calibration of wavelength, intensity, and duration of exposure, parameters that may vary widely between individuals and conditions. (2) More extensive clinical trials and research are needed to fully understand the efficacy and limitations of tNIR, and to establish standardized treatment protocols.

As research progresses, tNIR could become a significant tool in the neurotherapeutic arsenal, particularly for conditions where other treatments have limited effectiveness.

8.3 Transcranial Magnetic Stimulation (TMS)

Transcranial magnetic stimulation (TMS) leverages the principle of electromagnetic induction to generate transient magnetic fields, capable of traversing the cranial vault to elicit electric currents within the cerebral cortex. These induced currents have the potential to depolarize neuronal membranes, culminating in alterations in neuronal activity and connectivity patterns. This section is dedicated to providing an overview of TMS, encompassing its historical trajectory from conception to contemporary applications, the technological frameworks that underpin its functionality, its utility across various clinical domains, and the forefront of research innovations.

8.3.1 Historical Developments

Transcranial magnetic stimulation (TMS), a pivotal technique in neurostimulation, owes its conceptual foundation to the seminal experiments on electromagnetic induction by Michael Faraday in the 1830s. Faraday's discovery that a changing magnetic field could induce an electric current in a nearby conductor laid the groundwork for later applications in manipulating neural activity. However, it wasn't until the latter part of the twentieth century that TMS began to be shaped as a method for non-invasive brain stimulation. A landmark study conducted by Dr. Anthony Barker and his team at the University of Sheffield, UK, in 1985, marked the inaugural demonstration of TMS's ability to stimulate the human motor cortex non-invasively, showcasing the potential for direct cortical excitation without the need for surgical intervention (Barker et al., 1985). Since its inception, TMS technology has undergone substantial refinements, evolving from single-pulse paradigms to sophisticated repetitive TMS (rTMS) protocols, enabling both transient and lasting modulation of cortical excitability (Rossi et al., 2009).

Following its initial demonstration, TMS has undergone significant technological and methodological advancements, evolving from a novel research instrument to a clinically validated treatment modality. Its application has been broad, spanning from the elucidation of cortical function and neuroplastic mechanisms to the therapeutic intervention in psychiatric and neurological disorders. The U.S. Food and Drug Administration's (FDA) approval of TMS for treating major depressive disorder in 2008 was a landmark moment, highlighting its safety and effectiveness and paving the way for its integration into modern therapeutic practices. This approval was based on substantial evidence demonstrating TMS's capability to modulate neural activity in regions implicated in mood regulation, offering a non-pharmacological treatment option for patients with treatment-resistant depression.

Clinically, TMS has found applications in diagnosing and treating an array of neuropsychiatric conditions and neurological disorders such as stroke rehabilitation, Parkinson's disease, and chronic pain, showcasing its potential to enhance recovery and ameliorate symptoms (George et al., 2010; George & Post, 2011; Lefaucheur et al., 2014; McDonnell & Stinear, 2017; Buss et al., 2019; Chou et al., 2020).

The clinical adoption of TMS has been supported by a growing body of research that elucidates its mechanisms of action, optimal treatment parameters, and potential applications beyond depression, including but not limited to schizophrenia, obsessive-compulsive disorder, and chronic pain. Studies continue to explore the neurophysiological underpinnings of TMS, including its effects on neurotransmitter systems, neuroplasticity, and functional connectivity, thereby expanding our understanding of its therapeutic potential (George & Post, 2011; Lefaucheur et al., 2014). Innovations in neuronavigation technologies have improved the precision of target localization, while concurrent neuroimaging studies using fMRI and PET have begun to unravel the network-level changes induced by TMS, offering insights into its neuromodulatory mechanisms (Fox et al., 2014).

8.3.2 Technological Aspects

A transcranial magnetic stimulation (TMS) device consists of an electromagnetic coil that can be shaped into various configurations, including figure-eight, circular, or custom-tailored designs, which is connected to a pulse generator, often referred to as a stimulator. This coil, when strategically placed on the scalp, as shown in ▪ Fig. 8.2 (Wiseman et al., 2024), targets specific brain regions with the intent to modulate neural activity.

The operational mechanism of TMS involves the coil receiving transient, high-intensity electric currents from the pulse generator. These currents enable the coil to generate a focused magnetic field capable of penetrating the skull. This magnetic field, upon activation, dynamically interacts with the underlying brain tissue (Rossini et al., 1994), as shown in ▪ Fig. 8.3 (Tan et al., 2022). The specificity of the magnetic field's interaction with neuronal circuits allows for the targeted modulation of brain functions, affecting areas such as the prefrontal cortex which is often implicated in mood regulation and cognitive functions.

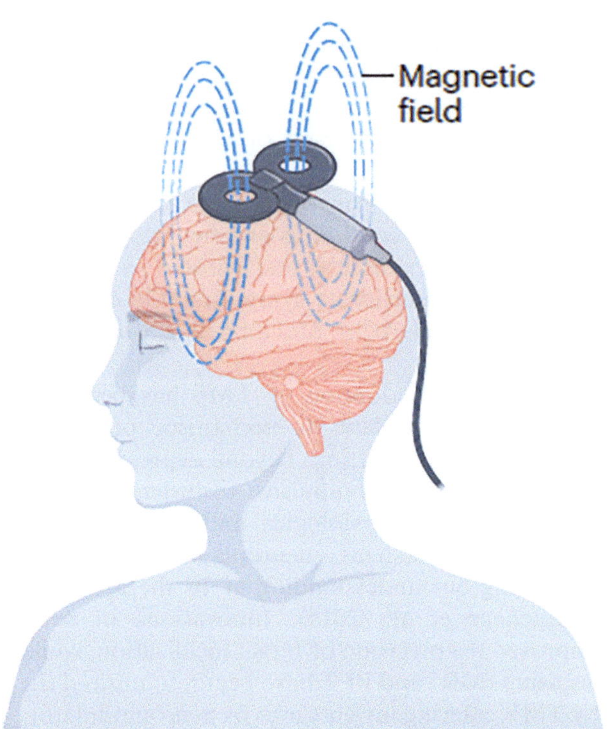

Fig. 8.2 A standard figure-eight TMS coil placed on the scalp; here, over dorsolateral prefrontal cortex. From Wiseman et al. (2024)

The magnetic pulses induced by the coil are of sufficient intensity and duration to depolarize neurons, leading to the modulation of neuronal activity. This process can enhance or suppress cortical excitability depending on the frequency and intensity of the stimulation, thereby offering a non-invasive means to influence brain activity (Chail et al., 2018). Such capabilities render TMS a valuable tool in both research and clinical settings, particularly for the treatment of various neuropsychiatric disorders and the exploration of brain function and interconnectivity (Chen et al., 2008; Rossi et al., 2009).

Given the intricate design and precision required in the configuration of the TMS apparatus, the effectiveness and safety of the stimulation are highly reliant on accurate placement and calibration of the coil relative to the targeted brain regions. This precision ensures that the induced magnetic fields are both effective in achieving the desired neural modulation and safe for the patient, minimizing unintended stimulation of adjacent neural tissue.

The effectiveness of TMS hinges significantly on the configuration and composition of its magnetic coils (Roth et al., 2002). Specific designs like the figure-eight coil are engineered to concentrate stimulation on precise brain areas, enhancing

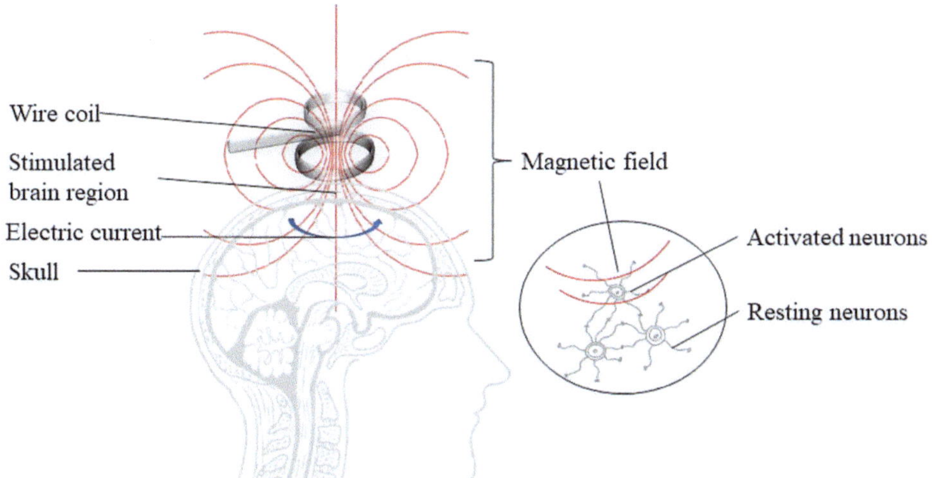

☑ **Fig. 8.3** In transcranial magnetic stimulation (TMS), a magnetic coil is strategically positioned on the patient's scalp, directly above the specific cortical area targeted for stimulation. This coil generates a rapidly fluctuating magnetic field that penetrates the skull and induces neuronal stimulation in the underlying brain tissue. The primary mechanism at work is electromagnetic induction, through which an electric current is created within the brain. This current leads to the focal depolarization of neurons, enabling targeted modulation of brain activity in the desired region. From Tan et al. (2022)

focal impact (Rossini et al., 1994). Coil materials also play a crucial role; for example, coils with a solid core as opposed to an air core are more efficient in magnetic field transmission, minimizing energy dissipation and facilitating more intense therapeutic sessions (Roth et al., 2002; Deng et al., 2014). The shape of the coil is another critical factor, influencing both the depth and specificity of stimulation: while double-cone and Hesed coils are tailored for deeper penetration into brain tissue, they typically provide a broader area of effect, reducing the focus (Zangen et al., 2005). Advances in coil technology, particularly the adoption of solid core materials, optimize the delivery of magnetic fields, significantly reducing energy loss and mitigating the risk of overheating during prolonged treatment sessions (Makaroff et al., 2023). These innovations collectively enhance the therapeutic potential of TMS by improving the precision and intensity of neural stimulation. ☑ Figure 8.4 shows different types of commonly used coils (Gutierrez et al., 2022).

Box 8.1 Transcranial Magnetic Stimulation (TMS)

A non-invasive brain stimulation technique that uses magnetic fields to induce electric currents in specific brain regions, influencing neuronal activity. TMS can be used to activate or inhibit neurons, depending on the frequency of stimulation.

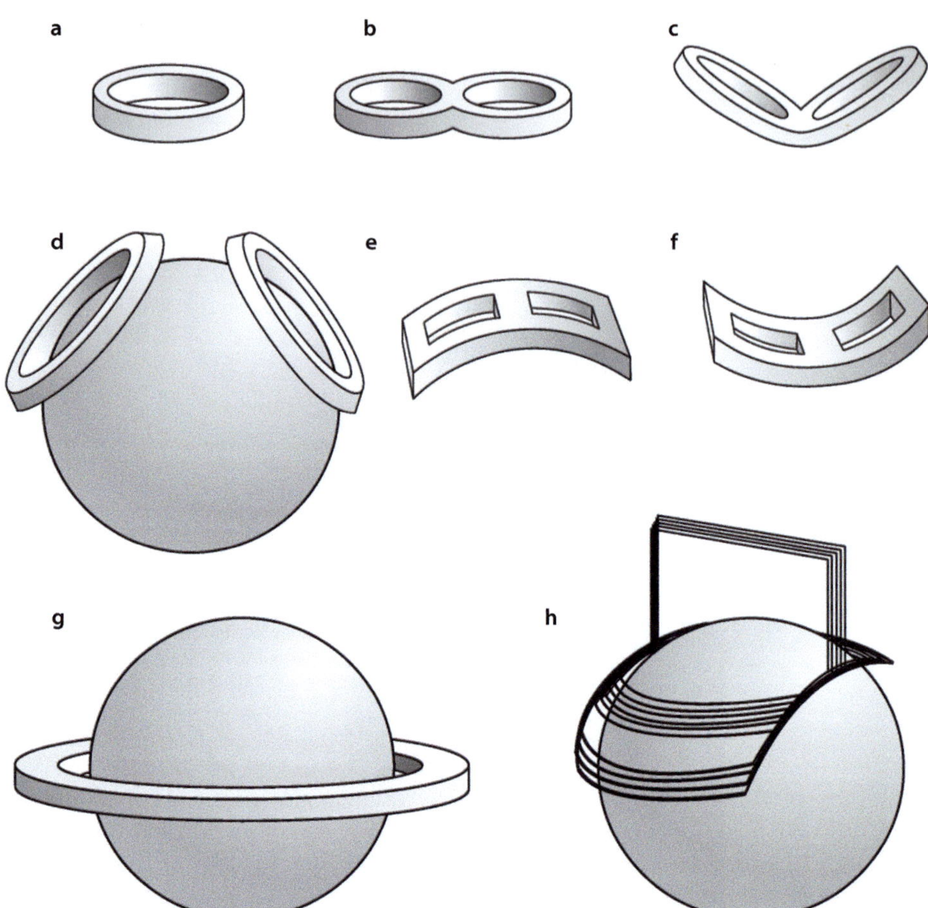

■ **Fig. 8.4** An illustrative comparison of the various types of coils commonly utilized in TMS. Each coil type is designed to meet specific therapeutic needs and research requirements by varying in shape and function. **a Circular coil**: This coil offers a broader field of stimulation, affecting larger brain areas simultaneously. While less targeted than the figure-eight, it is useful for general stimulation in experimental and clinical scenarios. **b Figure-eight coil**: This coil is favored for its ability to deliver focused magnetic fields, making it ideal for targeting specific cortical areas with precision. It is most commonly used in clinical settings for treating localized brain functions and disorders. **c Butterfly (V-shape) coil**: An angled electrode configuration ideal for targeting specific neural pathways or regions not accessible with straight electrodes, often used in cortical surface recordings. **d Double-cone coil**: Designed to reach deeper brain structures, the double-cone coil is suitable for stimulating areas that are not easily accessible with more superficial coil designs. Its structure allows it to better contour to the head's shape, facilitating deeper penetration. **e Downward-bending U-shape (DBU) coil**: A curved electrode designed to conform to the brain's contours, with openings to reduce impedance and improve contact, enhancing signal quality. **f Upward-bending U-shape (UBU) coil**: Electrodes arranged in a curved manner to target cylindrical regions, useful for deep brain stimulation and localized treatment of neurological disorders. **g Halo coil**: A ring electrode placed around a spherical surface, typically used for global brain stimulation or recording, providing a broad coverage area. **h Hesed coil (H-coil)**: Developed to target deeper brain regions without sacrificing the focus of stimulation. The H-coil is particularly valuable in research and treatment modalities requiring influence over deep-seated neurological pathways, such as in certain mood disorders. From Gutierrez et al. (2022) originally published under CC-BY 4.0

8.3.3 Neural Mechanism of TMS

The excitation of nerves by magnetic stimulation is primarily influenced by the voltage gradient aligned with the nerve fiber (Vucic & Kiernan, 2017). Due to the complex neural anatomy of the brain, excitation typically occurs at anatomical features like bends and branch points of nerve fibers, or more crucially at the axonal hillock—the juncture where the cell body transitions into the axon (Abdeen & Stuchly, 1994). The orientation of neurons relative to the induced electric field is therefore crucial in determining the level of neuronal activation and subsequently affecting the outcomes of Transcranial Magnetic Stimulation (TMS).

Although the specific neural circuits activated by TMS are not fully understood (Vucic et al., 2013; Siebner et al., 2022), it is known that TMS primarily stimulates the motor cortex at depths ranging from 1.5 to 2.1 cm below the skull surface (Rudiak & Marg, 1994). Research involving animal models has demonstrated that cortical stimulation induces a multifaceted corticomotoneuronal volley consisting of both direct (D) waves and several indirect (I) waves (Patton & Amassian, 1954), as illustrated in ◘ Fig. 8.5 (Vucic & Kiernan, 2017). This phenomenon has been further substantiated by human studies, which have employed cervical epidural recording techniques to detect the presence of D waves alongside multiple I waves, denoted as I1, I2, I3, and so forth, occurring at intervals ranging from 1.5 to 2.5 ms (Kaneko et al., 1997; Di Lazzaro et al., 1998). The complex interplay between the D and I waves is critical in shaping the parameters observed in transcranial magnetic stimulation (TMS) outcomes, indicating a fundamental aspect of how cortical stimulation impacts neuronal activity and connectivity (Siebner et al., 2022).

Several theoretical models have been developed to explain the phenomena of descending corticomotoneuronal volleys induced by TMS (Siebner et al., 2022). One early model suggests that the evoked D and I waves result from periodic stimulation of cortical output cells, such as Betz cells in layer V, by cortical interneurons with fixed temporal characteristics (Amassian et al., 1987). An alternate theory posits that I waves arise from independent sequences of interneuronal circuits, where each sequence is responsible for generating a distinct I wave (Day et al., 1989). A third hypothesis argues that magnetic stimulation triggers a widespread activation of neurons, which then fire repetitively in sync with their intrinsic membrane properties (Ziemann & Rothwell, 2000).

More recent computer simulation studies have introduced a feed-forward model that explains I-wave generation in terms of the location of interneuronal synapses on Betz cells, with later I waves emanating from synapses positioned further from the cell body and earlier ones from those closer to the soma (Rusu et al., 2014).

The direction of cortical current flow significantly influences the nature of the corticomotoneuronal volleys: I waves are optimally elicited by currents flowing in a posterior-anterior direction, while D waves are predominantly generated by lateral to medial currents (Kaneko et al., 1996; Sakai et al., 1997; Di Lazzaro et al., 2003). This suggests that D waves might represent direct activation of corticospinal axons, possibly at the axonal hillock, whereas I waves likely result from the complex interplay of cortical excitatory and inhibitory neurons (Di Lazzaro et al., 2008, 2012).

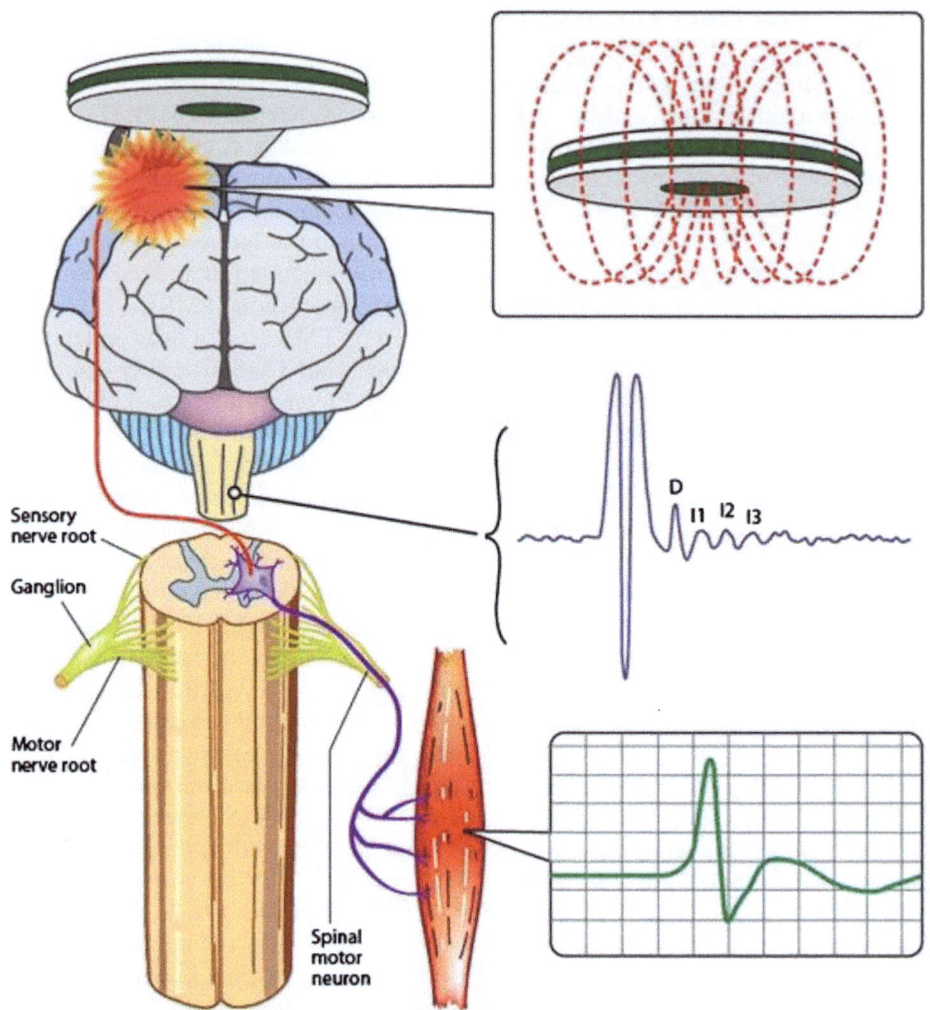

Fig. 8.5 Transcranial magnetic stimulation (TMS) induces a descending corticospinal volley that includes both direct (D) waves and several indirect (I) waves. The motor evoked potential (MEP), represented by a green curve, is subsequently recorded from a target muscle. This MEP serves as a critical biomarker for assessing the functionality of upper motor neurons, providing valuable insights into their operational state. From (Vucic & Kiernan, 2017)

In clinical settings, TMS has been instrumental in probing the functionality of cortical output cells and intracortical neuronal networks within the primary motor cortex (M1). This has led to profound insights into the pathophysiological mechanisms underlying neurodegenerative diseases such as ALS and has spurred the development of innovative diagnostic techniques (Vucic et al., 2013). The integrity of the motor cortex and corticospinal tract is typically assessed using TMS outcome measures like motor threshold (MT), motor evoked potential (MEP) amplitude, central motor conduction time (CMCT), duration of cortical silent period (CSP), as well as measures of short interval intracortical inhibition and intracortical facilitation. These parameters collectively provide valuable diagnostics in understanding and monitoring neurodegenerative conditions.

- MT is the minimum intensity required to evoke a MEP in a target muscle during 50% of TMS trials, according to the International Federation of Clinical Neurophysiology (Rossini et al., 1999). This threshold is defined as the lowest stimulator output needed to produce a small MEP response (over 50 μV). Advances in threshold-tracking TMS techniques have refined MT as the intensity necessary to consistently generate a target MEP of 0.2 mV ($\pm$20%) (Fisher et al., 2002; Vucic et al., 2006). MTs indicate the density of corticomotoneuronal projections to anterior horn cells, typically lower for dominant hand muscles due to higher projection density (Macdonell et al., 1991; Chen et al., 1998). Additionally, MTs reflect the excitability of cortical output cells, influenced by voltage-gated Na+ channels and the glutamatergic system (Boroojerdi et al., 2001; Lazzaro et al., 2003; Ziemann, 2004a).

- MEP amplitude reflects the cumulative effect of descending corticospinal volleys, composed of direct (D) and indirect (I) waves, on spinal and bulbar motor neurons, indicating the density of corticomotoneuronal projections (Lazzaro et al., 1998; Ziemann, 2004b). At threshold levels, TMS induces I waves at 1.5 ms intervals, with their frequency and amplitude escalating as stimulus intensity increases (Lazzaro et al., 1998). This relationship facilitates the construction of a stimulus-response curve. Unlike motor threshold, MEP amplitude assesses the function of higher-threshold motor cortex neurons farther from the TMS field center. Typically expressed as a percentage of the maximum compound muscle action potential (CMAP), MEP amplitude helps evaluate both upper and lower motor neuron contributions but varies significantly between individuals, which can limit its diagnostic utility for identifying cortical and corticomotoneuronal abnormalities (Hess et al., 1987).

- CMCT measures the duration required for a neural impulse to travel through the central nervous system and activate spinal or bulbar motor neurons, serving as an indicator of corticospinal tract integrity (Udupa & Chen, 2013). Various techniques, including F-wave and cervical nerve root stimulation, are employed to estimate CMCT (Claus, 1990), although these methods primarily provide approximations of the actual conduction time (Rossini et al., 1994).

- CSP is defined as the cessation of voluntary electromyography (EMG) activity in a target muscle following transcranial magnetic stimulation of the motor cortex (Cantello et al., 1992). It begins with the motor evoked potential (MEP) and lasts until voluntary EMG activity resumes (Chen et al., 2008; Vucic et al., 2013). CSP length varies with stimulus intensity but not with the size of the MEP or background EMG activity levels (Ziemann, 2004b). This period differs in physiological properties from the MEP, indicating distinct cortical outputs (Vucic & Kiernan, 2017). The underlying mechanisms of CSP involve cortical inhibitory neurons, which likely operate through GABA$_B$ receptor-mediated inhibitory postsynaptic potentials (Connors et al., 1988; Ziemann et al., 1993).

> **Box 8.2 Motor Threshold**
> The minimum intensity of stimulation necessary to produce a motor response (e.g., a twitch in a hand muscle) in a given percentage of trials. It is often used in TMS to calibrate the required stimulation intensity for an individual.

8.3.4 Types of TMS Modes

Transcranial magnetic stimulation (TMS) is characterized by three primary stimulation modes: single-pulse TMS (spTMS), paired-pulse TMS (ppTMS), and repetitive TMS (rTMS), as illustrated in ◘ Fig. 8.6 (Zhong et al., 2021).

The traditional use of TMS typically involves the application of spTMS, where a single magnetic pulse is administered to the brain (Rossini et al., 1994). This method is primarily employed for mapping cortical functions and evaluating cortical connectivity, offering insights into the structural and functional architecture of the brain (Klomjai et al., 2015). spTMS is particularly valuable in neurological examinations to assess the integrity and functioning of specific cortical areas (Chail et al., 2018).

In ppTMS, a preliminary conditioning stimulus is followed by a test stimulus, offering valuable information about intracortical inhibitory and excitatory circuits and their influence on motor cortex output (Rossini et al., 2015). By examining the interactions between these two stimuli, researchers can gain deeper understanding of the underlying neural mechanisms that regulate cortical excitability and motor function (Ziemann et al., 1996; Khammash et al., 2019).

rTMS, which operates in both low-frequency (<1 Hz) and high-frequency (>5 Hz) variants, delivers a series of magnetic pulses at set frequencies (Rossi et al., 2000), which can modulate neuronal excitability over prolonged intervals (Hallett, 2007), proving useful in therapeutic settings for conditions like depression and motor recovery after stroke (George & Post, 2011; Lefaucheur et al., 2014; Buss et al., 2019; Chou et al., 2020; Tan et al., 2022). rTMS alters cortical activity depending on the frequency of the magnetic pulses: frequencies above 5 Hz generally enhance activity, while those below 5 Hz suppress it (Fitzgerald et al., 2006). Standard rTMS often targets the prefrontal cortex using a figure-of-eight coil for about 37.5 min per session. Theta-burst stimulation (TBS), a more recent development, achieves comparable or greater effects in just 3 min (Suppa et al., 2016).

Low-frequency rTMS tends to have an inhibitory effect on the brain, reducing cortical excitability, brain metabolism, and neuronal activity, thus potentially calming hyperactive regions (Rollnik et al., 2002). High-frequency rTMS (HF-rTMS), however, increases cortical excitability; this elevation in neural activity can lead to improved motor symptoms and other functional enhancements by increasing the excitability of stimulated nerve cells (Klomjai et al., 2015; Lefaucheur et al., 2020). The effects of rTMS are not only immediate but also enduring, maintaining changes in cortical excitability long after the stimulation session has ended

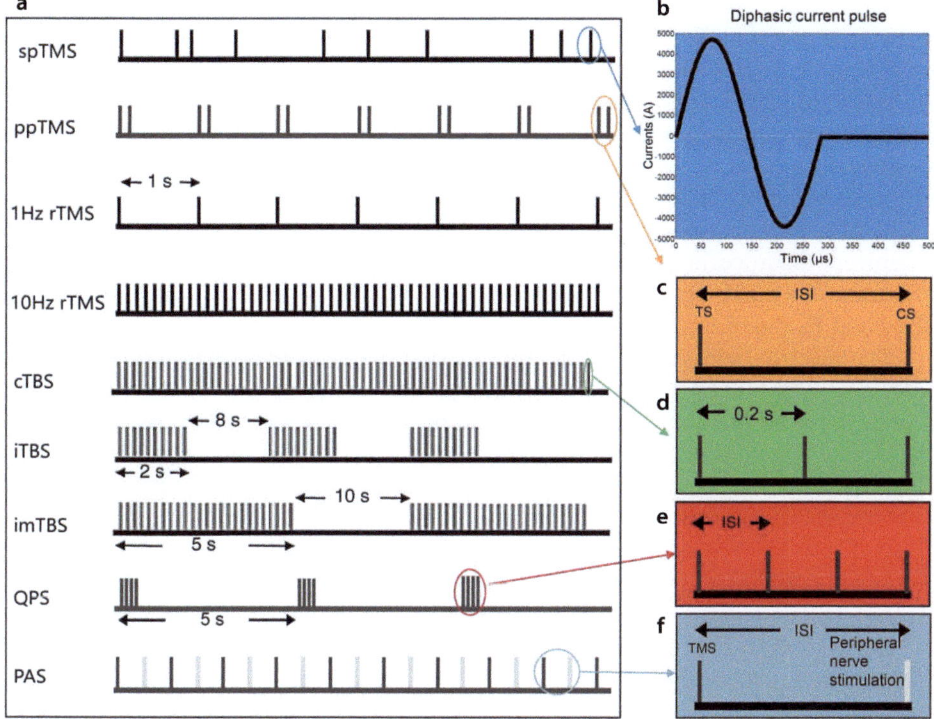

◘ **Fig. 8.6** Transcranial magnetic stimulation (TMS) is categorized into several distinct stimulation modes, each tailored for different research and clinical purposes. The diagram illustrates these various TMS modalities, showing how each mode differs in terms of pulse frequency, duration, and the inter-pulse intervals, providing a clear comparison of their operational frameworks. This differentiation helps in selecting the appropriate TMS mode based on the specific clinical or research requirements

a The primary modes of TMS include single-pulse TMS (spTMS), paired-pulse TMS (ppTMS), and repetitive TMS (rTMS). **b** Single-pulse TMS is used to probe the excitability and connectivity of cortical areas by delivering a single magnetic pulse. **c** ppTMS involves two pulses in quick succession to explore intracortical facilitation and inhibition dynamics. rTMS, on the other hand, involves the continuous application of magnetic pulses and is used for therapeutic purposes, influencing cortical plasticity and potentially altering brain function over longer periods. Within rTMS, Theta Burst Stimulation (TBS) represents a newer protocol that intensifies the effects of stimulation through a specific pattern of pulses. TBS can be divided into **d** continuous TBS (cTBS), which is inhibitory and typically reduces cortical excitability, and intermittent TBS (iTBS), which is excitatory and used to increase cortical activity. **e** Quadripulse TMS (QPS) and **f** paired associative stimulation (PAS) are two other emerging patterned modulation. From Zhong et al. (2021)

(Centonze et al., 2007), differing significantly from the transient effects observed with sTMS. Furthermore, rTMS has the capacity to influence the balance between excitatory and inhibitory neurons within the cortex, regulating neurotransmitter transmission and enhancing overall cortical function, which can result in marked clinical improvements.

Theta burst stimulation (TBS) represents an advanced form of rTMS that delivers bursts of stimulation at theta frequency. TBS is noted for its efficiency, requiring shorter periods of stimulation to induce lasting neurological changes (Suppa et al., 2016; Blumberger et al., 2018). There are two variants of TBS: intermittent TBS (iTBS), which is excitatory and can enhance cortical activity, and continuous TBS (cTBS), which is inhibitory and used to diminish cortical excitability (Blumberger et al., 2018). Both forms of TBS extend the therapeutic possibilities of rTMS, allowing for tailored treatments that adjust the cortical excitatory-inhibitory balance in patients. Quadripulse TMS (QPS) and paired associative stimulation (PAS) are two other emerging patterned modulation protocols as shown in ☐ Fig. 8.6.

The electric field generated by these techniques induces neuronal depolarization, triggering action potentials. The resultant modulation of neural activity, whether excitatory or inhibitory, fundamentally depends on various stimulation parameters, including the intensity, frequency, and pulse duration of the magnetic fields applied, enabling precise control over the therapeutic outcomes.

Box 8.3 Cortical Excitability

A term that describes the readiness of the brain cortex to respond to stimuli. NIBS techniques can alter cortical excitability, which is pivotal in studying and treating neurological disorders.

8.3.5 Applications

TMS is used both diagnostically to evaluate brain function and plasticity, and therapeutically to induce long-lasting changes in brain activity through repetitive sessions.

8.3.5.1 Use of TMS as a Diagnostic Tool

Transcranial magnetic stimulation (TMS) is valuable for studying movement disorders due to its ability to evaluate motor cortical physiology and connectivity by eliciting MEPs (Chen et al., 2008). TMS provides diagnostic insights into intra- and inter-hemispheric connectivity and corticospinal tract integrity, particularly after brain lesions. CMCT, a primary metric, measures corticospinal conduction from the primary motor cortex (M1) to a target muscle, identifying the quickest conduction time (Pellegrini et al., 2018). It tends to be prolonged in disorders such as Parkinson's disease or atypical Parkinsonism, such as multiple system atrophy or progressive supranuclear palsy. Additionally, motor thresholds and MEP metrics assess motor cortex excitability in these conditions (Chen et al., 2008; Schneider et al., 2008; Benussi et al., 2018a, 2018b).

While some studies show variable silent periods and intracortical inhibition in dystonia, others find them comparable to healthy controls due to high interindividual variability (Kang et al., 2011; Sadnicka et al., 2014). TMS metrics have also been explored in sensory areas, like the visual cortex using phosphenes to assess disease severity, and in Alzheimer's disease, where it complements traditional assessments and could parallel established biomarkers (Benussi et al., 2018a, 2018b; Padovani et al., 2018).

Thus, TMS offers reliable neurophysiological metrics for early diagnosis and monitoring of movement disorders, although it currently lacks the capability to ascertain the specific nature or cause of cortical lesions.

8.3.5.2 Use of TMS as a Therapeutic Tool

TMS is primarily used for treating neuropsychiatric disorders, with established evidence-based guidelines endorsing its efficacy in conditions like depression, pain, stroke, movement disorders, and schizophrenia (Lefaucheur et al., 2020). High-frequency (HF) rTMS targeting the left dorsolateral pre-frontal cortex (DLPFC) for depression and the M1 area contralateral to pain for analgesia are recommended with the highest level of efficacy. Other protocols like low-frequency (LF) rTMS of the right DLPFC for depression (Noda et al., 2015) and HF-rTMS for schizophrenia's negative symptoms hold probable efficacy. Additional applications for conditions like Parkinson's disease, post-stroke deficits, epilepsy, and post-traumatic stress disorder are considered possibly effective (Hasan et al., 2017; Zhou et al., 2017; Hanlon et al., 2018).

The therapeutic effects of TMS are supported by theories such as the repair model, which suggests TMS restructures brain dysfunctions, and the interactive model, which promotes natural brain adaptations to injury. Notably, large-scale studies and neuroimaging have confirmed the efficacy and safety of rTMS, particularly in depression, demonstrating changes in functional connectivity that predict treatment outcomes (Brunoni et al., 2017; Blumberger et al., 2018).

The left dorsolateral prefrontal cortex is frequently targeted in therapeutic rTMS due to its connections with limbic and subcortical structures like the caudate nucleus, globus pallidus, and thalamus. This region is crucial for executive functions such as attention, cognitive control, and set-shifting (Bonelli & Cummings, 2007; Fox et al., 2012; Panikratova et al., 2020). Stimulating this area can impact both clinical symptoms and executive abilities. ◘ Figure 8.7 (Wiseman et al., 2024) summarizes the cognitive effects of rTMS in treating disorders like major depressive disorder, obsessive-compulsive disorder, schizophrenia, Parkinson's disease, and Alzheimer's disease.

Beyond psychiatric applications, TMS is effective against neurological conditions like migraine, neuropathic pain, and post-stroke impairments. The FDA has approved specific TMS devices for treating migraine, and recent studies show potential in enhancing motor skills in stroke rehabilitation through tailored rTMS protocols. Although TMS's therapeutic reach is expanding, including applications in intraoperative neurophysiology and various neurological disorders, optimal stimulation parameters for routine clinical use are yet to be fully established.

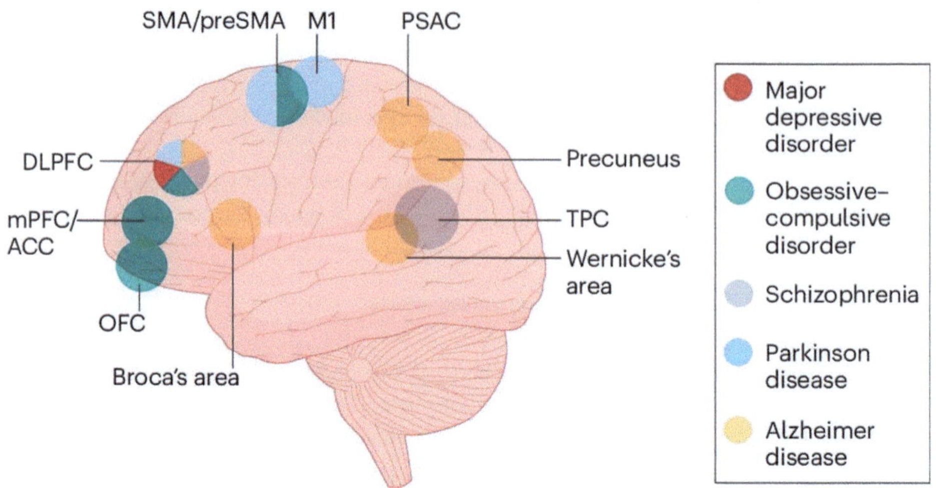

◘ Fig. 8.7 The sagittal view of the brain illustrates the specific cortical regions targeted by repetitive transcranial magnetic stimulation (rTMS) in the treatment of various neuropsychiatric disorders. Each region is color-coded according to the disorder it is associated with major depressive disorder (red), obsessive-compulsive disorder (green), schizophrenia (gray), Parkinson's disease (blue), and Alzheimer's disease (yellow)

The targeted areas include the anterior cingulate cortex (ACC), which is involved in mood regulation and cognitive processes; the dorsolateral prefrontal cortex (DLPFC), which is crucial for executive functions and is a common target in depression and schizophrenia therapies; the primary motor cortex (M1), which is frequently stimulated in movement disorders like Parkinson's disease; and the medial prefrontal cortex (mPFC) and orbital frontal cortex (OFC), both of which play roles in emotion regulation and decision-making and are targeted in OCD and depression treatments. Additional targeted regions include the presupplementary motor area (preSMA) and supplementary motor area (SMA), which are involved in motor control and cognitive planning and are stimulated in conditions like Parkinson's disease for motor symptom relief. The parietal somatosensory association cortices (PSAC) and temporoparietal cortex (TPC) are targeted in Alzheimer's disease and schizophrenia, reflecting their roles in sensory integration and cognitive processing. This detailed mapping underscores the therapeutic strategy of rTMS to modulate specific brain networks associated with each disorder, aiming to alleviate symptoms through precise magnetic stimulation of critical brain areas involved in each condition's pathology. From Wiseman et al. (2024)

8.3.6 **Latest Research Trends**

Contemporary research in the field of transcranial magnetic stimulation (TMS) is increasingly emphasizing the importance of personalizing treatment protocols to match the unique neuroanatomical and clinical characteristics of each patient. This approach involves adjusting critical parameters of TMS therapy, such as the positioning of the magnetic coil, the frequency of the magnetic pulses, and their intensity, to optimize therapeutic outcomes based on individual patient profiles (Fox et al., 2012). Moreover, there is a growing interest in integrating TMS with other therapeutic modalities, including cognitive-behavioral therapy (CBT) and pharmacological treatments, as a strategy to potentially enhance the effectiveness and sustainability of the therapeutic benefits (Donse et al., 2018).

The advancement of neuroimaging technologies, such as functional magnetic resonance imaging (fMRI) and positron emission tomography (PET), plays a pivotal role in complementing TMS treatments by providing deeper insights into the mechanisms of TMS-induced neuromodulation and identifying biological markers that could predict a patient's response to therapy (Siebner et al., 2009; Bergmann et al., 2016; Modak & Fitzgerald, 2021). Additionally, the realm of neurorehabilitation is witnessing an expanded use of TMS, especially in the context of stroke recovery and the rehabilitation of motor disorders. The potential of TMS to facilitate neuroplasticity, the brain's ability to reorganize itself by forming new neural connections, is a key focus of research in this area (Lefaucheur et al., 2014, 2020).

In summary, TMS represents a convergence of historical scientific inquiry and modern clinical neuroscience. From Faraday's foundational experiments to Barker's pioneering application and the subsequent clinical validation, TMS exemplifies the translational pathway from basic science to therapeutic innovation, offering new avenues for understanding and treating complex neurological and psychiatric conditions. Transcranial magnetic stimulation stands as a significant advancement in the repertoire of non-invasive brain stimulation techniques. Its capacity to modulate neural activity in a targeted manner, devoid of invasive procedures, renders TMS an invaluable tool for probing the complexities of brain function and for devising innovative therapeutic strategies for a spectrum of neurological and psychiatric conditions. In sum, TMS and its derivatives, rTMS and TBS, epitomize a confluence of neuroscientific innovation and clinical application, enabling non-invasive interrogation and modulation of brain function with remarkable specificity and adaptability.

Box 8.4 Objective of TMS

TMS utilizes magnetic fields to induce small electric currents in specific areas of the brain, thereby influencing neuronal activity. This technique can temporarily excite or inhibit particular brain regions depending on the parameters used.

■ **Key Mechanisms:**

Electromagnetic Induction: TMS works on the principle of electromagnetic induction. A coil placed near the scalp generates a rapidly changing magnetic field which penetrates the skull painlessly and induces currents in the brain tissue.

Neuronal Excitation and Inhibition: The induced currents can depolarize or hyperpolarize neurons, leading to excitation or inhibition of neuronal activity. The outcome depends on the frequency of stimulation—high-frequency TMS tends to excite neural activity, while low-frequency TMS can inhibit it.

Modulation of Cortical Networks: TMS not only affects the targeted cortical area but also the connected networks, which can result in changes in network activity across the brain. This is particularly important for its application in treating neurological disorders where dysfunctional network activity is a concern.

■ **Clinical Applications:**
TMS is widely used in both research and clinical settings. It helps in mapping brain function, studying the connectivity of different brain regions, and treating disorders such as major depressive disorder and migraine.

■ **Safety and Monitoring:**
TMS is considered safe when used according to established guidelines. Common side effects are mild and can include discomfort at the stimulation site or headache. However, it is contra-indicated in individuals with metal implants in the head or certain medical conditions.

■ **Future Directions:**
Ongoing research is exploring the use of TMS in cognitive enhancement, treatment of other psychiatric disorders, and recovery from brain injury, leveraging its ability to modulate cortical excitability and connectivity.

8.4 Transcranial Electrical Stimulation (tES)

Transcranial electrical stimulation (tES) is a sophisticated, non-invasive brain stimulation (NIBS) technique that has attracted widespread attention for its potential to alter cortical excitability, thereby offering promising avenues for both clinical treatments and neuroscientific research. tES is a method that involves the application of an electric field across the scalp to directly influence neuronal activity.

Depending on the current profile, as shown in ◘ Fig. 8.8 (Tavakoli & Yun, 2017), tES can be divided into (a) tDCS, transcranial direct current stimulation; (b) tACS, transcranial alternating current stimulation; (c) tRNS, transcranial random noise stimulation; (d) otDCS, oscillating tDCS. In this book you will learn about tDCS and tACS. The historical lineage of electrical stimulation of the brain reveals a progression from the use of high-intensity fields in treatments like electroconvulsive therapy (ECT) at currents of 60 mA, electro-anesthesia at 40 mA, and electro-sleep therapies employing currents between 3 and 10 mA, to the more nuanced application of lower intensities in contemporary tDCS protocols, typically between 1 and 2 mA (Priori, 2003; Nitsche et al., 2008). This evolution reflects a targeted effort to harness the therapeutic potentials of electrical brain stimulation while minimizing peripheral side effects, such as uncomfortable skin sensations and the visual phenomenon known as phosphenes.

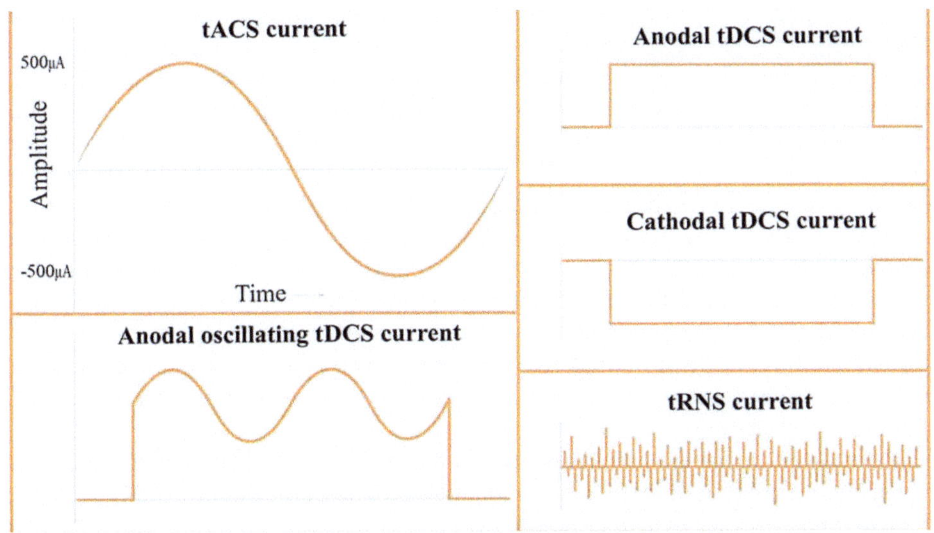

■ **Fig. 8.8** Transcranial electrical stimulation (tES) encompasses a variety of non-invasive protocols that utilize electrical currents to modulate neuronal activity within the brain. Here's a detailed description of each protocol. **Transcranial Direct Current Stimulation (tDCS)**: tDCS involves the application of a constant, low-intensity electrical current that is passed through two electrodes placed on the scalp. This method is designed to enhance or inhibit neuronal excitability, depending on the direction of the current. It influences cortical excitability and can have lasting effects on brain functions, useful in treating depression, chronic pain, and enhancing cognitive performance. **Transcranial Alternating Current Stimulation (tACS)**: Unlike tDCS, tACS applies a sinusoidal electric current which alternates at specific frequencies, aimed at entraining brain oscillations. This method is used to modulate rhythmic brain activity, potentially enhancing cognitive functions such as learning and memory, and is also being researched for its effects on sleep and mood disorders. **Transcranial Random Noise Stimulation (tRNS)**: This technique involves delivering a random electrical noise through electrodes placed on the scalp. tRNS is thought to increase cortical excitability more diffusely than tDCS by preventing the adaptation of neurons to stimulation. It has been shown to improve mathematical abilities, attention, and may benefit stroke rehabilitation. **Oscillating Transcranial Direct Current Stimulation (otDCS)**: A variant of tDCS, otDCS incorporates oscillating electrical currents at specific frequencies, combining elements of both tDCS and tACS. This approach is theorized to target and modulate the brain's natural oscillations more effectively, potentially improving the specificity and efficacy of the treatment for various neurological and psychiatric conditions. These protocols represent sophisticated tools in neuroscientific research and clinical applications, each providing unique mechanisms to influence brain function and offering potential therapeutic benefits for a range of neuropsychiatric conditions. Adapted from Tavakoli and Yun (2017) originally published under CC-BY

8.5 Transcranial Direct Current Stimulation (tDCS)

This section aims to provide an overview of tDCS, including its historical trajectory, technological nuances, clinical applications, and the forefront of ongoing research endeavors.

The operational mechanism of tDCS involves the application of a steady, unidirectional direct current across the scalp. This is achieved by positioning electrodes at specific locations on the head, whereby the current flows from an anodal (positively charged) electrode to a cathodal (negatively charged) electrode, as shown in Fig. 8.9 (Szymoniuk et al., 2023). This direct current stimulation aims to modulate neuronal membrane potentials, effectively increasing or decreasing neuronal excitability depending on the direction of the current flow, with anodal stimulation typically leading to an increase in excitability, and cathodal stimulation resulting in a decrease (Nitsche & Paulus, 2001; Stagg & Nitsche, 2011).

Recent advancements in the field have leveraged sophisticated neuroimaging and brain mapping technologies to refine electrode placement and stimulation parameters, thereby enhancing the precision and efficacy of tDCS. Clinical applications of tDCS have expanded significantly, covering a wide spectrum of conditions including, but not limited to, depression, chronic pain, stroke recovery, and cognitive enhancement in healthy individuals (Fregni & Pascual-Leone, 2007; Brunoni et al., 2011).

Moreover, ongoing research is actively exploring the integration of tDCS with other therapeutic modalities, such as cognitive–behavioral therapy and pharmacological interventions, to synergize and potentially amplify treatment outcomes. The exploration of individual variability in response to tDCS, as well as the development of personalized neuromodulation protocols, represents a critical frontier in the quest to optimize and tailor treatments to individual patient profiles (Kuo & Nitsche, 2012; Wurzman et al., 2016).

Thus, tDCS stands as a pivotal development in the field of non-invasive brain stimulation, characterized by its potential to modulate cortical excitability in a targeted and minimally invasive manner. Its application spans across therapeutic and research settings, embodying a significant contribution to the landscape of neuromodulation technologies.

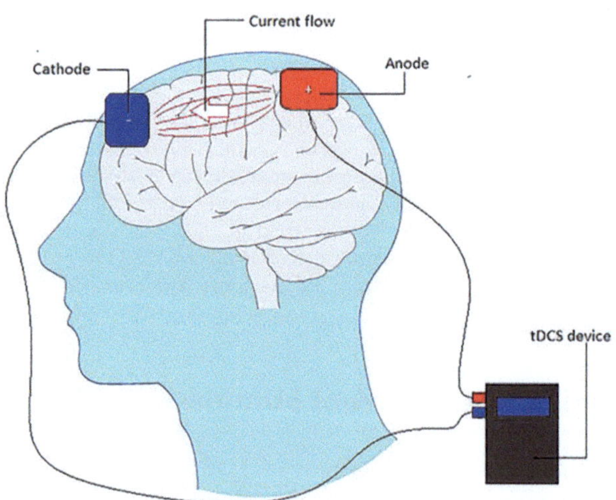

Fig. 8.9 The direct current predominantly passes through the brain, impacting the resting electrical charge located beneath the cathode, although there is some diversion through the skull. This shift in charge distribution modulates neuronal excitability and function. From Szymoniuk et al. (2023) originally published under CC-BY 4.0

8.5.1 Historical Development

The historical use of electrical phenomena for therapeutic purposes dates back to ancient civilizations, notably to ancient Rome, where electric fish was employed as a rudimentary form of electrotherapy to alleviate conditions such as headaches and pain. This early practice illustrates the longstanding human curiosity and experimentation with electrical methods for medical treatment.

Transitioning from these primal methods to the refined techniques of modern neuromodulation, the conceptual foundation for transcranial direct current stimulation (tDCS) was laid in the mid-twentieth century. Initial explorations into non-invasive electrical stimulation of the brain commenced, significantly diverging from the invasive procedures of earlier times. A pivotal moment in the preclinical development of tDCS occurred during the 1960s, with experiments demonstrating that the application of weak direct current (DC) could effectively alter neuronal activity in animal models. These findings were instrumental in establishing the theoretical and practical underpinnings of tDCS (Bindman et al., 1964).

The resurgence of tDCS research in the early twenty-first century, particularly in human subjects, is often attributed to the landmark studies by Nitsche and Paulus in 2000. Their research provided compelling evidence of tDCS's ability to modulate cortical excitability and induce lasting changes in neuroplasticity, marking a significant advancement in our understanding of the technique's therapeutic potential (Nitsche & Paulus, 2001). This period heralded the beginning of a new era in non-invasive brain stimulation, with subsequent studies exploring the efficacy and mechanisms of tDCS across a spectrum of cognitive, neurological, and psychiatric conditions.

Since these initial human trials, tDCS has emerged as a prominent research tool and a promising therapeutic intervention. Its application has been investigated extensively, demonstrating potential benefits in treating conditions such as depression, chronic pain, and stroke-induced motor deficits, as well as enhancing cognitive functions in healthy individuals (Fregni & Pascual-Leone, 2007; Brunoni et al., 2011). The evolution from ancient electrotherapy to contemporary tDCS highlights a remarkable journey of scientific and clinical innovation, leveraging the principles of electrical stimulation to modulate brain function in a precise and controlled manner.

8.5.2 Technological Aspects

Transcranial direct current stimulation (tDCS) is characterized by the administration of a low-intensity direct current through electrodes affixed to the scalp, typically not exceeding 2 milliamperes (mA). This process modulates the resting membrane potential of neurons within the targeted cerebral region, thereby altering their excitability, as shown in ◼ Fig. 8.10 (Green et al., 2020). The direction and magnitude of these changes in neuronal excitability are contingent on the polarity of the applied stimulation; anodal (positive) stimulation tends to augment

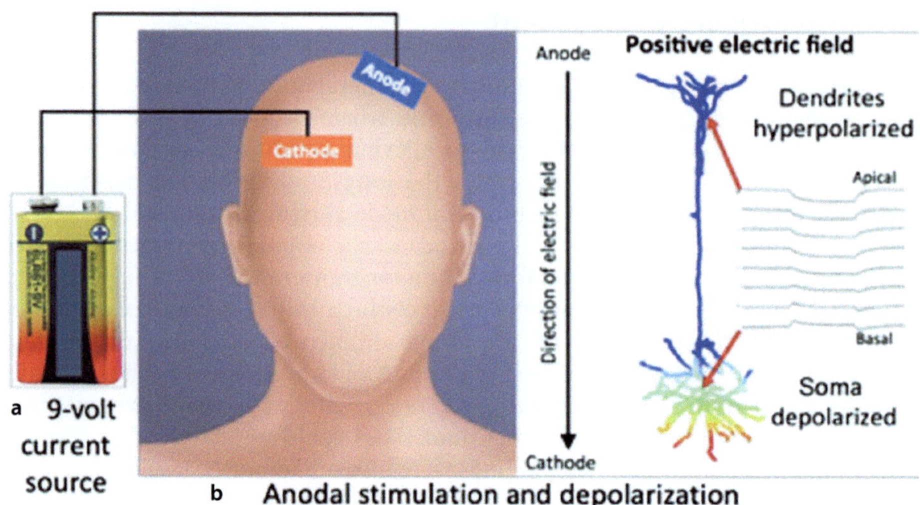

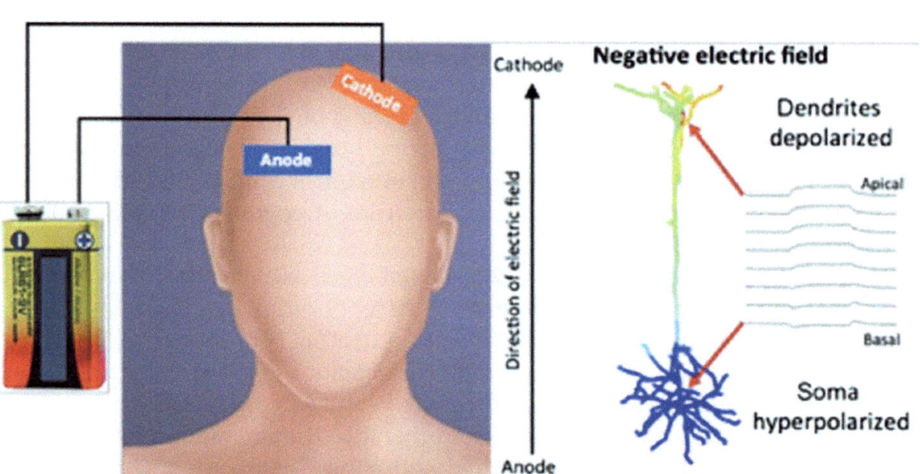

■ **Fig. 8.10** **a** The provided diagram showcases two distinct modalities of transcranial direct current stimulation (tDCS) application on the primary motor cortex (M1) and their differential effects on neuronal excitability. **b** The figure displays anodal stimulation applied to M1, with the cathodal electrode positioned over the contralateral supraorbital area. This setup facilitates the depolarization of neuronal membranes under the anodal electrode, thereby enhancing the excitability of neurons within M1. This increase in excitability is due to the positive charge of the anode, which reduces the resting membrane potential, making neurons more likely to fire. Conversely, the figure **c** illustrates a setup where the cathodal electrode is placed directly over M1, with the anodal electrode situated over the contralateral supraorbital area. In this configuration, the cathodal stimulation induces hyperpolarization of the neuronal membrane, reducing M1's excitability. The negative charge associated with the cathode increases the resting membrane potential, thus making it less likely for neurons to activate. The **right** section of the illustration depicts the impact of neuron orientation on the effectiveness of tDCS. It explains that the interaction between the electric field generated by tDCS and neuronal structures (such as the soma, dendrites, axon initial segment, and axon tree) varies depending on how the field aligns with the neuron's architectural orientation. This alignment significantly influences whether a neuron experiences net depolarization or hyperpolarization, which in turn affects neuronal excitability. For instance, an electric field oriented parallel to the length of a neuron may depolarize the axon's initial segment effectively, facilitating action potential initiation, whereas a perpendicular orientation might have less impact on excitability. This nuanced interaction underscores the importance of considering neuron morphology and circuit architecture in the effective application of TES. From Green et al. (2020)

---

neural activity, whereas cathodal (negative) stimulation usually attenuates it, as shown in ◘ Fig. 8.10.

The fundamental components of a tDCS system include a battery-operated stimulator connected to at least two electrodes positioned on the scalp, as shown in ◘ Fig. 8.11 (Rosa & Lisanby, 2012). These electrodes are designated as the anode and cathode, corresponding to their positive and negative charges, respectively. The electrical circuit is completed as current flows from the anode, traverses the skin, skull, and brain tissue, and finally arrives at the cathode. This flow induces excitatory effects under anodal stimulation and inhibitory effects under cathodal stimulation, affecting the cortical regions beneath the electrodes (Nitsche & Paulus, 2001).

During the administration of tDCS, the intensity of the current is maintained at a low level, ranging from 0.5 to 2 mA, over a session duration of approximately 10–30 min. The selection of such low-intensity currents ensures that tDCS is generally well-tolerated by participants, producing minimal adverse effects. The most commonly reported side effects include mild tingling or itching sensations at the electrode sites, underscoring the non-invasive and user-friendly nature of tDCS as a neuromodulation technique (Brunoni et al., 2011).

tDCS's mechanism of modulating neuronal excitability through such a non-invasive approach has garnered interest for its potential applications in various clinical and research contexts. The technique has been explored for its efficacy in enhancing cognitive functions, treating neuropsychiatric disorders, and facilitating rehabilitation post-stroke (Fregni & Pascual-Leone, 2007; Reato et al., 2013).

> **Box 8.5 Sham Stimulation**
> A placebo procedure used in clinical trials of brain stimulation devices. Sham stimulation mimics the physical sensation of the actual treatment without delivering the therapeutic level of brain stimulation.

8.5.3 Clinical Applications

Transcranial direct current stimulation (tDCS) has been rigorously explored as a potential therapeutic intervention for depression, demonstrating capacity for symptom alleviation, particularly through the modulation of neural activity in the dorsolateral prefrontal cortex. Review by Brunoni et al. (2012) provides substantial

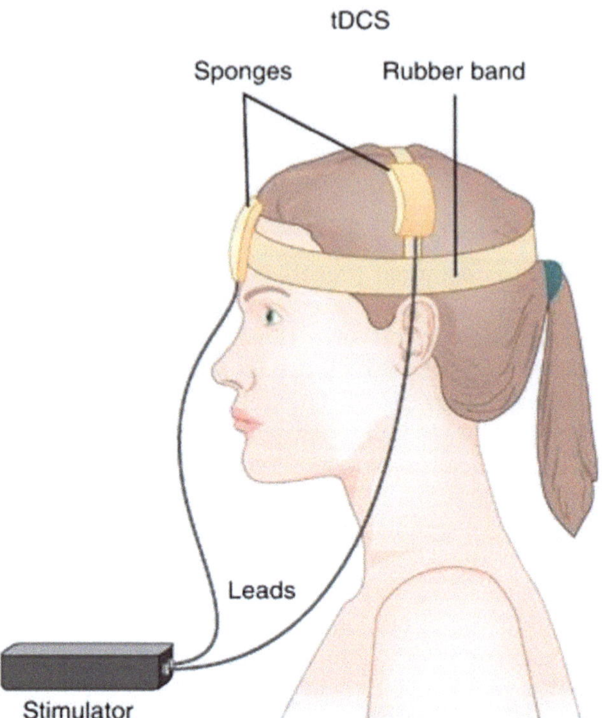

□ Fig. 8.11   The transcranial direct current stimulation (tDCS) setup involves the placement of two electrodes on the scalp to deliver a continuous, low-intensity electrical current to modulate cortical excitability. This setup includes the following components. **Electrodes**: Typically, two sponge electrodes soaked in a saline solution are used to ensure conductivity and comfort. These are positioned based on the brain regions targeted for stimulation or modulation. One electrode acts as the anode (positive electrode), which is generally placed over the targeted cortical area to enhance neuronal excitability by depolarizing the neuronal membranes. The other electrode, the cathode (negative electrode), is placed on another part of the scalp to provide a return path for the current, potentially leading to hyperpolarization and thus reduced excitability of the neurons under it. **Electrode Placement**: The specific location of the electrodes is crucial and is determined based on the neurophysiological or clinical objectives of the tDCS session. Common placements include the dorsolateral prefrontal cortex for cognitive enhancement or mood regulation and the primary motor cortex for studies involving motor function and rehabilitation.

Stimulator: A battery-powered device generates the direct current transmitted to the electrodes. The stimulator allows for the adjustment of current intensity and duration according to the protocol requirements. Typical current intensities range from 1 to 2 mA, and sessions can last from 20 to 30 min. **Current Flow**: The electrical current flows from the anode through the scalp and skull, penetrating the cortex to influence neuronal activity, and then returns to the cathode, completing the circuit. This flow influences the cortical areas beneath and between the electrodes, modulating neuronal excitability in targeted brain regions. **Safety and Comfort Measures**: The setup includes mechanisms to ensure participant comfort and safety, such as using conductive gel or saline to reduce skin resistance and irritation, and careful monitoring of the stimulation intensity and duration to avoid adverse effects like skin irritation or discomfort. The effectiveness of a tDCS session depends on these setup components working harmoniously to deliver precise, controlled electrical stimulation, tailored to achieve specific neuromodulatory outcomes. From Rosa and Lisanby (2012) evidence supporting the efficacy of tDCS in this regard, indicating its promise as an adjunct or alternative to conventional treatment modalities for depression.

Beyond its applications in mental health, tDCS has also been investigated for its therapeutic potential in addressing chronic pain syndromes, such as fibromyalgia, and enhancing outcomes in stroke rehabilitation. Fregni et al. (2006) underscored the utility of tDCS in mitigating pain associated with fibromyalgia, while Flöel (2014) highlighted its effectiveness in fostering motor function recovery and language rehabilitation post-stroke. These studies collectively point to the broad therapeutic spectrum of tDCS, encompassing both neuropsychiatric and neurorehabilitative domains.

The interest in tDCS extends into the realm of cognitive enhancement, targeting both healthy individuals and those experiencing cognitive deficits. Kuo and Nitsche (2012) have contributed to this body of research, investigating the potential of tDCS to augment various aspects of cognitive functioning. While findings in this area present a more nuanced picture, with some studies demonstrating positive outcomes and others indicating limited effects, the exploration of tDCS as a tool for cognitive enhancement continues to evolve. The variability in results underscores the complexity of brain stimulation interventions and highlights the importance of further research to delineate the parameters and conditions under which tDCS can be most beneficial.

8.5.4 Latest Research Trends

Contemporary research endeavors are deeply invested in elucidating the intricate mechanisms underpinning transcranial direct current stimulation (tDCS), with a particular focus on its role in facilitating neuroplasticity, influencing neurotransmitter systems, and modulating network connectivity within the brain (Stagg & Nitsche, 2011). Such understanding is crucial for harnessing the full therapeutic potential of tDCS. Investigations are currently directed towards refining stimulation protocols—including current density, precise electrode positioning, and optimal duration of stimulation—to both maximize the therapeutic benefits and minimize any adverse effects associated with the technique.

A significant development in this field is the movement towards personalized tDCS protocols, acknowledging the substantial variability in individual anatomical and neurophysiological characteristics. López-Alonso et al. (2014) emphasize the importance of tailoring tDCS approaches to accommodate these individual differences, which could significantly enhance the efficacy and safety of the interventions.

Moreover, the scope of tDCS application is expanding, with ongoing studies exploring its utility across a broader spectrum of neurological and psychiatric disorders. Investigations into the potential of tDCS in treating conditions such as schizophrenia, Alzheimer's disease, and substance addiction underscore the versatility of this neuromodulation technique (Brunoni et al. 2012). These efforts reflect the growing recognition of tDCS as a valuable, non-invasive tool for modulating brain activity, holding promise for diverse clinical applications.

In sum, tDCS represents a frontier of non-invasive brain stimulation with considerable promise for clinical neuroscience and neurorehabilitation. As research progresses in uncovering the mechanisms of action and optimizing the parameters of tDCS, it is poised to become an increasingly pivotal tool in the treatment and understanding of neurological and psychiatric conditions.

Box 8.6 Objective of tDCS

tDCS is a form of non-invasive brain stimulation that delivers a constant, low-intensity current to the brain area of interest via electrodes placed on the scalp. The goal is to modulate neuronal activity, either enhancing or inhibiting it, depending on the direction of the current.

■ **Key Mechanisms:**

Neuronal Modulation: tDCS influences neuronal excitability by altering the resting membrane potential of neurons. Anodal tDCS generally increases excitability by depolarizing neurons, while cathodal tDCS decreases excitability by hyperpolarizing them.

Long-lasting Effects: The effects of tDCS can extend beyond the stimulation period, influencing brain functions for minutes to hours, potentially due to changes in synaptic strength akin to long-term potentiation (LTP) and depression (LTD).

■ **Clinical Applications:**

tDCS has been explored for various applications, including enhancing cognitive functions such as learning and memory, treating mood disorders like depression, and aiding stroke recovery by enhancing neuroplasticity.

■ **Research and Efficacy:**

Studies have shown mixed results regarding the efficacy of tDCS, which can vary significantly depending on the specific parameters used (e.g., current intensity, electrode placement, and duration of stimulation).

■ **Safety and Considerations:**

tDCS is generally considered safe with mild side effects, the most common being slight itching, tingling, or discomfort at the electrode sites. However, protocols must be carefully designed to avoid skin irritation or other adverse effects.

■ **Future Directions:**

Ongoing research is aiming to optimize tDCS protocols, better understand the conditions under which it is most effective, and explore its potential in combination with other therapeutic modalities like physical therapy or cognitive training.

8.6   Transcranial Alternating Current Stimulation (tACS)

Transcranial alternating current stimulation (tACS) represents a pivotal development in the domain of non-invasive brain stimulation techniques, drawing considerable interest from the neuroscience community. This method entails the delivery of oscillatory electrical currents through the scalp to modulate the activity of cortical neurons in a rhythmic manner. The capacity of tACS to entrain brain oscillations to specific frequencies underlies its potential therapeutic and cognitive enhancement applications, including the augmentation of memory, attention, and various other cognitive processes.

The conceptual foundation of tACS traces back to early investigations into electrical brain stimulation, but it has only recently been refined and widely recognized within the neuroscience field. The technological framework of tACS involves sophisticated devices capable of generating and applying precise, oscillatory currents to targeted brain regions, aiming to synchronize neuronal firing patterns with external electrical oscillations (Antal et al., 2017; Paulus, 2011).

Clinically, tACS has been explored for its utility in treating a range of neurological and psychiatric conditions, with research probing into its efficacy in enhancing cognitive functions such as working memory, executive function, and language processing (Zaehle et al., 2010; Grover et al., 2023). The technique's non-invasiveness and specificity in modulating brain rhythms make it a promising tool for therapeutic interventions and cognitive enhancement in healthy individuals.

Recent research trends in tACS have focused on delineating the optimal stimulation parameters for various applications, understanding the underlying neural mechanisms, and exploring the technique's potential in personalized medicine. Studies have also investigated the synergistic effects of combining tACS with other non-invasive brain stimulation methods or cognitive training protocols to maximize cognitive benefits (Vossen et al., 2015; Santarnecchi et al. 2013).

8.6.1 Historical Development

Transcranial alternating current stimulation (tACS) represents an innovative advance within the realm of non-invasive brain stimulation, building upon centuries of exploration into electrical stimulation of the brain. The genesis of tACS is closely linked to the progressive refinement of transcranial electrical stimulation (tES) methodologies. This stimulation technique differentiates itself through its specific focus on modulating brain activity by applying oscillatory electrical currents, a concept that has evolved significantly alongside our deepening comprehension of the neural underpinnings of cognition.

The burgeoning interest in tACS can be traced back to the early twenty-first century, catalyzed by seminal research that illuminated the critical role of brain oscillations in mediating a wide array of cognitive functions, including but not limited to attention, memory, and perception. Pioneering studies, particularly those referenced by Paulus (2011), demonstrated that external electrical fields oscillating at frequencies congruent with those of natural brain rhythms could influence these rhythms in a manner that directly impacts cognitive processes. This insight opened up novel avenues for research, suggesting that tACS could be employed to either enhance or diminish specific cognitive functions by targeting associated brain oscillations.

The hypothesis driving tACS research posits that synchronized oscillations across neuronal assemblies are fundamental to cognitive operations. By aligning the external stimulation frequency with intrinsic brain frequencies related to specific cognitive domains, tACS offers a targeted approach to modulating the neural dynamics underlying these domains. This mechanism of action positions tACS as a promising tool for both investigating the functional significance of brain oscillations and for the therapeutic modulation of cognitive functions (Thut et al., 2011; Herrmann et al., 2013).

As the technique has evolved, subsequent research has explored a broad spectrum of applications, from enhancing learning and memory to improving motor function and mood regulation. The specificity of tACS, in terms of frequency targeting, allows for a nuanced approach to brain stimulation, providing a unique window into the oscillatory basis of cognition and behavior (Kuo & Nitsche, 2012; Reato et al., 2013).

Thus, tACS has emerged from the broader context of electrical brain stimulation as a distinct and sophisticated method for non-invasively modulating neural activity. Through the strategic application of oscillatory currents, tACS leverages our growing understanding of brain rhythms to explore and influence cognitive processes, marking a significant step forward in the field of neuromodulation.

8.6.2 **Technological Aspects**

Transcranial alternating current stimulation (tACS) is distinguished by its non-invasive approach to modulating neural activity, utilizing oscillatory electrical currents applied at specific frequencies directly to the scalp, as shown in ▪ Fig. 8.12 (Wu et al., 2021). This technique is fundamentally different from transcranial direct current stimulation (tDCS) which applies a constant direct current, thereby shifting the resting membrane potential of neurons. Instead, tACS leverages alternating current to dynamically interact with ongoing brain rhythms, aiming to entrain or synchronize the neural oscillations within the targeted cortical area to the frequency of the applied stimulation. This process of entrainment seeks to align the brain's natural oscillatory patterns—such as alpha (8–12 Hz), beta (13–30 Hz), or gamma (30–80 Hz) waves—with the externally applied current, potentially modifying the associated cognitive processes (Thut et al., 2011; Herrmann et al., 2013).

The apparatus for tACS comprises a stimulator unit that generates sinusoidal currents oscillating at a user-defined frequency, mirroring the spectrum of brain wave frequencies (e.g., delta <4 Hz, theta 4–7 Hz, alpha 8–12 Hz, beta 13–30 Hz, or gamma >30 Hz). These currents are delivered through electrodes strategically positioned over specific regions of the scalp to target underlying brain areas. The application of this current typically spans a duration of 10–30 min per session, with the precise frequency and duration of stimulation tailored to the desired neuromodulatory outcome.

Selecting the appropriate stimulation frequency allows researchers and clinicians to home in on distinct brain oscillations, facilitating a nuanced exploration of their contributions to various cognitive functions. Below, I will detail the primary types of tACS, emphasizing their mechanisms, applications, and potential cognitive and therapeutic impacts.

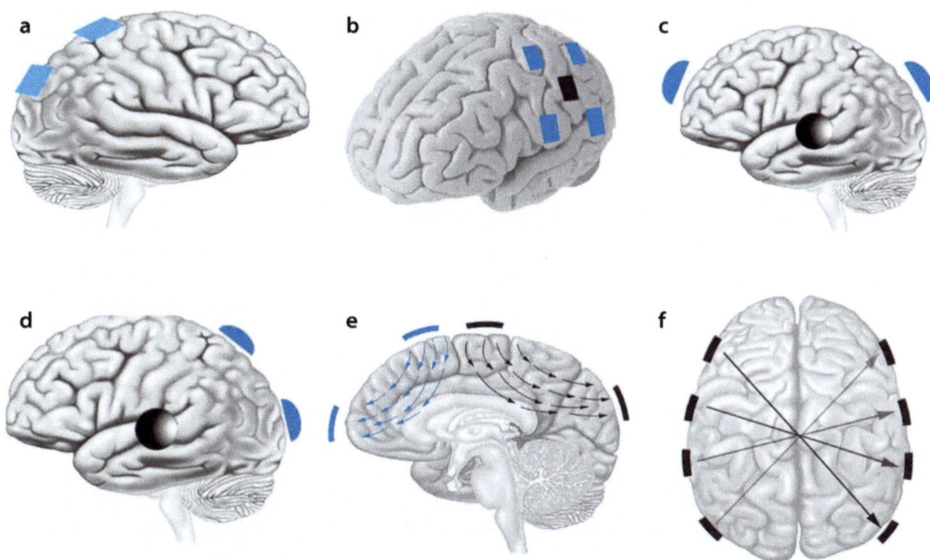

Fig. 8.12   In the implementation of transcranial alternating current stimulation (tACS), various configurations are employed to target specific neural mechanisms. **a Traditional tACS**: Here, electrode placement is strategically chosen based on the brain regions intended for stimulation. This method typically involves setting up electrodes over specific cortical areas that correspond with the desired neuromodulatory effects. **b High-Definition (HD)-tACS**: This advanced configuration uses multiple smaller electrodes (anodes) marked in blue, arranged around a central, larger electrode (cathode) shown in black. This setup allows for more focused and localized stimulation, enhancing the precision of the electric field distribution across the targeted cortical areas. **c Phase-Shifted tACS**: This variation introduces a phase difference between two blue anodes, while using a common black cathode. The phase shift between the anodes can modulate the interaction of the electrical fields, potentially leading to differential modulation of neural activity and enhancing the specificity of the desired cognitive or behavioral outcomes. **d Amplitude Modulated (AM)-tACS**: In this setup, the stimulation involves blue anodes and a common black cathode, with the key feature being variations in the stimulus amplitude over time. This modulation can influence the strength of the cortical impact, allowing researchers to study dynamic changes in brain activity in response to fluctuating electrical inputs. **e Temporally Interfering (TI) Stimulation**: This method utilizes two pairs of electrodes delivering high-frequency stimulations that independently would surpass the neural firing threshold but, when combined through interference, create a low-frequency stimulation envelope at the target site. This approach is designed to affect deeper brain structures without impacting the overlying cortex. **f Intersectional Short Pulse (ISP) Stimulation**: This technique employs pairs of electrodes that deliver brief pulses at varied timings. The differing pulse timings can create complex patterns of electrical activity, aiming to selectively stimulate particular neural populations or pathways by exploiting the temporal dynamics of neuronal excitability. Each of these tACS variants serves unique experimental and therapeutic purposes, allowing researchers and clinicians to tailor brain stimulation interventions more precisely according to the functional architecture of the brain and the specific neuropsychological objectives of the stimulation. From Wu et al. (2021) originally published under CC-BY

1. Standard tACS

Standard tACS involves the application of a sinusoidal current at a specific frequency to entrain neural oscillations within a targeted brain region. This type aims to synchronize neuronal activity with the external stimulation, potentially enhancing or inhibiting cognitive processes associated with the targeted oscillation frequency (Antal & Paulus, 2013). For instance, alpha-tACS (8–12 Hz) targets the alpha oscillations implicated in attention and relaxation, while gamma-tACS (>30 Hz) aims at gamma oscillations, associated with higher cognitive functions such as working memory and perception (Herrmann et al., 2013).

2. Amplitude-Modulated tACS (AM-tACS)

AM-tACS is a variant where the intensity of the oscillating current varies over time, allowing for the modulation of brain oscillations in a more dynamic manner. This approach can simulate the natural fluctuations in neural activity, potentially leading to more effective entrainment and cognitive modulation (Vossen et al., 2015). AM-tACS can be used to explore the effects of varying oscillatory dynamics on brain function and to investigate how changes in amplitude modulation influence cognitive outcomes.

3. Individualized tACS

Individualized tACS tailors the frequency and possibly other parameters of the stimulation to the individual's intrinsic brain oscillations, identified through baseline neurophysiological measurements such as EEG. This personalization aims to enhance the efficacy of tACS by aligning the stimulation more closely with the subject's natural neural rhythms, thereby optimizing the entrainment process (Soekadar et al., 2015). Individualized tACS is particularly promising for therapeutic applications, where the variability in brain oscillations across individuals may influence the response to stimulation (Kuo & Nitsche, 2012; Reato et al., 2013).

4. Multi-Frequency tACS

Multi-frequency tACS simultaneously applies currents at multiple frequencies, enabling the modulation of several neural oscillations at once. This approach can facilitate the investigation of interactions between different brain rhythms and their collective impact on cognition and behavior. Multi-frequency tACS has the potential to reveal complex neurodynamic processes and to offer novel therapeutic interventions targeting multiple dysregulated oscillatory patterns in neuropsychiatric disorders (Riecke et al., 2015).

5. Theta-Gamma Coupling tACS

Theta-Gamma coupling tACS specifically aims to enhance the coupling between theta and gamma oscillations, a phenomenon linked to memory processes and cognitive flexibility. By concurrently modulating theta and gamma frequencies in a coordinated manner, this type of tACS seeks to influence cognitive functions that rely on the integration of information across different brain regions and scales of neural activity (Axmacher et al., 2010).

Each type of tACS offers unique insights into the role of brain oscillations in cognitive processes and presents novel avenues for cognitive enhancement and therapeutic intervention. As research in this area progresses, further refinement of tACS methodologies and a deeper understanding of their neurophysiological effects will undoubtedly emerge, expanding the potential applications of this promising brain stimulation technique.

This targeted approach to brain stimulation, afforded by tACS, underscores its utility as a powerful tool in the investigation of the underlying neurophysiological mechanisms of cognitive functions and in the development of novel therapeutic strategies for neurological and psychiatric conditions.

8.6.3 Clinical Applications

Transcranial alternating current stimulation (tACS) has garnered considerable attention for its potential in cognitive enhancement, leveraging its ability to modulate brain oscillations through precisely targeted electrical currents. Research indicates that tACS can influence key aspects of cognitive functioning, including working memory and attention, by varying the frequency and specific location of the applied stimulation. A seminal study by Zaehle et al. (2010) demonstrated that tACS at alpha frequencies can significantly impact working memory performance, underscoring the technique's capacity to enhance cognitive processes through the modulation of brain rhythms.

Beyond cognitive enhancement, tACS is being investigated for its therapeutic potential in treating neuropsychiatric disorders characterized by aberrant brain oscillations, such as schizophrenia and depression. The work of Vossen et al. (2015) highlights the promise of tACS in this domain, suggesting that by normalizing dysregulated neural oscillations, tACS may offer a novel approach to managing these complex conditions.

Furthermore, emerging evidence supports the application of tACS in the field of neurorehabilitation, particularly for stroke recovery. Feurra et al. (2011) explored the role of tACS in promoting neural plasticity and functional reorganization, critical components of recovery in stroke patients. By facilitating the reestablishment of healthy neural oscillatory patterns, tACS may enhance the brain's intrinsic capacity for repair and adaptation following injury.

These areas of research collectively illustrate the diverse applications of tACS, from enhancing normal cognitive functions to addressing pathological neural dysregulation and aiding in neurological recovery. As our understanding of the neurophysiological mechanisms underpinning tACS continues to evolve, so does its potential as a versatile tool in cognitive neuroscience and neuropsychiatric treatment.

8.6.4 Latest Research Trends

Recent advancements in transcranial alternating current stimulation (tACS) research have been increasingly oriented towards personalizing stimulation protocols. This includes the customization of stimulation frequencies and patterns to optimize the effectiveness and response rates across diverse populations. Current research is heavily focused on uncovering the core mechanisms behind tACS, including its impact on neuroplasticity, network connectivity, and the regulation of neurotransmitter systems.

Furthermore, the integration of tACS with advanced neuroimaging techniques, such as functional magnetic resonance imaging (fMRI) and electroencephalography (EEG), represents a burgeoning research avenue. This synergistic approach enables a more detailed elucidation of the impacts of tACS on brain functionality and connectivity, providing insights into how oscillatory stimulation affects neural dynamics and cognitive processes (Antal & Paulus, 2013).

The exploration of tACS's therapeutic potential is also expanding, with a growing portfolio of clinical trials assessing its efficacy in addressing a wide array of conditions. Notably, research is exploring tACS's utility in modulating sleep patterns, alleviating mood disorders, and managing chronic pain, demonstrating the technique's versatility and potential as a therapeutic modality.

In summation, tACS emerges as a novel and non-invasive brain stimulation method with significant promise, despite its relatively nascent stage of development. Its unique capability to modulate brain oscillations in a frequency-specific manner opens new vistas for probing the neural underpinnings of cognition and for the formulation of innovative therapeutic strategies. Nonetheless, the field remains in an exploratory phase, with further research imperative to ascertain the most effective stimulation parameters, understand the long-term impacts, and unravel the mechanisms driving the observed effects of tACS. Future research endeavors, dedicated to refining stimulation protocols and deepening our grasp of tACS's neurophysiological implications, will be pivotal in fully leveraging its therapeutic potential.

Box 8.7 Objective of tACS

tACS is a non-invasive brain stimulation technique that applies a low-intensity alternating current through electrodes on the scalp. This method aims to influence brain activity by modulating the ongoing rhythmic electrical oscillations of the brain.

■ Key Mechanisms:

Oscillatory Synchronization: tACS synchronizes neuronal firing by aligning the phase of neuronal oscillations with the phase of the externally applied current, which can enhance or suppress specific brain rhythms depending on the frequency of stimulation.

Frequency-Specific Effects: The effects of tACS are largely frequency-dependent. For example, stimulating at alpha frequencies (8–12 Hz) may affect attention and relaxation, while beta frequencies (13–30 Hz) can influence alertness and motor functions.

■ Clinical Applications:

tACS has been explored for its potential to enhance cognitive functions, treat neuropsychiatric disorders, and improve motor rehabilitation. It is particularly noted for its ability to target specific neural rhythms, offering a tailored approach to brain stimulation.

- **Research and Efficacy:**
Research into tACS is relatively newer compared to other forms of brain stimulation. Initial studies have shown promising results in modulating cognitive states and enhancing performance, but further research is needed to establish its clinical efficacy and optimal stimulation parameters.

- **Safety and Considerations:**
tACS is generally well-tolerated with minimal side effects, most commonly mild sensations on the scalp such as tingling or itching during stimulation. Ensuring the correct electrode placement and stimulation settings is crucial for safety and effectiveness.

- **Future Directions:**
Future research aims to better understand how tACS affects long-term brain function and its potential as a therapeutic tool for chronic conditions. Combining tACS with neuroimaging techniques like EEG and fMRI could also enhance our understanding of its effects on brain dynamics.

Key Takeaways

1. **Non-Invasive Brain Stimulation (NIBS) Techniques Are Critical for Neuromodulation**: This chapter emphasizes the transformative potential of non-invasive brain stimulation (NIBS) techniques like transcranial magnetic stimulation (TMS), transcranial direct current stimulation (tDCS), and transcranial alternating current stimulation (tACS). These methods are integral in modulating neuronal activity, offering effective and safe therapeutic alternatives for various neuropsychiatric and neurological disorders.

2. **Mechanisms of Action and Technological Advancements**: Each NIBS technique operates through distinct mechanisms, such as electromagnetic induction (TMS), low-intensity electrical currents (tDCS, tACS), and mechanical waves (transcranial ultrasound stimulation—TUS). Recent technological advancements, including personalized stimulation protocols and integration with neuroimaging, have enhanced their precision and effectiveness in both research and clinical applications.

3. **Therapeutic Applications and Clinical Efficacy**: NIBS techniques have shown promising results in treating a range of conditions such as depression, chronic pain, and cognitive deficits following traumatic brain injuries. Techniques like TMS are FDA-approved for depression, and ongoing research is exploring broader applications, including Alzheimer's disease and stroke rehabilitation.

4. **Future Directions in Personalized Neuromodulation**: This chapter highlights the movement toward personalized brain stimulation protocols, tailoring treatments based on individual neurophysiological characteristics. This personalized approach, along with the integration of NIBS with other therapeutic modalities, is set to enhance the efficacy and application of these technologies in both cognitive enhancement and clinical treatment settings.

In conclusion, as we advance our understanding and improve the technologies associated with NIBS, we are not only expanding the frontiers of neuroscience but also enhancing our capacity to influence and heal the human brain. The integration of these techniques into routine clinical practice will undoubtedly transform the landscape of neurological and psychiatric treatment, offering new hope and improved outcomes for patients worldwide.

Conclusion

The exploration of non-invasive brain stimulation (NIBS) techniques, as detailed in this chapter, underscores their transformative potential in neuromodulation and neurotherapeutic applications. Techniques such as transcranial magnetic stimulation (TMS), transcranial direct current stimulation (tDCS), transcranial alternating current stimulation (tACS), transcranial ultrasound stimulation (TUS), and transcranial near-infrared (tNIR) stimulation represent a significant advancement in our ability to modulate brain function safely and effectively without the need for invasive procedures.

Each method discussed offers unique mechanisms by which neuronal activity can be influenced—be it through electromagnetic induction, electrical currents, mechanical waves, or light. These methods have been shown not only to affect neuronal excitability and brain oscillations but also to induce long-term plastic changes in the brain, which are pivotal for both understanding brain function and treating various neurological and psychiatric disorders.

The therapeutic potential of these techniques is vast, ranging from enhancing cognitive functions such as memory and attention to treating debilitating conditions like depression, chronic pain, and the cognitive deficits associated with traumatic brain injuries. The ongoing research and development of these technologies continue to refine their safety, efficacy, and application scope, promising to broaden their integration into clinical settings.

Moreover, the continuous innovation in NIBS is poised to revolutionize our approach to neurological and psychiatric care, making it more personalized, effective, and accessible. Future studies are expected to further elucidate the underlying mechanisms of these techniques, optimize their parameters, and expand their applications, potentially introducing new paradigms in the treatment and rehabilitation of complex brain disorders.

This section offers practical exercises designed to help apply the material learned to concrete scenarios.

Simulation of Transcranial Magnetic Stimulation (TMS)

Objective: To understand the principles of TMS and simulate its effects on neuronal excitability.

Task: Students will use simulation software to model the effect of different TMS protocols (high-frequency vs. low-frequency) on a virtual neuron model. They will observe changes in neuronal firing patterns and document their results.

Design a tDCS Experimental Protocol

Objective: To apply knowledge of tDCS in designing an experiment that investigates its effects on cognitive functions.

Task: Students will design a double-blind, placebo-controlled study to test the effectiveness of tDCS in enhancing working memory. They will specify the electrode placement, current strength, duration of stimulation, and methods for measuring working memory performance before and after stimulation.

Role-Play a Patient Consultation for tACS Therapy

Objective: To explore the clinical application of tACS and address patient concerns.

Task: In a role-playing exercise, one student acts as a neurologist explaining the benefits and risks of tACS to another student playing the role of a patient considering tACS for depression treatment.

Evaluate Safety and Ethics of NIBS

Objective: To critically assess the safety and ethical considerations of using NIBS.

Task: Students will review literature to compile a report on the known side effects of NIBS and discuss ethical concerns related to altering brain activity, especially in vulnerable populations.

Analyze Real Data from NIBS Studies

Objective: To analyze actual clinical data derived from NIBS studies.

Task: Using data sets available from public databases or collaborative research projects, students will perform statistical analysis to determine the efficacy of NIBS in treating neurological conditions, such as epilepsy or depression.

Create Educational Materials on NIBS for Patients

Objective: To develop communication skills and create patient-friendly educational materials.

Task: Students will create brochures or digital content that explains what NIBS is, how it works, its potential benefits, and risks in a format that is easily understandable by patients and the general public.

References

Abdeen, M. A., & Stuchly, M. A. (1994). Modeling of magnetic field stimulation of bent neurons. *IEEE Transactions on Biomedical Engineering, 41*(11), 1092–1095.

Amassian, V. E., Stewart, M., Quirk, G. J., & Rosenthal, J. L. (1987). Physiological basis of motor effects of a transient stimulus to cerebral cortex. *Neurosurgery, 20*(1), 74–93.

Antal, A., & Paulus, W. (2013). Transcranial alternating current stimulation (tACS). *Frontiers in Human Neuroscience, 7,* 317.

Antal, A., Alekseichuk, I., Bikson, M., Brockmöller, J., Brunoni, A. R., Chen, R., et al. (2017). Low intensity transcranial electric stimulation: Safety, ethical, legal regulatory and application guidelines. *Clinical Neurophysiology, 128*(9), 1774–1809.

Axmacher, N., Henseler, M. M., Jensen, O., Weinreich, I., Elger, C. E., & Fell, J. (2010). Cross-frequency coupling supports multi-item working memory in the human hippocampus. *Proceedings of the National Academy of Sciences, 107*(7), 3228–3233.

Barker, A. T., Jalinous, R., & Freeston, I. L. (1985). Non-invasive magnetic stimulation of human motor cortex. *The Lancet, 325*(8437), 1106–1107.

Benussi, A., Dell'Era, V., Cantoni, V., Ferrari, C., Caratozzolo, S., Rozzini, L., et al. (2018a). Discrimination of atypical parkinsonisms with transcranial magnetic stimulation. *Brain Stimulation, 11*(2), 366–373.

Benussi, A., Alberici, A., Ferrari, C., Cantoni, V., Dell'Era, V., Turrone, R., et al. (2018b). The impact of transcranial magnetic stimulation on diagnostic confidence in patients with Alzheimer disease. *Alzheimer's Research & Therapy, 10,* 1–10.

Bergmann, T. O., Karabanov, A., Hartwigsen, G., Thielscher, A., & Siebner, H. R. (2016). Combining non-invasive transcranial brain stimulation with neuroimaging and electrophysiology: Current approaches and future perspectives. *Neuroimage, 140,* 4–19.

Berman, M. H., Halper, J. P., Nichols, T. W., Jarrett, H., Lundy, A., & Huang, J. H. (2017). Photobiomodulation with near infrared light helmet in a pilot, placebo controlled clinical trial in dementia patients testing memory and cognition. *Journal of Neurology and Neuroscience, 8*(1), 176.

Bindman, L. J., Lippold, O. C. J., & Redfearn, J. (1964). The action of brief polarizing currents on the cerebral cortex of the rat (1) during current flow and (2) in the production of long-lasting after-effects. *The Journal of Physiology, 172*(3), 369.

Blumberger, D. M., Vila-Rodriguez, F., Thorpe, K. E., Feffer, K., Noda, Y., Giacobbe, P., et al. (2018). Effectiveness of theta burst versus high-frequency repetitive transcranial magnetic stimulation in patients with depression (THREE-D): A randomised non-inferiority trial. *The Lancet, 391*(10131), 1683–1692.

Bonelli, R. M., & Cummings, J. L. (2007). Frontal-subcortical circuitry and behavior. *Dialogues in Clinical Neuroscience, 9*(2), 141–151.

Boroojerdi, B., Battaglia, F., Muellbacher, W., & Cohen, L. G. (2001). Mechanisms influencing stimulus-response properties of the human corticospinal system. *Clinical Neurophysiology, 112*(5), 931–937.

Brunoni, A. R., Amadera, J., Berbel, B., Volz, M. S., Rizzerio, B. G., & Fregni, F. (2011). A systematic review on reporting and assessment of adverse effects associated with transcranial direct current stimulation. *International Journal of Neuropsychopharmacology, 14*(8), 1133–1145.

Brunoni, A. R., Nitsche, M. A., Bolognini, N., Bikson, M., Wagner, T., Merabet, L., et al. (2012). Clinical research with transcranial direct current stimulation (tDCS): Challenges and future directions. *Brain Stimulation, 5*(3), 175–195.

Brunoni, A. R., Chaimani, A., Moffa, A. H., Razza, L. B., Gattaz, W. F., Daskalakis, Z. J., & Carvalho, A. F. (2017). Repetitive transcranial magnetic stimulation for the acute treatment of major depressive episodes: A systematic review with network meta-analysis. *JAMA Psychiatry, 74*(2), 143–152.

Buss, S. S., Fried, P. J., & Pascual-Leone, A. (2019). Therapeutic noninvasive brain stimulation in Alzheimer's disease and related dementias. *Current Opinion in Neurology, 32*(2), 292–304.

Cantello, R., Gianelli, M., Civardi, C., & Mutani, R. (1992). Magnetic brain stimulation: The silent period after the motor evoked potential. *Neurology, 42*, 1951–1959.

Centonze, D., Koch, G., Versace, V., Mori, F., Rossi, S., Brusa, L., et al. (2007). Repetitive transcranial magnetic stimulation of the motor cortex ameliorates spasticity in multiple sclerosis. *Neurology, 68*(13), 1045–1050.

Chail, A., Saini, R. K., Bhat, P. S., Srivastava, K., & Chauhan, V. (2018). Transcranial magnetic stimulation: A review of its evolution and current applications. *Industrial Psychiatry Journal, 27*(2), 172–180.

Chávez-Martínez, A., Reyes-Villagrana, R. A., Rentería-Monterrubio, A. L., Sánchez-Vega, R., Tirado-Gallegos, J. M., & Bolivar-Jacobo, N. A. (2020). Low and high-intensity ultrasound in dairy products: Applications and effects on physicochemical and microbiological quality. *Foods, 9*(11), 1688.

Chen, R., Tam, A., Bütefisch, C., Corwell, B., Ziemann, U., Rothwell, J. C., & Cohen, L. G. (1998). Intracortical inhibition and facilitation in different representations of the human motor cortex. *Journal of Neurophysiology, 80*(6), 2870–2881.

Chen, R., Cros, D., Curra, A., Di Lazzaro, V., Lefaucheur, J. P., Magistris, M. R., et al. (2008). The clinical diagnostic utility of transcranial magnetic stimulation: Report of an IFCN committee. *Clinical Neurophysiology, 119*(3), 504–532.

Chou, Y. H., That, V. T., & Sundman, M. (2020). A systematic review and meta-analysis of rTMS effects on cognitive enhancement in mild cognitive impairment and Alzheimer's disease. *Neurobiology of Aging, 86*, 1–10.

Claus, D. (1990). Central motor conduction: Method and normal results. *Muscle & Nerve, 13*(12), 1125–1132.

Connors, B. W., Malenka, R. C., & Silva, L. R. (1988). Two inhibitory postsynaptic potentials, and GABAA and GABAB receptor-mediated responses in neocortex of rat and cat. *The Journal of Physiology, 406*(1), 443–468.

Day, B. L., Dressler, D., Maertens de Noordhout, A. C. D. M., Marsden, C. D., Nakashima, K., Rothwell, J. C., & Thompson, P. (1989). Electric and magnetic stimulation of human motor cortex: Surface EMG and single motor unit responses. *The Journal of Physiology, 412*(1), 449–473.

Deng, Z. D., Lisanby, S. H., & Peterchev, A. V. (2014). Coil design considerations for deep transcranial magnetic stimulation. *Clinical Neurophysiology, 125*(6), 1202–1212.

Di Lazzaro, V., Oliviero, A., Profice, P., Saturno, E., Pilato, F., Insola, A., et al. (1998). Comparison of descending volleys evoked by transcranial magnetic and electric stimulation in conscious humans. *Electroencephalography and Clinical Neurophysiology/Electromyography and Motor Control, 109*(5), 397–401.

Di Lazzaro, V., Oliviero, A., Pilato, F., Mazzone, P., Insola, A., Ranieri, F., & Tonali, P. A. (2003). Corticospinal volleys evoked by transcranial stimulation of the brain in conscious humans. *Neurological Research, 25*(2), 143–150.

Di Lazzaro, V., Ziemann, U., & Lemon, R. N. (2008). State of the art: Physiology of transcranial motor cortex stimulation. *Brain Stimulation, 1*(4), 345–362.

Di Lazzaro, V., Profice, P., Ranieri, F., Capone, F., Dileone, M., Oliviero, A., & Pilato, F. (2012). I-wave origin and modulation. *Brain Stimulation, 5*(4), 512–525.

Di Pino, G., Pellegrino, G., Assenza, G., Capone, F., Ferreri, F., Formica, D., et al. (2014). Modulation of brain plasticity in stroke: A novel model for neurorehabilitation. *Nature Reviews Neurology, 10*(10), 597–608.

Donse, L., Padberg, F., Sack, A. T., Rush, A. J., & Arns, M. (2018). Simultaneous rTMS and psychotherapy in major depressive disorder: Clinical outcomes and predictors from a large naturalistic study. *Brain Stimulation, 11*(2), 337–345.

Feurra, M., Bianco, G., Santarnecchi, E., Del Testa, M., Rossi, A., & Rossi, S. (2011). Frequency-dependent tuning of the human motor system induced by transcranial oscillatory potentials. *Journal of Neuroscience, 31*(34), 12165–12170.

Fisher, R. J., Nakamura, Y., Bestmann, S., Rothwell, J. C., & Bostock, H. (2002). Two phases of intracortical inhibition revealed by transcranial magnetic threshold tracking. *Experimental Brain Research, 143*, 240–248.

Fitzgerald, P. B., Fountain, S., & Daskalakis, Z. J. (2006). A comprehensive review of the effects of rTMS on motor cortical excitability and inhibition. *Clinical Neurophysiology, 117*(12), 2584–2596.

Flöel, A. (2014). tDCS-enhanced motor and cognitive function in neurological diseases. *Neuroimage, 85*, 934–947.

Fox, M. D., Buckner, R. L., White, M. P., Greicius, M. D., & Pascual-Leone, A. (2012). Efficacy of transcranial magnetic stimulation targets for depression is related to intrinsic functional connectivity with the subgenual cingulate. *Biological Psychiatry, 72*(7), 595–603.

Fox, M. D., Buckner, R. L., Liu, H., Chakravarty, M. M., Lozano, A. M., & Pascual-Leone, A. (2014). Resting-state networks link invasive and noninvasive brain stimulation across diverse psychiatric and neurological diseases. *Proceedings of the National Academy of Sciences, 111*(41), E4367–E4375.

Fregni, F., & Pascual-Leone, A. (2007). Technology insight: Noninvasive brain stimulation in neurology—Perspectives on the therapeutic potential of rTMS and tDCS. *Nature Clinical Practice Neurology, 3*(7), 383–393.

Fregni, F., Boggio, P. S., Lima, M. C., Ferreira, M. J., Wagner, T., Rigonatti, S. P., et al. (2006). A sham-controlled, phase II trial of transcranial direct current stimulation for the treatment of central pain in traumatic spinal cord injury. *PAIN®, 122*(1–2), 197–209.

George, M. S., & Post, R. M. (2011). Daily left prefrontal repetitive transcranial magnetic stimulation for acute treatment of medication-resistant depression. *American Journal of Psychiatry, 168*(4), 356–364.

George, M. S., Lisanby, S. H., Avery, D., McDonald, W. M., Durkalski, V., Pavlicova, M., et al. (2010). Daily left prefrontal transcranial magnetic stimulation therapy for major depressive disorder: A sham-controlled randomized trial. *Archives of General Psychiatry, 67*(5), 507–516.

Giordano, J., Bikson, M., Kappenman, E. S., Clark, V. P., Coslett, H. B., Hamblin, M. R., et al. (2017). Mechanisms and effects of transcranial direct current stimulation. *Dose-response, 15*(1), 1559325816685467.

Green, J., Jang, S., Choi, J., Jun, S. C., & Nam, C. S. (2020). Transcranial direct current stimulation (tDCS): A beginner's guide for neuroergonomists. In C. S. Nam (Ed.), *Neuroergonomics: Principles and practice* (pp. 77–101). Springer.

Grover, S., Fayzullina, R., Bullard, B. M., Levina, V., & Reinhart, R. M. (2023). A meta-analysis suggests that tACS improves cognition in healthy, aging, and psychiatric populations. *Science Translational Medicine, 15*(697), eabo2044.

Gutierrez, M. I., Poblete-Naredo, I., Mercado-Gutierrez, J. A., Toledo-Peral, C. L., Quinzaños-Fresnedo, J., Yanez-Suarez, O., & Gutierrez-Martinez, J. (2022). Devices and technology in transcranial magnetic stimulation: A systematic review. *Brain Sciences, 12*(9), 1218.

Hallett, M. (2007). Transcranial magnetic stimulation: A primer. *Neuron, 55*(2), 187–199.

Hanlon, C. A., Dowdle, L. T., & Henderson, J. S. (2018). Modulating neural circuits with transcranial magnetic stimulation: Implications for addiction treatment development. *Pharmacological Reviews, 70*(3), 661–683.

Hasan, A., Wobrock, T., Guse, B., Langguth, B., Landgrebe, M., Eichhammer, P., et al. (2017). Structural brain changes are associated with response of negative symptoms to prefrontal repetitive transcranial magnetic stimulation in patients with schizophrenia. *Molecular Psychiatry, 22*(6), 857–864.

Herrmann, C. S., Rach, S., Neuling, T., & Strüber, D. (2013). Transcranial alternating current stimulation: A review of the underlying mechanisms and modulation of cognitive processes. *Frontiers in Human Neuroscience, 7*, 279.

Hess, C. W., Mills, K. R., Murray, N. M., & Schriefer, T. N. (1987). Magnetic brain stimulation: Central motor conduction studies in multiple sclerosis. *Annals of Neurology, 22*(6), 744–752.

Hill, A. T., Fitzgerald, P. B., & Hoy, K. E. (2016). Effects of anodal transcranial direct current stimulation on working memory: A systematic review and meta-analysis of findings from healthy and neuropsychiatric populations. *Brain Stimulation, 9*(2), 197–208.

Kaneko, K., Kawai, S., Fuchigami, Y., Morita, H., & Ofuji, A. (1996). The effect of current direction induced by transcranial magnetic stimulation on the corticospinal excitability in human brain. *Electroencephalography and Clinical Neurophysiology/Electromyography and Motor Control, 101*(6), 478–482.

Kaneko, K., Fuchigami, Y., Morita, H., Ofuji, A., & Kawai, S. (1997). Effect of coil position and stimulus intensity in transcranial magnetic stimulation on human brain. *Journal of the Neurological Sciences, 147*(2), 155–159.

Kang, J. S., Terranova, C., Hilker, R., Quartarone, A., & Ziemann, U. (2011). Deficient homeostatic regulation of practice-dependent plasticity in writer's cramp. *Cerebral Cortex, 21*(5), 1203–1212.

Khammash, D., Simmonite, M., Polk, T. A., Taylor, S. F., & Meehan, S. K. (2019). Probing short-latency cortical inhibition in the visual cortex with transcranial magnetic stimulation: A reliability study. *Brain Stimulation, 12*(3), 702–704.

Klomjai, W., Katz, R., & Lackmy-Vallée, A. (2015). Basic principles of transcranial magnetic stimulation (TMS) and repetitive TMS (rTMS). *Annals of Physical and Rehabilitation Medicine, 58*(4), 208–213.

Kuo, M. F., & Nitsche, M. A. (2012). Effects of transcranial electrical stimulation on cognition. *Clinical EEG and Neuroscience, 43*(3), 192–199.

Lazzaro, V. D., Restuccia, D., Oliviero, A., Profice, P., Ferrara, L., Insola, A., et al. (1998). Magnetic transcranial stimulation at intensities below active motor threshold activates intracortical inhibitory circuits. *Experimental Brain Research, 119*, 265–268.

Lazzaro, V. D., Oliviero, A., Profice, P., Pennisi, M. A., Pilato, F., Zito, G., et al. (2003). Ketamine increases human motor cortex excitability to transcranial magnetic stimulation. *The Journal of Physiology, 547*(2), 485–496.

Lefaucheur, J. P., André-Obadia, N., Antal, A., Ayache, S. S., Baeken, C., Benninger, D. H., et al. (2014). Evidence-based guidelines on the therapeutic use of repetitive transcranial magnetic stimulation (rTMS). *Clinical Neurophysiology, 125*(11), 2150–2206.

Lefaucheur, J. P., Aleman, A., Baeken, C., Benninger, D. H., Brunelin, J., Di Lazzaro, V., et al. (2020). Evidence-based guidelines on the therapeutic use of repetitive transcranial magnetic stimulation (rTMS): An update (2014–2018). *Clinical Neurophysiology, 131*(2), 474–528.

Liebert, A., Bicknell, B., Laakso, E. L., Heller, G., Jalilitabaei, P., Tilley, S., , et al. (2021). Improvements in clinical signs of Parkinson's disease using photobiomodulation: A prospective proof-of-concept study. BMC Neurology, 21, 1-15.

López-Alonso, V., Cheeran, B., Río-Rodríguez, D., & Fernández-del-Olmo, M. (2014). Inter-individual variability in response to non-invasive brain stimulation paradigms. *Brain Stimulation, 7*(3), 372–380.

Macdonell, R. A. L., Shapiro, B. E., Chiappa, K. H., Helmers, S. L., Cros, D., Day, B. J., & Shahani, B. T. (1991). Hemispheric threshold differences for motor evoked potentials produced by magnetic coil stimulation. *Neurology, 41*(9), 1441–1441.

Makaroff, S. N., Nguyen, H., Meng, Q., Lu, H., Nummenmaa, A. R., & Deng, Z. D. (2023). Modeling transcranial magnetic stimulation coil with magnetic cores. *Journal of Neural Engineering, 20*(1), 016028.

McDonnell, M. N., & Stinear, C. M. (2017). TMS measures of motor cortex function after stroke: A meta-analysis. *Brain Stimulation, 10*(4), 721–734.

Meng, Y., Hynynen, K., & Lipsman, N. (2021). Applications of focused ultrasound in the brain: From thermoablation to drug delivery. *Nature Reviews Neurology, 17*(1), 7–22.

Modak, A., & Fitzgerald, P. B. (2021). Personalising transcranial magnetic stimulation for depression using neuroimaging: A systematic review. *The World Journal of Biological Psychiatry, 22*(9), 647–669.

Nitsche, M. A., & Paulus, W. (2001). Sustained excitability elevations induced by transcranial DC motor cortex stimulation in humans. *Neurology, 57*(10), 1899–1901.

Nitsche, M. A., Cohen, L. G., Wassermann, E. M., Priori, A., Lang, N., Antal, A., et al. (2008). Transcranial direct current stimulation: State of the art 2008. *Brain Stimulation, 1*(3), 206–223.

Nizamutdinov, D., Qi, X., Berman, M. H., Dougal, G., Dayawansa, S., Wu, E., et al. (2021). Transcranial near infrared light stimulations improve cognition in patients with dementia. *Aging and Disease, 12*(4), 954.

Noda, Y., Silverstein, W. K., Barr, M. S., Vila-Rodriguez, F., Downar, J., Rajji, T. K., et al. (2015). Neurobiological mechanisms of repetitive transcranial magnetic stimulation of the dorsolateral prefrontal cortex in depression: A systematic review. *Psychological Medicine, 45*(16), 3411–3432.

Padovani, A., Benussi, A., Cantoni, V., Dell'Era, V., Cotelli, M. S., Caratozzolo, S., et al. (2018). Diagnosis of mild cognitive impairment due to Alzheimer's disease with transcranial magnetic stimulation. *Journal of Alzheimer's Disease, 65*(1), 221–230.

Panikratova, Y. R., Vlasova, R. M., Akhutina, T. V., Korneev, A. A., Sinitsyn, V. E., & Pechenkova, E. V. (2020). Functional connectivity of the dorsolateral prefrontal cortex contributes to different components of executive functions. *International Journal of Psychophysiology, 151*, 70–79.

Patton, H. D., & Amassian, V. E. (1954). Single-and multiple-unit analysis of cortical stage of pyramidal tract activation. *Journal of Neurophysiology, 17*(4), 345–363.

Paulus, W. (2011). Transcranial electrical stimulation (tES–tDCS; tRNS, tACS) methods. *Neuropsychological Rehabilitation, 21*(5), 602–617.

Pellegrini, M., Zoghi, M., & Jaberzadeh, S. (2018). The effect of transcranial magnetic stimulation test intensity on the amplitude, variability and reliability of motor evoked potentials. *Brain Research, 1700*, 190–198.

Pitzschke, A., Lovisa, B., Seydoux, O., Zellweger, M., Pfleiderer, M., Tardy, Y., & Wagnières, G. (2015). Red and NIR light dosimetry in the human deep brain. *Physics in Medicine & Biology, 60*(7), 2921.

Polanía, R., Nitsche, M. A., & Ruff, C. C. (2018). Studying and modifying brain function with non-invasive brain stimulation. *Nature Neuroscience, 21*(2), 174–187.

Poyton, R. O., & Ball, K. A. (2011). Therapeutic photobiomodulation: Nitric oxide and a novel function of mitochondrial cytochrome c oxidase. *Discovery Medicine, 11*(57), 154–159.

Priori, A. (2003). Brain polarization in humans: A reappraisal of an old tool for prolonged non-invasive modulation of brain excitability. *Clinical Neurophysiology, 114*(4), 589–595.

Reato, D., Rahman, A., Bikson, M., & Parra, L. C. (2013). Effects of weak transcranial alternating current stimulation on brain activity—A review of known mechanisms from animal studies. *Frontiers in Human Neuroscience, 7*, 687.

Riecke, L., Formisano, E., Herrmann, C. S., & Sack, A. T. (2015). 4-Hz transcranial alternating current stimulation phase modulates hearing. *Brain Stimulation, 8*(4), 777–783.

Rollnik, J. D., Düsterhöft, A., Däuper, J., Kossev, A., Weissenborn, K., & Dengler, R. (2002). Decrease of middle cerebral artery blood flow velocity after low-frequency repetitive transcranial magnetic stimulation of the dorsolateral prefrontal cortex. *Clinical Neurophysiology, 113*(6), 951–955.

Rosa, M. A., & Lisanby, S. H. (2012). Somatic treatments for mood disorders. *Neuropsychopharmacology, 37*(1), 102–116.

Rossi, S., Pasqualetti, P., Rossini, P. M., Feige, B., Ulivelli, M., Glocker, F. X., et al. (2000). Effects of repetitive transcranial magnetic stimulation on movement-related cortical activity in humans. *Cerebral Cortex, 10*(8), 802–808.

Rossi, S., Hallett, M., Rossini, P. M., Pascual-Leone, A., & Safety of TMS Consensus Group. (2009). Safety, ethical considerations, and application guidelines for the use of transcranial magnetic stimulation in clinical practice and research. *Clinical Neurophysiology, 120*(12), 2008–2039.

Rossini, P. M., Barker, A. T., Berardelli, A., Caramia, M. D., Caruso, G., Cracco, R. Q., et al. (1994). Non-invasive electrical and magnetic stimulation of the brain, spinal cord and roots: Basic principles and procedures for routine clinical application. Report of an IFCN committee. *Electroencephalography and Clinical Neurophysiology, 91*(2), 79–92.

Rossini, P. M., Berardelli, A., Deuschl, G., et al. (1999). Applications of magnetic cortical stimulation. The International Federation of Clinical Neurophysiology. *Electroencephalography and clinical Neurophysiology Supplement, 52*, 171–185.

Rossini, P. M., Burke, D., Chen, R., Cohen, L. G., Daskalakis, Z., Di Iorio, R., et al. (2015). Non-invasive electrical and magnetic stimulation of the brain, spinal cord, roots and peripheral nerves: Basic principles and procedures for routine clinical and research application. An updated report from an IFCN Committee. *Clinical Neurophysiology, 126*(6), 1071–1107.

Roth, Y., Zangen, A., & Hallett, M. (2002). A coil design for transcranial magnetic stimulation of deep brain regions. *Journal of Clinical Neurophysiology, 19*(4), 361–370.

Rudiak, D., & Marg, E. (1994). Finding the depth of magnetic brain stimulation: A re-evaluation. *Electroencephalography and Clinical Neurophysiology/Evoked Potentials Section, 93*(5), 358–371.

Rusu, C. V., Murakami, M., Ziemann, U., & Triesch, J. (2014). A model of TMS-induced I-waves in motor cortex. *Brain Stimulation, 7*(3), 401–414.

Sadnicka, A., Hamada, M., Bhatia, K. P., Rothwell, J. C., & Edwards, M. J. (2014). A reflection on plasticity research in writing dystonia. *Movement Disorders, 29*(8), 980–987.

Sakai, K., Ugawa, Y., Terao, Y., Hanajima, R., Furubayashi, T., & Kanazawa, I. (1997). Preferential activation of different I waves by transcranial magnetic stimulation with a figure-of-eight-shaped coil. *Experimental Brain Research, 113*, 24–32.

Salehpour, F., Mahmoudi, J., Kamari, F., Sadigh-Eteghad, S., Rasta, S. H., & Hamblin, M. R. (2018). Brain photobiomodulation therapy: A narrative review. *Molecular Neurobiology, 55*, 6601–6636.

Santarnecchi, E., Polizzotto, N. R., Godone, M., Giovannelli, F., Feurra, M., Matzen, L., et al. (2013). Frequency-dependent enhancement of fluid intelligence induced by transcranial oscillatory potentials. *Current Biology, 23*(15), 1449–1453.

Schneider, S. A., Talelli, P., Cheeran, B. J., Khan, N. L., Wood, N. W., Rothwell, J. C., & Bhatia, K. P. (2008). Motor cortical physiology in patients and asymptomatic carriers of parkin gene mutations. *Movement Disorders, 23*(13), 1812–1819.

Siebner, H. R., Bergmann, T. O., Bestmann, S., Massimini, M., Johansen-Berg, H., Mochizuki, H., et al. (2009). Consensus paper: Combining transcranial stimulation with neuroimaging. *Brain Stimulation, 2*(2), 58–80.

Siebner, H. R., Funke, K., Aberra, A. S., Antal, A., Bestmann, S., Chen, R., et al. (2022). Transcranial magnetic stimulation of the brain: What is stimulated? A consensus and critical position paper. *Clinical Neurophysiology, 140*, 59–97.

Soekadar, S. R., Witkowski, M., Birbaumer, N., & Cohen, L. G. (2015). Enhancing Hebbian learning to control brain oscillatory activity. *Cerebral Cortex, 25*(9), 2409–2415.

Stagg, C. J., & Nitsche, M. A. (2011). Physiological basis of transcranial direct current stimulation. *The Neuroscientist, 17*(1), 37–53.

Suppa, A., Huang, Y. Z., Funke, K., Ridding, M. C., Cheeran, B., Di Lazzaro, V., et al. (2016). Ten years of theta burst stimulation in humans: Established knowledge, unknowns and prospects. *Brain Stimulation, 9*(3), 323–335.

Szymoniuk, M., Chin, J. H., Domagalski, Ł., Biszewski, M., Jóźwik, K., & Kamieniak, P. (2023). Brain stimulation for chronic pain management: A narrative review of analgesic mechanisms and clinical evidence. *Neurosurgical Review, 46*(1), 127.

Tan, S. W., Wu, A., Cheng, L. J., Wong, S. H., Lau, Y., & Lau, S. T. (2022). The effectiveness of transcranial stimulation in improving swallowing outcomes in adults with poststroke dysphagia: A systematic review and meta-analysis. *Dysphagia, 37*(6), 1796–1813.

Tavakoli, A. V., & Yun, K. (2017). Transcranial alternating current stimulation (tACS) mechanisms and protocols. *Frontiers in Cellular Neuroscience, 11*, 214.

Tedford, C. E., DeLapp, S., Jacques, S., & Anders, J. (2015). Quantitative analysis of transcranial and intraparenchymal light penetration in human cadaver brain tissue. *Lasers in Surgery and Medicine, 47*(4), 312–322.

Thut, G., Schyns, P. G., & Gross, J. (2011). Entrainment of perceptually relevant brain oscillations by non-invasive rhythmic stimulation of the human brain. *Frontiers in Psychology, 2*, 170.

Tufail, Y., Matyushov, A., Baldwin, N., Tauchmann, M. L., Georges, J., Yoshihiro, A., et al. (2010). Transcranial pulsed ultrasound stimulates intact brain circuits. *Neuron, 66*(5), 681–694.

Udupa, K., & Chen, R. (2013). Central motor conduction time. *Handbook of Clinical Neurology, 116*, 375–386.

Vossen, A., Gross, J., & Thut, G. (2015). Alpha power increase after transcranial alternating current stimulation at alpha frequency (α-tACS) reflects plastic changes rather than entrainment. *Brain Stimulation, 8*(3), 499–508.

Vucic, S., & Kiernan, M. C. (2017). Transcranial magnetic stimulation for the assessment of neurodegenerative disease. *Neurotherapeutics, 14*(1), 91–106.

Vucic, S., Howells, J., Trevillion, L., & Kiernan, M. C. (2006). Assessment of cortical excitability using threshold tracking techniques. *Muscle & Nerve, 33*(4), 477–486.

Vucic, S., Ziemann, U., Eisen, A., Hallett, M., & Kiernan, M. C. (2013). Transcranial magnetic stimulation and amyotrophic lateral sclerosis: Pathophysiological insights. *Journal of Neurology, Neurosurgery & Psychiatry, 84*(10), 1161–1170.

Wang, C. S. M., Chang, W. H., Yang, Y. K., & Cheng, K. S. (2023). Comparing transcranial direct current stimulation (tDCS) with other non-invasive brain stimulation (NIBS) in the treatment of Alzheimer's disease: A literature review. *Journal of Medical and Biological Engineering, 43*(4), 362–375.

Weber, M. J., Messing, S. B., Rao, H., Detre, J. A., & Thompson-Schill, S. L. (2014). Prefrontal transcranial direct current stimulation alters activation and connectivity in cortical and subcortical reward systems: A tDCS-fMRI study. *Human Brain Mapping, 35*(8), 3673–3686.

Wiseman, M., Sewell, I. J., Nestor, S. M., Giacobbe, P., Hamani, C., Lipsman, N., & Rabin, J. S. (2024). Cognitive effects of focal neuromodulation in neurological and psychiatric disorders. *Nature Reviews Psychology, 3*(4), 242–260.

Wu, L., Liu, T., & Wang, J. (2021). Improving the effect of transcranial alternating current stimulation (tACS): A systematic review. *Frontiers in Human Neuroscience, 15*, 652393.

Wurzman, R., Hamilton, R. H., Pascual-Leone, A., & Fox, M. D. (2016). An open letter concerning do-it-yourself users of transcranial direct current stimulation. *Annals of Neurology, 80*(1), 1.

Zaehle, T., Rach, S., & Herrmann, C. S. (2010). Transcranial alternating current stimulation enhances individual alpha activity in human EEG. *PLoS One, 5*(11), e13766.

Zangen, A., Roth, Y., Voller, B., & Hallett, M. (2005). Transcranial magnetic stimulation of deep brain regions: Evidence for efficacy of the H-coil. *Clinical Neurophysiology, 116*(4), 775–779.

Zhong, G., Yang, Z., & Jiang, T. (2021). Precise modulation strategies for transcranial magnetic stimulation: Advances and future directions. *Neuroscience Bulletin, 37*(12), 1718–1734.

Zhou, D. D., Wang, W., Wang, G. M., Li, D. Q., & Kuang, L. (2017). An updated meta-analysis: Short-term therapeutic effects of repeated transcranial magnetic stimulation in treating obsessive-compulsive disorder. *Journal of Affective Disorders, 215*, 187–196.

Ziemann, U. (2004a). TMS and drugs. *Clinical Neurophysiology, 115*(8), 1717–1729.

Ziemann, U. (2004b). Cortical threshold and excitability measurements. In *Handbook of clinical neurophysiology* (Vol. 4, pp. 317–335). Elsevier.

Ziemann, U., & Rothwell, J. C. (2000). I-waves in motor cortex. *Journal of Clinical Neurophysiology, 17*(4), 397–405.

Ziemann, U., Netz, J., Szelényi, A., & Hömberg, V. (1993). Spinal and supraspinal mechanisms contribute to the silent period in the contracting soleus muscle after transcranial magnetic stimulation of human motor cortex. *Neuroscience Letters, 156*(1–2), 167–171.

Ziemann, U. L. F., Rothwell, J. C., & Ridding, M. (1996). Interaction between intracortical inhibition and facilitation in human motor cortex. *The Journal of Physiology, 496*(3), 873–881.

Brain Stimulation Theory and Techniques

Comparative Analysis and Applications

Contents

U. Chaudhary, *Expanding Senses using Neurotechnology*, https://doi.org/10.1007/978-3-031-76081-5_9

Test your learning and check your understanding of this book's contents: use the "Springer Nature Flashcards" app to access questions using ▶ https://sn.pub/kmb-jyz. To use the app, please follow the instructions in ▶ Chap. 1.

This chapter provides a comparative overview of various noninvasive brain stimulation (NIBS) techniques, including Transcranial Magnetic Stimulation (TMS), Transcranial Direct Current Stimulation (tDCS), Transcranial Alternating Current Stimulation (tACS), and Deep Brain Stimulation (DBS). It explores these methods' distinct mechanisms, applications, and comparative efficacy in clinical and research settings. This chapter explores key theories like Network Modulation and Neuroplasticity, which form the basis for how these technologies affect neural activity and deliver therapeutic benefits. It also highlights the essential steps in designing and conducting brain stimulation studies, focusing on choosing the right stimulation parameters and accounting for individual differences in brain structure and function. Applications of brain stimulation across various neurological and psychiatric disorders are examined, highlighting its potential for enhancing cognitive functions and treating refractory conditions. This chapter concludes by considering future directions in brain stimulation research, including the integration of advanced imaging techniques and real-time feedback systems to enhance the specificity and efficacy of these interventions. This synthesis of current knowledge and emerging trends underscores the growing significance of brain stimulation in neuroscience and medicine, offering promising avenues for both understanding and treating complex brain disorders.

Learning Objectives

1. Comparative Analysis of Different Brain Stimulation Techniques: Gain a comparative understanding of various non-invasive brain stimulation (NIBS) methods including Transcranial Magnetic Stimulation (TMS), Transcranial Direct Current Stimulation (tDCS), Transcranial Alternating Current Stimulation (tACS), and the more invasive Deep Brain Stimulation (DBS).
2. Explore Mechanisms of Action: Learn about the distinct mechanisms through which each brain stimulation technique influences neuronal activity and induces therapeutic effects, focusing on their principles of operation such as electromagnetic induction and electrical current modulation.
3. Evaluate Ethical and Safety Considerations: Consider the ethical implications and safety concerns associated with the clinical use of brain stimulation techniques, particularly addressing issues of informed consent, privacy, equitable access, and the implications of cognitive enhancement.

4. Understand the Impact of Individual Variability on Treatment Outcomes: Recognize the role of individual anatomical and physiological differences in influencing the effectiveness of brain stimulation, underscoring the importance of personalized treatment approaches.

9.1 Comparison of Different Brain Stimulation Methods

Following an in-depth exploration of various brain stimulation techniques in ▶ Chaps. 7 and 8, it becomes essential to juxtapose these methods to elucidate their distinct characteristics, advantages, limitations, and clinical implications. In this comparative analysis, we concentrate on four principal brain stimulation modalities: Transcranial Magnetic Stimulation (TMS), Transcranial Direct Current Stimulation (tDCS), Transcranial Alternating Current Stimulation (tACS), and Deep Brain Stimulation (DBS). Each technique employs a unique approach to modulating brain activity, catering to different research and therapeutic needs. A comprehensive summary of these methods is encapsulated in ◘ Table 9.1, delineating their comparative aspects.

1. Transcranial Magnetic Stimulation (TMS)

   TMS utilizes electromagnetic induction to generate focal, transient magnetic fields, inducing electrical currents that modulate neuronal activity within targeted brain regions. TMS can be applied in single pulses (spTMS) for mapping brain function or in repetitive sessions (rTMS) for therapeutic purposes. Its noninvasive nature and high spatial resolution make TMS a valuable tool for both clinical and research settings, particularly in treating depression and elucidating cortical function and connectivity (Rossi et al., 2009; Lefaucheur et al., 2020).

2. Transcranial Direct Current Stimulation (tDCS)

   tDCS delivers a constant, low-intensity electrical current through the scalp to alter the resting membrane potential of neurons, thereby modulating their excitability. tDCS is celebrated for its simplicity, safety, and portability, offering potential benefits in cognitive enhancement and rehabilitation from neurological disorders such as stroke. However, its effects are generally more diffuse than TMS, and the optimal parameters for various applications are still under investigation (Nitsche et al., 2008).

3. Transcranial Alternating Current Stimulation (tACS)

   tACS applies oscillatory electrical currents to entrain brain oscillations in frequency-specific manners. This technique aims to modulate cognitive processes and brain rhythms associated with various functions, such as memory and attention. tACS's capacity to target specific neural oscillations presents novel research opportunities. However, its clinical efficacy and the mechanisms underlying its effects require further elucidation (Antal & Paulus, 2013).

■ **Table 9.1** Comparative analysis of different brain stimulation techniques

Feature/technique	TMS (Transcranial Magnetic Stimulation)	tDCS (Transcranial Direct Current Stimulation)	tACS (Transcranial Alternating Current Stimulation)	DBS (Deep Brain Stimulation)
Principle of operation	Uses magnetic fields to induce electric currents in the brain, stimulating neurons	Applies a constant, low-intensity direct current to stimulate neurons	Applies alternating current to modulate neuronal activity	Surgically implanted electrodes deliver electrical impulses to specific brain areas
Invasiveness	Noninvasive	Noninvasive	Noninvasive	Invasive (requires surgery)
Primary clinical applications	Depression, migraine, stroke rehabilitation, psychiatric disorders.	Depression, chronic pain, cognitive enhancement	Cognitive enhancement, mood disorders, sleep regulation	Parkinson's disease, essential tremor, dystonia, obsessive-compulsive disorder
Spatial resolution	High	Low	Moderate	Very high
Advantages	Noninvasive, targeted stimulation, adjustable parameters	Noninvasive, easy to administer, safe for prolonged use	Noninvasive, allows for frequency-specific modulation	Direct and precise targeting, adjustable stimulation, effective for drug-resistant conditions
Limitations	Temporary effects, requires repeated sessions, may cause discomfort	Subtle effects, dependent on electrode placement and individual variability	Limited research on long-term effects, optimal parameters not fully established	Surgical risks, potential for hardware-related complications, requires battery replacement
Typical duration of effects	Short-term, effects can last minutes to hours post-stimulation	Short to medium-term, effects can last for hours after stimulation	Short to medium-term, similar to tDCS in duration	Long term, with ongoing stimulation
Customization and adjustability	High (intensity, frequency, and location can be adjusted)	Moderate (intensity and location can be adjusted)	Moderate (frequency and intensity can be adjusted)	High (continuous adjustment of stimulation parameters post-surgery)

4. Deep Brain Stimulation (DBS)

DBS involves the surgical implantation of electrodes into specific brain areas, delivering continuous electrical impulses to modulate deep brain structures. Predominantly used for treating movement disorders like Parkinson's disease, DBS offers significant symptomatic relief when other treatments are ineffective. Despite its invasiveness, DBS's precise targeting and adjustable parameters provide a robust approach to managing severe neurological conditions (Wiseman et al., 2024).

In conclusion, each brain stimulation method for disrupting brain activity has unique strengths and limitations. TMS, tDCS, and tACS are noninvasive techniques, while DBS involves an invasive surgical procedure. The choice of technique depends on factors such as the targeted brain region, the desired effects, and the specific condition being treated. Overall, a comprehensive understanding of these brain stimulation methods can help clinicians and researchers select the most appropriate approach for their specific objectives. ◘ Figure 9.1 (Nasr et al., 2022) compares different NIBS techniques based on their spatial specificity, mechanistic specificity, and robustness. ◘ Figure 9.2 provides an overview of temporal and spatial resolution comparison of different NIBS techniques (Polanía et al., 2018).

Thus, these brain stimulation techniques offer diverse approaches to modulating neural activity, each with its distinct set of advantages, limitations, and clinical applications. As our understanding of the brain's complexities deepens, these methods will continue to evolve, enhancing their efficacy and expanding their therapeutic potentials. Now we will explore brain stimulation theory using the brain stimulation techniques we have learned so far.

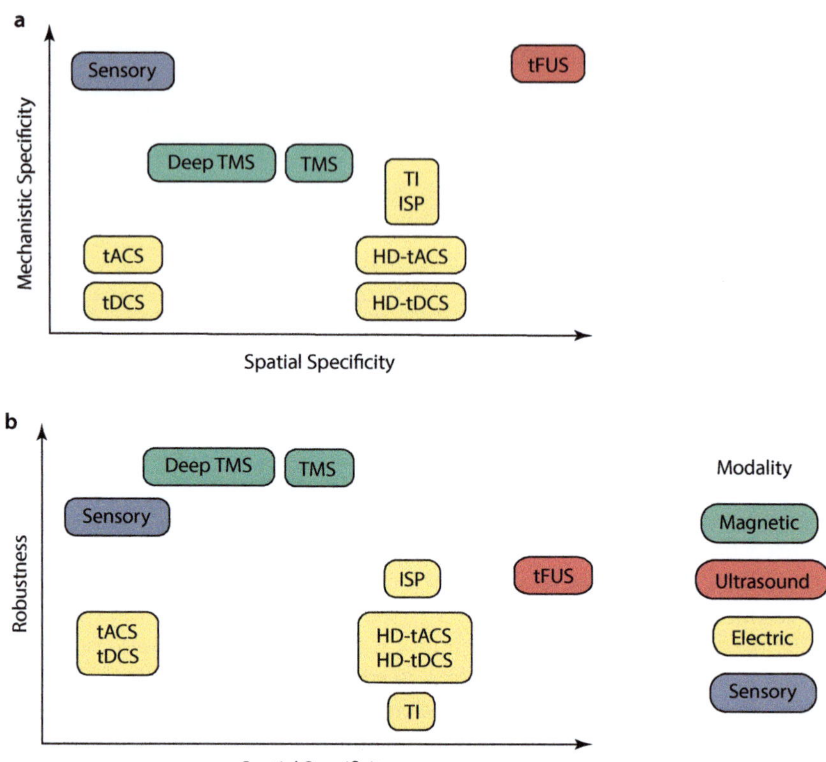

Fig. 9.1 Comparison of noninvasive brain stimulation methods—Noninvasive brain stimulation (NIBS) methods encompass several crucial aspects: spatial specificity, mechanistic specificity, and robustness. These figures provide an expanded analysis of these characteristics across different NIBS techniques. **a, b Spatial Specificity**: This parameter refers to the ability of a stimulation method to target precise brain regions. Transcranial magnetic stimulation (TMS) offers relatively high spatial specificity, typically influencing brain areas under the coil with focused magnetic fields, particularly when using figure-eight coils, which concentrate the field to a small cortical area. In contrast, transcranial direct current stimulation (tDCS) has lower spatial specificity due to the broader spread of electric fields under the electrode patches, which can affect larger and more diffuse brain regions. Transcranial alternating current stimulation (tACS) and transcranial random noise stimulation (tRNS) share a similar spatial profile to tDCS. Transcranial ultrasound stimulation (TUS) stands out with high spatial specificity, as it can target deep brain tissues with millimeter precision. **a Mechanistic Specificity**: This aspect evaluates how a method can modulate neuronal mechanisms specifically. TMS can selectively influence neuronal excitability and connectivity by inducing electric currents directly in neural tissue, which makes it highly mechanistically specific. tDCS modulates neuronal excitability through alterations in membrane potentials, offering a different level of mechanistic engagement, which broadly affects neuronal firing rates and patterns. tACS, interestingly, allows for the modulation of cortical oscillations by matching the frequencies of brain waves, providing a unique form of mechanistic specificity by entraining neural rhythms. **b Robustness**: Robustness refers to the consistency and reliability of the effects produced by each method. TMS is considered robust in terms of producing repeatable and measurable changes in brain activity and connectivity, as evidenced by its widespread use in clinical and research settings. tDCS and tACS provide less robust results which can vary significantly between individuals due to factors like skull thickness and inter-individual anatomical differences. Although highly specific spatially, TUS is still under investigation for its robust-

ness across different settings and populations. In summary, each NIBS method presents a unique profile of effectiveness based on spatial and mechanistic specificity and robustness, which should be carefully considered when choosing the appropriate technique for research or therapeutic purposes. This analysis not only aids in understanding each method's comparative strengths and limitations but also guides their application in clinical and experimental neuroscience. From Nasr et al. (2022)

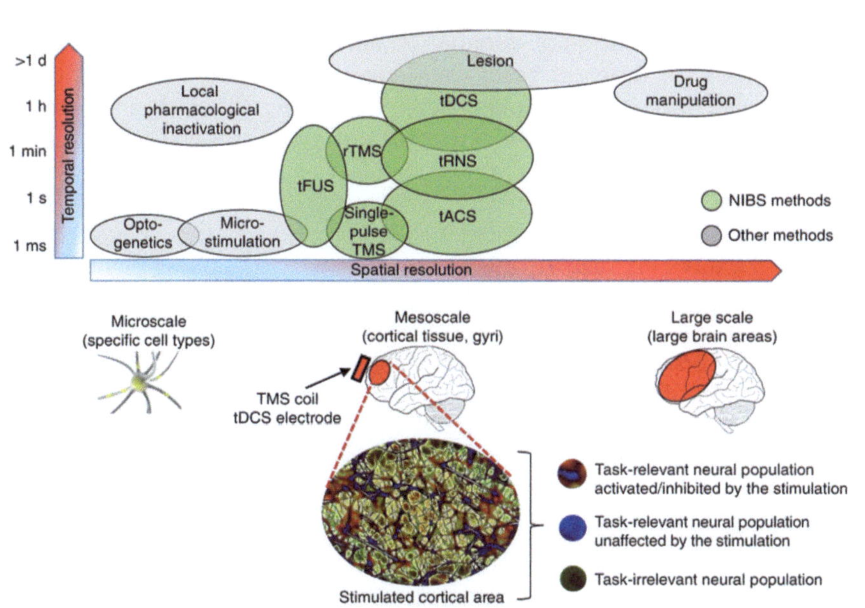

Fig. 9.2 NIBS exhibits distinct temporal and spatial resolution characteristics, contingent upon the specific methodologies employed. NIBS techniques, such as transcranial magnetic stimulation (TMS) and transcranial direct current stimulation (tDCS), typically operate at the mesoscale level. The temporal resolution of these techniques can vary significantly. For example, TMS can achieve high temporal resolution, while tDCS generally has lower temporal precision. The spatial resolution of NIBS methods is inherently limited by their mode of action. These techniques generally activate or inhibit relatively large populations of neurons, making it challenging to target specific neural circuits with high precision. The activation is often non-specific, affecting a broad area of the cortex. Despite this, NIBS studies frequently report apparent specificity in their outcomes. This perceived spatial resolution does not directly correlate with the precise anatomical targeting of neurons but rather with the broader disruption of neural activity across extensive cortical regions. Similarly, the temporal resolution of NIBS is influenced by the protocol used. High-frequency repetitive TMS (rTMS), for instance, can induce rapid changes in cortical excitability, thus offering relatively high temporal resolution. In contrast, tDCS induces more gradual changes in cortical excitability, leading to lower temporal resolution. The specificity observed in temporal dynamics during NIBS interventions often reflects the modulation of behaviorally relevant neural processes rather than precise temporal activation patterns. The specificity seen in NIBS studies, both spatial and temporal, is often a result of the disruption of behaviorally relevant operations performed by small, functionally interconnected groups of neurons within larger brain regions. These disruptions can manifest as changes in cognitive or motor functions, which researchers may misinterpret as evidence of precise anatomical or temporal targeting. However, the underlying mechanisms typically involve broader, more diffuse modulation of neural activity. In summary, while NIBS methods can provide valuable insights into brain function and hold therapeutic potential, their spatial and temporal resolutions are limited. The apparent specificity in NIBS outcomes more likely reflects the disruption of neural processes within larger brain networks rather than precise targeting of specific neural circuits. From Polanía et al. (2018)

9.2   Brain Stimulation Theory: Propagation Targeting

This section explores the fundamental principles underlying the transmission of electrical and magnetic fields through brain tissue, focusing on how these mechanisms influence the dispersion of stimulation throughout the brain. Critical factors such as tissue conductivity and electrode placement play pivotal roles in determining the distribution of stimulation energy within the cerebral landscape. The conductivity of different brain tissues (e.g., gray matter, white matter, cerebrospinal fluid) varies significantly, thereby affecting the path and strength of electrical currents or magnetic fields as they navigate through the brain (Wagner et al. 2004).

Furthermore, the precise positioning of electrodes or coils is paramount in targeting specific brain areas. The geometry and location of these elements directly impact the focality and intensity of the stimulation, necessitating meticulous planning and execution to achieve desired outcomes (Miranda et al., 2013).

The endeavor to accurately target specific brain regions is met with challenges, notably due to the complex anatomy of the brain and the variability in individual brain structures. Advances in neuroimaging and computational neuroscience have been instrumental in overcoming some of these obstacles. Computational models have emerged as powerful tools in the field, offering predictions of stimulation effects and guiding the optimization of stimulation parameters. These models incorporate detailed anatomical and physiological data, enabling the simulation of electrical or magnetic fields within the brain and their interaction with neural tissue (Dmochowski et al., 2011).

Such computational approaches are increasingly vital in the development of tailored and precise brain stimulation protocols. By simulating various scenarios, these models allow for the refinement of electrode placement and stimulation settings, aiming to maximize therapeutic effects while minimizing unwanted side effects. This strategy holds promise for enhancing brain stimulation interventions' specificity, efficacy, and safety, paving the way for more personalized treatment modalities (Ruffini et al. 2014).

In summary, the propagation of stimulation energy within the brain is governed by complex interactions between the applied fields and the heterogeneous tissue environment. Advances in computational modeling and neuroimaging are key to overcoming challenges in targeting and optimizing brain stimulation, driving progress toward more effective and personalized treatments.

9.2.1   Brain Stimulation Theory

Brain stimulation theory encompasses a range of principles and hypotheses underpinning the utilization of diverse methodologies to modulate neuronal activity within the brain. This body of theory is foundational to both academic research and clinical practices aimed at altering brain function to influence cognitive processes and behavior. At the core of this theory lies the premise that targeted interventions on neuronal activity can lead to significant modifications in brain function, which, in turn, have the potential to affect cognition and behavior in measurable ways (Dayan et al., 2013).

This exploration into brain stimulation theory will cover the foundational rationale for employing various brain stimulation techniques, elucidate the mechanisms through which these interventions exert their effects on neural activity, and examine the factors determining the effectiveness and outcomes of such stimulation endeavors.

Different brain stimulation techniques, including Transcranial Magnetic Stimulation (TMS), Transcranial Direct Current Stimulation (tDCS), Transcranial Alternating Current Stimulation (tACS), and Deep Brain Stimulation (DBS), are predicated on distinct physical principles and are designed to interact with neural circuits in specific ways. For instance, TMS utilizes magnetic fields to induce electrical currents in the brain, temporarily modulating neuronal excitability in targeted regions (Rossini et al., 2015). In contrast, tDCS applies a constant electrical current to modulate the resting membrane potential of neurons, thereby affecting neuronal activity over a broader area (Nitsche et al., 2008).

The mechanisms by which these techniques influence brain function is multifaceted, involving direct effects on neuronal excitability, synaptic plasticity, and network connectivity. For example, the application of TMS or tDCS can enhance or suppress cortical excitability in the stimulated area, leading to changes in local and network-level brain activity patterns. This modulation of brain activity can facilitate synaptic plasticity, the neural basis for learning and memory, thereby enabling potential therapeutic effects and cognitive enhancements (Ruffini et al. 2014).

Moreover, the outcome of brain stimulation is influenced by a variety of factors, including the specific parameters of the stimulation (e.g., intensity, frequency, duration), the targeted brain region, and individual differences in brain anatomy and physiology. Computational models and neuroimaging techniques are increasingly used to predict brain stimulation's effects and optimize stimulation protocols for individual patients, aiming to maximize therapeutic benefits while minimizing adverse effects (Miranda et al., 2013).

In summary, brain stimulation theory provides a conceptual framework for understanding and harnessing the potential of various brain stimulation techniques to modulate brain activity. By elucidating the mechanisms of action and optimizing stimulation parameters based on individual neuroanatomy and physiology, researchers and clinicians strive to improve the efficacy and safety of these interventions for therapeutic and research applications.

9.2.2 Rationale of Brain Stimulation

The foundational premise of brain stimulation emerges from the intricate understanding that cognitive functions and behaviors stem from the dynamic interplay among various brain regions and neural circuits. This complex neural orchestra's harmonious interactions between different components are crucial for maintaining normal brain function and behavioral outputs. However, in a multitude of neurological and psychiatric disorders, these essential interactions become perturbed, manifesting in altered patterns of brain activity and resultant functional impairments (Bassett & Sporns, 2017). Such disruptions can lead to a wide range of symptomatic expressions, including cognitive deficits, mood disturbances, and motor dysfunction, profoundly affecting an individual's quality of life.

Brain stimulation techniques, encompassing both invasive and noninvasive modalities, are predicated on the capability to modulate neuronal activity within specific brain areas or circuits selectively. These interventions are designed not merely to influence localized neural activity but to recalibrate the broader network dynamics, aiming to rectify the dysregulated patterns of brain function characteristic of various pathologies. By precisely targeting the neural substrates underlying dysfunctional processes, brain stimulation holds the potential to restore more normalized activity patterns, thereby ameliorating symptoms and enhancing functional outcomes (Dayan et al., 2013; Polanía et al., 2018).

For instance, Deep Brain Stimulation (DBS), a surgical intervention involving the implantation of electrodes in specific brain regions, has been demonstrated to offer significant symptom relief in conditions like Parkinson's disease and obsessive-compulsive disorder by modulating the activity of neural circuits implicated in these disorders (Lozano et al., 2019). Similarly, noninvasive approaches such as Transcranial Magnetic Stimulation (TMS) and Transcranial Direct Current Stimulation (tDCS) have shown promise in the treatment of depression and cognitive rehabilitation post-stroke, respectively, by targeting relevant cortical areas to induce beneficial changes in neural activity and connectivity (Brunoni et al., 2012; Lefaucheur et al., 2020).

These brain stimulation interventions aim to enhance the patient's quality of life by restoring functional integrity and reducing symptomatology. The continuous advancement in our understanding of brain network dynamics and the development of more sophisticated stimulation techniques are pivotal in refining these therapeutic strategies, making them more effective and tailored to individual patient needs.

9.2.3 Mechanisms of Action

The mechanisms through which various brain stimulation techniques exert their effects are intricately tied to the modulation of neuronal excitability. This modulation typically involves altering the membrane potentials of neurons, leading to either an upregulation or downregulation of neural activity. The specific outcome—whether an increase or decrease in activity—hinges on a variety of factors, including the stimulation parameters (such as intensity and frequency) and the particular brain region being targeted.

Two pivotal models stand out among the theoretical frameworks that elucidate the principles behind brain stimulation: Network Modulation Theory and Neuroplasticity Theory.

1. Network Modulation Theory

    Network Modulation Theory posits that the effects of brain stimulation extend beyond the immediate site of application, influencing a broad network of interconnected brain regions. This theory is premised on the understanding that cognitive functions and behaviors emerge from the complex interactions within these networks. Brain stimulation can induce widespread modulatory effects by targeting specific nodes or connections within these networks, potentially restoring disrupted neural communication and enhancing cognitive and motor functions. ◘ Figure 9.3 (To et al. 2018) depicts the brain networks as a graph. This perspective is supported by neuroimaging studies that demonstrate how localized stimulation can lead to functional and structural changes across distant but interconnected brain areas (Bassett & Sporns, 2017).

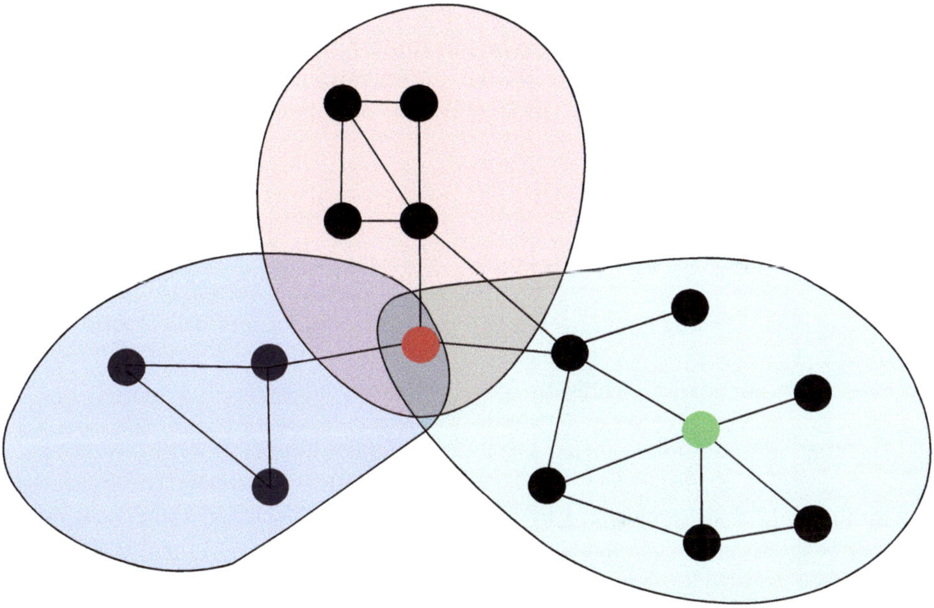

◘ **Fig. 9.3** Brain networks are intricately structured in a manner analogous to graphs, where nodes (represented as black dots) and edges (illustrated as black lines interlinking the nodes) form the fundamental components. Within this graphical framework, certain nodes exhibit more intense and frequent interactions among themselves than with nodes outside their group. These densely interconnected nodes are recognized as modules, visually distinguished in diagrams as colored communities within the broader network. In this modular architecture, provincial hubs play a critical role. These hubs are nodes that boast a high degree of connectivity, primarily linking nodes within the same module. For example, a node depicted in green within a module might serve as a provincial hub, signifying its central role in fostering intra-module communication and maintaining the coherence of that module's functional activities. In contrast, connector hubs are pivotal for their role in bridging disparate modules. These nodes are characterized by their extensive connections that span multiple modules, facilitating inter-module communication and integration. Such nodes are crucial for the overall connectivity of the brain network, allowing for complex interactions across different functional areas. A node-colored red might be an example of a connector hub, indicating its broader connectivity profile that extends beyond its immediate module, linking diverse and functionally distinct areas of the brain network. Understanding the roles and characteristics of provincial and connector hubs enhances our comprehension of the brain's functional topology. This conceptualization not only aids in mapping the structural and functional landscape of the brain but also in understanding how information is processed and integrated across various cognitive and neurological functions. From To et al. (2018) originally published under CC-BY

2. Neuroplasticity Theory

Neuroplasticity Theory focuses on the capacity of brain stimulation to induce lasting changes in the brain's structure and function. According to this theory, appropriately timed and targeted stimulation can strengthen or weaken synaptic connections, a process known as synaptic plasticity. This synaptic modification is thought to underlie learning, memory, and recovery from brain injury. Brain stimulation techniques, by facilitating synaptic plasticity, can promote adaptive changes in neural circuits, potentially leading to improvements in various neurological and psychiatric conditions (Fritsch et al., 2010).

Both theories underscore the potential of brain stimulation as a tool for probing the neural basis of cognition and behavior and as a therapeutic intervention. By elucidating the mechanisms underpinning brain stimulation, these theoretical models contribute to the optimization of stimulation protocols, guiding the development of more effective and personalized approaches to treatment.

Box 9.1 Objective of Propagation Targeting

Propagation targeting in brain stimulation is the practice of optimizing the delivery of stimulation energy to specific brain areas. It involves understanding the pathways through which electrical and magnetic fields disperse within brain tissue, influenced by factors such as tissue conductivity and electrode placement.

■ Key Concepts:

Tissue Conductivity: Different brain tissues (e.g., gray matter, white matter, and cerebrospinal fluid) have varying levels of conductivity, which affects how stimulation energy moves through the brain. Effective targeting requires knowledge of these properties to predict and control the spread of stimulation.

Electrode Placement: The precise location and geometry of stimulation electrodes or coils are crucial for focusing the energy on desired brain regions while minimizing effects on non-targeted areas. Accurate placement enhances the focality and intensity of the stimulation, which is vital for both safety and efficacy.

■ Challenges in Effective Targeting:

Complex Brain Anatomy: The intricate structure of the human brain poses significant challenges for targeting specific brain areas. Variability in brain anatomy across individuals further complicates this task.

Advanced Techniques for Precision: Utilization of neuroimaging tools like MRI and computational models is essential for mapping brain structures and simulating the effects of stimulation, thereby improving targeting accuracy.

■ Applications and Implications:

Enhanced Efficacy and Safety: Improved targeting can lead to more effective stimulation protocols by directly influencing the neural circuits involved in specific disorders, thus enhancing therapeutic outcomes and reducing side effects.

Future Directions: Ongoing advancements in imaging and computational neuroscience continue to refine our ability to target and stimulate brain regions more precisely, promising even greater improvements in brain stimulation therapies.

9.2.4 Fundamentals of Network Modulation Theory

The Network Modulation Theory within the domain of brain stimulation marks a significant evolution from a localized perspective to a systemic conceptualization of brain function and its modulation via stimulation techniques. This theory asserts that both invasive and noninvasive forms of brain stimulation impact not merely the directly targeted neurons but also the broader, interconnected neural networks.

9.2.4.1 Conceptual Basis

Traditional approaches to brain stimulation predominantly focused on the immediate effects at the stimulation site. Network Modulation Theory, however, proposes that the brain functions as an integrated network, where stimulation of one node can influence distant regions within the network, underscoring a shift towards understanding brain function through the lens of distributed processing and network dynamics (Sporns, 2011).

9.2.4.2 Mechanisms of Network Modulation

At the heart of this theory is the notion that altering neural activity in a specific area can propagate through synaptic connections, engendering a series of functional modifications across the network. For instance, Deep Brain Stimulation (DBS) in Parkinson's disease not only targets the subthalamic nucleus but also impacts the basal ganglia-thalamocortical circuit, leading to symptomatic improvement (McIntyre et al., 2004).

9.2.4.3 Evidence from Neuroimaging

Technological advancements in neuroimaging, such as functional magnetic resonance imaging (fMRI) and positron emission tomography (PET), have been pivotal in substantiating the effects of network modulation. These techniques have identified changes in regions far from the stimulation site, signaling network-level impacts (Lozano & Lipsman, 2013).

9.2.4.4 Applications in Clinical Practice

Neurological Disorders: DBS's application in conditions like Parkinson's disease leverages network modulation to counterbalance pathological network dynamics, offering symptom relief (Deuschl et al., 2006).

Psychiatric Disorders: Network theories guide the use of Transcranial Magnetic Stimulation (TMS) in depression, targeting crucial nodes within mood-regulating networks (Fox et al., 2012).

9.2.4.5 Challenges and Future Directions

1. Mapping Individual Networks: Variability in neural network architectures across individuals poses a challenge, highlighting the need for personalized mapping to refine stimulation protocols (Bassett & Sporns, 2017).
2. Understanding Long-term Effects: The enduring impact of network modulation on neural plasticity and network organization remains an area of active investigation (Reithler et al., 2011).
3. Technological Advances: Progress in neurostimulation technologies and computational models is anticipated to enhance the precision and adaptability of network modulation techniques, potentially enriching therapeutic outcomes.
4. Network Modulation Theory has profoundly enriched our comprehension of brain stimulation's systemic effects, transitioning the focus from localized impacts to the intricacies of network-wide dynamics. This theoretical framework fosters novel therapeutic strategies for a spectrum of neurological and psychiatric conditions, steering the field towards more tailored and efficacious stimulation interventions.

9

Box 9.2   Objective of Network Modulation Theory

This theory explores how brain stimulation extends beyond localized effects, influencing broader neural networks that underpin cognitive functions and behaviors. It asserts that stimulating one node within these networks can impact distant interconnected regions, potentially normalizing or enhancing network function.

tially correcting dysfunctional interactions.

Strategic Node Targeting: Successful brain stimulation involves identifying and targeting strategic nodes—regions that significantly influence overall network behavior. This requires a deep understanding of the functional connectivity within the brain. Please look at Box ▶ 9.3 for the definition of functional connectivity.

■ **Key Concepts:**

Connectivity and Interaction: Cognitive behaviors emerge from the complex interplay within neural networks. Targeting specific nodes with brain stimulation can influence the entire network, altering its dynamics and poten-

■ **Clinical Applications:**

Broad Therapeutic Impact: Network Modulation Theory is critical for designing treatments for neurological and psychiatric disorders where network dysfunctions are prevalent, such as in depression or Parkinson's disease.

By modulating key network nodes, clinicians can achieve significant improvements in symptom management.

■ Challenges and Considerations:

Mapping Network Dynamics: One of the main challenges is accurately mapping the complex web of neural connections and understanding how they contribute to specific disorders.

Advances in neuroimaging and computational modeling are crucial in addressing these challenges.

Individual Variability: The effectiveness of targeting network nodes varies between individuals due to differences in brain network architecture. Personalized approaches, possibly informed by detailed neuroimaging data, are essential for optimizing outcomes.

Box 9.3   Functional Connectivity

Definition: Functional connectivity refers to the statistical dependence between distinct brain regions or neural elements that share functional properties. It is measured through the correlation of physiological signals, such as neuronal activity or hemodynamic responses, across different parts of the brain during resting state or while engaged in a task.

■ Key Aspects:

Measurement Tools: Functional connectivity is commonly assessed using neuroimaging techniques such as functional Magnetic Resonance Imaging (fMRI) or Electroencephalography (EEG), which detect synchronous activity patterns between brain areas. It has also been used with other brain signal recording techniques outlined in this book.

Applications: Understanding functional connectivity helps in mapping the brain's network organization and is crucial for studying brain functions, diagnosing neurological and psychiatric conditions, and developing targeted treatments.

Interpretation: High functional connectivity between two regions indicates that they are likely to be involved in similar or related processes or that one region may influence the activity of the other.

9.2.5   Fundamental Concepts of Neuroplasticity

Neuroplasticity Theory, within the context of brain stimulation, elucidates the brain's intrinsic ability to undergo structural and functional reorganization by forming new neural connections across the lifespan. This adaptability is pivotal for understanding how brain stimulation therapies can effectuate lasting changes in brain function, particularly in reaction to injury, disease, or therapeutic interventions.

9.2.5.1 Definition and Scope

Neuroplasticity encompasses the nervous system's resilience and adaptability, allowing for the reconfiguration of its structure, functionality, and connections in response to varying stimuli. This concept extends to both synaptic plasticity, which involves the modulation of synaptic connection strength, and structural plasticity, which pertains to alterations in neural pathways and synapses (Pascual-Leone et al., 2005).

9.2.5.2 Mechanisms of Plasticity

The brain's plastic changes are propelled by mechanisms such as neurogenesis (the birth of new neurons), synaptic strengthening or weakening (through long-term potentiation or depression, respectively), and alterations in dendritic structure (branching and pruning). These processes are modulated by environmental exposures, learning experiences, and instances of brain injury (Cramer et al., 2011). ◘ Figure 9.4 (Huang et al., 2017) depicts how NIBS affects synaptic plasticity.

9.2.5.3 Plasticity and Learning

Learning and memory epitomize neuroplasticity, with repeated neural pathway activation reinforcing these pathways, underpinning the mechanisms of learning and skill development (Kleim & Jones, 2008).

9.2.5.4 Neuroplasticity in Brain Stimulation

Stimulation-Induced Plasticity: Techniques such as TMS and tDCS can instigate targeted plastic changes, enhancing or mitigating neural activity in specific regions. These modifications are central to the therapeutic efficacy of brain stimulation in treating depression, aiding stroke rehabilitation, and managing chronic pain (Fregni & Pascual-Leone, 2007).

Homeostatic Plasticity: This refers to the brain's capacity to preserve neural equilibrium amidst stimulation-induced changes, ensuring that excitatory and inhibitory network balances are maintained to avert maladaptive outcomes, such as epilepsy (Turrigiano, 2012).

9.2.5.5 Clinical Applications

Rehabilitation After Stroke: Post-stroke rehabilitation leverages neuroplasticity, employing brain stimulation to foster neural network reorganization and the restitution of motor and cognitive functions (Hummel & Cohen, 2006).

Treatment of Disorders: In conditions like major depressive disorder, neuroplastic changes induced by TMS are posited to mitigate symptoms by reconfiguring dysfunctional neural circuits (George et al., 2013).

9.2.5.6 Challenges and Future Directions

1. Individual Variability: Personal differences in neuroplastic responses necessitate tailored brain stimulation protocols, incorporating genetic, neuroimaging, and behavioral data (Cash et al., 2021).

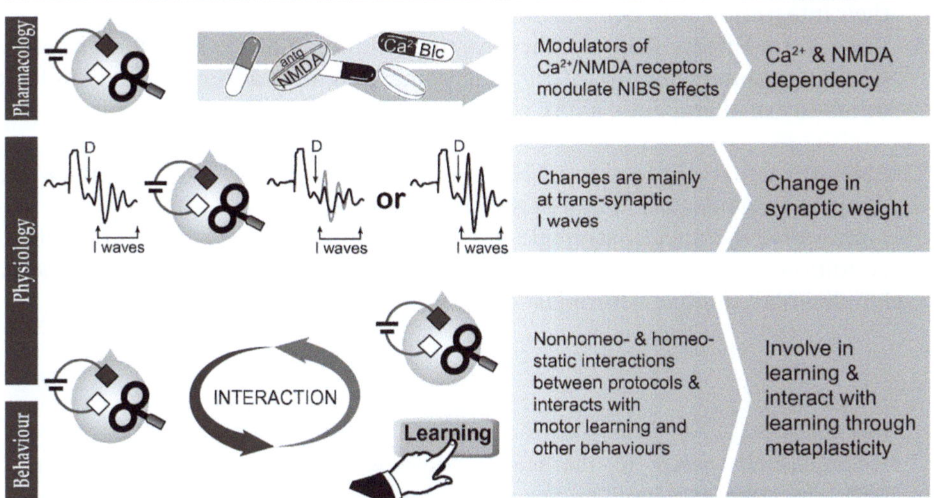

☐ **Fig. 9.4** Three primary lines of evidence robustly support the hypothesis that Noninvasive Brain Stimulation (NIBS) exerts its therapeutic and modulatory effects through mechanisms closely associated with synaptic plasticity. **Pharmacological Influence on NIBS:** The effectiveness of NIBS is significantly influenced by pharmacological agents that modify the function of key receptors and channels pivotal for synaptic plasticity. For instance, drugs that interact with calcium channels and NMDA receptors have been found to alter the outcomes of NIBS. This interaction highlights the essential role of these molecular components in mediating the synaptic plasticity effects induced by NIBS. Such findings suggest that NIBS may facilitate synaptic changes through the modulation of intracellular calcium levels and NMDA receptor activity, which are crucial for long-term potentiation (LTP) and long-term depression (LTD), foundational processes of synaptic plasticity. **Impact on I-Waves in Epidural Recordings:** In the context of Transcranial Magnetic Stimulation (TMS), a form of NIBS, epidural recordings of descending neural volleys have shown that NIBS predominantly affects I-waves rather than D-waves. I-waves are believed to be generated through trans-synaptic activation of cortical neurons, whereas D-waves are thought to result from direct activation of axons. NIBS's selective modulation of I-waves implies that its effects are mediated through trans-synaptic mechanisms rather than through direct neuronal activation. This distinction underscores the role of synaptic networks and their plasticity in the functional impact of NIBS. **Synergy with Learning and Practice:** The efficacy of NIBS is not only impacted by its standalone application but is also significantly enhanced when combined with motor practice and cognitive learning activities. This synergistic effect suggests that NIBS may optimize the brain's plastic response to learning and practicing new skills. Such findings indicate that NIBS likely facilitates improvements in motor and cognitive functions by engaging and perhaps reinforcing the neural circuits involved in learning and memory, thereby promoting synaptic plasticity. This aspect of NIBS is particularly compelling, as it points to its potential use in rehabilitative and educational settings to enhance the learning outcomes and recovery processes. These lines of evidence collectively affirm the role of synaptic plasticity as a fundamental mechanism through which NIBS achieves its effects, offering valuable insights into how NIBS could be optimized for therapeutic and cognitive enhancement applications. From Huang et al. (2017)

2. Optimizing Stimulation Protocols: Research continues to define the most effective stimulation parameters and to investigate novel stimulation modalities and their integration with behavioral therapies (Dayan et al., 2013).
3. Long-Term Effects: A comprehensive understanding of the long-term implications of induced neuroplasticity, including potential adverse effects and the durability of therapeutic gains, remains crucial.

The Neuroplasticity Theory furnishes a foundational framework for tapping into the brain's reorganizational and adaptive capabilities through stimulation. It heralds innovative therapeutic strategies for a spectrum of neurological and psychiatric conditions, underscoring the necessity of personalized treatments tailored to each individual's unique neuroplastic potential.

Box 9.4 Objective of Neuroplasticity

Neuroplasticity refers to the brain's ability to reorganize itself by forming new neural connections throughout life. This adaptability is essential for learning, memory, and recovery from brain injury, and is a key mechanism through which brain stimulation therapies exert their effects.

■ **Key Aspects:**

Synaptic Plasticity: This involves changes in the strength of synapses, which are the connections between neurons. Enhancements (long-term potentiation) and reductions (long-term depression) in synaptic strength are fundamental for learning and adapting to new information or environments.

Structural Plasticity: It refers to the brain's ability to actually change its physical structure through growth and reorganization of dendrites, axons, and synapses. This is often a response to learning or injury.

■ **Clinical Relevance:**

Rehabilitation: Neuroplasticity is the foundation of many rehabilitation strategies following neurological damage, such as after a stroke, where retraining the brain can lead to regaining function.

Therapeutic Interventions: Understanding neuroplasticity allows for the development of targeted therapies in psychiatric and neurological disorders, aiming to harness the brain's plastic properties to bring about beneficial changes.

■ **Challenges and Considerations:**

Critical Periods: While neuroplasticity occurs throughout life, there are critical periods in development when the brain is particularly receptive to change. Outside these periods, inducing plasticity can require more effort or different strategies.

Individual Differences: Variability in neuroplastic potential among individuals can affect the outcomes of interventions aimed at inducing plasticity, necessitating personalized approaches.

- **Future Directions:**

Enhancing Plasticity: Research is ongoing into methods to enhance neuroplasticity via pharmacological agents, lifestyle changes, and technological interventions such as brain stimulation, to improve outcomes in therapy and rehabilitation.

Monitoring Changes: Advanced imaging techniques are increasingly used to visualize changes in brain connectivity and structure in real-time, offering deeper insights into how neuroplastic changes unfold.

9.2.6 Factors Influencing Brain Stimulation Outcomes

The efficacy of brain stimulation is contingent upon a multitude of factors that need to be meticulously considered to achieve desired therapeutic or cognitive outcomes. Some of the critical elements are:

1. **Stimulation Parameters**

   The selection of stimulation parameters—including intensity, frequency, duration, and waveform—is paramount in influencing the outcomes of brain stimulation interventions. These parameters must be strategically chosen, taking into account the target brain region and the specific effects intended. This precision is crucial as these parameters directly influence the neural response, with the potential to either upregulate or downregulate neural activity in the targeted area. ▣ Figure 9.5 (He et al., 2020) shows noninvasive brain stimulation protocols commonly used. Research demonstrates that variations in these parameters can lead to significantly different outcomes. For instance, in the case of rTMS, frequencies above 5 Hz (high-frequency) tend to increase cortical excitability. In contrast, frequencies at or below 1 Hz (low-frequency) are associated with a decrease in excitability, highlighting the importance of parameter selection in achieving the desired therapeutic effect (Rossi et al., 2009).

2. **Targeted Brain Region**

   The specificity of targeting the correct brain region is paramount, as it determines the neural circuits impacted by the stimulation. An in-depth understanding of the neuroanatomy and functional connectivity pertinent to the disorder or cognitive process under investigation guides the identification of target regions. This is exemplified in the application of TMS for depression, where targeting the dorsolateral prefrontal cortex is based on its role in mood regulation and its connectivity with other mood-related brain regions (Fox et al., 2012).

3. **Individual Variability**

   Individual responses to brain stimulation can vary significantly and be influenced by unique anatomical, physiological, and genetic factors. This variability underscores the need for a personalized approach to brain stimulation, where parameters are adjusted to the individual's specific neurobiological profile.

Commonly used NIBS Protocols

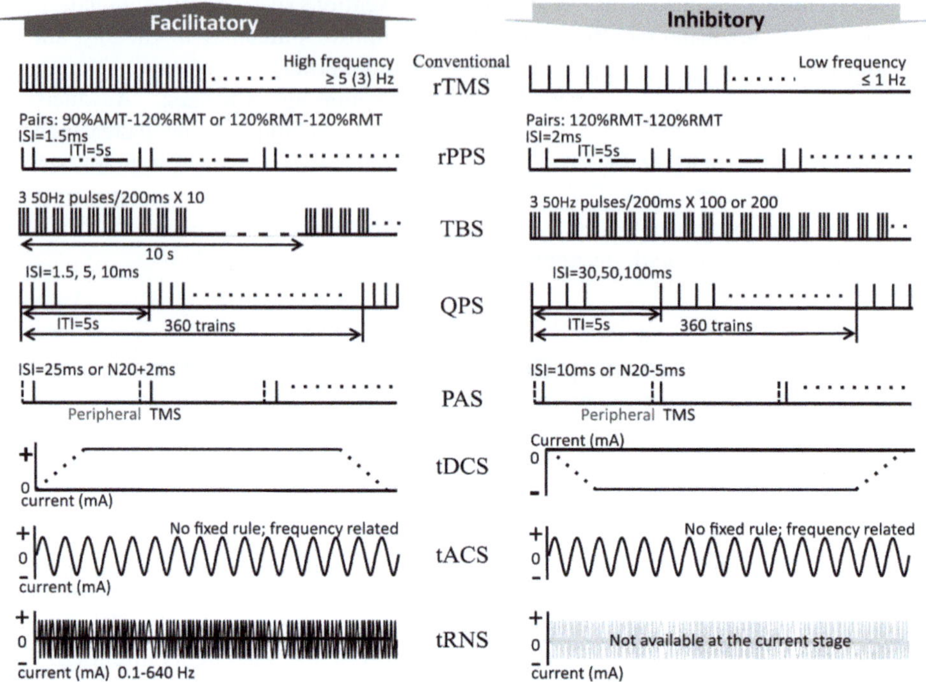

■ Fig. 9.5 The illustration provides an overview of commonly used noninvasive brain stimulation protocols, distinguishing between facilitatory and inhibitory methods. On the left panel, facilitatory protocols are listed, which are primarily designed to enhance cortical excitability and are used to stimulate neural activity. These include various Transcranial Magnetic Stimulation (TMS) protocols such as Repetitive TMS **(rTMS)**: Delivers sequences of pulses at different frequencies. Theta Burst Stimulation **(TBS)**: A form of rTMS that delivers bursts of stimulation at theta frequency. Pair Associative Stimulation **(PAS)**: Combines peripheral nerve stimulation with TMS. Quadri-pulse Stimulation **(QPS)**: Involves four closely spaced pulses to potentiate neuronal circuits. Additionally, the panel includes traces from Transcranial Electrical Stimulation (TES) protocols: transcranial direct current stimulation **tDCS** (anodal and cathodal): Uses direct current to either increase (anodal) or decrease (cathodal) neuronal excitability. The vertical axis represents the direction of current flow, with anodal stimulation generally depicted as upward for facilitatory effects and cathodal as downward for inhibitory effects. Transcranial Alternating Current Stimulation **(tACS)**: Applies alternating current at specific frequencies to either synchronize or desynchronize neural oscillations, depending on the frequency used. Transcranial Random Noise Stimulation **(tRNS)**: Delivers random frequency currents that can increase cortical excitability through stochastic resonance effects. Each protocol depicted in the diagrams is accompanied by a graphical trace representing the methodological application, either enhancing or suppressing neuronal function, which is indicated along the vertical axis showing the direction of current or stimulus intensity. This clear categorization helps in understanding the operational mechanisms and potential applications of each brain stimulation protocol, from therapeutic interventions to cognitive enhancement strategies. From He et al. (2020)

Techniques such as neuro navigated brain stimulation, which integrates MRI images for precise targeting, and the consideration of genetic polymorphisms affecting neural plasticity are examples of how individual variability is being addressed in clinical practice (Cheeran et al., 2008; Ridding & Ziemann, 2010).

4. **State-Dependency**

   The brain's state at the time of stimulation—whether at rest or engaged in a task—can significantly affect brain stimulation outcomes. This is due to the state-dependent nature of neural excitability and plasticity. For example, tDCS-induced modulation of cortical excitability has been shown to be more pronounced when participants are engaged in tasks that activate the targeted brain regions, emphasizing the need to consider the cognitive state of individuals during stimulation (Miniussi et al., 2013).

5. **Interactions with Pharmacological Interventions**

   The concurrent use of pharmacological treatments can modify the effects of brain stimulation, necessitating a careful consideration of drug-stimulation interactions. Some medications may enhance the effects of stimulation by increasing neural plasticity, while others might reduce its efficacy. Understanding these interactions is crucial for optimizing combination therapies, especially in complex conditions like treatment-resistant depression, where multiple modalities are often employed (Brunoni et al., 2011).

   Thus, achieving optimal outcomes in brain stimulation therapy demands a comprehensive approach that considers the intricacies of stimulation parameters, targeted brain regions, individual variability, state-dependency, and the potential interactions with pharmacological interventions. As our understanding of these factors deepens, so does our capacity to tailor brain stimulation therapies to meet the unique needs of each patient, enhancing efficacy and safety. ◘ Figure 9.6 (Polanía et al., 2018) depicts the factors determining the variability of neurophysiological and behavioral effects in NIBS research.

9.2.7 Challenges and Future Directions

While the field of brain stimulation has made significant strides, it confronts a spectrum of challenges that necessitate further investigation and innovation:

1. **Optimization of Stimulation Parameters and Protocols**

   Determining the most effective stimulation parameters and protocols for particular conditions or cognitive functions is crucial. This endeavor often demands extensive, well-designed clinical trials that are sufficiently powered to elucidate subtle differences in outcomes. For instance, large-scale randomized controlled trials (RCTs) are required to establish the efficacy and optimal parameters of interventions like Transcranial Magnetic Stimulation (TMS) for depression or cognitive enhancement (George et al., 2013; Brunoni et al., 2016).

2. **Precision in Targeting Brain Regions and Circuits**

   Advancing the accuracy of targeting specific brain regions and circuits remains a priority. This challenge calls for the integration of sophisticated neuroimaging techniques, such as functional MRI (fMRI) and diffusion tensor

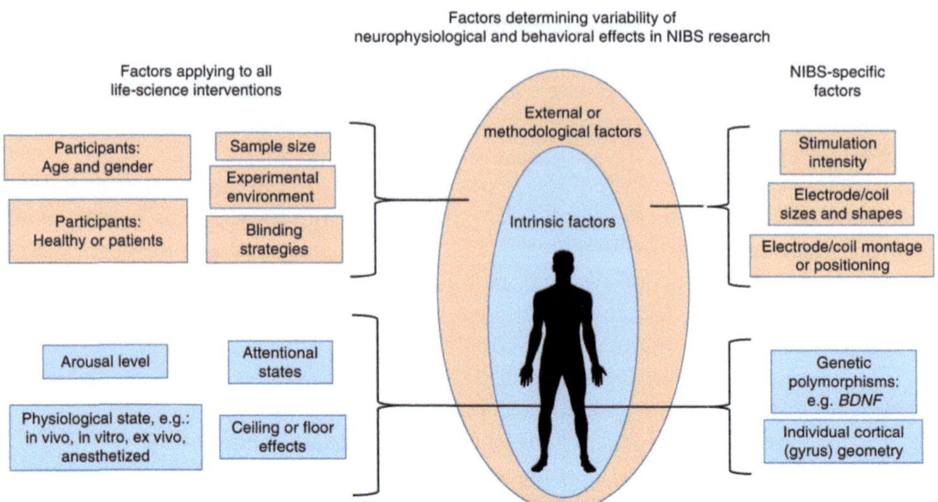

Fig. 9.6 Numerous factors contribute to the variability observed in the effects of Noninvasive Brain Stimulation (NIBS), mirroring the complexity found in other experimental interventions across the life sciences. These include general factors such as participant age, sex, neurological diversity, and baseline neurophysiological state, which can influence the outcome of any biological intervention. In addition to these general variables, there are specific factors intrinsic to NIBS that warrant careful consideration: **Stimulation Parameters:** Variations in intensity, frequency, duration, and the type of stimulation (e.g., tDCS, TMS, tACS) can significantly affect outcomes. Each parameter can modulate neural activity differently, and slight alterations can lead to divergent effects across studies. **Electrode or Coil Placement:** The location and positioning of the stimulating electrode or coil are crucial for targeting specific brain regions. Misplacement or inconsistent positioning between sessions or subjects can lead to variability in results. **Individual Neuroanatomy:** Variations in skull thickness, scalp conductivity, and brain morphology can affect the efficacy of the stimulation, altering the electric field distribution and the resultant neuromodulation. **State Dependency:** The neural state of an individual at the time of stimulation, including factors like alertness, cognitive load, or emotional state, can influence the effects of NIBS. **Circadian Rhythms:** The time of day when the stimulation is administered can also impact its effectiveness, as neural excitability varies throughout the day. From Polanía et al. (2018)

imaging (DTI), with computational models to enhance the precision of stimulation. These tools can help map the brain's functional and structural connectivity, thereby guiding more focused and individualized stimulation approaches (Fox et al., 2013; Opitz et al., 2015).

3. **Long-Term Effects and Safety**

   Investigating the long-term effects and safety profiles of brain stimulation techniques, especially noninvasive methods like tDCS and tACS, is imperative. As these technologies gain popularity in both clinical and research domains, understanding their prolonged use implications is vital for ensuring patient safety and the integrity of research outcomes. Systematic reviews and meta-analyses of longitudinal studies are instrumental in assessing these aspects (Bikson et al., 2016; Kekic et al., 2016).

4. **Unraveling Underlying Neural Mechanisms**

   Elucidating the neural mechanisms mediating brain stimulation's effects is essential for refining these interventions. This includes exploring how different stimulation modalities influence neuroplasticity, neural oscillations, and network dynamics. Advances in this area can lead to the development of more targeted and effective stimulation strategies, potentially opening new therapeutic avenues (Bestmann et al., 2015; Polanía et al., 2018).

As the theoretical and empirical understanding of brain stimulation continues to evolve, so does the potential to leverage these techniques for probing cognitive processes and devising innovative treatments for neurological and psychiatric conditions. Future research, grounded in interdisciplinary collaboration and technological innovation, will play a pivotal role in overcoming these challenges and expanding the utility of brain stimulation therapies.

Box 9.5 Understanding Variability in Brain Stimulation

The effectiveness of brain stimulation techniques such as TMS, tDCS, tACS, and DBS can vary significantly due to a range of factors that influence how stimulation impacts the brain. Recognizing these factors is crucial for optimizing treatment protocols and achieving consistent therapeutic results.

■ **Critical Influencing Factors:**

Stimulation Parameters: Choices regarding the intensity, frequency, duration, and waveform of the stimulation directly affect its efficacy. Tailoring these parameters to the specific needs of the condition being treated can enhance therapeutic outcomes.

Targeted Brain Region: Accurate targeting of brain regions that are functionally relevant to the disorder or desired outcome is essential. This requires precise knowledge of brain anatomy and function, often guided by neuroimaging techniques.

Individual Variability: Anatomical and physiological differences between individuals, such as skull thickness, brain morphology, and neurochemical background, can alter the effects of brain stimulation. Personalized treatment approaches, potentially informed by diagnostic imaging, are therefore necessary.

State-Dependency: The state of the brain at the time of stimulation (e.g., whether a person is resting, engaged in a task, or experiencing particular emotions) can influence the effectiveness of the stimulation, as neural responsiveness varies with brain state.

Interactions with Pharmacological Treatments: Concurrent medications can modulate the effects of brain stimulation, either enhancing or inhibiting its impact depending on the drugs' actions on neural plasticity or excitability.

■ **Clinical Implications:**

Personalization of Therapy: Understanding and integrating these factors enable clinicians to tailor brain stimulation treatments to individual patients, enhancing both safety and effectiveness.

Research and Development: Ongoing research into these factors also informs the development of new stimulation devices and protocols, driving advances in both technology and therapeutic methodologies.

9.3   Planning Stimulation: Closed-Loop Systems for Psychiatric Disorders

This section explores the innovative concept of closed-loop brain stimulation systems, which use real-time brain activity monitoring and symptom tracking to dynamically adjust stimulation settings. Powered by advanced algorithms, these systems fine-tune stimulation based on continuous feedback from brain activity and clinical responses. This adaptive approach represents a major step forward in delivering personalized and more effective treatments for psychiatric conditions like depression, anxiety, and schizophrenia.

9.3.1   Closed-Loop Brain Stimulation Systems

Closed-loop brain stimulation systems represent a paradigm shift in neuromodulation therapies. Unlike traditional, open-loop systems that deliver constant stimulation without regard for the patient's neural state or symptom changes, closed-loop systems leverage real-time data to optimize therapeutic effects dynamically. These systems utilize various biosensors to monitor neural signals, such as electroencephalography (EEG) or local field potentials (LFPs), alongside clinical symptom tracking, as shown in ◙ Fig. 9.7 (Bergmann, 2018). The integration of this data through advanced computational algorithms allows for the automatic adjustment of stimulation parameters, ensuring that the therapy is responsive to the patient's current neurological and psychological state (Rouse et al., 2011; Scangos et al., 2021).

9.3.2   Components of Closed-Loop Brain Stimulation Systems

The architecture of closed-loop brain stimulation systems is fundamentally composed of three integral components:

9.3.2.1   Sensing

This component is tasked with the ongoing surveillance of neural signals or biomarkers that are indicative of the psychiatric disorder being targeted. The acquisition of these signals can be accomplished through various methods, including electroencephalography (EEG) for noninvasive monitoring, local field potentials (LFPs) via intracranial electrodes for a more direct measure of brain activity, or other physiological markers associated with the disorder's pathology. The choice of sensing technique is crucial, as it needs to reliably reflect the disorder's neural underpinnings (Parastarfeizabadi & Kouzani, 2017).

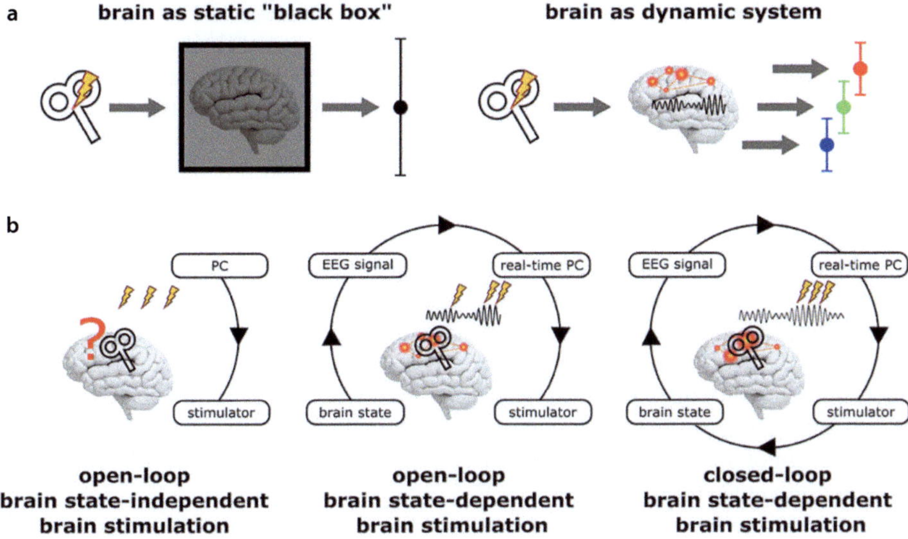

Fig. 9.7 The principal scenarios of brain stimulation with respect to the current brain state are crucial for understanding the variability and efficacy of Noninvasive Brain Stimulation (NIBS) effects. **a** Static vs. Dynamic Brain Conceptualization—**Static "Black Box" Approach**: Traditional NIBS methods often treat the brain as a static entity, where stimulation parameters do not account for the ongoing internal state or its fluctuations. This perspective, illustrated on the left, may lead to inconsistent outcomes across different sessions or individuals due to the natural variability in brain states. **Dynamic System Approach**: In contrast, recognizing the brain as the dynamic system it truly is (shown on the right) involves considering the current state of brain activity during stimulation. This approach acknowledges that brain states can influence the effects of stimulation, potentially leading to more consistent and state-dependent outcomes. **b** Brain State-Independent vs. State-Dependent Stimulation—**Open-loop, Brain State-Independent Stimulation**: This method does not take into account the current brain state; it applies a preset stimulation protocol without real-time adjustments based on neuroimaging or other neurophysiological feedback. This approach does not require the integration of concurrent neuroimaging or real-time systems to control the stimulation, simplifying the setup but potentially overlooking important variations in brain state that could influence effectiveness. **Open-loop, Brain State-Dependent Brain Stimulation (BSDBS)**: This more nuanced approach utilizes concurrent neuroimaging techniques, such as EEG, to monitor the brain state in real-time and adjusts or triggers stimulation based on specific neurophysiological markers (e.g., the amplitude or phase of EEG oscillations). This method allows for more tailored stimulation based on the immediate brain state, enhancing the intervention's relevance and effectiveness. However, it does not aim to alter the brain state being monitored systematically but to utilize its current state to optimize stimulation timing and parameters. **Closed-loop BSDBS**: This advanced scenario extends open-loop BSDBS by monitoring and actively modifying the targeted brain state. It requires a real-time feedback system that adjusts the stimulation parameters based on the desired changes in brain activity, thereby exerting a controlled influence on specific neural patterns. For instance, stimulation could be designed to enhance or suppress particular brain oscillations, effectively steering brain activity towards a therapeutic goal. Each of these scenarios represents a different level of integration between brain stimulation technology and our understanding of brain dynamics. The transition from open-loop state-independent methods to closed-loop state-dependent systems marks a significant evolution in the field, promising more precise and effective interventions tailored to the dynamic conditions of the human brain. This progression underscores the need for advanced neuroimaging integration and real-time data processing capabilities to fully exploit the potential of NIBS in both research and clinical applications. From Bergmann (2018) originally published under CC-BY

9.3.2.2 Signal Processing and Decision-Making

The data captured by the sensing component undergoes real-time processing and analysis. This process uses sophisticated algorithms and computational models to identify significant features and discerns patterns within the neural activity that are pertinent to the disorder. Based on this analysis, the system evaluates the current state of neural activity and makes informed decisions on the necessity and nature of adjustments to the stimulation parameters, aiming to normalize aberrant neural dynamics (Rosin et al., 2011).

The core of closed-loop systems lies in their adaptive algorithms, designed to effectively interpret complex neural and clinical data to modulate stimulation protocols. These algorithms employ machine learning and signal processing techniques to identify patterns in the monitored signals that correlate with symptom fluctuations or changes in brain activity associated with psychiatric disorders. By recognizing these patterns, the system can adjust the intensity, frequency, duration, and location of stimulation in real-time, aiming to normalize aberrant neural activity and alleviate symptoms (Opitz et al. 2015).

9.3.2.3 Stimulation

Following the decision-making process, the stimulation component is responsible for delivering targeted electrical impulses to specific brain regions or circuits implicated in the disorder, as shown in ▶ Fig. 8.12. This can be achieved using various neuromodulation techniques such as Deep Brain Stimulation (DBS) for invasive, focused stimulation, Transcranial Magnetic Stimulation (TMS), or Transcranial Electrical Stimulation (tES) for noninvasive interventions. The system adaptively modifies stimulation parameters, including intensity, frequency, and duration, in real-time, guided by the preceding analysis to ensure optimal therapeutic outcomes (Opitz et al. 2015).

9.3.3 Applications in Psychiatric Disorders

Recent advancements in closed-loop brain stimulation underscore its transformative potential for treating various neurological and psychiatric disorders. This adaptive approach tailors treatment to individual patient needs, enhancing efficacy and minimizing side effects through real-time adjustment of stimulation parameters based on specific neural signals or biomarkers. Key findings in several disorders illustrate the promise of closed-loop systems:

1. Major Depressive Disorder (MDD)

    In the context of MDD, closed-loop stimulation systems are designed to target neural signals associated with mood regulation. This includes monitoring specific oscillatory patterns or assessing functional connectivity within the brain's mood-regulating networks. By dynamically adjusting stimulation in response to detected neural activity, these systems aim to modulate the neural circuits implicated in depression, potentially leading to alleviating mood symptoms. Research indicates that targeting areas like the dorsolateral prefrontal cortex, which is involved in mood regulation, with adaptive stimulation could enhance therapeutic outcomes for patients with depression (Drysdale et al., 2017; Scangos et al., 2021).

2. Obsessive-Compulsive Disorder (OCD)

   For OCD, closed-loop systems offer the possibility of directly influencing brain circuits associated with compulsive behavior and obsessions. By monitoring neural activity indicative of these patterns and adaptively modifying stimulation parameters, it is conceivable to mitigate the severity of obsessive thoughts and compulsive behaviors. Such interventions could significantly improve the quality of life for individuals with OCD, providing a novel avenue for treatment-resistant cases (Figee et al., 2013). Early results suggest that adaptive DBS targeting areas like the nucleus accumbens can yield significant symptom improvements and lower side effects compared to conventional DBS, offering new hope for individuals with this challenging condition (Denys et al., 2010; Widge et al., 2019).

3. Post-Traumatic Stress Disorder (PTSD)

   Although research on closed-loop stimulation for PTSD is in the preliminary stages, it could potentially detect neural signatures associated with fear and anxiety responses, enabling the modulation of brain circuits relevant to the disorder. Targeting neural activity related to hyperarousal and intrusive memories and adjusting stimulation to normalize this activity offers a promising strategy for reducing PTSD symptoms such as flashbacks and heightened stress responses (Koenigs and Grafman 2009; Koek et al., 2024).

4. Epilepsy

   Closed-loop systems like Responsive Neurostimulation (RNS) have marked a significant advance in epilepsy treatment, particularly for drug-resistant cases. By detecting early signs of seizure activity and delivering precise electrical stimulation to prevent seizure propagation, these systems have been shown to reduce seizure frequency substantially. Studies report significant improvements in seizure control and quality of life among patients using RNS systems (Morrell, 2011; Heck et al., 2014).

5. Parkinson's Disease

   In Parkinson's disease, closed-loop Deep Brain Stimulation (DBS) systems that adjust stimulation parameters based on neural biomarkers, such as beta-band oscillations in the subthalamic nucleus, have demonstrated enhanced motor symptom management. Compared to traditional open-loop DBS, adaptive DBS offers improved control of motor symptoms with fewer stimulation-induced side effects, underscoring the benefits of a more tailored therapeutic approach (Little et al., 2013; Arlotti et al., 2018).

9.3.4 Advantages and Challenges

Closed-loop brain stimulation systems represent a paradigm shift in neuromodulation therapy, offering significant advantages over traditional open-loop systems.

These benefits stem from their dynamic adaptability to the patient's neural activity, which facilitates personalized, efficient, and safer treatment modalities. However, the implementation of these systems, especially in psychiatric disorders, faces several technical and clinical challenges.

9.3.4.1 Advantages of Closed-Loop Systems

1. Personalized Treatment

   Closed-loop systems adjust stimulation parameters in real-time based on ongoing brain activity, enabling a highly personalized treatment approach. This adaptability ensures that each patient receives stimulation tailored to their unique neural patterns, enhancing treatment efficacy (Sitaram et al., 2017).

2. Improved Efficiency

   The capability to dynamically modulate stimulation parameters allows closed-loop systems to target specific neural circuits more precisely. This precision can potentially shorten the duration and decrease the intensity of stimulation needed to elicit therapeutic effects, making the treatment more efficient (Arlotti et al., 2018).

3. Reduced Side Effects

   By providing stimulation only when necessary and fine-tuning the stimulation parameters to the patient's current state, closed-loop systems aim to reduce unwanted side effects. This approach improves the safety and tolerability of brain stimulation treatments (Widge et al., 2019).

9.3.4.2 Challenges in Development and Implementation

1. Identifying Reliable Neural Biomarkers

   The efficacy of closed-loop systems relies heavily on accurately detecting neural signals associated with specific psychiatric conditions. However, pinpointing reliable biomarkers is challenging due to the complexity and variability inherent in psychiatric disorders. The search for biomarkers that can consistently predict treatment response or symptom fluctuations remains ongoing.

2. Developing Advanced Signal Processing and Decision-Making Algorithms

   Closed-loop systems necessitate sophisticated algorithms capable of real-time signal processing and decision-making. These algorithms must efficiently analyze complex neural data to determine the appropriate adjustments to stimulation parameters. Developing such algorithms requires a deep understanding of both the neural underpinnings of psychiatric disorders and advanced computational techniques (Rouse et al., 2011).

   In conclusion, closed-loop brain stimulation systems are emerging as a transformative approach in treating psychiatric disorders, promising personalized, adaptive therapies that enhance efficacy and minimize side effects. Despite existing challenges, their potential to revolutionize psychiatric care through targeted, symptom-specific interventions is substantial. As ongoing research addresses these limitations, the future of psychiatric treatment with closed-loop systems appears increasingly promising, marking a significant step forward in managing complex neurological and psychiatric conditions.

Box 9.6 Objective of Closed-Loop Systems

Closed-loop brain stimulation systems represent a significant advancement in neuromodulation for treating psychiatric disorders. These systems dynamically adjust stimulation parameters based on real-time feedback from the patient's brain activity, optimizing therapeutic outcomes.

■ **Key Components:**

Real-Time Monitoring: These systems continuously monitor brain activity using techniques like EEG or local field potentials to assess the current state of neural functioning.

Adaptive Stimulation: Based on the monitored data, the system adjusts the stimulation parameters (such as intensity and frequency) in real time to tailor the therapy according to the patient's immediate needs.

Feedback Integration: Integration of computational algorithms allows these systems to learn from the patient's responses to stimulation, improving the precision of treatment over time.

■ **Clinical Applications:**

Depression and Anxiety: Closed-loop systems can modulate areas involved in mood regulation, potentially adjusting depressive and anxious symptoms more effectively than static stimulation methods.

Schizophrenia: Targeting specific neural circuits that are dysregulated in schizophrenia might help in managing symptoms like hallucinations and disorganized thinking.

■ **Benefits:**

Personalized Treatment: By adjusting to changes in brain activity, closed-loop systems offer a personalized approach, potentially increasing the efficacy and reducing side effects compared to traditional stimulation methods.

Enhanced Efficiency: These systems provide stimulation only when necessary and with optimal parameters, which can conserve battery life in implanted devices and reduce patient exposure to unnecessary stimulation.

■ **Challenges:**

Complexity in Implementation: The technology requires sophisticated algorithms and integration of various sensor inputs, making it complex and expensive to implement.

Understanding Individual Differences: Effective use of closed-loop systems requires a deep understanding of individual differences in brain structure and function, which can be challenging to standardize across a diverse patient population.

9.4 Ethical Considerations and Future Directions in Brain Stimulation

Ethical considerations take center stage in this exploration of brain stimulation, especially when techniques are employed to disrupt ongoing brain activity. These concerns are paramount to ensuring that the application of brain stimulation technologies adheres to principles of respect, justice, and beneficence.

9.4.1 **Ethical Considerations**

1. Informed Consent

    A cornerstone of ethical research and treatment involves ensuring participants and patients are fully informed about the procedures they will undergo. This includes a comprehensive understanding of potential risks and benefits, allowing individuals to make educated decisions regarding their participation. Informed consent is particularly crucial in brain stimulation studies due to the invasive nature of some techniques and the potential for unknown long-term effects (Fins et al., 2011).

2. Privacy and Data Security

    The sensitive nature of brain data collected during stimulation procedures mandates stringent privacy and data security measures. Protecting this information from unauthorized access is essential to maintaining patient confidentiality and trust in medical research and practice (Illes et al., 2010).

3. Equitable Access

    Equity in access to brain stimulation technologies is a significant ethical concern. Disparities in availability based on socioeconomic status, geographic location, or other factors could exacerbate existing health inequities. Ensuring fair access to these potentially life-changing treatments is an ethical imperative.

4. Cognitive Enhancement and Fairness

    The use of brain stimulation for cognitive enhancement in healthy individuals raises questions about competition, fairness, and societal expectations. The potential for creating disparities in cognitive abilities through access to enhancement technologies warrants careful consideration of ethical and social implications (Hamilton et al., 2011).

9.4.2 **Future Directions and Emerging Trends**

1. Novel Stimulation Techniques

    Innovations in brain stimulation, such as focused ultrasound, are under development. These new methods promise more targeted and less invasive options for modulating brain activity, expanding the toolkit available to researchers and clinicians (Tyler, 2011).

2. Personalized Brain Stimulation

    Advances in understanding individual brain structure and function enhance the potential for personalized brain stimulation interventions. Tailoring treatments to individual differences could significantly improve efficacy and reduce side effects (Lozano et al., 2019).

3. Integration with Other Therapies

    Combining brain stimulation with other treatments, such as medication, psychotherapy, or behavioral interventions, may offer synergistic benefits. This integrated approach could lead to more comprehensive treatment strategies for neurological and psychiatric disorders (Bergmann et al., 2016).

4. Brain-Computer Interfaces (BCIs)

The advancement of BCIs is paving the way for sophisticated closed-loop brain stimulation systems. These technologies hold the promise of revolutionizing treatment approaches for various conditions and enhancing human cognition in ways previously unimagined (Lebedev & Nicolelis, 2017).

By conscientiously navigating ethical considerations and actively engaging with emerging trends, the field of brain stimulation can continue to evolve, offering profound insights into brain function and novel treatments for an array of conditions while potentially unlocking new realms of human cognitive capability.

Box 9.7 Ethical Considerations in Brain Stimulation

Brain stimulation technologies raise several ethical concerns, particularly as they relate to consent, privacy, and the potential for cognitive enhancement.

Informed Consent: Ensuring that patients fully understand the risks and benefits of brain stimulation, along with the potential for unknown long-term effects, is crucial. This is especially important given the complexity and novelty of these interventions.

Privacy and Data Security: Brain data is highly sensitive. Protecting this data from unauthorized access is essential to maintaining trust and confidentiality.

Equitable Access: Addressing disparities in access to brain stimulation technologies is critical to prevent exacerbating existing health inequities.

Cognitive Enhancement: The use of brain stimulation for enhancing cognitive functions in healthy individuals raises questions about fairness, social pressure, and the potential widening of social inequalities.

Future Directions in Brain Stimulation: Advancements in brain stimulation are expected to continue, driven by technological innovations and deeper understanding of brain function.

Novel Stimulation Techniques: Techniques such as focused ultrasound and personalized electric field modulation are emerging, promising less invasive and more targeted approaches.

Precision Medicine: Increased precision in targeting and intervention, informed by individual genetic, anatomical, and functional profiles, could enhance both efficacy and safety.

Integration with Other Therapies: Combining brain stimulation with pharmacological treatments and behavioral therapies may lead to synergistic effects, offering comprehensive approaches to treatment.

Brain-Computer Interfaces (BCIs): Advances in BCIs could further refine stimulation techniques, leading to more sophisticated closed-loop systems that adjust interventions in real-time based on direct brain activity feedback.

Key Takeaways

1. **Diverse Mechanisms of Brain Stimulation Techniques**: This chapter highlights the unique mechanisms of various brain stimulation techniques, including Transcranial Magnetic Stimulation (TMS), Transcranial Direct Current Stimulation (tDCS), Transcranial Alternating Current Stimulation (tACS), and Deep Brain Stimulation (DBS). Each method influences neuronal activity differently, catering to both non-invasive and invasive therapeutic needs, ranging from modulating cortical excitability to targeting deep brain structures.

2. **Applications Across Neurological and Psychiatric Disorders**: Brain stimulation techniques are applied in treating a wide spectrum of conditions. TMS and tDCS are effective in addressing depression, cognitive rehabilitation, and chronic pain management, while DBS shows remarkable results in movement disorders like Parkinson's disease and emerging potential in psychiatric conditions such as obsessive-compulsive disorder (OCD) and treatment-resistant depression.

3. **Challenges in Precision and Individual Variability**: A significant challenge highlighted is the variability in patient outcomes, influenced by factors such as brain anatomy, stimulation parameters, and individual neurophysiological differences. Personalized brain stimulation approaches, incorporating advanced neuroimaging and computational modeling, are essential for optimizing therapeutic outcomes and minimizing adverse effects.

4. **Future Directions and Emerging Technologies**: This chapter emphasizes the importance of advancing brain stimulation technologies, particularly in integrating closed-loop systems and adaptive stimulation protocols that dynamically adjust based on real-time neural feedback. Future developments focus on enhancing precision, optimizing patient-specific treatments, and expanding applications to a broader range of neuropsychiatric conditions.

In conclusion, the field of brain stimulation is marked by rapid advancements and the ongoing integration of new technologies. Future directions include refining stimulation protocols to be more personalized and adjusting parameters in real-time based on continuous feedback from brain activity monitoring. Such advancements could significantly enhance the efficacy and safety of brain stimulation therapies, making them a cornerstone of modern neurological and psychiatric treatment regimes. As we continue to deepen our understanding of brain function and refine our technological tools, the potential applications of brain stimulation are bound to expand, offering new hope and improved outcomes for patients with complex neurological conditions.

Conclusion

In this chapter, we have performed a comparative analysis of various brain stimulation methods, including Transcranial Magnetic Stimulation (TMS), Transcranial Direct Current Stimulation (tDCS), Transcranial Alternating Current Stimulation (tACS), and Deep Brain Stimulation (DBS). Each of these techniques leverages unique mechanisms to influence neuronal activity, offering a range of therapeutic and research applications. Their comparative efficacy and suitability depend significantly on the specific clinical or research requirements, such as the need for precision, depth of stimulation, and the particular neurological condition being targeted.

Brain stimulation is underpinned by several critical theories that provide a framework for understanding its mechanisms and effects. Network Modulation Theory suggests that stimulation extends beyond localized effects, influencing interconnected brain networks and potentially restoring or enhancing their functionality. Neuroplasticity Theory supports the idea that brain stimulation can induce long-term changes in neural connections, which is fundamental for learning, memory, and recovery from neurological damage. These theories highlight the potential of brain stimulation to not only probe the physiological basis of brain function but also to modify it in ways that can lead to significant improvements in cognitive and motor functions.

Planning a brain stimulation study requires meticulous design and consideration of numerous factors that can influence outcomes. Researchers must carefully select stimulation parameters such as intensity, frequency, and duration while also considering the individual variability in anatomy and physiology. The use of neuroimaging and ncuronavigation technologies can enhance the precision of targeting and the effectiveness of the intervention. Additionally, understanding the state-dependency of brain responses to stimulation can lead to more effective and personalized treatment protocols.

The application of brain stimulation extends beyond the treatment of disease. It holds promise for enhancing cognitive abilities, investigating the foundations of neural processes, and even modifying behavior. Clinically, these techniques are being applied to treat a wide array of conditions, from depressive disorders and epilepsy to chronic pain and movement disorders. Each method offers distinct advantages, such as the non-invasiveness of TMS, tDCS, and tACS, or DBS's direct and robust effects.

This section offers practical exercises designed to help apply the material learned to concrete scenarios.

Design a Brain Stimulation Protocol

Objective: Apply knowledge of different brain stimulation techniques to design a protocol for a specific neurological condition.

Task: Students choose a condition (e.g., depression, Parkinson's disease) and decide which stimulation technique (TMS, tDCS, tACS, or DBS) would be most appropriate. They must justify their choice based on the technique's mechanism of action, targeted brain regions, and expected outcomes.

Simulate Network Modulation

Objective: Understand how stimulating different brain areas can influence neural networks.

Task: Using software simulations, students apply virtual electrodes at various cortical or subcortical sites and observe changes in neural network activity. This exercise helps illustrate concepts like network modulation and functional connectivity.

Ethical Debate on Cognitive Enhancement:

Objective: Explore the ethical implications of using brain stimulation for cognitive enhancement in healthy individuals.

Task: Organize a debate where students discuss the potential social, ethical, and psychological implications of enhancing cognitive function through non-invasive brain stimulation, considering issues of equity, consent, and long-term effects.

Role-Play Patient-Doctor Interaction for DBS Treatment

Objective: Develop communication skills and deepen understanding of the clinical considerations involved in DBS.

Task: One student plays the role of a neurologist explaining the DBS procedure, risks, benefits, and post-surgical care to another student who acts as a patient with Parkinson's disease. This exercise emphasizes patient education and informed consent.

Create a Brain Stimulation Safety Protocol

Objective: Learn about the safety concerns associated with brain stimulation.

Task: Students develop a comprehensive safety protocol for a hypothetical new brain stimulation lab. The protocol must address equipment safety, patient screening, session monitoring, and emergency procedures.

Closed-Loop System Design Challenge

Objective: Conceptualize a closed-loop brain stimulation system for a psychiatric disorder.

Task: Students design a schematic of a closed-loop system tailored for a disorder like major depressive disorder or OCD, describing the types of sensors and stimulation devices used, how feedback is processed, and how stimulation parameters are adjusted in real-time.

References

Antal, A., & Paulus, W. (2013). Transcranial alternating current stimulation (tACS). *Frontiers in Human Neuroscience, 7*, 317.

Arlotti, M., Marceglia, S., Foffani, G., Volkmann, J., Lozano, A. M., Moro, E., et al. (2018). Eight-hours adaptive deep brain stimulation in patients with Parkinson disease. *Neurology, 90*(11), e971–e976.

Bassett, D. S., & Sporns, O. (2017). Network neuroscience. *Nature Neuroscience, 20*(3), 353–364.

Bergmann, T. O. (2018). Brain state-dependent brain stimulation. *Frontiers in Psychology, 9*, 422698.

Bergmann, T. O., Karabanov, A., Hartwigsen, G., Thielscher, A., & Siebner, H. R. (2016). Combining noninvasive transcranial brain stimulation with neuroimaging and electrophysiology: Current approaches and future perspectives. *Neuroimage, 140*, 4–19.

Bestmann, S., de Berker, A. O., & Bonaiuto, J. (2015). Understanding the behavioural consequences of noninvasive brain stimulation. *Trends in Cognitive Sciences, 19*(1), 13–20.

Bikson, M., Grossman, P., Thomas, C., Zannou, A. L., Jiang, J., Adnan, T., et al. (2016). Safety of transcranial direct current stimulation: Evidence based update 2016. *Brain Stimulation, 9*(5), 641–661.

Brunoni, A. R., Ferrucci, R., Bortolomasi, M., Vergari, M., Tadini, L., Boggio, P. S., et al. (2011). Transcranial direct current stimulation (tDCS) in unipolar vs. bipolar depressive disorder. *Progress in Neuro-Psychopharmacology and Biological Psychiatry, 35*(1), 96–101.

Brunoni, A. R., Nitsche, M. A., Bolognini, N., Bikson, M., Wagner, T., Merabet, L., et al. (2012). Clinical research with transcranial direct current stimulation (tDCS): Challenges and future directions. *Brain Stimulation, 5*(3), 175–195.

Brunoni, A. R., Moffa, A. H., Fregni, F., Palm, U., Padberg, F., Blumberger, D. M., et al. (2016). Transcranial direct current stimulation for acute major depressive episodes: Meta-analysis of individual patient data. *The British Journal of Psychiatry, 208*(6), 522–531.

Cash, R. F., Cocchi, L., Lv, J., Fitzgerald, P. B., & Zalesky, A. (2021). Functional magnetic resonance imaging–guided personalization of transcranial magnetic stimulation treatment for depression. *JAMA Psychiatry, 78*(3), 337–339.

Cheeran, B., Talelli, P., Mori, F., Koch, G., Suppa, A., Edwards, M., et al. (2008). A common polymorphism in the brain-derived neurotrophic factor gene (BDNF) modulates human cortical plasticity and the response to rTMS. *The Journal of Physiology, 586*(23), 5717–5725.

Cramer, S. C., Sur, M., Dobkin, B. H., O'Brien, C., Sanger, T. D., Trojanowski, J. Q., et al. (2011). Harnessing neuroplasticity for clinical applications. *Brain, 134*(6), 1591–1609.

Dayan, E., Censor, N., Buch, E. R., Sandrini, M., & Cohen, L. G. (2013). Noninvasive brain stimulation: From physiology to network dynamics and back. *Nature Neuroscience, 16*(7), 838–844.

Denys, D., Mantione, M., Figee, M., Van Den Munckhof, P., Koerselman, F., Westenberg, H., et al. (2010). Deep brain stimulation of the nucleus accumbens for treatment-refractory obsessive-compulsive disorder. *Archives of General Psychiatry, 67*(10), 1061–1068.

Deuschl, G., Schade-Brittinger, C., Krack, P., Volkmann, J., Schäfer, H., Bötzel, K., et al. (2006). A randomized trial of deep-brain stimulation for Parkinson's disease. *New England Journal of Medicine, 355*(9), 896–908.

Dmochowski, J. P., Datta, A., Bikson, M., Su, Y., & Parra, L. C. (2011). Optimized multi-electrode stimulation increases focality and intensity at target. *Journal of Neural Engineering, 8*(4), 046011.

Drysdale, A. T., Grosenick, L., Downar, J., Dunlop, K., Mansouri, F., Meng, Y., et al. (2017). Resting-state connectivity biomarkers define neurophysiological subtypes of depression. *Nature Medicine, 23*(1), 28–38.

Figee, M., Luigjes, J., Smolders, R., Valencia-Alfonso, C. E., Van Wingen, G., De Kwaasteniet, B., et al. (2013). Deep brain stimulation restores frontostriatal network activity in obsessive-compulsive disorder. *Nature Neuroscience, 16*(4), 386–387.

Fins, J. J., Mayberg, H. S., Nuttin, B., Kubu, C. S., Galert, T., Sturm, V., et al. (2011). Misuse of the FDA's humanitarian device exemption in deep brain stimulation for obsessive-compulsive disorder. *Health Affairs, 30*(2), 302–311.

Fox, M. D., Buckner, R. L., White, M. P., Greicius, M. D., & Pascual-Leone, A. (2012). Efficacy of transcranial magnetic stimulation targets for depression is related to intrinsic functional connectivity with the subgenual cingulate. *Biological Psychiatry, 72*(7), 595–603.

Fox, M. D., Liu, H., & Pascual-Leone, A. (2013). Identification of reproducible individualized targets for treatment of depression with TMS based on intrinsic connectivity. *Neuroimage, 66*, 151–160.

Fregni, F., & Pascual-Leone, A. (2007). Technology insight: Noninvasive brain stimulation in neurology—perspectives on the therapeutic potential of rTMS and tDCS. *Nature Clinical Practice Neurology, 3*(7), 383–393.

Fritsch, B., Reis, J., Martinowich, K., Schambra, H. M., Ji, Y., Cohen, L. G., & Lu, B. (2010). Direct current stimulation promotes BDNF-dependent synaptic plasticity: Potential implications for motor learning. *Neuron, 66*(2), 198–204.

George, M. S., Taylor, J. J., & Short, E. B. (2013). The expanding evidence base for rTMS treatment of depression. *Current Opinion in Psychiatry, 26*(1), 13–18.

Hamilton, R., Messing, S., & Chatterjee, A. (2011). Rethinking the thinking cap: Ethics of neural enhancement using noninvasive brain stimulation. *Neurology, 76*(2), 187–193.

He, W., Fong, P. Y., Leung, T. W. H., & Huang, Y. Z. (2020). Protocols of noninvasive brain stimulation for neuroplasticity induction. *Neuroscience Letters, 719*, 133437.

Heck, C. N., King-Stephens, D., Massey, A. D., Nair, D. R., Jobst, B. C., Barkley, G. L., et al. (2014). Two-year seizure reduction in adults with medically intractable partial onset epilepsy treated with responsive neurostimulation: Final results of the RNS System Pivotal trial. *Epilepsia, 55*(3), 432–441.

Huang, Y. Z., Lu, M. K., Antal, A., Classen, J., Nitsche, M., Ziemann, U., et al. (2017). Plasticity induced by noninvasive transcranial brain stimulation: A position paper. *Clinical Neurophysiology, 128*(11), 2318–2329.

Hummel, F. C., & Cohen, L. G. (2006). Noninvasive brain stimulation: A new strategy to improve neurorehabilitation after stroke? *The Lancet Neurology, 5*(8), 708–712.

Illes, J., Moser, M. A., McCormick, J. B., Racine, E., Blakeslee, S., Caplan, A., et al. (2010). Neurotalk: Improving the communication of neuroscience research. *Nature Reviews Neuroscience, 11*(1), 61–69.

Kekic, M., Boysen, E., Campbell, I. C., & Schmidt, U. (2016). A systematic review of the clinical efficacy of transcranial direct current stimulation (tDCS) in psychiatric disorders. *Journal of Psychiatric Research, 74*, 70–86.

Kleim, J. A., & Jones, T. A. (2008). Principles of experience-dependent neural plasticity: Implications for rehabilitation after brain damage.

Koek, R. J., Avecillas-Chasin, J., Krahl, S. E., Chen, J. W., Sultzer, D. L., Kulick, A. D., et al. (2024). Deep brain stimulation of the amygdala for treatment-resistant combat post-traumatic stress disorder: Long-term results. *Journal of Psychiatric Research.*

Koenigs, M., & Grafman, J. (2009). Posttraumatic stress disorder: The role of medial prefrontal cortex and amygdala. *The Neuroscientist, 15*(5), 540–548.

Lebedev, M. A., & Nicolelis, M. A. (2017). Brain-machine interfaces: From basic science to neuroprostheses and neurorehabilitation. *Physiological Reviews, 97*(2), 767–837.

Lefaucheur, J. P., Aleman, A., Baeken, C., Benninger, D. H., Brunelin, J., Di Lazzaro, V., et al. (2020). Evidence-based guidelines on the therapeutic use of repetitive transcranial magnetic stimulation (rTMS): An update (2014–2018). *Clinical Neurophysiology, 131*(2), 474–528.

Little, S., Pogosyan, A., Neal, S., Zavala, B., Zrinzo, L., Hariz, M., et al. (2013). Adaptive deep brain stimulation in advanced Parkinson disease. *Annals of Neurology, 74*(3), 449–457.

Lozano, A. M., & Lipsman, N. (2013). Probing and regulating dysfunctional circuits using deep brain stimulation. *Neuron, 77*(3), 406–424.

Lozano, A. M., Lipsman, N., Bergman, H., Brown, P., Chabardes, S., Chang, J. W., et al. (2019). Deep brain stimulation: Current challenges and future directions. *Nature Reviews Neurology, 15*(3), 148–160.

McIntyre, C. C., Savasta, M., Kerkerian-Le Goff, L., & Vitek, J. L. (2004). Uncovering the mechanism (s) of action of deep brain stimulation: Activation, inhibition, or both. *Clinical Neurophysiology, 115*(6), 1239–1248.

Miniussi, C., Harris, J. A., & Ruzzoli, M. (2013). Modelling noninvasive brain stimulation in cognitive neuroscience. *Neuroscience & Biobehavioral Reviews, 37*(8), 1702–1712.

Miranda, P. C., Mekonnen, A., Salvador, R., & Ruffini, G. (2013). The electric field in the cortex during transcranial current stimulation. *Neuroimage, 70*, 48–58.

Morrell, M. J. (2011). Responsive cortical stimulation for the treatment of medically intractable partial epilepsy. *Neurology, 77*(13), 1295–1304.

Nasr, K., Haslacher, D., Dayan, E., Censor, N., Cohen, L. G., & Soekadar, S. R. (2022). Breaking the boundaries of interacting with the human brain using adaptive closed-loop stimulation. *Progress in Neurobiology, 216*, 102311.

Nitsche, M. A., Cohen, L. G., Wassermann, E. M., Priori, A., Lang, N., Antal, A., et al. (2008). Transcranial direct current stimulation: State of the art 2008. *Brain Stimulation, 1*(3), 206–223.

Opitz, A., Paulus, W., Will, S., Antunes, A., & Thielscher, A. (2015). Determinants of the electric field during transcranial direct current stimulation. *Neuroimage, 109*, 140–150.

Parastarfeizabadi, M., & Kouzani, A. Z. (2017). Advances in closed-loop deep brain stimulation devices. *Journal of Neuroengineering and Rehabilitation, 14*, 1–20.

Pascual-Leone, A., Amedi, A., Fregni, F., & Merabet, L. B. (2005). The plastic human brain cortex. *Annual Review of Neuroscience, 28*, 377–401.

Polanía, R., Nitsche, M. A., & Ruff, C. C. (2018). Studying and modifying brain function with non-invasive brain stimulation. *Nature Neuroscience, 21*(2), 174–187.

Reithler, J., Peters, J. C., & Sack, A. T. (2011). Multimodal transcranial magnetic stimulation: Using concurrent neuroimaging to reveal the neural network dynamics of noninvasive brain stimulation. *Progress in Neurobiology, 94*(2), 149–165.

Ridding, M. C., & Ziemann, U. (2010). Determinants of the induction of cortical plasticity by noninvasive brain stimulation in healthy subjects. *The Journal of Physiology, 588*(13), 2291–2304.

Rosin, B., Slovik, M., Mitelman, R., Rivlin-Etzion, M., Haber, S. N., Israel, Z., et al. (2011). Closed-loop deep brain stimulation is superior in ameliorating parkinsonism. *Neuron, 72*(2), 370–384.

Rossi, S., Hallett, M., Rossini, P. M., Pascual-Leone, A., & Safety of TMS Consensus Group. (2009). Safety, ethical considerations, and application guidelines for the use of transcranial magnetic stimulation in clinical practice and research. *Clinical Neurophysiology, 120*(12), 2008–2039.

Rossini, P. M., Burke, D., Chen, R., Cohen, L. G., Daskalakis, Z., Di Iorio, R., et al. (2015). Noninvasive electrical and magnetic stimulation of the brain, spinal cord, roots and peripheral nerves: Basic principles and procedures for routine clinical and research application. An updated report from an IFCN Committee. *Clinical Neurophysiology, 126*(6), 1071–1107.

Rouse, A. G., Stanslaski, S. R., Cong, P., Jensen, R. M., Afshar, P., Ullestad, D., et al. (2011). A chronic generalized bi-directional brain–machine interface. *Journal of Neural Engineering, 8*(3), 036018.

Ruffini, G., Fox, M. D., Ripolles, O., Miranda, P. C., & Pascual-Leone, A. (2014). Optimization of multifocal transcranial current stimulation for weighted cortical pattern targeting from realistic modeling of electric fields. *Neuroimage, 89*, 216–225.

Scangos, K. W., Khambhati, A. N., Daly, P. M., Makhoul, G. S., Sugrue, L. P., Zamanian, H., et al. (2021). Closed-loop neuromodulation in an individual with treatment-resistant depression. *Nature Medicine, 27*(10), 1696–1700.

Sitaram, R., Ros, T., Stoeckel, L., Haller, S., Scharnowski, F., Lewis-Peacock, J., et al. (2017). Closed-loop brain training: The science of neurofeedback. *Nature Reviews Neuroscience, 18*(2), 86–100.

Sporns, O. (2011). *Networks of the brain*. Massachusetts.

To, W. T., De Ridder, D., Hart, J., Jr., & Vanneste, S. (2018). Changing brain networks through noninvasive neuromodulation. *Frontiers in Human Neuroscience, 12*, 128.

Turrigiano, G. (2012). Homeostatic synaptic plasticity: Local and global mechanisms for stabilizing neuronal function. *Cold Spring Harbor Perspectives in Biology, 4*(1), a005736.

Tyler, W. J. (2011). Noninvasive neuromodulation with ultrasound? A continuum mechanics hypothesis. *The Neuroscientist, 17*(1), 25–36.

Wagner, T. A., Zahn, M., Grodzinsky, A. J., & Pascual-Leone, A. (2004). Three-dimensional head model simulation of transcranial magnetic stimulation. *IEEE Transactions on Biomedical Engineering, 51*(9), 1586–1598.

Wiseman, M., Sewell, I. J., Nestor, S. M., Giacobbe, P., Hamani, C., Lipsman, N., & Rabin, J. S. (2024). Cognitive effects of focal neuromodulation in neurological and psychiatric disorders. *Nature Reviews Psychology*, 1–19.

Neurofeedback Basics and Applications

Techniques, Benefits, and Ethical Considerations

Contents

Test your learning and check your understanding of this book's contents: use the "Springer Nature Flashcards" app to access questions using ▶ https://sn.pub/kmb-jyz. To use the app, please follow the instructions in ▶ Chap. 1.

This chapter explores the innovative realm of neurofeedback, a sophisticated branch of biofeedback that empowers individuals to modulate their own brain activity using operant conditioning principles and advanced brain imaging technologies. Neurofeedback utilizes real-time feedback from EEG and fMRI to enable conscious adjustment of brain functions, tapping into the brain's neuroplastic capabilities to enhance cognitive and emotional processing. The technique is presented as a potent tool for clinical applications, effectively addressing a range of neurological and psychiatric disorders such as attention-deficit/hyperactivity disorder (ADHD), epilepsy, anxiety, and depression, and offering a viable alternative or complement to traditional drug therapies. Additionally, neurofeedback is explored for its potential in cognitive enhancement and mental wellness among healthy individuals, highlighting its benefits in improving cognitive abilities, promoting relaxation, and enhancing overall well-being. This chapter sets the stage for a deeper exploration of neurofeedback's foundational concepts, neurophysiological mechanisms, and broader implications for mental health and cognitive enhancement in Volume 2.

Learning Objectives

1. Understand Neurofeedback Principles: Grasp the foundational principles of neurofeedback, including the role of operant conditioning and real-time feedback in enabling individuals to consciously adjust their brain functions.
2. Explore Neurofeedback Technologies: Learn about the technologies that enable neurofeedback, particularly EEG and fMRI, and understand how these tools capture and feed back brain activity to the user.
3. Assess Clinical Applications: Understand how neurofeedback leverages the brain's neuroplastic capabilities to aid in rehabilitation, enhancing cognitive functions and compensating for neurological deficits.
4. Discuss Ethical and Future Implications: Consider the ethical implications of neurofeedback, especially in terms of consent and its use in cognitive enhancement and discuss the future potential of integrating neurofeedback with other therapeutic modalities.

Neurofeedback, a sophisticated branch of biofeedback, is a groundbreaking non-invasive approach designed to empower individuals to control their own brain activity. This method relies heavily on the principles of operant conditioning, a learning process where behaviors are shaped through rewards and punishments to encourage desirable changes. Neurofeedback integrates state-of-the-art real-time brain imaging technologies, such as EEG and fMRI, to provide users with immediate visual or auditory feedback on their brain activity. This feedback loop enables individuals to consciously adjust their brain functions, tapping into the brain's neuroplasticity to enhance cognitive and emotional processing (Sterman, 2000; Hammond, 2005).

The applications of neurofeedback are extensive, offering significant therapeutic potential for a range of neurological and psychiatric disorders. It has been recognized as a viable option for managing conditions like attention-deficit/hyperactivity disorder (ADHD), epilepsy, anxiety, and depression, often serving as an adjunct or alternative to traditional drug therapies. Research indicates that neurofeedback can substantially alleviate symptoms of these disorders, improving life quality and functional outcomes for many patients (Arns et al., 2009; Micoulaud-Franchi et al., 2014).

Beyond clinical applications, neurofeedback is also explored for its potential in cognitive enhancement and mental wellness among healthy individuals. Studies have shown its effectiveness in improving cognitive abilities such as memory, attention, and executive functions. This expands its applicability to enhancing mental performance in various settings, from academic to professional environments (Gruzelier, 2014). Additionally, neurofeedback facilitates greater mind–body integration, promoting relaxation, reducing stress, and enhancing overall well-being, thus serving as a tool for personal development and mental health maintenance (Schoenberg & David, 2014).

In the following chapters, we will explore the foundational concepts of neurofeedback, and the neurophysiological mechanisms that facilitate this technique. This discussion aims to provide a thorough understanding of neurofeedback's role in neuroscience and mental health, setting the stage for further detailed exploration in Volume 2.

10

Box 10.1 Brain Wave

Definition: A brain wave refers to the rhythmic electrical activity generated by the neurons in the brain. These oscillations are detected using electroencephalography (EEG) and are characterized by their frequency, amplitude, and phase. Brain waves are typically classified based on their frequency into several categories, each associated with different types of brain activity and states of consciousness.

■ **Key Types:**

Delta Waves: The slowest type of brain waves, predominantly present during deep sleep.

Theta Waves: Associated with light sleep, relaxation, and creativity, these waves are slower than alpha waves but faster than delta waves.

Alpha Waves: Common during states of wakeful relaxation with closed eyes, these waves are slower than beta waves and are associated with calm, non-aroused states.

Beta Waves: Dominant during waking states involving active thinking, problem-solving, and active concentration.

Gamma Waves: The fastest type of brain waves, associated with high-level cognitive functioning and information processing.

Significance: Brain waves are crucial for understanding brain activity and can be indicative of a person's mental state, alertness, and overall brain health. Changes in normal brain wave patterns can be used to diagnose neurological conditions, such as epilepsy and sleep disorders.

10.1 Historical Context

The historical roots of neurofeedback trace back to seminal research conducted in the mid-twentieth century, marking a period of groundbreaking discoveries that set the stage for the development of neurofeedback as a scientific discipline. In the late 1950s and early 1960s, two parallel streams of research unfolded, each contributing uniquely to the foundational knowledge and application of neurofeedback.

One of the earliest pivotal moments in the field's history was the work of Joe Kamiya at the University of Chicago. Kamiya's pioneering research, published in 1968, revealed a remarkable human capability: individuals could learn to exert conscious control over their alpha brainwave activity. Alpha waves, oscillating between 8 and 12 Hz, are commonly associated with states of relaxation and calmness. Through operant conditioning techniques, participants in Kamiya's studies were able to increase or decrease their alpha wave activity by receiving immediate auditory feedback. This discovery not only challenged prevailing notions about the involuntary nature of brainwave patterns but also opened new vistas for exploring the potential of mental self-regulation for enhancing mental health and well-being (Kamiya, 1969).

Concurrently, at the University of California, Los Angeles (UCLA), Barry Sterman and his colleagues embarked on a series of experiments that would further cement the therapeutic potential of neurofeedback. Sterman's initial studies, motivated by an interest in sleep and arousal patterns, inadvertently led to a landmark finding. In 1969, Sterman and his team trained cats to increase their sensorimotor rhythm (SMR) brainwaves, a type of brain activity linked to physical relaxation while maintaining mental alertness. Following this conditioning, the cats demonstrated a remarkable resistance to chemically induced seizures. This observation not only highlighted the profound impact of brainwave modulation on neurological function but also laid the groundwork for neurofeedback's applications in treating seizure disorders and other neurological conditions (Sterman et al., 1974).

Together, the work of Kamiya and Sterman represents the dual pillars upon which the edifice of neurofeedback was constructed. Their research not only demonstrated the feasibility of brainwave control through feedback mechanisms but also illuminated the therapeutic implications of such control, paving the way for subsequent developments in the field. These early experiments underscored the potential of neurofeedback to revolutionize approaches to mental health treatment, cognitive enhancement, and the broader understanding of human brain function.

10.2 Theoretical Basis

Neurofeedback represents an advanced approach for influencing brain activity, deeply ingrained in the principles of operant conditioning—a pivotal theory in behavioral psychology. Initially detailed by B.F. Skinner in 1938, this learning theory explains how behaviors are molded and altered through carefully applied rewards and punishments. Skinner's seminal work on operant conditioning established a foundation for understanding behavioral acquisition and modification based on the consequences of actions, highlighting the importance of external stimuli in driving behavioral changes (Skinner, 1938).

Incorporating operant conditioning into neurofeedback enables precise control over brain activity, presenting a contemporary method to enhance cognitive abilities and address a variety of neurological and psychiatric conditions. Utilizing the brain's natural plasticity and capacity for learned behavior, neurofeedback provides a distinctive method for improving mental health and well-being. This approach builds upon the fundamental behavioral psychology concepts introduced by Skinner, enriched by decades of subsequent research that has expanded our understanding of behavior modification (Sterman, 2000; Hammond, 2005).

Neurofeedback operates on the idea that individuals can learn to alter their own brain activity by leveraging real-time feedback. This section explores the core principles underpinning neurofeedback, such as operant conditioning, neuroplasticity, self-regulation, and the feedback loop mechanism, providing a framework for understanding how this technique facilitates active participation in one's own cognitive and psychological health enhancement.

10.3   Operant Conditioning

Operant conditioning operates on the premise that behaviors followed by positive outcomes are more likely to recur, while those followed by negative outcomes are less likely to be repeated. This principle is applied in neurofeedback through the use of real-time feedback signals that reinforce desirable brainwave patterns or discourage undesirable ones. Key to this process are the concepts of reinforcement, which aims to increase the frequency of a desired behavior, and punishment, which seeks to decrease the likelihood of an unwanted behavior. In the context of neurofeedback, reinforcement might involve providing positive feedback when an individual successfully increases a specific beneficial brainwave pattern, thereby encouraging the continuation of that behavior. Conversely, the absence of feedback or the provision of negative feedback when undesirable brainwave patterns are detected acts as a form of punishment, discouraging those patterns (Sterman, 2000; Hammond, 2005).

10.3.1   Reinforcement and Punishment

Operant conditioning, a foundational theory within the field of behavioral psychology, essentially relies on the strategic application of two primary mechanisms: reinforcement, which is designed to increase the likelihood of a behavior's occurrence, and punishment, which aims to decrease that likelihood. This conceptual framework, rigorously developed by B.F. Skinner in the mid-twentieth century, posits that behavior is a function of its consequences, thus behaviors that are followed by favorable outcomes are more likely to be repeated, whereas those followed by unfavorable outcomes are less likely to recur (Skinner, 1938).

Reinforcement is further subdivided into positive reinforcement, where a desirable stimulus is presented in response to a behavior, thereby increasing the probability of that behavior being repeated, and negative reinforcement, where an unpleasant stimulus is removed following a behavior, also increasing the likelihood of the behavior's recurrence. Positive reinforcement might involve giving a reward to reinforce a desired action, while negative reinforcement could entail the removal of an aversive condition as a result of the desired behavior (Skinner, 1953).

Conversely, punishment operates by introducing an adverse stimulus (positive punishment) or removing a desirable one (negative punishment) following an undesired behavior, with the intent of reducing the frequency of that behavior. For instance, positive punishment could include the application of a negative consequence after an undesired action, whereas negative punishment involves taking away something favorable to decrease the likelihood of an undesired behavior repeating (Skinner, 1953).

1. **Positive Reinforcement**

   The strategy of enhancing the probability of a behavior's recurrence through the introduction of a desirable stimulus subsequent to that behavior is termed positive reinforcement. This technique is predicated on the principle that behaviors followed by rewarding or favorable outcomes are more likely to be repeated in the future. B.F. Skinner, a pioneer in the study of operant conditioning, extensively documented this phenomenon, illustrating how positive reinforcement could be effectively utilized to shape and modify behaviors (Skinner, 1957).

   A quintessential example of positive reinforcement can be observed in educational and parenting practices, where a child receives a reward for exhibiting a desired behavior. For instance, presenting a child with a small treat or privilege as a reward for diligently completing their homework serves as a tangible form of positive reinforcement. The treat acts as a reinforcing stimulus, increasing the child's motivation to engage in the desired behavior of completing homework assignments promptly in anticipation of future rewards. This method not only encourages the repetition of the behavior but also fosters a positive association with the task at hand, thereby enhancing the learning experience and behavioral outcomes (Skinner, 1957; Kazdin, 1982).

   The application of positive reinforcement is grounded in the concept of behavior modification, a core aspect of operant conditioning theory. Through the strategic use of reinforcements, individuals can be encouraged to adopt more adaptive behaviors, thereby facilitating behavioral change and improvement in various settings, including schools, homes, and therapeutic environments. This approach underscores the importance of incentives in influencing behavior and provides a valuable tool for educators, parents, and clinicians in promoting desired behaviors and skills (Kazdin, 1982).

2. **Negative Reinforcement**

The concept of negative reinforcement involves the removal or avoidance of an aversive stimulus to enhance the frequency of a desired behavior. This principle operates on the foundation that the elimination of an unfavorable condition following a specific behavior can increase the likelihood of that behavior being elicited again. B.F. Skinner, a seminal figure in behavioral psychology, extensively discussed this mechanism in his work, highlighting its effectiveness in behavior modification and learning processes (Skinner, 1953).

A classic illustration of negative reinforcement can be found in daily routines, such as the use of an alarm clock. When an alarm emits a loud sound to awaken an individual, the sound is perceived as an aversive stimulus. The act of turning off the alarm—to cease the unpleasant noise—upon waking functions as negative reinforcement. The removal of the loud sound reinforces the action of waking up at the predetermined time. This process not only encourages the individual to wake up when the alarm sounds in the future but also utilizes the principle of avoidance learning, where the motivation to eliminate the aversive stimulus (the alarm sound) leads to the repeated performance of the desired behavior (waking up at a specific time) (Skinner, 1953; Catania, 1992).

Negative reinforcement is distinguished from punishment in operant conditioning; while both involve aversive stimuli, negative reinforcement aims to increase desirable behaviors by removing or avoiding negative conditions, whereas punishment seeks to decrease undesirable behaviors by introducing or removing stimuli. The nuanced application of negative reinforcement provides a powerful tool in shaping behavior, with broad implications across educational, therapeutic, and everyday settings, enabling individuals to adapt and modify behaviors in response to their environment (Catania, 1992).

3. **Positive Punishment**

The implementation of positive punishment involves the presentation of an adverse outcome or stimulus subsequent to an undesirable behavior, with the intent of reducing the frequency of that behavior. This approach is grounded in the principle that the introduction of an unfavorable consequence in response to a specific action will decrease the likelihood of that action being repeated in the future. The concept of positive punishment is distinct from negative reinforcement; while negative reinforcement seeks to increase desired behaviors by removing aversive stimuli, positive punishment aims directly at diminishing unwanted behaviors through the addition of unpleasant stimuli. Azrin and Holz's (1966) research in the realm of behavior modification provides a foundational understanding of how positive punishment can be effectively utilized to alter behavior patterns.

A commonly cited example of positive punishment in both theoretical discussions and practical applications involves corrective measures in response to misbehavior. When a child engages in inappropriate or unacceptable behavior, the imposition of additional chores serves as a form of positive punishment. The extra chores act as an unfavorable stimulus, directly linked to the misbehavior, with the intention of discouraging such actions in the future. This method of behavioral correction is based on the premise that the association between the undesired behavior and the immediate negative consequence will lead to a reduction in the occurrence of that behavior (Azrin & Holz, 1966).

The use of positive punishment is a subject of considerable debate, particularly concerning its effectiveness and ethical implications in various settings, including child rearing, education, and behavior therapy. While positive punishment can be an immediate and powerful deterrent for undesirable behaviors, its long-term effectiveness and potential psychological impacts imply critical considerations. The application of positive punishment strategies requires careful consideration of the context, the severity of the punishment, and the specific behavior being addressed to ensure that it is used appropriately and effectively (Kazdin, 1982; Skinner, 1953).

4. **Negative Punishment**

The strategy of negative punishment involves the withdrawal of a favorable or positive stimulus with the aim of reducing the likelihood of a specific behavior's recurrence. This behavioral modification technique is predicated on the principle that the removal of a desirable condition or reward in response to an undesirable behavior acts as a deterrent, thereby decreasing the frequency of that behavior. Negative punishment is distinct from positive punishment; while the latter involves introducing an aversive stimulus to suppress unwanted behaviors, negative punishment focuses on the removal of pleasant stimuli as a consequence of those behaviors. This approach is elaborated in the work of Redmon (1991), which discusses the application of negative punishment in behavioral interventions.

◘ Figure 10.1 depicts operant conditioning as a mechanism of pain with reinforcement and punishment (Bąbel, 2020).

A practical application of negative punishment can be observed in parenting or educational contexts, where privileges or rewards are contingent on certain behaviors or performance criteria. For instance, a teenager who receives poor grades may experience the withdrawal of video game privileges as a form of negative punishment. The underlying logic is that by removing access to a valued activity in response to unsatisfactory educational performance, the teenager is incentivized to improve their grades to regain their privileges. This scenario exemplifies how negative punishment can be employed to encourage behavioral change by establishing a clear consequence for failing to meet established expectations (Redmon, 1991).

The use of negative punishment raises important considerations regarding its effectiveness and the potential for unintended consequences, such as resentment or reduced motivation. As with all behavioral modification techniques, the implementation of negative punishment should be thoughtful and strategic, taking into account the individual's needs, the specific behavior being targeted, and the overall goals of the intervention. Careful application of negative punishment can be a valuable component of a comprehensive behavior management strategy, promoting desirable behaviors through the structured removal of rewards (Kazdin, 1982; Skinner, 1953).

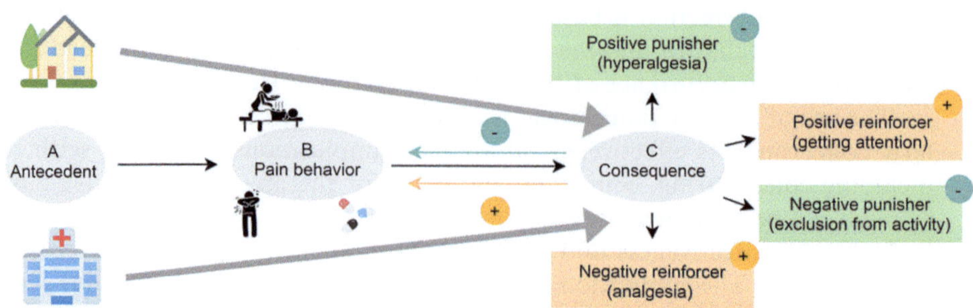

☐ **Fig. 10.1** Operant conditioning serves as a fundamental mechanism in the modulation of pain behaviors. These behaviors, such as rubbing a painful area, taking pain medication, or seeking therapeutic massage, can be influenced by both positive and negative reinforcements or punishments that occur in specific contexts, like at home or in a clinical setting. Positive reinforcements might include receiving attention and care from loved ones, which can increase the likelihood of engaging in certain pain behaviors when similar situations arise. Conversely, negative reinforcements could involve the relief of pain through analgesic medications, promoting the repetition of such behaviors to avoid discomfort. Furthermore, pain behaviors can also be subject to punishments that either encourage or discourage these actions. Positive punishment might involve an increase in pain after certain activities, known as hyperalgesia, which discourages the behavior. On the other hand, negative punishment could involve being excluded from activities due to visible pain behaviors, potentially decreasing these behaviors over time. In essence, the presence of an antecedent stimulus (also called discriminative stimulus) acts as a cue that triggers pain behaviors, which are then either reinforced or punished. Reinforced behaviors are more likely to recur under similar stimuli, while those that are punished tend to diminish. This dynamic interaction between behaviors, reinforcements, punishments, and contextual stimuli highlights the complexity of pain management through the lens of operant conditioning. From Bąbel (2020)

10.3.2 Shaping

Shaping represents a sophisticated technique for the gradual development of complex behaviors through the systematic reinforcement of successive approximations toward the target behavior. This method hinges on the concept that by selectively reinforcing behaviors that are progressively closer to the desired outcome, an individual or animal can be guided to exhibit the target behavior that would not naturally occur to such a precise degree. The foundational principle of shaping involves breaking down the learning process into incremental steps, each step bringing the subject closer to the final desired behavior, thereby facilitating learning that might otherwise be too challenging to achieve in a single step (Skinner, 1953). To this purpose, the antecedent stimulus—often called in behavioral psychology *discriminative stimulus*—aimed at triggering a specific behavioral response, can be represented by a physical signal perceivable in one sensory domain (e.g., the frequency of a target tone, or the bar level in a virtual high jump task, or the target cycling speed of an avatar in a virtual bike competition). The physical signal also indicates the availability of reinforcement or punishment. Gradually, smoothly and successively modulating such physical signal by small steps, according to the performance of the test subject, the shaping takes place until the desired behavioral effect.

The utility of shaping extends across various domains, notably in animal training, where it plays a pivotal role in teaching animals to perform tasks or exhibit behaviors that are outside their normal repertoire. Peterson (2004) highlights the effectiveness of shaping in animal training, illustrating how trainers can guide animals through a series of small, manageable steps toward the execution of complex or unnatural tasks. For example, a trainer might reward a dolphin for initially touching a ball with its nose, gradually requiring more specific interactions with the ball, such as pushing it through a hoop, before the final behavior is fully mastered and consistently performed in exchange for reinforcement.

Shaping is not limited to animal training; it also finds application in therapeutic settings, education, and skill development, where incremental learning and behavior modification are essential. This technique underscores the importance of positive reinforcement in learning and the capacity for operant conditioning to facilitate the acquisition of new skills and behaviors through structured, step-by-step reinforcement strategies (Kazdin, 2012).

The concept of shaping reaffirms the adaptive nature of behavior under the influence of environmental contingencies and the potential for operant conditioning principles to guide the learning process. By applying shaping techniques, complex behaviors can be taught and refined, demonstrating the profound impact of systematic reinforcement on behavior modification (Peterson, 2004; Skinner, 1953).

10.3.3 Extinction

Extinction, a critical concept within the operant conditioning framework, refers to the process by which a previously reinforced behavior decreases in frequency or ceases altogether due to the discontinuation of reinforcement. This phenomenon underscores the fundamental principle that the persistence of learned behaviors is closely tied to their consequences. As articulated by B.F. Skinner in his seminal work on behavioral psychology, behaviors that are not reinforced will gradually diminish over time, leading to extinction. The principle of extinction is instrumental in understanding how behaviors can be modified or eliminated through the manipulation of reinforcement schedules (Skinner, 1957).

The process of extinction plays a vital role in both theoretical and applied settings, offering insights into how learned behaviors can be altered. For instance, in therapeutic contexts, extinction principles are applied to reduce or eliminate maladaptive behaviors by systematically withdrawing the reinforcements that maintain them. This approach can be particularly effective in treating phobias, addictions, and other behavior-related disorders, by gradually reducing the undesirable behaviors through the careful management of reinforcement (Kazdin, 2012).

Moreover, the concept of extinction has implications for understanding habit formation, skill acquisition, and the maintenance of behavioral change. It highlights the importance of consistent reinforcement for the continuation of desired behaviors and provides a mechanism for behavior modification through the strategic removal of reinforcement. However, it's noteworthy that the process of extinction can be accompanied by a temporary increase in the behavior's frequency or intensity, known as an extinction burst, before the behavior diminishes, reflecting the complexity of behavior change processes (Lerman & Iwata, 1995).

Extinction emphasizes the dynamic nature of behavior and its susceptibility to environmental influences, reinforcing the significance of reinforcement patterns in shaping and modifying behavior over time (Skinner, 1957).

10.3.4 Schedules of Reinforcement

In the exploration of operant conditioning, B.F. Skinner and his colleague, C.B. Ferster, delineated a series of reinforcement schedules that play a pivotal role in shaping the learning process. These schedules, characterized by distinct patterns and timings of reinforcement delivery, critically influence the rate and strength of behavior acquisition. The foundational work presented in "Schedules of Reinforcement" (Ferster & Skinner, 1957) elaborates on four primary types of reinforcement schedules: fixed-ratio, variable-ratio, fixed-interval, and variable-interval. Each schedule imposes a unique structure on how and when reinforcements are administered, thereby affecting the dynamics of learning and behavior modification.

1. **Fixed-Ratio Schedule**: This schedule delivers reinforcement after a specified number of responses. For instance, a reward might be given after every fifth response. This schedule tends to produce a high rate of responding, with a brief pause following the delivery of reinforcement, reflecting the direct link between the behavior's frequency and the receipt of rewards.
2. **Variable-Ratio Schedule**: Under this schedule, reinforcement is provided after a random number of responses, with the number varying around a specific average. This unpredictability leads to a high and steady rate of responding, as the next reinforcement could occur at any time, making it particularly effective in maintaining behaviors over time.
3. **Fixed-Interval Schedule**: Reinforcement is available only after a set period has elapsed since the last reinforcement, regardless of how many responses occur during that interval. This schedule typically results in a "scalloped" pattern of responses, with a slow start after reinforcement followed by an increase in rate as the time for the next reinforcement approaches.
4. **Variable-Interval Schedule**: Here, reinforcement becomes available after varying intervals of time, with the intervals fluctuating around an average. This schedule produces a moderate, steady rate of responding, as the first response after the interval has passed is reinforced, without the predictability of when the reinforcement will be available.

The strategic application of these programs in various contexts, from educational settings to therapy and behavior modification programs, demonstrates their utility in effectively managing and shaping behavior. The differences in response patterns elicited by these schedules underscore the nuanced ways in which reinforcement can be optimized to achieve specific learning goals and behavioral outcomes (Ferster & Skinner, 1957).

10.3.5 **Discriminative Stimulus**

The concept of a discriminative stimulus plays a crucial role in the learning and behavioral modification processes. A discriminative stimulus is a specific cue or environmental physical signal that indicates the availability of reinforcement or punishment, effectively guiding the individual to understand under what conditions a particular behavior will lead to positive or negative consequences. This stimulus enables the organism to distinguish between situations in which a behavior will be reinforced (leading to a reward) or punished (resulting in an adverse outcome), thereby facilitating more targeted and adaptive behavioral responses.

The discriminative stimulus functions as a critical component in the learning process by marking the contextual boundaries for behavior reinforcement. This signaling mechanism aids in the development of more complex behavioral repertoires by teaching the individual to modulate their behavior based on the presence or absence of particular cues in the environment. For example, a traffic light serves as a discriminative stimulus in everyday life; the green light signals the availability of reinforcement (proceeding through the intersection without penalty), while the red light indicates that proceeding would be punished (potentially receiving a citation for running the red light).

The theoretical underpinnings of discriminative stimuli and their role in shaping behavior were extensively explored by Skinner and other behavioral psychologists, who demonstrated how these cues could be systematically manipulated to enhance learning and behavior modification outcomes. By selectively reinforcing behaviors in the presence of specific stimuli, individuals can learn to perform certain actions more reliably when those cues are present, optimizing their behavior to align with the contingencies of their environment (Skinner, 1938; Skinner, 1953).

The application of discriminative stimuli extends beyond laboratory settings to practical applications in education, therapy, and behavior management, offering a powerful tool for encouraging desirable behaviors and discouraging undesirable ones. Through the strategic use of these stimuli, it is possible to facilitate more precise and effective learning experiences, underscoring the importance of environmental cues in the operant conditioning process.

Operant conditioning, a cornerstone concept in behavioral psychology, finds widespread application across diverse domains such as education, psychotherapy, animal training, and behavior modification programs. ◘ Table 10.1 summarizes the different operant conditioning principles. In the innovative field of neurofeedback, operant conditioning is ingeniously applied to encourage specific brainwave patterns, leveraging the brain's inherent neuroplastic abilities. Neurofeedback sessions typically involve monitoring brain activity via EEG and providing real-time feedback to individuals based on their brainwave patterns.

For example, in a program aimed at increasing alpha wave frequency, which is associated with states of relaxation, the generation of alpha waves by the participant is positively reinforced. This reinforcement, often in the form of visual or auditory feedback (e.g., a visual level meter, or an auditory signal that changes in magnitude or frequency), encourages the brain to produce more alpha waves. The

◼ Table 10.1 Different operant conditioning principles

Principle	Type	Description	Example
Reinforcement	Positive	Introduces a pleasant stimulus to increase a behavior	Giving a child praise for doing homework
	Negative	Removes an unpleasant stimulus to increase a behavior	Turning off an alarm when waking up early
Punishment	Positive	Adds an unpleasant stimulus to decrease a behavior	Scolding a child for running into the street
	Negative	Removes a pleasant stimulus to decrease a behavior	Taking away a toy when a child misbehaves
Shaping	–	Reinforces closer approximations of a desired behavior gradually	Training a dog to roll over by rewarding incremental steps
Extinction	–	Behavior decreases when reinforcement is withheld	Ignoring a tantrum so that it decreases over time
Schedules of reinforcement	–	Different schedules influence how and when behaviors are reinforced	Giving a treat to a pet after it performs a trick (fixed-ratio)
Discriminative stimulus	–	A stimulus in the presence of which a particular response will be reinforced	A whistle blown before mealtime makes a dog expect food

success of this approach is underpinned by the brain's capacity for neuroplasticity, the remarkable ability to reorganize and adapt in response to new experiences, training, or environmental changes. Research by Pascual-Leone and colleagues underscores the transformative potential of neuroplasticity, highlighting how targeted interventions, such as neurofeedback, can induce lasting changes in brain function and structure (Pascual-Leone et al., 2005).

In this context, lasting changes in brain function and structure correspond to learning effects. From a neuro-functional point of view, learning needs as prerequisites intact neural processing within the cortico-thalamic-striatal loop (Hinterberger et al., 2005) and preserved plasticity of the N-methyl-D-aspartate (NMDA) receptor-dependent synaptic connections of this system (Koralek et al. 2012). Research by Birbaumer et al. (1990) highlighted that NMDA synapses become strengthened through learning as the post-synaptic membrane potential should vary depending on prior cell activity history. In addition, specific neural substrates, such as the reticular nucleus of the thalamus, are thought to operate as reward-regulatory components (Birbaumer et al., 1990). Two exemplary applications showing the necessity of these two prerequisites (intact cortico-thalamic-striatal processing and NMDA-plasticity) for skill learning consist of a neurofeedback task in humans, where modulation of slow cortical potentials was implemented (Hinterberger et al., 2005) and of a neuro-prosthetic control in animals (Koralek et al. 2012). As you read in ▸ Chap. 7, the neural processes that modulate the synaptic strength are called long-term potentiation and long-term depression, tested in vitro already in the end of last century (Bi & Poo, 1998). They represent essential mechanisms for neuroplasticity.

> **Box 10.2 Objective of Operant Conditioning in Neurofeedback**
>
> Operant conditioning is a fundamental principle in neurofeedback, where it is used to train individuals to voluntarily control their brain activity. This method relies on the concept of rewards and punishments to reinforce desirable brainwave patterns, thereby promoting beneficial changes in brain function.
>
> ■ **Key Mechanisms:**
>
> **Reinforcement and Punishment:** In neurofeedback, desirable brainwave patterns are typically reinforced by positive feedback or rewards, while less desirable patterns may be discouraged through the absence of rewards or through negative feedback. This training encourages the brain to produce more of the beneficial patterns.
>
> **Feedback Mechanisms:** Real-time feedback is critical in operant conditioning within neurofeedback. This feedback can be visual, auditory, or even tactile, depending on the system, and it provides immediate information about the individual's brainwave state.
>
> ■ **Applications:**
>
> **Behavioral and Cognitive Improvements:** By reinforcing specific brainwave patterns, neurofeedback can be used to improve focus, reduce anxiety, enhance sleep, and mitigate symptoms of ADHD.
>
> **Therapeutic Uses:** Operant conditioning through neurofeedback is also employed in therapeutic settings to help manage symptoms of epilepsy, depression, and other neurological conditions, by training patients to alter their own brain activity in beneficial ways.
>
> ■ **Challenges:**
>
> **Consistency of Training:** Effective operant conditioning requires consistent and repeated training sessions. The individual must be able to reliably reproduce the rewarded brain activity over time to achieve lasting changes.
>
> **Measurement of Progress:** Accurately measuring and quantifying changes in brain activity due to operant conditioning are essential to determine the effectiveness of neurofeedback and to make necessary adjustments to the training protocol.
>
> ■ **Future Directions:**
>
> **Enhanced Learning Algorithms:** Advancements in machine learning could improve the customization and effectiveness of feedback mechanisms, potentially speeding up the learning process and making treatments more efficient.

10.4 **Neuroplasticity**

Neuroplasticity, also widely recognized as brain plasticity, encapsulates the brain's remarkable capacity to undergo structural and functional changes in response to various experiences throughout an individual's life. This principle constitutes a foundational pillar of contemporary neuroscience, challenging earlier notions of a rigid, unchangeable brain and propelling the understanding that the brain is indeed a dynamic and malleable organ. The concept of neuroplasticity underscores the brain's ability to reorganize itself by forming new neural connections, thereby facilitating continuous learning and adaptation to new information, environments, and experiences (Buonomano & Merzenich, 1998).

The implications of neuroplasticity extend across the lifespan, from the rapid developmental changes observed in infancy and childhood to the adaptability displayed in adulthood and even into older age. This adaptive capacity of the brain plays a critical role in learning new skills, recovering from brain injuries, and compensating for lost functions. Research in this field has demonstrated that neuroplastic changes can result from a variety of experiences, including education, environmental influences, physical activity, and cognitive therapies.

Pioneering studies by researchers such as Merzenich and colleagues have provided empirical evidence supporting the concept of neuroplasticity, showing that the brain's cortical and subcortical regions can reorganize in response to training and rehabilitation efforts (Merzenich et al., 1984). Similarly, work by Pascual-Leone et al. has illustrated how the human brain cortex adapts to changes in sensory inputs and motor outputs, further highlighting the brain's plastic nature (Pascual-Leone et al., 2005).

These discoveries have profound implications for therapeutic practices, suggesting that interventions designed to harness the brain's plasticity can lead to significant improvements in cognitive and physical functions. As such, neuroplasticity is not merely a theoretical concept but a practical framework guiding the development of innovative neurorehabilitation techniques, educational strategies, and mental health interventions, marking a paradigm shift in our understanding of the brain's capacity for change and adaptation.

Neuroplasticity encompasses several key dimensions that elucidate the brain's adaptability and responsiveness to experiences and environmental changes. These aspects highlight the multifaceted nature of brain plasticity, demonstrating how structural and functional alterations underpin learning, recovery, and adaptation processes.

1. **Structural Plasticity**: Structural plasticity refers to modifications in the brain's physical architecture that emerge through learning and experiential engagement. Two primary mechanisms exemplify this aspect: neurogenesis and synaptic plasticity. Neurogenesis, the birth of new neurons, expands the brain's neural network, while synaptic plasticity, involving changes in the efficacy of synaptic connections, strengthens or weakens these connections based on activity levels. This dynamic interplay facilitates the brain's ability to restructure itself in response to ongoing learning and environmental demands (Kolb & Whishaw, 1998).
2. **Functional Plasticity**: Functional plasticity is the brain's capacity to redistribute functions from damaged regions to intact ones, a critical adaptive response

following injury. This form of plasticity is evident in recovery processes where, post-injury, unaffected brain areas assume the roles of compromised regions, thereby (partially) preserving functionality. Such adaptability underscores the brain's resilience and its ability to maintain operations despite localized damage (Nudo, 2006).

3. **Critical Periods**: Critical periods represent windows of heightened sensitivity during early developmental stages, wherein the brain exhibits an extraordinary degree of plasticity. In these intervals, the neural substrate is especially responsive to environmental stimuli, facilitating rapid learning and development. The concept of critical periods highlights the temporal aspects of neuroplasticity, emphasizing the influence of timing on the brain's developmental trajectory (Hensch, 2004).

4. **Long-Term Potentiation (LTP) and Long-Term Depression (LTD)**: LTP and LTD are molecular processes, as you read in ▶ Chap. 7, that respectively enhance or diminish synaptic strength, playing pivotal roles in learning and memory formation. LTP increases synaptic efficiency, facilitating stronger neural connections, whereas LTD reduces synaptic strength, allowing the brain to "unlearn" associations. These processes are integral to the dynamic modulation of synaptic networks, supporting the brain's ability to encode and erase memories (Malenka & Bear, 2004).

5. **Experience-Dependent Plasticity**: Experience-dependent plasticity underscores the brain's ability to reconfigure its synaptic connections based on experiential input. This form of plasticity is foundational to skill acquisition and adaptation to novel environments, illustrating how experiences shape the brain's structure and function. Through experience-dependent plasticity, the brain continually evolves, reflecting the cumulative impact of individual experiences on neural architecture (Buonomano & Merzenich, 1998).

Together, these dimensions of neuroplasticity illuminate the complex processes by which the brain evolves and adapts, underscoring the intrinsic connection between neural mechanisms and experiential factors. The exploration of these aspects provides profound insights into the brain's capacity for change, with significant implications for education, rehabilitation, and the understanding of human cognition and behavior.

In the context of rehabilitation, neuroplasticity serves as the foundational principle behind therapeutic strategies aimed at recuperating lost neurological functions following strokes or traumatic brain injuries. Kleim and Jones (2008) emphasize the pivotal role of neuroplasticity in rehabilitation, where targeted exercises and tasks are designed to stimulate affected brain areas, promoting the reestablishment of neural connections and the recruitment of alternative pathways to regain lost abilities. This approach underscores the potential of rehabilitation therapies to facilitate significant recovery by engaging the brain's plastic capabilities.

Furthermore, the intersection of neurofeedback and neuroplasticity represents a compelling area of exploration within neuroscience. Neurofeedback, a technique that utilizes real-time feedback to enable individuals to consciously modulate their brain activity, capitalizes on the principles of neuroplasticity to induce lasting changes in brain function. By providing individuals with the possibility to modulate their own neural patterns in specific contexts, neurofeedback offers a unique method for enhancing cognitive performance, managing psychiatric symptoms, and improving overall mental well-being.

The foundation of neurofeedback's effectiveness lies in its ability to engage neuroplastic mechanisms. Through repeated neurofeedback sessions, individuals learn to modulate their brainwave patterns, leading to sustained changes in brain function that reflect the principles of experience-dependent neuroplasticity. This process is supported by a growing body of research indicating that neurofeedback can induce lasting changes in neural activity patterns, thereby improving symptoms associated with various neurological and psychiatric conditions (Mahncke et al., 2006; Merzenich et al., 2014; Gruzelier, 2014; Thibault et al., 2016).

The convergence of neurofeedback and neuroplasticity not only highlights the adaptability of the brain but also underscores the potential for targeted interventions to facilitate meaningful and lasting changes in brain activity. As our understanding of neuroplasticity expands, so too does the potential for neurofeedback contributes to the development of personalized and effective treatments that harness the brain's natural capacity for change and growth.

10.5 Self-Regulation and the Brain

This section explores the mechanisms by which neurofeedback enhances the self-regulation of brain functions, thereby improving cognitive and emotional control through more effective management of brain activity. Self-regulation forms the bedrock of human cognition, essential for adaptively managing thoughts, emotions, and actions to achieve goals and respond to environmental demands. Neurofeedback, as a form of biofeedback, utilizes real-time displays of brain activity to foster this capacity. By providing individuals with direct, real-time insights into their own brainwave patterns, neurofeedback enables the conscious modulation of these patterns (Hammond, 2005). This empowerment through biofeedback can significantly enhance an individual's cognitive performance, boost emotional resilience, and promote overall mental health. The process involves learning to alter one's brainwaves to states associated with optimal function and well-being, thereby potentially alleviating symptoms of various neurological and psychiatric conditions. This method stands out by offering a unique, non-invasive pathway to mental healthcare that emphasizes active patient engagement and self-efficacy in managing brain function.

The core mechanism underlying neurofeedback involves the establishment of a feedback loop that integrates real-time monitoring of brain activity with instantaneous feedback provided to the individual. This dynamic interaction is pivotal for allowing conscious modulation of brainwave patterns, thereby empowering individuals to deliberately influence their neural activity (Gruzelier, 2014).

In practice, neurofeedback leverages advanced neuroimaging technologies, predominantly electroencephalography (EEG), to provide continuous, real-time observation of an individual's brainwave activities (Thibault et al., 2015). This data is then presented back to the user through visual or auditory feedback, depicting the current state of their neural dynamics. Such immediate feedback is instrumental in teaching individuals how to consciously adjust their brain activity, targeting specific brainwave patterns linked to desired cognitive or emotional outcomes (Heinrich et al., 2007). ■ Figure 10.2 (Enriquez-Geppert et al., 2019) depicts the neurofeedback experimental design.

10.5.1 **Neurofeedback and Cognitive Control**

Neurofeedback's effectiveness in enhancing cognitive control is anchored in its capacity to train individuals to modify distinct brainwave frequencies that correlate with different cognitive states. For example, beta waves, which are closely linked with attention and memory processes, are frequently targeted in neurofeedback protocols. Research conducted by Egner and Gruzelier (2004) provided a significant contribution to this field by demonstrating that neurofeedback training designed to modulate increase beta wave activity could lead to marked improvements in attention and working memory (Egner & Gruzelier, 2004). This manipu-

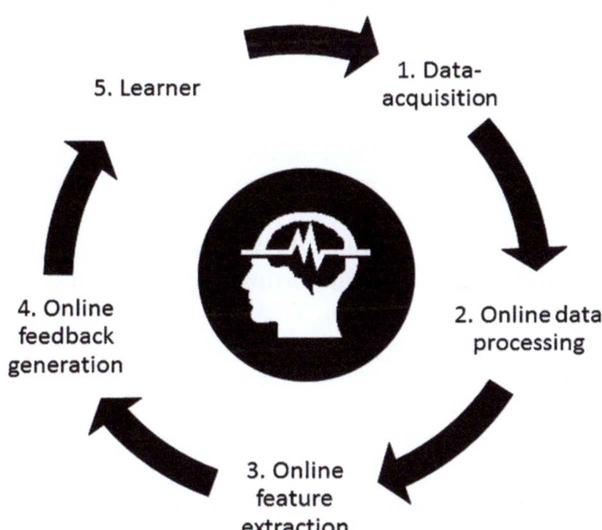

■ **Fig. 10.2** The diagram offers a detailed visualization of experimental setups in neurofeedback research. In these studies, participants engage in a specific task designed to elicit certain brain activity patterns. This brain activity is captured in real time, typically using technologies such as electroencephalography (EEG) or functional magnetic resonance imaging (fMRI). The data collected is then processed immediately, and the results are converted into a user-friendly format, such as a visual gauge displayed on a screen. Participants receive instructions on how to alter the levels of the gauge, which serves as a direct visual representation of their brain activity. The objective for participants is to consciously adjust their brainwave patterns by focusing on specific mental states or cognitive strategies that influence the gauge's readings. This process provides a dynamic and interactive platform for participants to learn how to modulate their neural functions through direct feedback, effectively linking their cognitive efforts with changes in brain activity. This schematic encapsulation not only details the flow of information and interaction in neurofeedback experiments but also underscores the practical applications of such research in teaching individuals to gain control over their cognitive and emotional states through self-regulation of brain activity. From Enriquez-Geppert et al. (2019)

lation of brainwave frequencies directly impacts cognitive functions, offering a potent method for enhancing capacities vital for self-regulation.

Their study exemplifies how neurofeedback can be strategically used to enhance cognitive functions by teaching individuals to control and adjust their brain activity consciously (Egner & Gruzelier, 2004). By focusing on specific frequencies like alpha, neurofeedback taps into the neural underpinnings of cognitive processes, enabling individuals to optimize their mental states through targeted brainwave modulation (Zoefel et al., 2011). This approach not only underscores the adaptive capabilities of the brain but also highlights neurofeedback's potential as a transformative tool for cognitive enhancement and the development of higher-order cognitive skills essential in various every day and professional contexts (Enriquez-Geppert et al., 2017).

10.5.2 Emotional Regulation Through Neurofeedback

Beyond the scope of cognitive improvements, neurofeedback has also demonstrated significant efficacy in enhancing emotional regulation, a critical component of overall self-regulation (Johnston et al., 2010). The technique involves training individuals to modulate their alpha wave frequencies, which are commonly associated with states of calmness and relaxation. This modulation is aimed at reducing symptoms of anxiety and depression, thereby promoting greater emotional stability. Hammond's seminal work in 2005 laid the groundwork for understanding how increased alpha activity can aid in managing emotional responses (Hammond, 2005).

Recent empirical studies, including those conducted by Cheon et al. (2015), have provided additional evidence supporting the effectiveness of neurofeedback in the domain of emotional regulation. These studies underscore neurofeedback's capacity to ameliorate symptoms of emotional dysregulation by reinforcing adaptive neural activity patterns. Through the targeted increase of alpha waves, neurofeedback enables individuals to achieve a more relaxed and stable emotional state, offering a non-pharmacological option for enhancing emotional well-being (Cheon et al., 2015).

These findings highlight neurofeedback's role as a versatile intervention that not only improves cognitive abilities but also equips individuals with the skills necessary to better manage their emotional states. This dual capability makes neurofeedback a valuable tool in both clinical and therapeutic settings, where it can be used to help individuals achieve a more balanced and effective regulation of their emotional and cognitive functions (Enriquez-Geppert et al., 2017).

10.5.3 Neurofeedback in Clinical Rehabilitation

The application of neurofeedback is notably impactful in the realm of clinical rehabilitation, especially for individuals recuperating from neurological injuries or disorders (Birbaumer et al., 2009). This advanced technique capitalizes on the brain's inherent neuroplastic capabilities, facilitating the retraining and recovery of brain functions that may have been impaired. Neurofeedback achieves this by guiding the brain to reorganize and strengthen neural connections, effectively compensating for lost functionalities (Haller et al., 2013).

Kleim and Jones, in their 2008 review, have highlighted the significant role of neurofeedback in the rehabilitation process, particularly in the context of stroke recovery. Their research points out how neurofeedback can be instrumental in facilitating motor recovery in stroke patients. The technique does so by specifically reinforcing neural pathways that are crucial for motor function. Through repeated sessions, patients are able to engage and strengthen these pathways, promoting the recovery of motor skills that were affected by the stroke (Kleim & Jones, 2008).

This approach underscores neurofeedback's pivotal role in harnessing the principles of neuroplasticity for functional restoration. By enabling the brain to adapt and modify its activity in response to feedback, neurofeedback provides a powerful tool for recovery and rehabilitation in clinical settings. The method not only helps in restoring physical abilities but also contributes to the overall resilience and adaptability of the brain, offering renewed hope and improved quality of life for patients dealing with the aftermath of neurological impairments (Renton et al., 2017).

10.5.4 Enhancing Executive Functions

Neurofeedback's profound impact on self-regulation is particularly visible in its capacity to enhance executive functions, including inhibitory control, cognitive flexibility, and planning. These higher-order cognitive processes are critical for effective self-regulation, as they enable individuals to adeptly manage and respond to the complex demands of social interactions and environmental challenges (Zimmerman & Schunk, 2004; Roebers, 2017).

Different research has been performed on neurofeedback training that specifically targets the prefrontal cortex, a crucial brain area responsible for executive functions. This region orchestrates a range of cognitive activities that govern decision-making, problem-solving, and behavior moderation. The findings from this research indicate that neurofeedback can significantly improve performance on tasks that measure executive functions. The training facilitates changes in brain activity that bolsters cognitive processes essential for planning, shifting between tasks, and controlling impulses. This suggests a viable neurobiological pathway through which neurofeedback enhances the brain's ability to perform these essential functions (Scheinost et al., 2013; Zotev et al., 2013; Johnston et al. 2010; Paret et al., 2016; Mennella et al., 2017).

By reinforcing and optimizing neural activity in the prefrontal cortex, neurofeedback provides a strategic intervention that strengthens the neural substrates underpinning self-regulation. This enhancement in executive functioning not only aids in daily cognitive tasks but also has broader implications for improving overall mental health and adaptive behavior (Zotev et al., 2014).

10.5.5   Lifelong Neuroplasticity and Adaptation

The widespread effectiveness of neurofeedback at various life stages underscores the universal and enduring nature of neuroplasticity, highlighting its capacity for facilitating cognitive and emotional adaptations throughout a person's lifetime. Neurofeedback is a dynamic tool that can be effectively applied across a diverse age spectrum—from young children grappling with developmental challenges to adults aiming for cognitive enhancement (Monastra et al., 2002), and even older adults addressing age-related cognitive declines. For children, neurofeedback has been used to mitigate symptoms of ADHD and autism, enhancing their ability to concentrate and interact socially (Arns et al., 2013; Sampedro Baena et al., 2021). In adults, the technique has been employed to refine cognitive abilities like memory and attention, which are crucial for professional and personal efficacy (Wang & Hsieh, 2013; Jiang et al., 2017).

Furthermore, in the elderly, neurofeedback training has shown promise in staving off the effects of cognitive aging, such as memory lapses and decreased processing speed, thus improving their quality of life. This versatility not only demonstrates neurofeedback's role in promoting self-regulation but also showcases its potential to enhance neural adaptability and resilience across the lifespan. Through targeted neural training, neurofeedback leverages the brain's inherent plasticity, offering a tailored approach to bolster cognitive and emotional health in individuals at any age (Sitaram et al., 2017).

Thus, neurofeedback synergistically combines operant conditioning, neuroplasticity, self-regulation, and the feedback loop to optimize brain function and behavior. Operant conditioning underlies the method, utilizing reinforcement to encourage desirable brainwave patterns, thus guiding the learning process inherent in neurofeedback. This process is supported by neuroplasticity, the brain's capacity to form new neural connections, allowing for lasting changes in brain activity and behavior as a result of neurofeedback training. Self-regulation is a key outcome, with individuals learning to control their neural processes and, by extension, their cognitive and emotional states, through real-time feedback. This feedback loop—monitoring brain activity, receiving feedback, and making adjustments—enables a dynamic interaction between the individual and the neurofeedback system. Together, these elements create a powerful framework for neurofeedback, promoting adaptive neural functioning and enhancing overall well-being (Sitaram et al., 2017).

Neurofeedback represents a significant leap forward in the quest to enhance self-regulation, offering a window into the dynamic interplay between the brain's electrical activity and our cognitive and emotional lives. Through the targeted modulation of brainwave patterns, individuals can achieve greater control over their mental states, leading to improvements in cognitive performance, emotional resilience, and adaptive behavior. As neuroscientific research continues to uncover the intricacies of neuroplasticity and self-regulation, neurofeedback stands poised to offer increasingly sophisticated approaches to enhancing human potential.

> **Box 10.3 Objective of Self-regulation in Neurofeedback**
>
> Self-regulation through neurofeedback involves training individuals to consciously control their own brain activity. This capability is fundamental for managing emotions, behaviors, and cognitive processes.
>
> ■ **Key Concepts:**
>
> **Neural Basis of Self-regulation:** Self-regulation is underpinned by several brain regions, including the prefrontal cortex, which plays a crucial role in decision-making, emotional regulation, and impulse control.
>
> **Mechanisms of Action:** Through neurofeedback, individuals learn to monitor and adjust their brainwave patterns in real-time. This practice enhances neural pathways involved in self-regulation, leading to improved cognitive functions and emotional control.
>
> ■ **Benefits:**
>
> **Enhanced Cognitive Control:** Training in self-regulation can help individuals better manage attentional processes, reduce impulsivity, and enhance their capacity to focus on complex tasks.
>
> **Emotional Stability:** Improved self-regulation also aids in managing anxiety, depression, and stress by fostering a greater capacity to modulate emotional responses.
>
> ■ **Challenges:**
>
> **Skill Acquisition:** Learning to control one's brain activity effectively requires time and practice, and the level of mastery can vary significantly between individuals.
>
> **Transfer of Training:** Ensuring that gains in self-regulatory capacity achieved during neurofeedback sessions transfer to everyday life situations remains a significant challenge.
>
> ■ **Clinical Implications:**
>
> **Therapeutic Applications:** Self-regulation skills acquired through neurofeedback have significant therapeutic potential, particularly in treating ADHD, anxiety disorders, and other conditions where self-regulation deficits are prevalent.
>
> ■ **Future Directions:**
>
> Integration with behavioral interventions: Combining neurofeedback with cognitive-behavioral strategies may enhance overall treatment efficacy, helping individuals apply their new self-regulation skills more broadly.

10.6 Technological Aspect of Neurofeedback

The technological foundation of neurofeedback is built upon advanced neuroimaging and signal processing tools that enable the precise monitoring, analysis, and feedback of brain activity, as shown in ◘ Fig. 10.3 (Mano et al., 2017). This section explores the key technologies that facilitate neurofeedback, highlighting how they contribute to the field's efficacy and versatility.

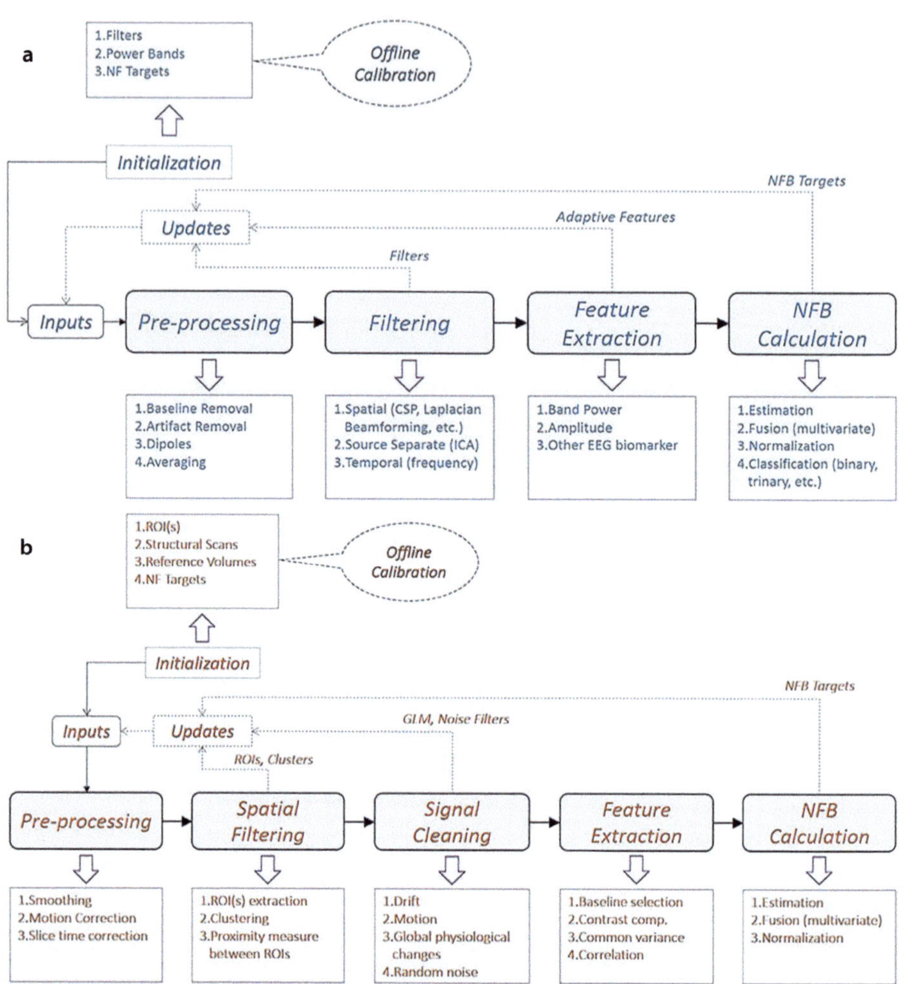

◘ Fig. 10.3 **a** Neurofeedback (NFB) using EEG. **b** NFB using fMRI. The implementation of neurofeedback involves a series of methodical steps that begin with the capture of brain signals and extend through to the delivery of feedback based on specific neural metrics. **Brain recording:** The initial step in neurofeedback involves recording brain activity using neuroimaging technologies such as electroencephalography (EEG). This involves placing sensors on the scalp to capture electrical signals produced by brain activity, which are crucial for identifying patterns associated with various cognitive and emotional states. **Signal Preprocessing**: Once brain signals are acquired, they undergo preprocessing to ensure accuracy and usability. This phase includes filtering noise and artifacts that might distort the data, such as electrical interference or signals caused by muscle movements. The preprocessing also involves amplifying the brain signals to enhance their clarity and distinguishability. **Data Analysis**: After preprocessing, the data are analyzed to extract relevant features, such as specific brainwave frequencies (e.g., alpha, beta, theta). This analysis helps to identify the patterns that neurofeedback training will target. For example, increasing beta waves for improved concentration or enhancing alpha waves to promote relaxation. **Neurofeedback Metrics**: Based on the analysis, neurofeedback metrics are established. These metrics define the parameters that users are trained to control during neurofeedback sessions. The metrics are typically customized to the individual's therapeutic or enhancement goals and are influenced by their baseline brain activity patterns. **Feedback Mechanism**: The core of neurofeedback training involves providing real-time feedback to the individual about their brainwave patterns. This feedback can be visual, such as a moving graph or changing colors on a screen, or auditory, such as a tone that changes pitch or volume depending on the person's brain activity. The immediate feedback helps users learn to voluntarily adjust their brainwaves to match desired states associated with specific cognitive or emotional outcomes. **Iteration and Adaptation**: Neurofeedback sessions are iterative, often requiring multiple sessions over a period of time. The feedback mechanism is adapted based on the user's progress and changing needs. Adjustments to the feedback algorithm may be made to optimize training effectiveness and maintain engagement. **Evaluation and Refinement**: Continuous evaluation of the effectiveness of neurofeedback is critical. This involves assessing changes in the brain's activity patterns and any corresponding improvements in cognitive or emotional functions. Based on these evaluations, further refinements to the neurofeedback protocol may be implemented to enhance its efficacy. This structured approach to neurofeedback ensures that the intervention is both personalized and adaptive, capitalizing on the brain's capacity for plasticity and self-regulation to achieve specific therapeutic and developmental goals. From Mano et al. (2017) originally published under CC-BY

10.6.1 Initial Recording of Brain Signals

The first step in neurofeedback implementation involves the recording of brain activity, using brain signal acquisition techniques you read in ▶ Chap. 2.

Electroencephalography (EEG) remains the foundational technology in neurofeedback, offering real-time monitoring of the brain's electrical activity through electrodes placed on the scalp. As you read in ▶ Chap. 2, this non-invasive method captures the electrical impulses generated by neuronal activity, allowing for the observation and analysis of various brainwave patterns. Niedermeyer and da Silva (2005) emphasize EEG's crucial role in neurofeedback, providing the data necessary for understanding and influencing brain function through feedback mechanisms (Gruzelier, 2014; Enriquez-Geppert et al., 2017).

Functional Magnetic Resonance Imaging (fMRI) represents a more recent advancement in neurofeedback technology (Sulzer et al., 2013; Watanabe et al., 2017), allowing for the visualization of brain activity in much greater detail than EEG. fMRI neurofeedback provides feedback based on blood oxygen level-dependent (BOLD) signals, reflecting changes in localized brain activity with high spatial resolution. This method enables participants to regulate specific brain regions and networks, offering new possibilities for neurofeedback applications (Sulzer et al., 2013; Sitaram et al., 2017).

Weiskopf et al. (2003) highlight fMRI neurofeedback's potential in addressing conditions like depression and anxiety by targeting neural circuits involved in these disorders, showcasing the technology's capacity for precise and targeted brain modulation. A setup of fMRI-based neurofeedback is shown in ◘ Fig. 10.4 (Thibault et al., 2018).

Along with EEG and fMRI, fNIRS and MEG have been used for neurofeedback studies. fNIRS neurofeedback has shown promise in enhancing cognitive functions and emotional regulation by targeting prefrontal cortex activity (Naseer & Hong, 2015). MEG neurofeedback allows for the modulation of specific brain activ-

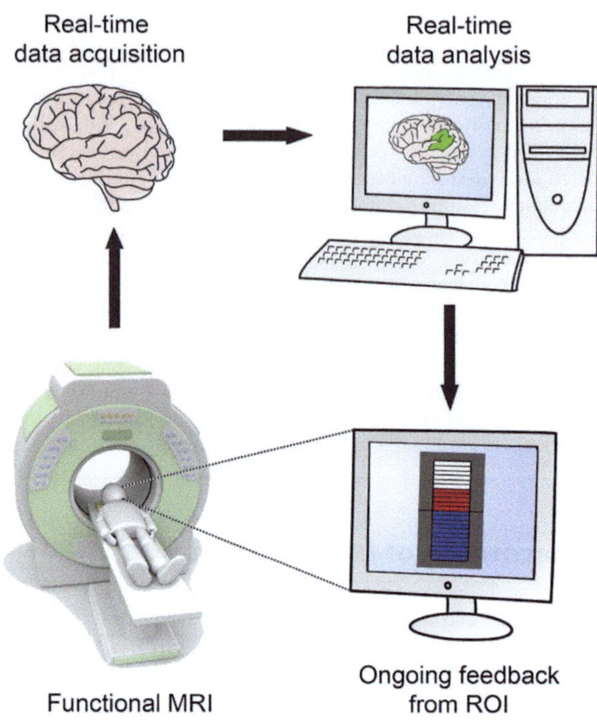

10

▫ Fig. 10.4   Functional magnetic resonance imaging (fMRI) neurofeedback utilizes advanced imaging technology to provide real-time feedback based on brain activity, often visualized through a standard thermometer feedback display. The steps involved are **fMRI Scanning**: fMRI neurofeedback begins with participants undergoing an fMRI scan. This non-invasive imaging technique measures brain activity by detecting changes associated with blood flow. When an area of the brain is more active, it consumes more oxygen and blood flow to that area increases, which can be detected and mapped by fMRI. **Data Extraction and Processing**: During the neurofeedback session, the fMRI data is processed in real time to extract specific signals of interest, typically from regions of the brain that are targeted for regulation. These regions are chosen based on their relevance to the participant's therapeutic goals, such as areas involved in emotion regulation or attention. **Thermometer Display**: The extracted brain signals are then converted into a visual format, commonly a thermometer display. This display provides a straightforward, intuitive visual feedback mechanism. The height of the mercury or similar indicator within the thermometer increases or decreases in response to the participant's brain activity level relative to a predetermined target state. **Real-Time Feedback**: Participants view the thermometer display while in the MRI scanner and attempt to control their brain activity to adjust the level of the thermometer. For example, if the goal is to reduce anxiety by enhancing activity in a relaxation-associated brain area, the participant would try to raise the thermometer level by increasing activity in that area. **Training and Adaptation**: The process involves repeated trials where the participant uses the visual feedback to learn how to alter their brain activity. The system may adjust the difficulty or sensitivity of the feedback based on the participant's performance to optimize the learning curve and maintain engagement. **Application and Outcomes**: Over time, participants learn to control specific brain regions' activity without real-time feedback, applying their learned skills to manage psychological or cognitive challenges. The effectiveness of the training is typically evaluated through behavioral assessments and follow-up scans to measure lasting changes in brain activity. Using a thermometer display in fMRI neurofeedback simplifies complex neural data into an easily understandable format, helping participants learn to modulate their brain activity effectively. This method is widely used for its intuitive nature, making it accessible for participants across various ages and conditions. From Thibault et al. (2018)

ities, including sensory processing and motor functions, by providing precise localization of brain activity (Thibault et al., 2016). Invasive neurofeedback has been explored in the context of treating intractable epilepsy, Parkinson's disease, and other conditions where precise neural modulation is required (Leuthardt et al. 2004).

10.6.2 Preprocessing Procedures

Once brain signals are recorded, they undergo a series of preprocessing procedures to ensure their quality and usability for neurofeedback. These procedures include steps described in ▶ Chap. 4 and also as outlined in ◘ Fig. 10.3.

10.6.3 Application of Neurofeedback Metrics

With clean and processed brain data, neurofeedback metrics are applied to guide the feedback provided to the individual. These metrics may include:
1. Threshold Setting: Determining the thresholds for desired brainwave activity levels, which participants aim to achieve or maintain during neurofeedback sessions (Vernon et al., 2003).
2. Real-Time Feedback: Real-Time Feedback is the cornerstone of neurofeedback, where processed brain activity is presented back to the user through visual, auditory, or haptic interfaces. This feedback enables individuals to learn how to modulate their brainwave patterns consciously (Sitaram et al., 2017). Advanced neurofeedback systems may employ Machine Learning and Artificial Intelligence (AI) to optimize feedback protocols and personalize treatment, enhancing the adaptability and effectiveness of interventions (Ros et al., 2020).
3. Progress Monitoring: Utilizing session-by-session data to track changes in brainwave patterns over time, assessing the individual's progress and adjusting neurofeedback protocols as needed (Arns et al., 2014).

These implementation steps, from the initial recording of brain signals through preprocessing to the application of neurofeedback metrics, form the backbone of neurofeedback interventions. Each step is designed to enhance the precision and personalization of the neurofeedback process, aiming to optimize cognitive and emotional regulation through targeted brainwave modulation.

10.7 Clinical Applications

Neurofeedback's clinical applications are broad and diverse, with substantial evidence supporting its efficacy in treating a variety of conditions:
1. Attention Deficit Hyperactivity Disorder (ADHD): Neurofeedback has been shown to improve symptoms of ADHD, including inattention, hyperactivity, and impulsivity, by modulating theta and beta waves (Arns et al., 2014; Arns & Kenemans, 2014).

2. Anxiety: Hammond (2005) reported that neurofeedback could reduce symptoms of anxiety by altering brainwave patterns associated with stress and relaxation.
3. Depression: Cheon et al. (2015) found neurofeedback to be effective in alleviating symptoms of depression, offering a promising alternative to traditional treatments.
4. Post-Traumatic Stress Disorder (PTSD): Gapen et al. (2016) demonstrated neurofeedback's potential in reducing PTSD symptoms, including flashbacks and hyperarousal.
5. Epilepsy: Sterman and Egner (2006) highlighted neurofeedback's role in reducing seizure frequency and severity in epilepsy patients by stabilizing brainwave activity.
6. Autism Spectrum Disorders (ASD): Kouijzer et al. (2009) reported improvements in social behavior and cognitive functions in individuals with ASD through neurofeedback training.

Beyond its clinical utility, neurofeedback has been explored for its potential in cognitive enhancement. Vernon et al. (2003) found that neurofeedback training could lead to significant improvements in memory, attention, and executive functioning. By targeting specific brainwave frequencies, individuals can achieve heightened cognitive states, improving their mental agility and processing speed.

Neurofeedback has also been applied to enhance performance in sports and artistic fields, where focus, creativity, and mental resilience are crucial. Wilson et al. (2006) discussed how neurofeedback could optimize focus and concentration among athletes, leading to improved performance outcomes. Similarly, artists have utilized neurofeedback to enhance creativity and innovation, leveraging the technique's ability to foster optimal brainwave states conducive to creative thought and expression.

Despite its promising applications, the field of neurofeedback faces several challenges, including issues related to placebo effects, methodological variations across studies, and the need for standardized protocols. In the research conducted by Schabus et al. (2017), as shown in ◘ Fig. 10.5 (Thibault et al., 2017), individuals were provided with instantaneous feedback on their brain activity: the successful amplification of the designated neural signal resulted in a greater rotation of the needle displayed on the monitor before them. The participants engaged in 12 sessions of genuine neurofeedback, followed by a three-month washout period, after which they participated in 12 sessions of placebo (sham) neurofeedback, or experienced the sequence in reverse order. The study found that while the ability to regulate neural activity was enhanced in the group receiving genuine feedback, neither the genuine nor the sham neurofeedback sessions led to improvements in objective sleep quality metrics. Furthermore, subjective assessments reported similar levels of improvement following both genuine and sham feedback sessions. Schönenberg et al. (2017) address these challenges, emphasizing the importance of rigorous research designs, placebo-controlled studies, and the standardization of neurofeedback protocols to enhance the reliability and validity of findings in the field. These challenges highlight the need for ongoing research and methodological refinement to fully harness neurofeedback's potential.

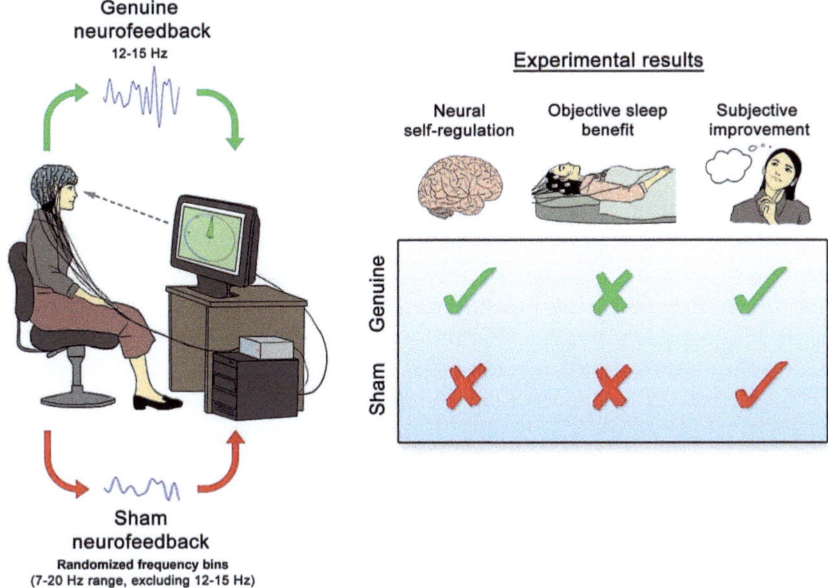

Fig. 10.5 Comparing genuine and sham neurofeedback involves evaluating the effectiveness and psychological impact of real neurofeedback sessions against placebo-like controls, where feedback is not based on the participant's actual brain activity. An overview of how this comparison is typically structured and analyzed is as follows. **Genuine Neurofeedback**: In genuine neurofeedback sessions, participants receive real-time feedback that accurately reflects their brain activity. This feedback is typically generated from EEG or fMRI data, which monitors electrical patterns or blood flow changes in the brain. The feedback aims to train participants to self-regulate specific neural processes associated with cognitive or emotional functions. **Sham Neurofeedback**: Sham neurofeedback serves as the control condition. In these sessions, the feedback provided to the participants is not based on their brain activity but is either randomized or based on pre-recorded data from other sessions or individuals. This placebo feedback is designed to be indistinguishable from genuine neurofeedback to the participant, ensuring that any psychological effects can be attributed to the belief in receiving true feedback, rather than the feedback itself. **Experimental Design**: Studies typically use a double-blind design, where neither the participants nor the experimenters interacting with them know whether the feedback is genuine or sham. This approach helps to eliminate bias and isolate the effects of the neurofeedback training itself. **Outcome Measures**: Both types of sessions are assessed using a variety of outcome measures to evaluate their efficacy. These measures may include behavioral tests, psychological assessments, and physiological measurements before and after the training sessions. Neuroimaging tools may also be used to observe any changes in brain structure or function. **Participant Experience**: Researchers often gather data on participants' perceptions of the session to assess whether they could discern between genuine and sham feedback. Participant blinding effectiveness is crucial for interpreting the results of neurofeedback studies. **Statistical Analysis**: The effectiveness of genuine versus sham neurofeedback is analyzed statistically to determine significant differences in the outcomes. These analyses help to clarify whether changes in participants' cognitive or emotional states can be attributed to actual neurofeedback or are due to placebo effects. **Implications for Research and Therapy**: The comparison between genuine and sham neurofeedback is vital for validating the effectiveness of neurofeedback as a therapeutic tool. If genuine neurofeedback shows significantly greater improvements than sham feedback, it supports the efficacy of neurofeedback training. Conversely, if no significant differences are found, it may suggest that placebo effects or other nonspecific factors are at play. This comparative approach ensures that the benefits attributed to neurofeedback are due to the intervention itself and not to psychological expectations or experimental bias, thereby reinforcing the validity of neurofeedback in clinical and psychological research. From Thibault et al. (2017)

10.8 Limitations of Neurofeedback

While neurofeedback holds promise as a non-invasive method for modulating brain activity and treating various neurological and psychological conditions, it is not without its limitations. Researchers and practitioners must address these challenges through rigorous scientific investigation, standardization of protocols, and continuous evaluation of the technique's safety and efficacy.

Despite numerous studies supporting the efficacy of neurofeedback, the scientific community has not reached a consensus on its effectiveness. The variability in research findings is partly due to differences in methodologies, sample sizes, and protocols. High-quality, large-scale randomized controlled trials are still needed to establish definitive evidence for neurofeedback's efficacy across different conditions. The placebo effect poses a significant challenge in neurofeedback research. Participants' expectations and beliefs about the treatment can influence outcomes, making it difficult to isolate the specific effects of neurofeedback. Additionally, blinding in neurofeedback studies is challenging because participants are often aware of the training process, which can introduce bias. The effectiveness of neurofeedback can vary widely among individuals. Factors such as age, baseline neurophysiological characteristics, and psychological state can influence how well a person responds to neurofeedback training. This variability makes it difficult to predict which individuals will benefit most from the intervention.

There is no standardized protocol for neurofeedback training, leading to significant variability in the implementation of the technique. Differences in the number of sessions, duration, frequency, and specific neurofeedback protocols (e.g., frequency bands targeted) can affect the outcomes. This lack of standardization complicates the comparison of results across studies and the generalization of findings. The accuracy and reliability of neurofeedback are dependent on the quality of the equipment and the expertise of the practitioner. EEG systems can be susceptible to artifacts from muscle movements, eye blinks, and external electrical interference. These artifacts can obscure the true neural signals and impact the effectiveness of the training.

Neurofeedback requires specialized equipment and trained professionals, making it relatively expensive and less accessible compared to other therapeutic interventions. The cost can be prohibitive for many individuals, and insurance coverage for neurofeedback is often limited or non-existent. While some studies have demonstrated short-term benefits of neurofeedback, there is limited evidence regarding its long-term efficacy. The sustainability of treatment effects remains unclear, and there is a need for long-term follow-up studies to determine whether the benefits of neurofeedback persist over time.

As with any intervention that involves brain modulation, there are ethical and safety considerations. The long-term effects of altering brain activity through neurofeedback are not fully understood, and there is a need for ongoing monitoring and research to ensure that the technique is safe. Ethical concerns also arise regarding the use of neurofeedback in vulnerable populations, such as children and individuals with severe neurological conditions.

10.9 **Ethical Implications and Future Research**

Ethical concerns in the practice of neurofeedback are multifaceted and chiefly involve the imperative of informed consent, particularly when engaging vulnerable populations such as children. It is essential that both participants and their guardians are thoroughly informed about the procedure, including its nature, potential benefits, and associated risks. Moreover, the ethical debate surrounding neurofeedback also touches on its potential misuse, especially when used for non-therapeutic enhancements which could provoke issues regarding fairness, privacy, and the nature of personal achievement. It is necessary to establish stringent ethical guidelines to manage these complexities effectively, ensuring that neurofeedback is practiced with utmost integrity and respect for individual autonomy.

The evolution of neurofeedback research is progressively aimed at discovering new applications and bolstering the evidence base through comprehensive, controlled studies. Sitaram et al. (2017) underscore the importance of exploring the extensive capabilities of neurofeedback, ranging from treating a variety of psychological and neurological disorders to augmenting cognitive and physical functions. Such pioneering research is vital for reinforcing the scientific underpinnings of neurofeedback, refining protocols, and confirming its efficacy across varied populations and settings.

There is also growing enthusiasm for enhancing therapeutic outcomes in neurofeedback by integrating it with other treatment modalities, such as pharmacotherapy and psychotherapy. This integrative approach proposes a more holistic treatment strategy that could potentially magnify benefits and tackle the complex aspects of numerous disorders more effectively. Arns and Kenemans (2014) advocate for this type of synergy, suggesting that combining neurofeedback with conventional treatments could produce enhanced therapeutic effects, thereby elevating overall efficacy and improving patient outcomes. This approach points towards a future where neurofeedback is part of a broader, more nuanced therapeutic landscape, aligning with personalized treatment plans to optimize health outcomes.

> **Key Takeaways**
> 1. **Neurofeedback as a Tool for Enhancing Cognitive and Emotional Control**: This chapter highlights neurofeedback's potential to enable individuals to consciously regulate brain activity using real-time feedback mechanisms. This non-invasive method utilizes EEG or fMRI to improve cognitive functions, emotional regulation, and mental well-being by leveraging the brain's neuroplasticity. It is particularly useful in treating conditions like ADHD, anxiety, and depression.
> 2. **Operant Conditioning as the Mechanism Behind Neurofeedback**: Neurofeedback is deeply rooted in operant conditioning principles, wherein individuals learn to modify brainwave patterns through positive reinforcement or negative feedback. By reinforcing desirable brain states and discouraging undesirable patterns, neurofeedback allows for targeted improvements in attention, relaxation, or mood regulation.

3. **Diverse Clinical Applications and Efficacy**: Neurofeedback has demonstrated efficacy across various clinical domains, including treating ADHD, epilepsy, PTSD, and autism spectrum disorders. It has been shown to alleviate symptoms, enhance cognitive performance, and offer a complementary or alternative solution to traditional therapies, making it a versatile tool for neurological and psychological interventions.

4. **Ethical Considerations and Future Directions**: This chapter also addresses the ethical concerns surrounding neurofeedback, particularly regarding informed consent and its use for cognitive enhancement in healthy individuals. It emphasizes the need for further research, standardization, and the integration of neurofeedback with other therapeutic modalities to enhance its efficacy and ensure safe, ethical applications in both clinical and non-clinical settings.

In essence, neurofeedback represents a pivotal chapter in the evolving narrative of neuroscience and technology's convergence. As we move forward, the fusion of neurofeedback principles with BCI systems promises to propel us into new realms of understanding and capability, heralding a future where the full potential of the human brain can be realized and harnessed in ways we are just beginning to imagine.

Conclusion

In this chapter, we have taken a deep dive into neurofeedback, covering its foundations, techniques, clinical uses, and the extensive research supporting its effectiveness. Neurofeedback operates by providing individuals with instantaneous feedback on their brain activity. Neurofeedback methods, ranging from traditional EEG-based methods to advanced fMRI and fNIRS techniques, exemplify the field's versatility and its ability to tailor interventions to individual needs and conditions. This innovative approach empowers individuals to consciously modulate their brainwave patterns, a process intricately linked to the brain's inherent neuroplastic capabilities. By engaging in this feedback loop, participants exploit their brain's adaptability, learning to adjust their neural activity in real-time. The essence of neurofeedback lies in its ability to condition the brain towards desirable neural activity patterns through consistent reinforcement. ◘ Figure 10.6 summarizes neurofeedback procedure (Sitaram et al., 2017).

The clinical applications of neurofeedback are vast, demonstrating significant potential in addressing ADHD, anxiety, depression, PTSD, epilepsy, and autism spectrum disorders. Beyond clinical settings, neurofeedback holds promise for enhancing cognitive functions, such as memory, attention, and executive functioning, as well as for boosting sports performance and artistic creativity. These applications are supported by a solid body of research, including efficacy studies and investigations into neurofeedback's mechanisms of action, underscoring its potential as a transformative tool in mental health and cognitive development. In Volume 2 of this book, we will deep dive into various applications of neurofeedback.

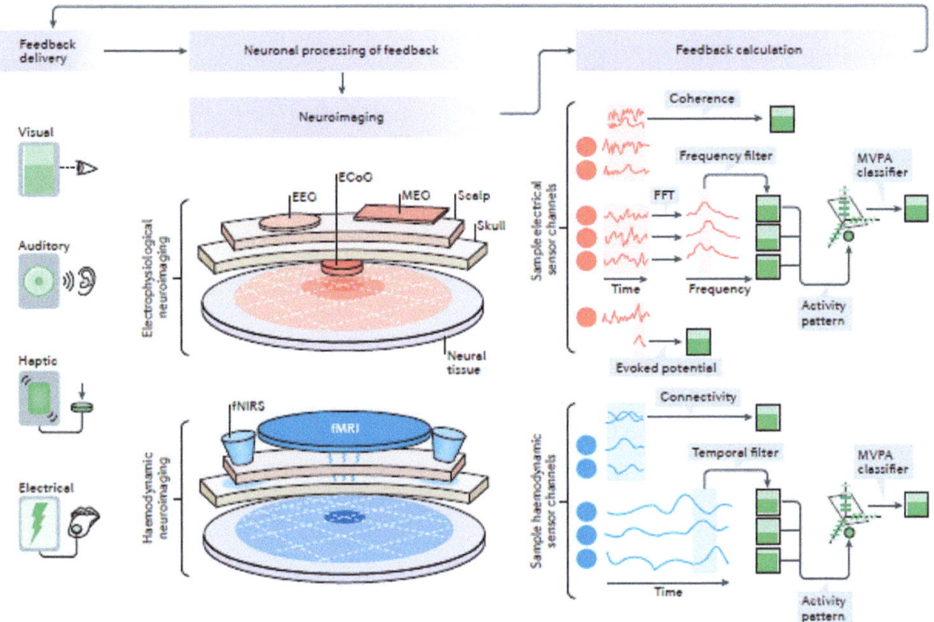

■ **Fig. 10.6** Neurofeedback is a sophisticated technique that starts by monitoring brain activity using various advanced neuroimaging and electrophysiological methods. These methods include electroencephalography (EEG), which captures electrical activity through scalp sensors; magnetoencephalography (MEG), which detects magnetic fields generated by neural activity; and invasive electrocorticography (ECoG), which involves placing electrodes directly on the brain surface. Hemodynamic imaging techniques such as functional magnetic resonance imaging (fMRI) and functional near-infrared spectroscopy (fNIRS) are also employed to measure brain activity based on blood flow changes associated with neural activity. These diverse technologies are used to map brain activity onto standard brain models, providing a comparative analysis of spatial resolutions. Additionally, the temporal resolution differences between electrophysiological methods (which provide real-time data) and hemodynamic techniques (which have slower response times due to reliance on vascular responses) are explored through the examination of sample signals from these sensors. The neurofeedback process involves several signal processing techniques. Univariate methods analyze data from single channels or specific brain regions to identify evoked potentials or similar features. Bivariate methods, such as coherence or connectivity analyses, are used to evaluate the functional connectivity between two channels. More complex analyses, such as multivariate pattern analysis (MVPA), are applied to sensor arrays to detect patterns across multiple variables, such as power in specific frequency ranges or levels of brain activation. Once the brain signals are processed, they are fed back to the user through various modalities, including visual displays, auditory signals, haptic feedback, or direct electrical stimulation. This feedback is designed to help individuals consciously modulate their brain activity, effectively closing the feedback loop. The brain processes this input, potentially leading to functional changes over time. This intricate feedback system not only allows for real-time interaction with brain processes but also empowers users to influence their cognitive functions and mental states actively. From Sitaram et al. (2017)

However, the journey of neurofeedback from a promising intervention to a widely recognized therapeutic modality is not without challenges. Ethical considerations, particularly concerning informed consent and the potential for misuse, alongside the need for standardized protocols and methodological rigor in research, are pivotal areas that must be addressed. Furthermore, the integration of neurofeedback with other therapies offers a promising avenue for enhancing therapeutic outcomes, suggesting a collaborative future that leverages the strengths of multiple treatment modalities.

As we conclude this exploration of neurofeedback, it is imperative to recognize its role as a critical component of the broader landscape of brain–computer interface (BCI) systems. Neurofeedback not only stands as a potent intervention in its own right but also forms an essential backbone of any BCI system, a topic that readers will encounter in Volume 2 of this book. The principles of neurofeedback—capturing, analyzing, and feeding back brain activity information—mirror the core functionalities of BCIs, which rely on direct communication pathways between the brain and external devices.

This symbiotic relationship highlights the transformative potential of neurofeedback within the BCI domain, where the understanding and manipulation of brain signals can lead to groundbreaking applications in rehabilitation, communication, and enhanced human-computer interaction. As we venture further into the integration of neurofeedback and BCI technologies, we stand on the cusp of a new frontier in neuroscience and technology, poised to unlock unprecedented possibilities for augmenting human capabilities and addressing complex neurological conditions.

This section offers practical exercises designed to help apply the material learned to concrete scenarios.

Simulate a Neurofeedback Session

Objective: Understand the process and experience of neurofeedback.

Task: Using simulation software, students will experience what it is like to participate in a neurofeedback session. They will attempt to control their simulated brainwave activity based on real-time feedback provided by the software.

Design a Neurofeedback Protocol for ADHD

Objective: Develop a neurofeedback treatment protocol for a specific disorder.

Task: Students will design a neurofeedback protocol aimed at enhancing concentration in individuals with ADHD. They should specify the type of brain waves to be targeted, the type of feedback to be used, session length, and total number of sessions.

Analyze Neurofeedback Data

Objective: Learn how to interpret EEG data from neurofeedback sessions.

Task: Students will analyze EEG data collected before and after a series of neurofeedback sessions to identify changes in brainwave patterns and assess the effectiveness of the training.

Ethical Debate on Neurofeedback

Objective: Explore the ethical implications of neurofeedback.

Task: Students will engage in a debate on the ethical considerations of using neurofeedback for enhancing cognitive abilities in healthy individuals versus its therapeutic use in clinical populations.

Create a Neurofeedback Awareness Campaign

Objective: Educate others about the benefits and limitations of neurofeedback.

Task: Students will create informational materials (e.g., brochures, presentations, social media posts) aimed at educating the public about what neurofeedback is, how it works, its potential benefits, and its limitations.

Develop a Research Proposal

Objective: Propose a study to investigate a novel application of neurofeedback.

Task: Students will write a research proposal outlining a study designed to explore a new application of neurofeedback, such as its potential use in enhancing artistic creativity or improving athletic performance. The proposal should include the study's background, objectives, methodology, expected outcomes, and ethical considerations.

References

Arns, M., & Kenemans, J. L. (2014). Neurofeedback in ADHD and insomnia: Vigilance stabilization through sleep spindles and circadian networks. *Neuroscience & Biobehavioral Reviews, 44*, 183–194.

Arns, M., De Ridder, S., Strehl, U., Breteler, M., & Coenen, A. (2009). Efficacy of neurofeedback treatment in ADHD: The effects on inattention, impulsivity and hyperactivity: A meta-analysis. *Clinical EEG and Neuroscience, 40*(3), 180–189.

Arns, M., Conners, C. K., & Kraemer, H. C. (2013). A decade of EEG theta/beta ratio research in ADHD: A meta-analysis. *Journal of Attention Disorders, 17*(5), 374–383.

Arns, M., Heinrich, H., & Strehl, U. (2014). Evaluation of neurofeedback in ADHD: The long and winding road. *Biological Psychology, 95*, 108–115.

Azrin, N. H., & Holz, W. C. (1966). Punishment. In W. K. Honig (Ed.), *Operant behavior: Areas of research and application* (pp. 380–447). Appleton-Century-Crofts.

Bąbel, P. (2020). Operant conditioning as a new mechanism of placebo effects. *European Journal of Pain, 24*(5), 902–908.

Bi, G.-Q., & Poo, M.-M. (1998). Synaptic modifications in cultured hippocampal neurons: Dependence on spike timing, synaptic strength, and postsynaptic cell type. *The Journal of Neuroscience, 18*(24), 10464–10472.

Birbaumer, N., Elbert, T., Canavan, A. G., & Rockstroh, B. (1990). Slow potentials of the cerebral cortex and behavior. *Physiological Reviews, 70*(1), 1–41.

Birbaumer, N., Murguialday, A. R., Weber, C., & Montoya, P. (2009). Neurofeedback and brain–computer interface: Clinical applications. *International Review of Neurobiology, 86*, 107–117.

Buonomano, D. V., & Merzenich, M. M. (1998). Cortical plasticity: From synapses to maps. *Annual Review of Neuroscience, 21*(1), 149–186.

Catania, A. C. (1992). *Learning*. Prentice Hall.

Cheon, E. J., Koo, B. H., Seo, W. S., Lee, J. Y., Choi, J. H., & Song, S. H. (2015). Effects of neurofeedback on adult patients with psychiatric disorders in a naturalistic setting. *Applied Psychophysiology and Biofeedback, 40*, 17–24.

Egner, T., & Gruzelier, J. H. (2004). EEG biofeedback of low beta band components: Frequency-specific effects on variables of attention and event-related brain potentials. *Clinical Neurophysiology, 115*(1), 131–139.

Enriquez-Geppert, S., Huster, R. J., & Herrmann, C. S. (2017). EEG-neurofeedback as a tool to modulate cognition and behavior: A review tutorial. *Frontiers in Human Neuroscience, 11*, 51.

Enriquez-Geppert, S., Smit, D., Pimenta, M. G., & Arns, M. (2019). Neurofeedback as a treatment intervention in ADHD: Current evidence and practice. *Current Psychiatry Reports, 21*, 1–7.

Ferster, C. B., & Skinner, B. F. (1957). *Schedules of reinforcement*. Appleton-Century-Crofts.

Gapen, M., van der Kolk, B. A., Hamlin, E., Hirshberg, L., Suvak, M., & Spinazzola, J. (2016). A pilot study of neurofeedback for chronic PTSD. *Applied Psychophysiology and Biofeedback, 41*, 251–261.

Gruzelier, J. H. (2014). EEG-neurofeedback for optimising performance. I: A review of cognitive and affective outcome in healthy participants. *Neuroscience & Biobehavioral Reviews, 44*, 124–141.

Haller, S., Kopel, R., Jhooti, P., Haas, T., Scharnowski, F., Lovblad, K. O., et al. (2013). Dynamic reconfiguration of human brain functional networks through neurofeedback. *NeuroImage, 81*, 243–252.

Hammond, D. C. (2005). Neurofeedback with anxiety and affective disorders. *Child and Adolescent Psychiatric Clinics, 14*(1), 105–123.

Heinrich, H., Gevensleben, H., & Strehl, U. (2007). Annotation: Neurofeedback–train your brain to train behaviour. *Journal of Child Psychology and Psychiatry, 48*(1), 3–16.

Hensch, T. K. (2004). Critical period regulation. *Annual Review of Neuroscience, 27*, 549–579.

Hinterberger, T., Veit, R., Wilhelm, B., Weiskopf, N., Vatine, J.-J., & Birbaumer, N. (2005). Neuronal mechanisms underlying control of a brain–computer interface. *European Journal of Neuroscience, 21*(11), 3169–3181.

Jiang, Y., Abiri, R., & Zhao, X. (2017). Tuning up the old brain with new tricks: Attention training via neurofeedback. *Frontiers in Aging Neuroscience, 9*, 52.

Johnston, S. J., Boehm, S. G., Healy, D., Goebel, R., & Linden, D. E. (2010). Neurofeedback: A promising tool for the self-regulation of emotion networks. *NeuroImage, 49*(1), 1066–1072.

Kamiya, J. (1969). Operant control of the EEG alpha rhythm and some of its reported effects on consciousness. In C. Tart (Ed.), *Altered states of consciousness* (pp. 519–529). Wiley.

Kazdin, A. E. (1982). *The token economy: A review and evaluation*. Plenum Press.

Kazdin, A. E. (2012). *Behavior modification in applied settings*. Waveland Press.

Kleim, J. A., & Jones, T. A. (2008). Principles of experience-dependent neural plasticity: Implications for rehabilitation after brain damage.

Kolb, B., & Whishaw, I. Q. (1998). Brain plasticity and behavior. *Annual Review of Psychology, 49*(1), 43–64.

Koralek, A. C., Jin, X., Long Ii, J. D., Costa, R. M., & Carmena, J. M. (2012). Corticostriatal plasticity is necessary for learning intentional neuroprosthetic skills. *Nature, 483*(7389), 331–335.

Kouijzer, M. E., de Moor, J. M., Gerrits, B. J., Congedo, M., & van Schie, H. T. (2009). Neurofeedback improves executive functioning in children with autism spectrum disorders. *Research in Autism Spectrum Disorders, 3*(1), 145–162.

Lerman, D. C., & Iwata, B. A. (1995). Prevalence of the extinction burst and its attenuation during treatment. *Journal of Applied Behavior Analysis, 28*(1), 93–94.

Leuthardt, E. C., Schalk, G., Wolpaw, J. R., Ojemann, J. G., & Moran, D. W. (2004). A brain–computer interface using electrocorticographic signals in humans. *Journal of Neural Engineering, 1*(2), 63.

Mahncke, H. W., Bronstone, A., & Merzenich, M. M. (2006). Brain plasticity and functional losses in the aged: Scientific bases for a novel intervention. *Progress in Brain Research, 157*, 81–109.

Malenka, R. C., & Bear, M. F. (2004). LTP and LTD: An embarrassment of riches. *Neuron, 44*(1), 5–21.

Mano, M., Lécuyer, A., Bannier, E., Perronnet, L., Noorzadeh, S., & Barillot, C. (2017). How to build a hybrid neurofeedback platform combining EEG and fMRI. *Frontiers in Neuroscience, 11*, 249134.

Mennella, R., Patron, E., & Palomba, D. (2017). Frontal alpha asymmetry neurofeedback for the reduction of negative affect and anxiety. *Behaviour Research and Therapy, 92*, 32–40.

Merzenich, M. M., Nelson, R. J., Stryker, M. P., Cynader, M. S., Schoppmann, A., & Zook, J. M. (1984). Somatosensory cortical map changes following digit amputation in adult monkeys. *Journal of Comparative Neurology, 224*(4), 591–605.

Merzenich, M. M., Van Vleet, T. M., & Nahum, M. (2014). Brain plasticity-based therapeutics. *Frontiers in Human Neuroscience, 8*, 385.

Micoulaud-Franchi, J. A., Geoffroy, P. A., Fond, G., Lopez, R., Bioulac, S., & Philip, P. (2014). EEG neurofeedback treatments in children with ADHD: An updated meta-analysis of randomized controlled trials. *Frontiers in Human Neuroscience, 8*, 906.

Monastra, V. J., Monastra, D. M., & George, S. (2002). The effects of stimulant therapy, EEG biofeedback, and parenting style on the primary symptoms of attention-deficit/hyperactivity disorder. *Applied Psychophysiology and Biofeedback, 27*, 231–249.

Naseer, N., & Hong, K. S. (2015). fNIRS-based brain-computer interfaces: A review. *Frontiers in Human Neuroscience, 9*, 3.

Niedermeyer, E., & da Silva, F. L. (Eds.). (2005). *Electroencephalography: Basic principles, clinical applications, and related fields*. Lippincott Williams & Wilkins.

Nudo, R. J. (2006). Mechanisms for recovery of motor function following cortical damage. *Current Opinion in Neurobiology, 16*(6), 638–644.

Paret, C., Ruf, M., Gerchen, M. F., Kluetsch, R., Demirakca, T., Jungkunz, M., et al. (2016). fMRI neurofeedback of amygdala response to aversive stimuli enhances prefrontal–limbic brain connectivity. *NeuroImage, 125*, 182–188.

Pascual-Leone, A., Amedi, A., Fregni, F., & Merabet, L. B. (2005). The plastic human brain cortex. *Annual Review of Neuroscience, 28*, 377–401.

Peterson, G. B. (2004). A day of great illumination: BF Skinner's discovery of shaping. *Journal of the Experimental Analysis of Behavior, 82*(3), 317–328.

Redmon, W. K. (1991). Pinpointing the technological fault in applied behavior analysis. *Journal of Applied Behavior Analysis, 24*(3), 441.

Renton, T., Tibbles, A., & Topolovec-Vranic, J. (2017). Neurofeedback as a form of cognitive rehabilitation therapy following stroke: A systematic review. *PLoS One, 12*(5), e0177290.

Roebers, C. M. (2017). Executive function and metacognition: Towards a unifying framework of cognitive self-regulation. *Developmental Review, 45*, 31–51.

Sampedro Baena, L., Fuente, G. A. C. D. L., Martos-Cabrera, M. B., Gómez-Urquiza, J. L., Albendín-García, L., Romero-Bejar, J. L., & Suleiman-Martos, N. (2021). Effects of neurofeedback in children with attention-deficit/hyperactivity disorder: A systematic review. *Journal of Clinical Medicine, 10*(17), 3797.

Schabus, M., Griessenberger, H., Gnjezda, M. T., Heib, D. P., Wislowska, M., & Hoedlmoser, K. (2017). Better than sham? A double-blind placebo-controlled neurofeedback study in primary insomnia. *Brain, 140*(4), 1041–1052.

Scheinost, D., Stoica, T., Saksa, J., Papademetris, X., Constable, R. T., Pittenger, C., & Hampson, M. (2013). Orbitofrontal cortex neurofeedback produces lasting changes in contamination anxiety and resting-state connectivity. *Translational Psychiatry, 3*(4), e250.

Schoenberg, P. L., & David, A. S. (2014). Biofeedback for psychiatric disorders: A systematic review. *Applied Psychophysiology and Biofeedback, 39*, 109–135.

Schönenberg, M., Wiedemann, E., Schneidt, A., Scheeff, J., Logemann, A., Keune, P. M., & Hautzinger, M. (2017). Confusion regarding operant conditioning of the EEG–authors' reply. *The Lancet Psychiatry, 4*(12), 897–898.

Sitaram, R., Ros, T., Stoeckel, L., Haller, S., Scharnowski, F., Lewis-Peacock, J., et al. (2017). Closed-loop brain training: The science of neurofeedback. *Nature Reviews Neuroscience, 18*(2), 86–100.

Skinner, B. F. (1938). *The behavior of organisms: An experimental analysis.* Appleton-Century.

Skinner, B. F. (1953). *Science and human behavior.* Macmillan.

Skinner, B. F. (1957). *Verbal behavior.* Copley Publishing Group.

Sterman, M. B. (2000). Basic concepts and clinical findings in the treatment of seizure disorders with EEG operant conditioning. *Clinical Electroencephalography, 31*(1), 45–55.

Sterman, M. B., & Egner, T. (2006). Foundation and practice of neurofeedback for the treatment of epilepsy. *Applied Psychophysiology and Biofeedback, 31*, 21–35.

Sterman, M. B., Macdonald, L. R., & Stone, R. K. (1974). Biofeedback training of the sensorimotor electroencephalogram rhythm in man: Effects on epilepsy. *Epilepsia, 15*(3), 395–416.

Sulzer, J., Haller, S., Scharnowski, F., Weiskopf, N., Birbaumer, N., Blefari, M. L., et al. (2013). Real-time fMRI neurofeedback: Progress and challenges. *NeuroImage, 76*, 386–399.

Thibault, R. T., Lifshitz, M., Birbaumer, N., & Raz, A. (2015). Neurofeedback, self-regulation, and brain imaging: Clinical science and fad in the service of mental disorders. *Psychotherapy and Psychosomatics, 84*(4), 193–207.

Thibault, R. T., Lifshitz, M., & Raz, A. (2016). The self-regulating brain and neurofeedback: Experimental science and clinical promise. *Cortex, 74*, 247–261.

Thibault, R. T., Lifshitz, M., & Raz, A. (2017). Neurofeedback or neuroplacebo? *Brain, 140*(4), 862–864.

Thibault, R. T., MacPherson, A., Lifshitz, M., Roth, R. R., & Raz, A. (2018). Neurofeedback with fMRI: A critical systematic review. *NeuroImage, 172*, 786–807.

Vernon, D., Egner, T., Cooper, N., Compton, T., Neilands, C., Sheri, A., & Gruzelier, J. (2003). The effect of training distinct neurofeedback protocols on aspects of cognitive performance. *International Journal of Psychophysiology, 47*(1), 75–85.

Wang, J. R., & Hsieh, S. (2013). Neurofeedback training improves attention and working memory performance. *Clinical Neurophysiology, 124*(12), 2406–2420.

Watanabe, T., Sasaki, Y., Shibata, K., & Kawato, M. (2017). Advances in fMRI real-time neurofeedback. *Trends in Cognitive Sciences, 21*(12), 997–1010.

Weiskopf, N., Veit, R., Erb, M., Mathiak, K., Grodd, W., Goebel, R., & Birbaumer, N. (2003). Physiological self-regulation of regional brain activity using real-time functional magnetic resonance imaging (fMRI): Methodology and exemplary data. *NeuroImage, 19*(3), 577–586.

Wilson, V. E., Peper, E., & Moss, D. (2006). "The Mind Room" in Italian Soccer training: The use of biofeedback and neurofeedback for optimum performance. *Biofeedback, 34*(3), 79–81.

Zimmerman, B. J., & Schunk, D. H. (2004). Self-regulating intellectual processes and outcomes: A social cognitive perspective. In *Motivation, emotion, and cognition* (pp. 337–364). Routledge.

Zoefel, B., Huster, R. J., & Herrmann, C. S. (2011). Neurofeedback training of the upper alpha frequency band in EEG improves cognitive performance. *NeuroImage, 54*(2), 1427–1431.

Zotev, V., Phillips, R., Young, K. D., Drevets, W. C., & Bodurka, J. (2013). Prefrontal control of the amygdala during real-time fMRI neurofeedback training of emotion regulation. *PLoS One, 8*(11), e79184.

Zotev, V., Phillips, R., Yuan, H., Misaki, M., & Bodurka, J. (2014). Self-regulation of human brain activity using simultaneous real-time fMRI and EEG neurofeedback. *NeuroImage, 85*, 985–995.

Neurotechnology for Communication

ALS Patient Case Study and BCI Development

Contents

Test your learning and check your understanding of this book's contents: use the "Springer Nature Flashcards" app to access questions using ▶ https://sn.pub/kmb-jyz. To use the app, please follow the instructions in ▶ Chap. 1.

This chapter ties all the concepts learned so far by focusing on a case study involving patient P11. The process encompasses the design and implementation of an auditory brain–computer interface (BCI) that utilizes eye movements to facilitate communication independent of visual capabilities. The chapter outlines the experimental design, including signal acquisition, preprocessing, feature extraction, and the deployment of machine learning algorithms such as Support Vector Machine (SVM) to classify responses. Real-world trials illustrate the challenges and complexities involved, emphasizing the need for accuracy, reliability, and real-time feedback in the communication system. Through a detailed discussion of the methodology and results, the chapter not only demonstrates the practical application of theoretical knowledge in neurotechnology but also highlights the profound impact of these systems on enhancing the quality of life for individuals with severe communication barriers. This exploration serves as a foundational study for numerous studies like those discussed extensively in Volume 2.

Learning Objectives

1. **Understand the Application of Neurotechnology in Communication**: Students will learn how to apply the concepts learned so far to develop communication systems for individuals with severe physical limitations like ALS.
2. **Explore the Development and Implementation of a Communication System**: Gain insight into the step-by-step process of designing, implementing, and refining a communication system that uses eye movements to facilitate communication for a patient without any means of communication. This includes the setup, feature extraction, model training, and real-time feedback mechanisms.

You may be asking yourself why we covered so many concepts in the earlier chapters of this book. The answer becomes clear in this chapter, where we will harness all that knowledge to address a challenge affecting thousands of people globally. Here, we transition from theory to practical application, using the principles and techniques we have discussed to develop solutions for real-world problems. This approach not only reinforces our learning but also demonstrates the direct impact of our skills in making meaningful changes in the lives of many.

In February 2018, Prof. Niels Birbaumer and I had our initial meeting with an amyotrophic lateral sclerosis (ALS) patient we refer to as P11 in our publications, which you will read in this chapter. To clarify, this was not our first experience with ALS; Prof. Birbaumer has engaged with hundreds of ALS patients over his career, and I have worked with dozens. This visit, however, was explicitly our first encounter with P11.

> **Box 11.1**
>
> *Amyotrophic Lateral Sclerosis*, commonly referred to as ALS or Lou Gehrig's disease, is a progressive neurodegenerative disorder that affects motor neurons in the brain and spinal cord. These neurons are responsible for controlling voluntary muscle movement, and their degeneration leads to muscle weakness, atrophy, and spasticity. Over time, individuals with ALS experience an increased loss of mobility in the limbs, and difficulties in speaking, swallowing, and breathing. ALS does not typically impair cognitive function or the senses (sight, smell, taste, hearing, and touch). The exact cause of ALS is unknown, and there is currently no cure, but treatments are available to help manage symptoms and improve quality of life. The disease is characterized by its rapid progression, although the rate can vary significantly from person to person. Different communication systems for patients with ALS is discussed extensively in Volume 2.

The need for our visit arose after P11's primary caregivers visited our brain–computer interface (BCI) lab at the University of Tübingen. They reported that P11 could not communicate using traditional eye-tracker systems, a technology often utilized by patients with severe physical limitations. For those unfamiliar with this technology, I have included an explanation in Box ▶ 11.2. P11's caregivers sought our expertise to explore alternative communication methods.

Upon our arrival, contrary to what the caregivers had believed, I observed that although P11 had lost gaze-fixation ability, P11 still retained some ability to move eyes. Typically, an eye-tracker requires gaze-fixation ability to function, as detailed in Box ▶ 11.2. Thus, P11 was unable to use the eye-tracker for communication. Therefore, P11's caregivers ingeniously deployed a method using a simple sheet of paper divided into four color-coded sectors with letters, as shown in ▪ Fig. 11.1 (Tonin et al., 2020). They would hold this sheet in front of P11, who communicated by moving eyes to indicate "yes" or by not moving eyes to signal "no." This binary system allowed P11 to choose letters and effectively communicate, albeit with significant assistance required for interpretation.

Recognizing the potential to enhance and automate this process, I suggested to Prof. Birbaumer that we could develop a system that would allow P11 to communicate more independently without continuous caregiver intervention. We proposed to utilize our BCI expertise and develop a system that would record P11's eye movement using electrooculography (EOG), about which you learned in ▶ Chap. 2, and translate them into communication actions.

We left P11's family with a promise to return with a prototype system within a month, hopeful that we could significantly improve P11's ability to communicate and interact with the environment.

🔹 **Fig. 11.1**   The speller schema utilized for P11 is an ingenious adaptation created by P11's caretakers to facilitate communication despite P11's severe limitations. In this system, the alphabet is segmented into color-coded sections on a simple sheet of paper. Each segment corresponds to a different group of letters, organized by color to make them distinct and easier for P11 to identify and select through eye movements. This method allows P11 to indicate choices by moving and not moving the eyes. The caretakers use this visual cue to spell out colors and characters. For instance, the caretaker says "Yellow" and observes P11's eye movement. If P11 moves the eye, it means yes, and the caretaker spells the characters inside the sector "Yellow". When no eye movement was observed, the caretaker moved to the next color, and so on. By breaking down the alphabet into these manageable, color-coded blocks, the speller schema optimizes the use of limited eye movements, enabling effective communication despite the challenges posed by ALS. This approach not only enhances P11's ability to express but also provides a model that could be adapted for other individuals with similar communication barriers. (From Tonin et al., 2020; originally published under CC BY 4.0)

11

> ### Box 11.2 Eye-Tracker-Based Communication Systems
>
> An eye-tracker-based communication system is a sophisticated technology designed to enable individuals who are unable to speak or use their hands to communicate through eye movements. This system utilizes a device that tracks where and how the eyes move, translating these movements into selections or commands on a computer screen.
>
> **■ Key Components:**
>
> *Eye-Tracking Device*: A camera or sensor that captures the eye movement. It is typically mounted on a computer or tablet.
>
> *Interface*: The visual display on which letters, words, or symbols are arrayed. Users select these by looking at them for a specified period, often indicated by a blink or a fixed stare.
>
> **■ How It Works:**
>
> The device calibrates the user's eye movements, learning to distinguish between intentional stares and normal eye blinks.
>
> The user looks at specific items on the screen to select them, effectively

"typing" out words or controlling a cursor.

Customizable settings allow for adjustment of dwell time (the time a user must fixate to make a selection) and sensitivity to suit individual capabilities and preferences.

- **Applications:**

Communication: Allows individuals with conditions like ALS, cerebral palsy, or severe paralysis to spell out words and sentences.

Environmental Control: Can be integrated with systems to control lights, televisions, and other home appliances.

Computer Access: Enables browsing the Internet, accessing email, and using most computer applications.

- **Benefits:**

Provides a vital communication lifeline for those with severe physical limitations.

Empowers users by giving them a degree of independence in daily interactions.

- **Limitations:**

Requires the ability to control one's gaze with some degree of precision.

May not be suitable for individuals whose eye control diminishes over time.

This technology not only helps in basic communication but also enhances the quality of life by broadening social interactions and access to information technology.

11.1 Problem Statement

You might find it puzzling when I say we need to develop a communication system for an ALS patient who cannot use a traditional eye-tracker and requires a method based on eye movement but independent of visual capabilities. What exactly do I mean by "independent of vision," and why is such independence crucial? For the moment, I ask you to accept this premise at face value. The detailed explanations and justifications for this approach will be thoroughly explored in Volume 2 of this book.

For now, let us focus on constructing a system that captures the eye movements of P11. This system will interpret P11's specific eye movements as binary signals, encoding them as 1's and 0's. These binary codes will then be integrated into a predefined speller schema, as illustrated in ◘ Fig. 11.1. The system will use a sound output to audibly present these options to P11, allowing P11 to make selections using eye movements alone. This process will enable P11 to formulate words and sentences autonomously without needing a caretaker's constant assistance to communicate.

Let us take a moment to review the foundational knowledge we have built throughout this book, which will be crucial for addressing the communication challenges faced by individuals like P11:

1. Acquiring Brain Signals
    ▶ Chapter 2: We explored various noninvasive methods to capture brain signals, focusing on techniques like EEG and fNIRS that do not require surgical interventions.
    ▶ Chapter 3: We deep-dived into invasive techniques that are irrelevant to this problem for now.
2. Processing Brain Signals
    ▶ Chapter 4: Here, we covered the essential steps for refining raw brain data into usable signals, detailing filtering, noise reduction, and signal enhancement techniques.
3. Feature Selection and Model Building
    ▶ Chapter 5: This chapter introduced the process of selecting meaningful features from processed signals and using these features to build predictive models with machine learning algorithms. These models are pivotal for interpreting brain activity in real time.
    Skipping over the chapters dedicated to brain stimulation, we reached:
4. Neurofeedback
    ▶ Chapter 10: We learned about neurofeedback mechanisms, which train individuals to modify their brain activity. This involves acquiring brain signals during a task, analyzing these signals, and providing real-time feedback. This forms a closed-loop system where the interface responds dynamically to the user's brain patterns.
    Remember that ▶ Chaps. 7, 8, and 9 also expanded on the applications of closed-loop systems in various settings.
    All the concepts and techniques from these chapters converge to help us design and implement a communication system for P11. In Volume 2, we will further integrate these with brain stimulation technologies to create advanced systems that, until recently, were the stuff of science fiction. Now, let us return to the pressing challenge we have before us.

11.2 Experimental Paradigm: The Solution

Once we were back in our lab, my team members got together, and we developed an experimental paradigm to assess and provide communication using the remnant eye-movement capabilities. This paradigm includes several key components:

Participant Setup: We decided to record both the EOG and EEG using the EEG system available in our lab.

Task Design: We designed specific tasks that P11 could perform while we recorded the EOG and EEG signals. The task was to move the eye and not move the eye in response to auditory stimuli. The design of these tasks is critical as it directly influences the type and quality of signals we can capture.

Data Acquisition: This involved recording EOG and EEG signals during the task performance. Data acquisition quality impacts subsequent processing stages' effectiveness, making it a critical component of our experimental paradigm.

Signal Processing: Once we have recorded the data, it underwent rigorous pre-processing to filter out noise and enhance the quality of the signals, as previously discussed.

Communication Interface: Finally, the processed signals were interfaced with a communication interface developed using the schema shown in ▣ Fig. 11.1. This interface used the binary "yes" and "no" to allow P11 to convey his thoughts or choices, essentially enabling communication.

By understanding this experimental paradigm, you can better appreciate how we used the remnant eye-movement capabilities of P11 to create functional communication systems for individuals who might otherwise be unable to express themselves. Now, let us look at each step in detail.

11.3 Task Design

We structured the study around four distinct session types, designed to train and empower P11 to use an oculomotor strategy for controlling a spelling system, as shown in ▣ Fig. 11.2 (Jaramillo-Gonzalez et al., 2021). Each session type had specific goals and procedures as follows.

1. **Training Sessions**

   Objective: Introduce and practice the basic response strategy.

   Procedure: P11 listened to 20 personal questions—10 expected to elicit a "yes" response and 10 a "no," presented in a pseudo-random order. P11 responded by moving eyes to indicate "yes" and refraining from any eye movement to signify "no." The length of the response time varied based on P11's ease with eye movements.

   Feedback: At the end of each response, P11 received a neutral auditory feedback, "Danke" (Thank you in English), indicating the completion of the trial.

2. **Feedback Sessions**

   Objective: Provide real-time feedback on the accuracy of the responses.

   Procedure: Similar to the training sessions, but at the end of each response window, the system provided specific feedback indicating whether the response was correctly recognized as "yes" or "no."

   Feedback: This session used affirmative feedback ("yes" or "no") based on the system's recognition of the patient's response.

3. **Copy Spelling Sessions**

   Objective: Practice spelling predefined sentences.

   Procedure: P11 was instructed to spell specific sentences using the spelling system. This session focused on the application of learned responses to structured tasks.

   Feedback: The system monitored and recorded the accuracy of the spelled sentences, providing insights into P11 proficiency.

4. **Free Spelling Sessions**

   Objective: Enable free-form spelling to assess independent use of the system.

11

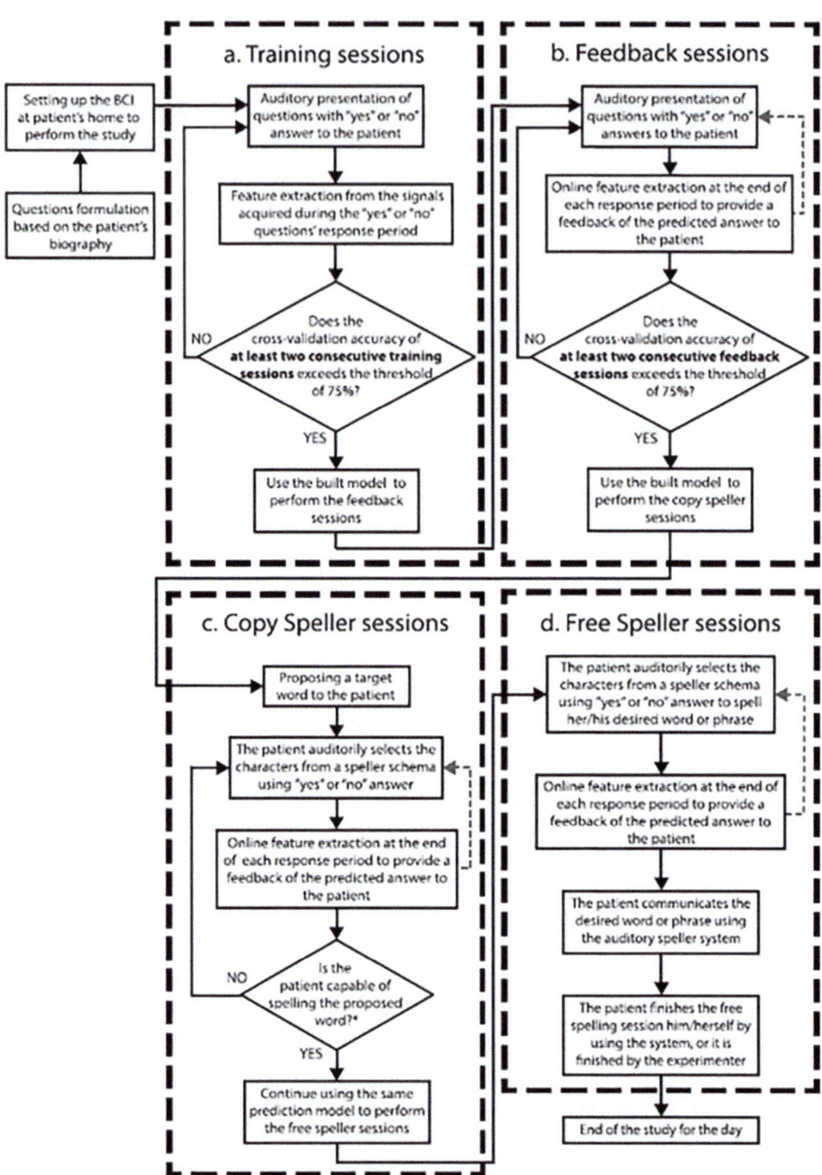

◘ Fig. 11.2 This figure illustrates the structured training approach using four session types designed to enable communication through eye movements. Training sessions introduce patients to the system with 20 personal questions and provide neutral feedback ("Thank you"). Feedback Sessions build on this by offering direct feedback ("yes" or "no") on the accuracy of responses. Copy spelling sessions involve spelling predefined sentences to practice and refine the learned skills. At the same time, free spelling sessions allow patients to spell freely, testing their independent use of the system. Each session type is critical in progressively enabling effective communication, with feedback mechanisms tailored to enhance learning and proficiency. (From Jaramillo-Gonzalez et al., 2021; originally published under CC BY 4.0)

Procedure: Following the structured practice of the copy spelling session, P11 was given the freedom to spell using the spelling system. This session tested P11's ability to apply the spelling strategy in a more open-ended, natural communication context.

Feedback: As with the copy spelling sessions, feedback was provided based on the system's recognition of the intended spelling.

Each session is depicted in the illustrative ◘ Fig. 11.2, showing the flow and integration of these activities. During these sessions, P11 developed strategies to respond to auditory questions by using eye movements to communicate effectively. This system used a binary response mechanism where moving the eyes indicated a "yes" and not moving them indicated a "no." Each type of session—training, feedback, copy spelling, and free spelling—played a critical role in helping P11 master the use of the oculomotor-based communication system.

11.4 Signal Acquisition

Once we had outlined and finalized our design, we started recording signals from P11. We used a 16-channel EEG amplifier (V-Amp DC, Brain Products, Germany) with silver/silver chloride (Ag/AgCl) active electrodes. This setup was critical for capturing both electrooculographic (EOG) and electroencephalographic (EEG) data. Specifically, we utilized four EOG electrodes to monitor eye movements—two placed at SO1 and IO1 to detect vertical movements and two at LO1 and LO2 to capture horizontal movements, as shown in ◘ Fig. 11.3a (Jaramillo-Gonzalez et al., 2021). This arrangement is beneficial for analyzing the different axes along which the eyes can move.

Additionally, during some of the sessions, we recorded data from at least seven EEG channels, as shown in ◘ Fig. 11.3b. These channels were not primarily used for communication purposes but were included for comprehensive brain activity analysis. The EEG and EOG channels were referenced to a common electrode on the right mastoid—a bony area behind the ear. They were also grounded to an electrode placed at the FPz location, which is located at the front of the scalp, just above the forehead, as shown in ◘ Fig. 11.3a.

The setup was carefully calibrated to ensure that electrode impedances did not exceed 10 kΩ, as higher impedances can reduce signal data quality. The sampling frequency for recording the brain signals was set at 500 Hz, which is suitable for effectively capturing a wide range of neural and ocular activities.

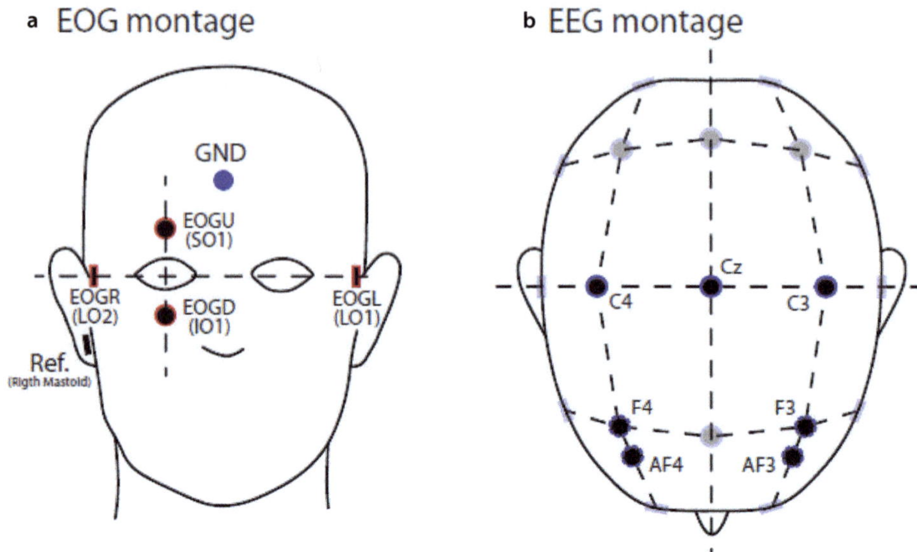

Fig. 11.3 In our study, we meticulously designed the setup for recording electrooculographic (EOG) and electroencephalographic (EEG) data, crucial for capturing detailed eye and brain activity. Let us look into how these setups were arranged: **a EOG Setup**: Our EOG setup utilized a minimum of four channels to monitor eye movements. These channels were strategically placed to capture horizontal and vertical movements of the eyes accurately. **Horizontal eye movement**: Two electrodes were positioned at the cantus of each eye—LO1 at the left cantus and LO2 at the right cantus. These locations are essential for detecting the side-to-side movements of the eyes. **Vertical eye movement**: Two additional electrodes were placed in relation to the orbit of the eye—SO1 above the superior orbit and IO1 below the inferior orbit. These positions help in measuring the up-and-down movements of the eyes. In our notation for the study, we labeled these electrodes as EOGL (Left), EOGR (Right), EOGU (Up), and EOGD (Down), corresponding to their positions and the direction of the eye movements they monitor. **b EEG Setup**: For the EEG recording, our focus was on a select number of electrodes that would provide crucial data from areas associated with motor control and cognitive functions. **Central motor areas**: We included electrodes at positions C3, Cz, and C4. These are key sites on the scalp overlying the primary motor cortex, involved in the planning, control, and execution of voluntary movements. Position Cz is located at the midline of the head directly above the central sulcus, while C3 and C4 lie symmetrically on either side of Cz, over the left and right hemispheres, respectively. **Prefrontal areas**: These regions are critical for cognitive functions such as decision-making, problem-solving, and moderating social behavior. Monitoring these areas helps in studying how brain activity in these regions correlates with different tasks or states during the sessions. (From Jaramillo-Gonzalez et al., 2021; originally published under CC BY 4.0)

11.5 Signal Preprocessing

After acquiring the signals, we moved to preprocessing them. We utilized a series of data preprocessing steps to ensure that the time-series signals captured from the EEG and EOG setups were of high quality and suitable for further analysis. Here is a detailed breakdown of these preprocessing procedures.

11.5.1 Filtering with a Digital Finite Impulse Response (FIR) Filter

Purpose and Application: We applied a digital FIR filter to remove unwanted frequencies from the signal, enhancing signal clarity and focusing on the frequencies of interest.

Passband Specification: The FIR filter was configured with a passband from 0.1 to 35 Hz. This range is crucial as it includes the delta, theta, alpha, and beta waves, which are significant for most neurological studies.

Notch Filter at 50 Hz: A notch filter specifically targeting 50 Hz was employed to eliminate electrical noise commonly caused by power lines. This step is essential for minimizing interference that can distort the signal of interest.

11.5.2 Elimination of Initial Data Points

Initial Data Removal: The first 50 data points were discarded from each time series. This procedure is standard practice in signal processing to remove transitory effects that often occur at the start of a signal due to the initiation of filtering. These initial fluctuations can skew initial analyses and lead to misleading interpretations.

11.5.3 Standardization of the Channels

Normalization Process: After filtering, all channels were standardized to ensure uniformity across all datasets. Standardization involves adjusting the data so that each channel has a mean of zero and a standard deviation of one.

Importance of Standardization: This step is crucial as it normalizes the signal amplitude across different recording sessions and subjects, facilitating more accurate comparisons and analyses. It helps to mitigate any biases or variations in signal strength due to individual differences or electrode placement inconsistencies.

The outcome of these preprocessing steps is shown on the truncated sample signal in ◘ Fig. 11.4.

◘ Figure 11.5 shows the average eye-movement responses corresponding to trials with "yes" and "no" answers.

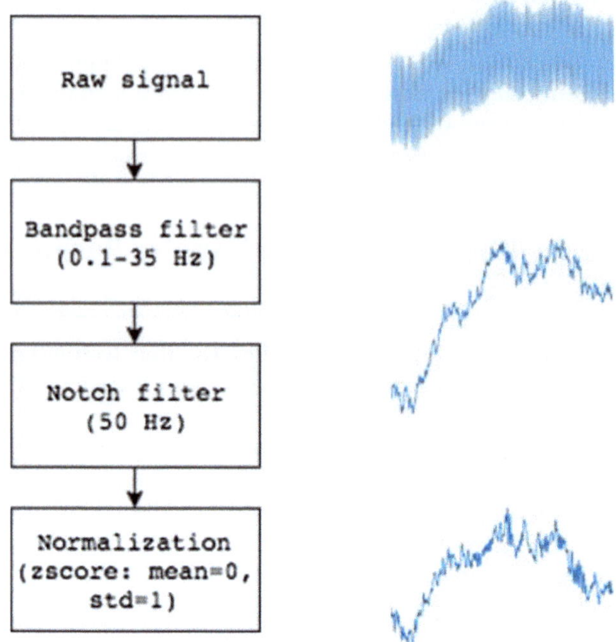

 **Fig. 11.4** Preprocessing of EEG and EOG time-series data. This figure illustrates the sequential steps involved in preprocessing EEG and EOG data recorded from P11. Each box and the associated trace represents a key stage in the data preparation. **Raw data acquisition**: Displays a sample raw signal captured through the EEG and EOG channels before any preprocessing is applied. **Application of FIR filter**: This shows the signal after applying a finite impulse response (FIR) filter with a passband from 0.1 to 35 Hz alongside a 50 Hz notch filter to remove electrical noise. **Standardization of signal**: Depicts the standardized signals with each channel adjusted to have a mean of zero and a standard deviation of one, ensuring consistency across the dataset for further analysis. (Unpublished figure from my erstwhile laboratory)

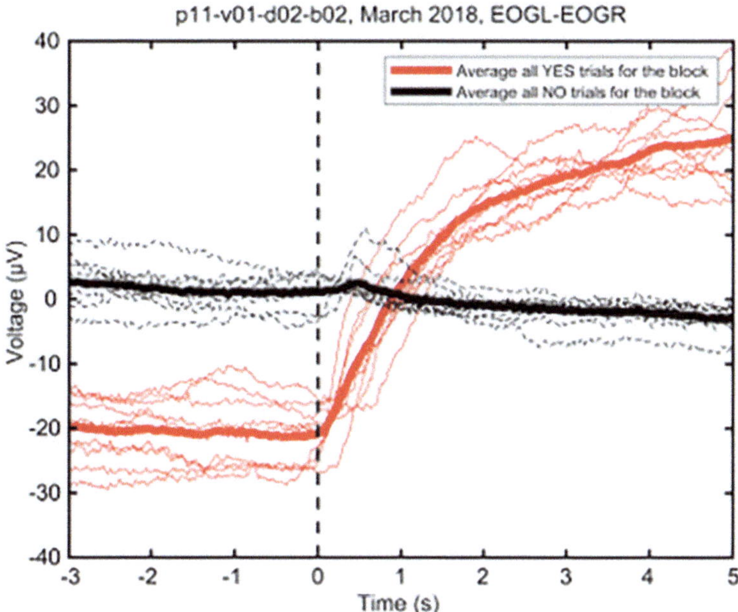

◘ Fig. 11.5 This figure displays the average eye-movement responses for trials with "yes" in red and "no" in black answers, respectively. The thin traces in the diagram represent the eye movements from individual trials, capturing the variability across different sessions. The thick traces, on the other hand, depict each category's average response, providing a clear visual summary of the typical eye-movement patterns associated with "yes" and "no" responses. This aggregation helps to illustrate the general trends in eye movements. It enhances the visibility of consistent patterns across numerous trials. (Unpublished figure from my erstwhile laboratory)

11.6 Feature Extraction and Model Building

The initial training sessions, as detailed in ▶ Sect. 11.3, were critical for collecting data from which we could identify and extract reliable features for building a machine-learning model. This model is essential for implementing real-time feedback in subsequent sessions, such as feedback, copy, and free spelling sessions. It enables the participant to receive instantaneous feedback on their responses, which is crucial for effectively adjusting and learning the communication strategy.

In the training sessions for patient P11, we carried out an initial feature extraction analysis, illustrated in ◘ Fig. 11.6. During this phase, we identified several potential features, including the maximum and minimum amplitudes of the eye movements, the precise times these amplitudes occurred, and the amplitude range, which is the difference between the maximum and minimum values. Among these, the maximum and minimum amplitudes and their timing were selected as the most promising features for modeling.

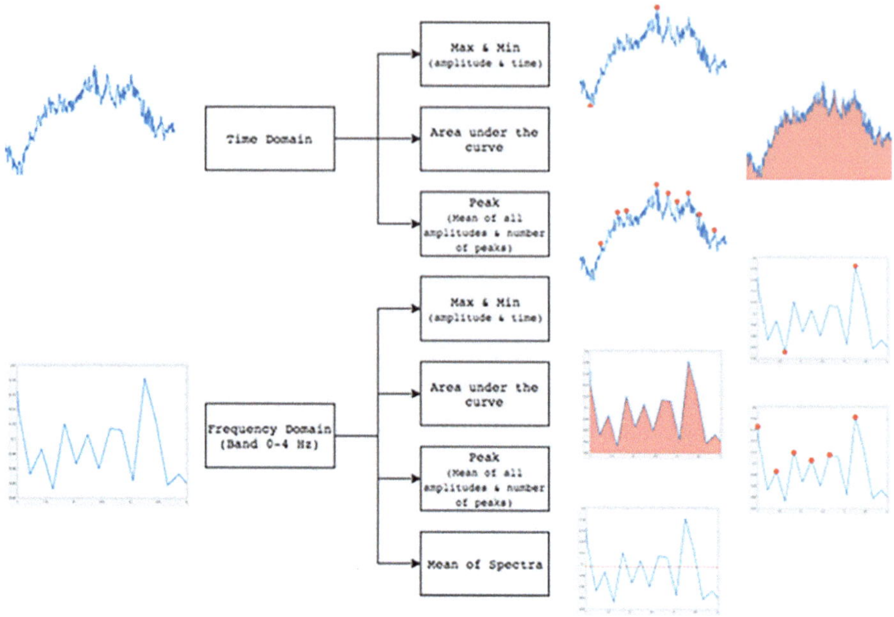

◘ Fig. 11.6 Illustrates the various features that were investigated during the analysis phase. These features include the maximum and minimum amplitudes of eye movements, the specific moments these extremes were recorded, and the amplitude range, which measures the difference between the maximum and minimum values. Other potential features, such as the area under the curve and the mean of spectra, were also considered to ensure a comprehensive understanding of the eye-movement patterns. Each feature was evaluated for its relevance and reliability in distinguishing between different types of responses, aiding in the development of a robust model for interpreting the P11's communication signals. (Unpublished figure from my erstwhile laboratory)

These selected features were then used to train a Support Vector Machine (SVM), a type of classification algorithm, to differentiate between "yes" and "no" responses based on the patterns identified in P11's eye movements. We performed fivefold cross-validation on this SVM to ensure its reliability and accuracy.

Box 11.3

A Support Vector Machine (SVM) is a powerful supervised machine learning algorithm used primarily for classification and regression tasks. It operates by finding a hyperplane that best divides a dataset into classes. The SVM algorithm seeks to maximize the margin between the data points of the different classes, which are nearest to the hyperplane. These points are known as support vectors. SVMs are particularly effective in high-dimensional spaces and are versatile in that they can be adapted for nonlinear classification using various kernel functions, such as polynomial or radial basis function (RBF). This adaptability allows SVMs to handle both linear and complex nonlinear relationships between data points.

11.7 **Communication**

After successfully capturing the signal and constructing a model that encapsulates P11's learned communication strategy, the subsequent phase involved enabling P11 to utilize this model to select characters and construct words and sentences. This process was carried out through online sessions, the framework of which is detailed in ◘ Fig. 11.2.

Post training, P11's eye-movement response signals were processed in real time during all interactive sessions. This process included extracting predetermined features relevant to the task, which were then fed into a Support Vector Machine (SVM) classifier. This classifier was tasked with interpreting P11's responses by differentiating between "yes" and "no" answers based on eye movements.

The performance of the SVM classifier was constantly evaluated to ensure its effectiveness. The classifier was only employed in feedback and copy spelling sessions when its accuracy consistently surpassed a specifically set threshold that was well above the chance level. This rigorous standard was necessary to ensure that the feedback and communication facilitated by the system were both precise and dependable.

The criteria and parameters used to determine when and how the SVM model was utilized are further detailed in ◘ Fig. 11.2. For curious minds interested in a deeper dive into the technical and procedural nuances of the communication system developed for P11, I recommend reviewing the detailed descriptions and outcomes presented in the scientific publications by Tonin et al. (2020) and Jaramillo-Gonzalez et al. (2021). These studies provide extensive insights into the methodologies and the effectiveness of the communication schema implemented.

A summary of the process described for the communication system described in this chapter is shown in ◘ Fig. 11.7 (Chaudhary et al., 2021).

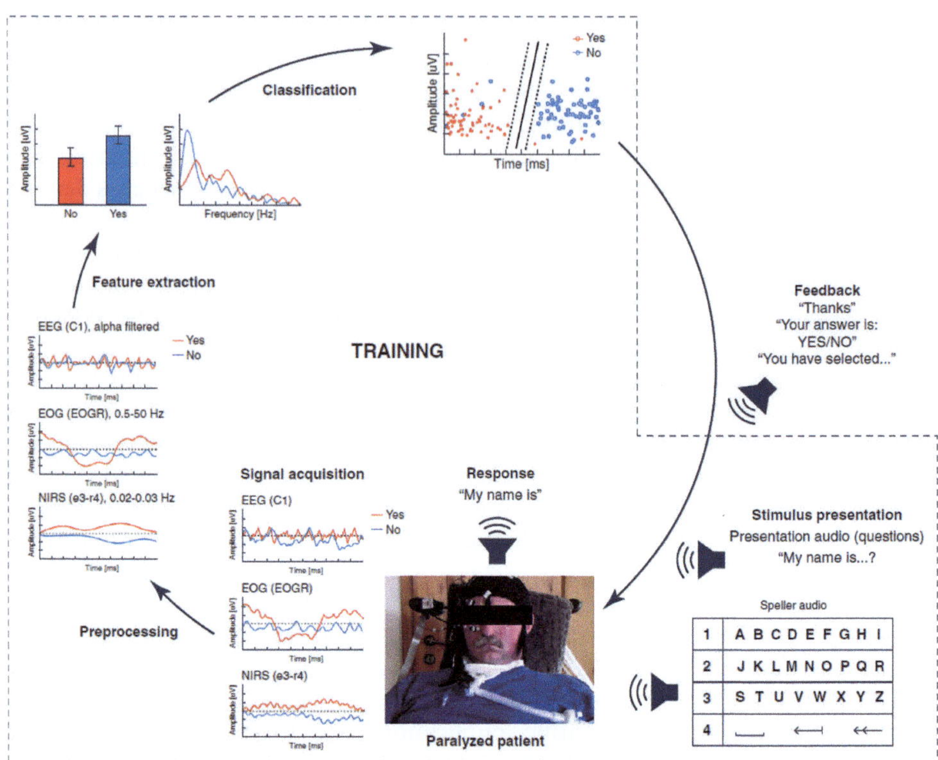

Fig. 11.7   This schematic represents the layout of an auditory noninvasive brain–computer interface (BCI) designed for communication by patients in a completely locked-in state (CLIS). The system integrates multiple types of neurophysiological data: EEG (electroencephalogram), EOG (electrooculogram), and NIRS (near-infrared spectroscopy). These signals are initially recorded and then subjected to preprocessing, which includes filtering to improve the signal quality. From these preprocessed signals, specific features that effectively distinguish between "yes" and "no" responses are extracted. This is critical as the precision of feature extraction directly influences the performance of the subsequent steps. Once the relevant features are identified, a machine learning algorithm—such as a Support Vector Machine (SVM) or Linear Discriminant Analysis (LDA)—is used to train a classifier. This classifier is refined through multiple training trials to maximize its ability to differentiate between "yes" and "no" answers with significant accuracyDuring the "feedback" trials that follow the training phase, the patient receives real-time feedback based on the classifier's predictions of their responses. This immediate feedback loop is essential for reinforcing the learning and adjustment of the patient's response strategies. The selection of letters from an auditory speller is integrated into this process but is contingent on achieving a stable and correct classification of "yes" and "no" responses with an accuracy exceeding 70%. Only when this level of reliability is reached does the system enable the patient to select letters and thereby communicate effectively, ensuring that the communication interface is both functional and reliable. (From Chaudhary et al., 2021; originally published under CC BY 4.0)

> **Key Takeaways**
> 1. **Development of Innovative Communication Systems for ALS Patients**: The chapter details how neurotechnology, specifically brain-computer interfaces (BCIs), can revolutionize communication for patients with severe physical limitations like Amyotrophic Lateral Sclerosis (ALS). By utilizing electrooculography (EOG) and auditory-based interfaces, individuals who are "locked-in" can regain some level of communication autonomy, enabling them to express choices and thoughts despite their physical incapacities.
> 2. **Eye Movement as a Viable Communication Modality**: A central theme is the use of residual eye movement as a means to facilitate communication. Even when traditional eye-tracking systems fail, eye movements detected via EOG offer an alternative, noninvasive approach to interpret binary responses such as "yes" or "no," significantly improving the quality of life for patients.
> 3. **Machine Learning in BCIs for Real-Time Communication**: The chapter emphasizes the integration of machine learning algorithms, such as Support Vector Machine (SVM), which classify eye-movement patterns to create real-time feedback mechanisms. These systems enhance the accuracy and reliability of communication systems, allowing patients to participate more actively in daily interactions with minimal caregiver intervention.
> 4. **Broader Implications and Future Directions**: The successful implementation of these neurotechnological systems for ALS patients opens up future possibilities for broader applications in other neurological disorders. By combining advanced signal processing, real-time feedback, and personalized system designs, BCIs offer a promising future for patient-centered solutions in communication and beyond.

In conclusion, the journey from theoretical understanding to practical application, as outlined in this chapter, not only challenges our technical and scientific acumen but also enriches our ability to make a profound and compassionate impact on society. The intersection of neurotechnology and patient-centric design stands as a beacon of hope, symbolizing a future where technology bridges gaps created by human frailty.

Conclusion

In this chapter, we applied most of the concepts we learned in this book to develop a communication system for a patient without any means of communication. The case study involving patient P11 has provided valuable insights into the practical challenges and potential solutions associated with developing neurotechnology-based communication systems. We saw how the signal we acquired is processed and employed to solve real-world problems; this is just the tip of the iceberg.

The methodologies and findings discussed in this chapter lay a robust foundation for further exploration as we look to the future. Moreover, the lessons learned from this project can inform similar applications across different neurological conditions, potentially transforming the lives of many more individuals around the world. The concepts you learned in this book form the backbone of a fascinating technology called brain–computer interface (BCI) or brain–machine interface (BMI), as shown in ◘ Fig. 11.8 (Chaudhary et al., 2016). Volume 2 of this book series focuses exclusively and extensively on BCIs/BMIs.

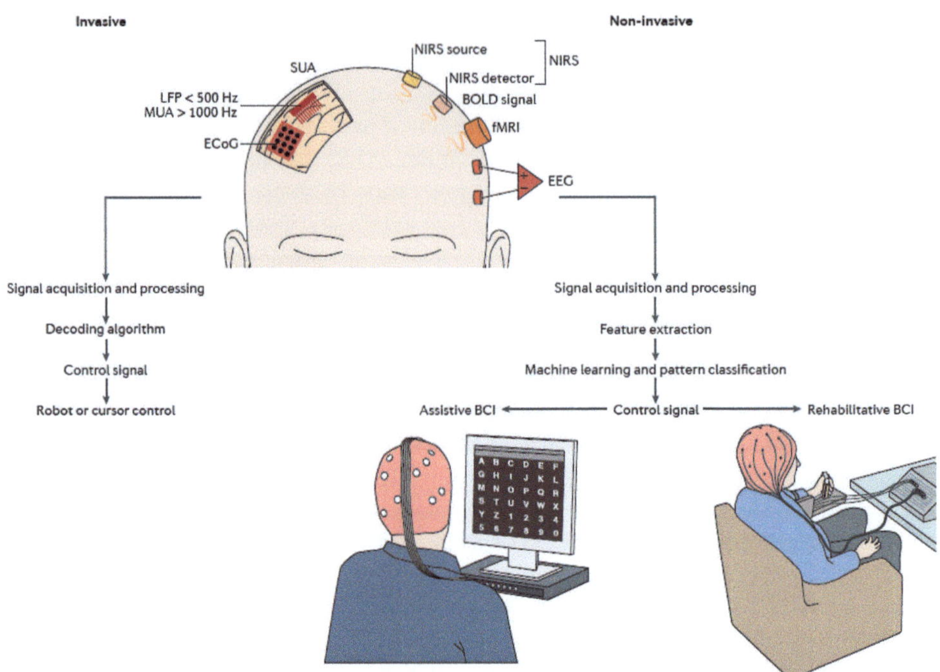

Fig. 11.8 Overview of invasive and noninvasive brain–computer interface (BCI) systems. This figure illustrates the workflow of both invasive and noninvasive BCI systems, highlighting the different methods of signal acquisition, processing, and application. **Invasive BCI—signal acquisition**: Invasive BCIs use electrodes implanted directly into the brain. These include: Single-unit activity (SUA)—Captures signals from individual neurons. Multi-unit activity (MUA)—Records high-frequency signals (>1000 Hz) from multiple neurons. Local field potentials (LFP)—Measures low-frequency signals (<500 Hz) from a population of neurons. Electrocorticography (ECoG)—Records electrical activity from the cortical surface. **Signal processing**: The acquired signals are processed and decoded using specialized algorithms to generate control signals. **Applications**: The control signals are used to operate external devices such as robotic arms or computer cursors, providing direct and precise control for the user. **Noninvasive BCI—signal acquisition**: Noninvasive BCIs use external sensors to record brain activity. These include: Electroencephalography (EEG)—Measures electrical activity through electrodes placed on the scalp. Functional Near-Infrared Spectroscopy (fNIRS)—Uses near-infrared light to measure changes in blood oxygenation levels. Functional Magnetic Resonance Imaging (fMRI)—Detects blood-oxygen-level-dependent (BOLD) signals to map brain activity. **Signal processing**: The acquired signals undergo feature extraction followed by machine learning and pattern classification to generate control signals. **Applications**: The control signals are utilized in assistive BCIs (e.g., spelling devices for communication) and rehabilitative BCIs (e.g., devices for motor rehabilitation). **Key steps—signal acquisition and processing**: Initial step involving the collection and preprocessing of neural signals. **Feature extraction**: Identifying relevant features from the processed signals. **Machine learning and pattern classification**: Applying algorithms to classify the signals and translate them into control commands. **Control signal generation**: Producing signals that can control external devices based on the user's neural activity. This figure provides an overview of the distinct pathways and technologies employed in invasive and noninvasive BCIs, illustrating their applications in enhancing communication and rehabilitation for users with neurological conditions. (From Chaudhary et al., 2016)

References

Chaudhary, U., Birbaumer, N., & Ramos-Murguialday, A. (2016). Brain–computer interfaces for communication and rehabilitation. *Nature Reviews Neurology, 12*(9), 513–525.

Chaudhary, U., Mrachacz-Kersting, N., & Birbaumer, N. (2021). Neuropsychological and neurophysiological aspects of brain-computer-interface (BCI) control in paralysis. *The Journal of Physiology, 599*(9), 2351–2359.

Jaramillo-Gonzalez, A., Wu, S., Tonin, A., Rana, A., Ardali, M. K., Birbaumer, N., & Chaudhary, U. (2021). A dataset of EEG and EOG from an auditory EOG-based communication system for patients in locked-in state. *Scientific Data, 8*(1), 8.

Tonin, A., Jaramillo-Gonzalez, A., Rana, A., Khalili-Ardali, M., Birbaumer, N., & Chaudhary, U. (2020). Auditory electrooculogram-based communication system for ALS patients in transition from locked-in to complete locked-in state. *Scientific Reports, 10*(1), 8452.